TC 7-100.2

Opposing Force Tactics

December 2011

DISTRIBUTION RESTRICTION:
Approved for public release; distribution is unlimited.

**HEADQUARTERS
DEPARTMENT OF THE ARMY**

This publication is available at
Army Knowledge Online (www.us.army.mil);
General Dennis J. Reimer Training and Doctrine
Digital Library (http://www.train.army.mil).

Training Circular
No. 7-100.2

TC 7-100.2

Headquarters
Department of the Army
Washington, D.C., 9 December 2011

Opposing Force Tactics

Contents

Page

PREFACE ix

INTRODUCTION x
    Operational Environments x
    Opposing Force xi
    The COE and OPFOR Continue to Evolve xv

Chapter 1   STRATEGIC AND OPERATIONAL FRAMEWORK 1-1
    National-Level Organization 1-1
    National Security Strategy 1-4
    Strategic Campaign 1-7
    Operational-Level Organization 1-8
    Operational Designs 1-10
    Types of Offensive and Defensive Action 1-12
    Systems Warfare 1-13
    The Role of Paramilitary Forces in Operations 1-16

Chapter 2   COMMAND AND CONTROL 2-1
    Concept and Principles 2-1
    Command and Support Relationships 2-2
    Tactical-Level Organizations 2-3
    Organizing the Tactical Battlefield 2-11
    Functional Organization of Forces and Elements 2-15
    Headquarters, Command, and Staff 2-17
    Command Posts 2-28
    Command and Control Systems 2-31

Chapter 3   OFFENSE 3-1
    Purpose of the Offense 3-1
    Planning the Offense 3-3
    Preparing for the Offense 3-8
    Executing the Offense 3-8
    Types of Offensive Action—Tactical Groups, Divisions, and Brigades 3-9
    Tactical Offensive Actions—Detachments, Battalions, and Below 3-20

Distribution Restriction: Approved for public release; distribution is unlimited.

Contents

Chapter 4	**DEFENSE**	**4-1**
	Purpose of the Defense	4-1
	Planning the Defense	4-2
	Preparing for the Defense	4-7
	Executing the Defense	4-9
	Types of Defensive Action—Tactical Groups, Divisions, and Brigades	4-10
	Tactical Defensive Actions—Detachments, Battalions, and Below	4-18
Chapter 5	**BATTLE DRILLS**	**5-1**
	Purpose of Battle Drills	5-1
	Actions on Contact	5-1
	Breaking Contact	5-4
	Situational Breach	5-5
	Fire and Maneuver	5-7
	Fixing	5-9
Chapter 6	**OTHER COMBINED ARMS ACTIONS**	**6-1**
	Actions of the Disruption Force	6-1
	Counterreconnaissance	6-2
	Antilanding Actions	6-6
	Urban Combat	6-7
Chapter 7	**INFORMATION WARFARE**	**7-1**
	Tactical-Level INFOWAR	7-1
	Elements of INFOWAR	7-3
Chapter 8	**RECONNAISSANCE**	**8-1**
	Combined Arms Mission	8-1
	Concept	8-2
	Assets	8-4
	Reconnaissance Planning	8-7
	Reconnaissance Elements	8-12
	Reconnaissance Methods	8-16
Chapter 9	**INDIRECT FIRE SUPPORT**	**9-1**
	Fire Support Concepts	9-1
	Indirect Fire Support Weapons	9-3
	Command and Control	9-4
	Fire Support Planning	9-11
	Targeting	9-12
	Target Acquisition and Reconnaissance	9-15
	Methods of Fire	9-18
	Fire Support of Maneuver Operations	9-19
	Tactical Deployment	9-20
	Logistics	9-25
Chapter 10	**AVIATION**	**10-1**
	Command and Control	10-1
	Airspace Management	10-5
	Missions	10-10
	Planning and Preparation	10-19

Contents

	Flight Tactics	10-27
Chapter 11	**AIR DEFENSE**	**11-1**
	Air Defense System	11-1
	Command and Control	11-4
	Air Surveillance	11-7
	Tactical Assets	11-10
	Employment	11-13
	Nonlethal Air Defense Measures	11-27
Chapter 12	**ENGINEER SUPPORT**	**12-1**
	Adaptive Engineer Support	12-1
	Missions and Tasks	12-1
	command and control	12-4
	Support to Information Warfare	12-5
	Engineer Reconnaissance	12-5
	Mobility	12-7
	Countermobility	12-19
	Survivability	12-30
Chapter 13	**CBRN AND SMOKE**	**13-1**
	Weapons of Mass Distruction	13-1
	Preparedness	13-1
	Staff Responsibility	13-2
	Chemical Warfare	13-2
	Biological Warfare	13-6
	Radiological Weapons	13-8
	Nuclear Warfare	13-9
	CBRN Protection	13-11
	Smoke	13-17
Chapter 14	**LOGISTICS**	**14-1**
	Stratecic and Operational Logistics Support	14-1
	Tactical Staff Responsibilities	14-1
	Tactical Logistics Concepts	14-3
	Logistics Missions	14-3
	Tailored Logistics Units	14-4
	Materiel Support	14-7
	Maintenance	14-9
	Transportation	14-9
	Personnel Support	14-12
	Medical Support	14-13
	Support of Combat Actions	14-14
	Post-Combat Support	14-16
Chapter 15	**SPECIAL-PURPOSE FORCES AND COMMANDOS**	**15-1**
	SECTION I – SPECIAL-PURPOSE FORCES	**15-1**
	Command and Control	15-1
	Missions	15-2
	Organization for combat	15-5

Contents

    SPF Brigades, Battalions, Companies, and Teams 15-6
    Tactics, Techniques, and Procedures 15-18
    Equipment 15-21
    Personnel 15-22
    Logistics 15-23
    SECTION II – COMMANDOS 15-25
    Command and Control 15-25
    Missions 15-25
    Organization for Combat 15-27
    Commando Brigades and Battalions 15-27
    Tactics, Techniques, and Procedures 15-33
    Personnel 15-34

Chapter 16   **MARKSMEN AND SNIPERS 16-1**
    Similarities and Differences 16-1
    Marksmen 16-1
    Snipers 16-4
    Antimateriel Role 16-9
    Role in Information Warfare 16-11
    Differences Between Regular Military Snipers and Irregular Force Snipers .. 16-11

    GLOSSARY Glossary-1
    REFERENCES References-1
    INDEX Index-1

Figures

Figure 1-1. National Command Authority 1-2
Figure 1-2. The State's armed forces 1-3
Figure 1-3. Conceptual framework for implementing the State's national security
    strategy 1-4
Figure 1-4. Example of a strategic campaign 1-7
Figure 1-5. State and OPFOR planning framework 1-9
Figure 1-6. Operational designs 1-11
Figure 1-7. Combat system 1-14
Figure 2-1. Possible DTG organization (example) 2-5
Figure 2-2. Possible BTG organization (example) 2-5
Figure 2-3. Motorized infantry-based DTG symbol 2-6
Figure 2-4. Mechanized infantry-based DTG symbol 2-6
Figure 2-5. Tank-based BTG symbol 2-6
Figure 2-6. Battalion example 2-7
Figure 2-7. Company (battery) example 2-7
Figure 2-8. Battalion-size detachment (BDET) example 2-8
Figure 2-9. Company-size detachment (CDET) example 2-8

Contents

Figure 2-10. Heliborne infantry-based BDET symbol 2-9
Figure 2-11. Attack helicopter-based BDET symbol 2-9
Figure 2-12. Mechanized infantry-based CDET symbol 2-10
Figure 2-13. Tank-based task-organized platoon symbol 2-10
Figure 2-14. Motorized infantry-based task-organized squad symbol 2-10
Figure 2-15. Linear AOR (example 1) 2-12
Figure 2-16. Linear AOR (example 2) 2-12
Figure 2-17. Nonlinear AOR (example 1) 2-13
Figure 2-18. Nonlinear AOR (example 2) 2-13
Figure 2-19. DTG or division command group and staff 2-18
Figure 2-20. Battalion command section and staff 2-27
Figure 3-1. Integrated attack (example 1) 3-10
Figure 3-2. Integrated attack (example 2) 3-11
Figure 3-3. Integrated attack (example 3) 3-12
Figure 3-4. Dispersed attack (example 1) 3-14
Figure 3-5. Dispersed attack (example 2) 3-15
Figure 3-6. Dispersed attack (example 3) 3-16
Figure 3-7. Spoiling attack (example) 3-18
Figure 3-8. Counterattack (example) 3-19
Figure 3-9. Assault (example 1) 3-21
Figure 3-10. Assault (example 2) 3-22
Figure 3-11. Assault (example 3) 3-23
Figure 3-12. Annihilation ambush (example 1) 3-30
Figure 3-13. Annihilation ambush (example 2) 3-30
Figure 3-14. Annihilation ambush (example 3) 3-31
Figure 3-15. Annihilation ambush using infantry antiarmor (hunter-killer) teams (example 4) 3-32
Figure 3-16. Harassment ambush (example) 3-33
Figure 3-17. Containment ambush (example 1) 3-34
Figure 3-18. Containment ambush (example 2) 3-35
Figure 3-19. Raid (example 1) 3-37
Figure 3-20. Raid (example 2) 3-37
Figure 3-21. Reconnaissance attack (example) 3-42
Figure 4-1. Maneuver defense (example 1) 4-11
Figure 4-2. Maneuver defense (example 2) 4-13
Figure 4-3. Area defense (example 1) 4-15
Figure 4-4. Area defense (example 2) 4-16
Figure 4-5. Area defense (example 3) 4-17
Figure 4-6. Simple and complex battle positions 4-19
Figure 4-7. Combat security outpost (example) 4-21
Figure 4-8. CDET in an SBP (example 1) 4-22

Contents

Figure 4-9. CDET in an SBP (example 2) ... 4-23
Figure 4-10. Reconnaissance support to an SBP (example) ... 4-25
Figure 4-11. Defense of a CBP (example) ... 4-30
Figure 4-12. CSOPs in the disruption zone supporting a CBP (example 1) ... 4-32
Figure 4-13. CSOPs in the disruption zone supporting a CBP (example 2) ... 4-33
Figure 4-14. Reconnaissance support to a CBP (example) ... 4-34
Figure 5-1. Actions on contact (example) ... 5-3
Figure 5-2. Breaking contact (example) ... 5-4
Figure 5-3. Situational breach (example) ... 5-6
Figure 5-4. Fire and maneuver (example) ... 5-8
Figure 5-5. Fixing (example) ... 5-10
Figure 6-1. Counterreconnaissance detachment (example) ... 6-4
Figure 6-2. Execution of counterreconnaissance (example) ... 6-6
Figure 6-3. Multidimensional battlefield ... 6-8
Figure 6-4. Urban detachment attacking enemy-controlled building complex (example) ... 6-11
Figure 8-1. Effective ranges of example reconnaissance assets ... 8-4
Figure 8-2. Zone of reconnaissance responsibility (example) ... 8-8
Figure 9-1. Possible IFC components in a DTG ... 9-5
Figure 9-2. IFC headquarters ... 9-6
Figure 9-3. Target report flow ... 9-14
Figure 9-4. Observation posts in the battle formation of an artillery battalion (example) ... 9-16
Figure 9-5. Artillery battalion and battery disposition (example) ... 9-21
Figure 10-1. Example of aviation in a DTG IFC ... 10-3
Figure 10-2. Example of DTG-level aviation other than in the IFC ... 10-4
Figure 10-3. Example of aviation in a BTG ... 10-5
Figure 10-4. Airspace procedural control measures (example) ... 10-9
Figure 10-5. Immediate DAS request process ... 10-24
Figure 12-1. Covert breach (example) ... 12-10
Figure 12-2. Mechanized breach (example) ... 12-11
Figure 12-3. Engineer support of a mechanized infantry battalion crossing (example) ... 12-18
Figure 12-4. AT minefield configuration with track-attack mines (example) ... 12-23
Figure 12-5. AT minefield configuration with full-width-attack mines (example) ... 12-23
Figure 12-6. AP minefield configuration (example) ... 12-24
Figure 12-7. Mixed minefield with blast AP rows between AT rows (example) ... 12-25
Figure 12-8. Example mixed minefield with an AP minefield leading to a full-width AT minefield ... 12-25
Figure 12-9. Mechanical AT minelaying sequence (example) ... 12-27
Figure 13-1. Decontamination trucks using echelon-left formation ... 13-16
Figure 13-2. Checkerboard area smokescreen (example) ... 13-22
Figure 13-3. Ring area smokescreen (example) ... 13-23
Figure 13-4. Mixed area smokescreen (example) ... 13-24

Figure 13-5. Smoke in the offense against an enemy in defensive positions (example 1) 13-26
Figure 13-6. Smoke in the offense against an enemy in defensive positions (example 2) 13-27
Figure 13-7. Smoke in the offense against an enemy on the move (example) 13-28
Figure 13-8. Smoke in the defense (example) 13-29
Figure 13-9. Smoke in an opposed water obstacle crossing (example) 13-30
Figure 13-10. Alternating blinding smoke and illuminating lines against the enemy at night (example) 13-31
Figure 14-1. Resources section 14-2
Figure 14-2. Integrated support command headquarters 14-5
Figure 14-3. DTG ISG (example) 14-7
Figure 15-1. SPF brigade (example) 15-7
Figure 15-2. SPF battalion (example) 15-8
Figure 15-3. SPF company (example) 15-10
Figure 15-4. SPF company with specialized teams (example) 15-11
Figure 15-5. Commando brigade (example) 15-28
Figure 15-6. Air infiltration company (example) 15-29
Figure 15-7. Reconnaissance company (example) 15-30
Figure 15-8. INFOWAR company (example) 15-30
Figure 15-9. Commando battalion (example).. 15-31
Figure 15-10. Commando company (example) 15-31
Figure 15-11. Weapons company (example) 15-32
Figure 15-12. INFOWAR platoon (example) 15-32
Figure 15-13. High-mobility reconnaissance platoon, commando battalion (example) 15-32

Tables

Table 2-1. Command and support relationships 2-2
Table 2-2. Command post system 2-28
Table 4-1. First-priority preparation tasks for a battalion or BDET battle position 4-27
Table 4-2. Second-priority preparation tasks for a battalion or BDET battle position 4-28
Table 4-3. Third-priority preparation tasks for a battalion or BDET battle position 4-29
Table 7-1. INFOWAR elements, objectives, and targets 7-4
Table 10-1. Classification of attack targets 10-13
Table 10-2. Calculation of aircraft sorties (example) 10-21
Table 10-3. Levels of combat readiness 10-26
Table 11-1. Example security measures 11-30
Table 12-1. Engineer support for preparation and conduct of the offense 12-3
Table 12-2. Engineer support for preparation and conduct of the defense 12-3
Table 12-3. Engineer support for preparation and conduct of tactical movement 12-7

Contents

Table 12-4. Preferred water obstacle-crossing methods 12-15
Table 13-1. Electro-optical and other systems defeated by obscurants 13-18
Table 13-2. Tactical employment of smoke and other obscurants from various
    sources 13-25
Table 14-1. Levels of medical care 14-13
Table 16-1. Differences between regular military snipers and irregular force snipers 16-12

Preface

This training circular (TC) is one of a series that describes an opposing force (OPFOR) for training U.S. Army commanders, staffs, and units. See the References section for a list of other TCs in this series. (Other publications in the former Field Manual [FM] 7-100 series will be converted to TCs as well.) Together, these TCs outline an OPFOR than can cover the entire spectrum of military and paramilitary capabilities against which the Army must train to ensure success in any future conflict.

Applications for this series of TCs include field training, training simulations, and classroom instruction throughout the Army. All Army training venues should use an OPFOR based on these TCs, except when mission rehearsal or contingency training requires maximum fidelity to a specific country-based threat or enemy. Even in the latter case, trainers should use appropriate parts of the OPFOR TCs to fill information gaps in a manner consistent with what they do know about a specific threat or enemy.

This publication applies to the Active Army, the Army National Guard (ARNG) /Army National Guard of the United States (ARNGUS), and the United States Army Reserve (USAR) unless otherwise stated.

Headquarters, U.S. Army Training and Doctrine Command (TRADOC) is the proponent for this publication. The preparing agency is the Contemporary Operational Environment and Threat Integration Directorate (CTID), TRADOC G-2 Intelligence Support Activity (TRISA)-Threats. Send comments and suggested improvements on DA Form 2028 (Recommended Changes to Publications and Blank Forms) directly to CTID at the following address: Director, CTID, TRISA-Threats, ATTN: ATIN-T (Bldg 53), 700 Scott Avenue, Fort Leavenworth, KS 66027-1323.

This publication is available at Army Knowledge Online (AKO) at http://www.us.army.mil and on the General Dennis J. Reimer Training and Doctrine Digital Library (RDL) at http://www.adtdl.army.mil. Readers should monitor those sites and also the TRADOC G2-TRISA Web sites listed below for the status of this TC and information regarding updates. The TC is also available (after AKO login) in AKO files under Organizations/DoD Organizations/Army/Army Command/TRADOC/HQ Staff/DCS, G-2 (Intelligence)/TRISA/TRISA-CTID/Hybrid Threat Doctrine at https://www.us.army.mil/suite/files/11318389 (for TRISA-CTID folder) or https://www.us.army.mil/suite/files/30837459 (for Hybrid Threat Doctrine folder). Periodic updates, subject to the normal approval process, will occur as a result of the normal production cycle. The date on the cover and title page of the electronic version will reflect the latest update.

Unless this publication states otherwise, masculine nouns or pronouns do not refer exclusively to men.

Introduction

This training circular (TC), as part of the TC 7-100 series, describes an opposing force (OPFOR) that exists for the purpose of training U.S. forces for potential combat operations. This OPFOR reflects a composite of the characteristics of military and/or paramilitary forces that may be present in actual operational environments (OEs) in which U.S. forces might become involved in the near- and mid-term. Like those actual threats or enemies, the OPFOR will continue to present new and different challenges for U.S. forces. The nature of OEs is constantly changing, and it is important for U.S. Army training environments to keep pace with real-world developments.

OPERATIONAL ENVIRONMENTS

The DOD officially defines an *operational environment* as "a composite of the conditions, circumstances, and influences that affect the employment of capabilities and bear on the decisions of the commander" (JP 3-0). This definition applies to an OE for a specific operation, at any level of command. In planning a training scenario and its road to war, trainers need to take into consideration the entire OE and its impact on the OPFOR's operations and tactics.

CONTEMPORARY OPERATIONAL ENVIRONMENT

The *Contemporary Operational Environment* (COE) **is the collective set of conditions, derived from a composite of actual worldwide conditions, that pose realistic challenges for training, leader development, and capabilities development for Army forces and their joint, intergovernmental, interagency, and multinational partners.** COE is a collective term for the related aspects of contemporary OEs that exist or could exist today or in the near- and mid-term future (next 10 years). It is not a totally artificial construct created for training. Rather, it is a representative composite of all the operational variables and actors that create the conditions, circumstances, and influences that can affect military operations in various actual OEs in this contemporary timeframe. This composite can, therefore, provide realistic and relevant conditions necessary for training and leader development.

Why It Is Called Contemporary

The COE is "contemporary" in the sense that it does not represent conditions that existed only in the past or that might exist only in the remote future, but rather those conditions that exist today and in the near- and mid-term future. This composite COE consists not only of the military and/or paramilitary capabilities of potential real-world adversaries, but also of the manifestations of the seven other operational variables that help define any OE.

Training Applications

The COE is particularly valuable in training. Its flexible composite should be capable of addressing the qualities of virtually any OE in which the units or individuals being trained might be called upon to operate. In training environments, an OE is created to approximate the demands of actual OEs that U.S. forces might encounter and to set the conditions for producing desired training outcomes. This involves the appropriate combination of an OPFOR (with military and/or paramilitary capabilities representing a composite of a number of potential adversaries) and other variables of the OE in a realistic, feasible, and plausible manner. See TC 7-101 for more detail on the incorporation of the COE into the design of training exercises.

The Army trains as it will fight. It trains and educates its members to develop agile leaders and organizations able to adapt to any situation and operate successfully in any OE. A training objective consists of task, conditions, and standard. The "conditions" for Army training events must include an OE that is realistic, relevant, and challenging to the ability of the training unit to accomplish the same kinds of mission-essential tasks that would be required of it in an actual OE for an actual operation. As much as possible, a combination of live, virtual, constructive, and gaming training enablers can help replicate conditions representative of an actual OE. (See FM 7-0.)

> **Conditions.** *Those variables of an operational environment or situation in which a unit, system, or individual is expected to operate and may affect performance.*
>
> JP 1-02

In predeployment training, the OE created for a training exercise should represent as closely as possible the conditions of the anticipated OE for the actual mission. Otherwise, the OE for training may represent a composite of the types of conditions that might exist in a number of actual OEs.

OPERATIONAL VARIABLES

All military operations will be significantly affected by a number of variables in the OE beyond simply military forces. Analysis of any OE, including the composite OE created for training purposes, focuses on eight interrelated operational variables:

- **Political.** Describes the distribution of responsibility and power at all levels of governance—formally constituted authorities, as well as informal or covert political powers.
- **Military.** Explores the military and/or paramilitary capabilities of all relevant actors (enemy, friendly, and neutral) in a given OE.
- **Economic.** Encompasses individual and group behaviors related to producing, distributing, and consuming resources.
- **Social.** Describes the cultural, religious, and ethnic makeup within an OE and the beliefs, values, customs, and behaviors of society members.
- **Information.** Describes the nature, scope, characteristics, and effects of individuals, organizations, and systems that collect, process, disseminate, or act on information.
- **Infrastructure.** Is composed of the basic facilities, services, and installations needed for the functioning of a community or society.
- **Physical Environment.** Includes the geography and man-made structures as well as the climate and weather in the area of operations.
- **Time.** Describes the timing and duration of activities, events, or conditions within an OE, as well as how the timing and duration are perceived by various actors in the OE.

The memory aid for these variables is PMESII-PT.

An assessment of these eight operational variables and their relationships helps to understand any OE and its impact on a particular operation. The operational variables form the basis for determining the conditions under which a unit will not only operate but also under which it will train. (See TC 7-101 for guidance on use of the operational variables in creating an appropriate OE for a training exercise.) Just as in an actual operation, commanders and staffs must seek to develop an understanding of the particular OE they face in a training event.

The OPFOR represents a major part of the military variable in training exercises. As such, it must fit in with the characteristics of the other seven operational variables that are selected for that exercise.

OPPOSING FORCE

AR 350-2, which establishes policies and procedures for the Army's Opposing Force (OPFOR) Program, defines an *opposing force* as "a plausible, flexible military and/or paramilitary force representing a composite of varying capabilities of actual worldwide forces, used in lieu of a specific threat force for

Introduction

training and developing U.S. forces." The TC 7-100 series describes the doctrine, organizations, and equipment of such an OPFOR and how to combine it with other operational variables to portray the qualities of a full range of conditions appropriate to Army training environments. As a training tool, the OPFOR must be a challenging, uncooperative sparring partner capable of stressing any or all warfighting functions and mission-essential tasks of the U.S. force.

> *Note.* Although the OPFOR is primarily a training tool, it may be used for other purposes. For example, some capability development activities that do not require simulation of a specific real-world potential adversary may use an OPFOR to portray the "threat" or "enemy."

When U.S. forces become involved in a particular country or region, they must take into account the presence and influence of various types of threats and other actors. In a training environment, an OPFOR can represent a composite of those nation-state or non-state actors that constitute military and/or paramilitary forces that could present a threat to the United States, its friends, or its allies. As in actual OEs, the OE used in training environments will also include various types of other, nonmilitary actors that are not part of the OPFOR, but could be part of the OE. The OPFOR employs tactics that can either mitigate or exploit the OE.

The commander of a U.S. unit plans and conducts training based on the unit's mission-essential task list and priorities of effort. The commander establishes the conditions in which to conduct training to standards. These conditions should include an OPFOR that realistically challenges the ability of the U.S. unit to accomplish its tasks. Training requirements will determine whether the OPFOR's capabilities are fundamental, sophisticated, or a combination of these.

THE HYBRID THREAT FOR TRAINING

In exercise design (see TC 7-101), the type(s) of forces making up the OPFOR will depend upon the conditions determined to be appropriate for accomplishing training objectives. In some cases, the OPFOR may only need to reflect the nature and capabilities of a regular military force, an irregular force, or a criminal organization. However, in order to be representative of the types of threats the Army is likely to encounter in actual OEs, the OPFOR will often need to represent the capabilities of a hybrid threat.

A *hybrid threat* **is the diverse and dynamic combination of regular forces, irregular forces, terrorist forces, and/or criminal elements unified to achieve mutually benefitting effects.** See TC 7-100 for more information on the nature of hybrid threats. However, TC 7-100.2 will focus on the representation of the tactics of such hybrid threats in training exercises. In that context, the force that constitutes the enemy, adversary, or threat for an exercise is called *the Hybrid Threat*, with the acronym HT. Whenever the acronym is used, readers should understand that as referring to the Hybrid Threat. The HT is a realistic and representative composite of actual hybrid threats. This composite constitutes the enemy, adversary, or threat whose military and/or paramilitary forces are represented as an OPFOR in training exercises.

The OPFOR, when representing a hybrid threat, must be a challenging, uncooperative adversary or enemy. It must be capable of stressing any or all warfighting functions and mission-essential tasks of the U.S. armed force being trained.

Military forces may have paramilitary forces acting in loose affiliation with them, or acting separately from them within the same training environment. These relationships depend on the scenario, which is crafted based on the training requirements and conditions of the Army unit being trained.

The OPFOR tactics described in TC 7-100.2 are appropriate for use by an OPFOR that consists either entirely or partly of regular military forces. Some of these tactics, particularly those carried out by smaller organizations, can also be used by irregular forces or even by criminal elements. Even those tactics carried out primarily by regular military forces may involve other components of the HT acting in some capacity. When either acting alone or in concert with other components of the HT, irregular forces and/or criminal elements can also use other tactics, which are outlined in other parts of the TC 7-100 series.

Introduction

BASELINE OPFOR

This TC introduces the baseline tactical doctrine of a flexible, thinking, adaptive OPFOR that applies its doctrine with considerable flexibility, adaptability, and initiative. It is applicable to the entire training community, including the OPFORs at all of the combat training centers (CTCs), the TRADOC schools, and units in the field. It provides an OPFOR that believes that, through adaptive use of all available forces and capabilities, it can create opportunities that, properly leveraged, can allow it to fight and win, even against an opponent such as the United States.

As a baseline for developing specific OPFORs for specific training environments, this TC describes an OPFOR that is representative of the forces of contemporary nation-states. This composite of the characteristics of real-world military forces (possibly combined with irregular forces and/or criminal elements) provides a framework for the realistic and relevant portrayal of capabilities and actions that U.S. armed forces might face in actual OEs.

THE STATE

TC 7-100.2 outlines the tactical-level doctrine of an OPFOR that primarily represents the armed forces of a nation-state. For this composite of real-world nation-state threats, the TC 7-100 series refers to the country to which the regular military forces belong as "the State." The general characteristics of State's doctrine and strategy could fit a number of different types of potential adversaries in a number of different scenarios.

> *Note.* In specific U.S. Army training environments, the generic name of the State may give way to other fictitious country names. (See guidance in AR 350-2.)

The OPFOR exists for the purpose of opposing U.S. forces in training exercises. However, like most countries in the world, the State typically does not design its forces just to fight the United States or its allies. It may design them principally to deal with regional threats and to take advantage of regional opportunities. At the same time, the State is aware that aggressive pursuit of its regional goals might lead to intervention by a major power, such as the United States, from outside the region. To the extent possible, therefore, it might invest in technologies and capabilities that have utility against both regional and extraregional opponents. The basic force structure of the OPFOR is the same for conflict with either type of opponent.

The State must go to war—or continue the war after extraregional intervention—with whatever forces and capabilities it had going into the war. However, it can adapt how it uses those forces and capabilities to fit the nature of the conflict and its opponent(s). Either on its own or as part of the HT for training, the State can employ adaptive strategy, operations, and tactics.

At the strategic level, the State's ability to challenge U.S. interests includes not only the military and paramilitary forces of the State, but also the State's diplomatic-political, informational, and economic instruments of power. Rarely would any country engage the United States or a U.S.-led coalition with purely military means. Trainers need to consider the total OE and all instruments of power at the disposal of the State—not just the military element. It is also possible that the State could be part of an alliance or coalition, in which case the OPFOR could include multinational forces. These nation-state forces may also operate in conjunction with non-state actors such as irregular forces or criminal elements as part of the HT for training.

FLEXIBILITY

The OPFOR must be flexible enough to fit various training requirements. It must be scalable and tunable. Depending on the training requirement, the OPFOR may be a large, medium, or small force. Its technology may be state-of-the-art, relatively modern, obsolescent, obsolete, or an uneven combination of those categories. Its ability to sustain operations may be limited or robust.

Introduction

In the OPFOR baseline presented in this TC, the authors often say that the State or the OPFOR "may" be able to do something or "might" or "could" do something. They often use the progressive forms of verbs to say that the State "is developing" a capability or "is continually modernizing." The State participates in the global market, which can allow it to acquire things it cannot produce domestically. Such descriptions give scenario writers considerable flexibility in determining what the State or the OPFOR actually has at a given point in time or a given place on the battlefield—in a particular scenario.

Thinking

This TC describes how the OPFOR thinks, especially how it thinks about fighting its regional neighbors and/or the United States. This thinking determines basic OPFOR tactics—as well as strategy and operations, which are the subject of FM 7-100.1). It drives OPFOR organizational structures and equipment acquisition or adaptation. It also determines how the nation-state OPFOR that represents the armed forces of the State would interact with other, non-state actors that may be present in the COE.

Just because the U.S. force knows something about how the OPFOR has fought in the past does not mean that the OPFOR will always continue to fight that way. A thinking OPFOR will learn from its own successes and failures, as well as those of its potential enemies. It will adapt its thinking, its makeup, and its way of fighting to accommodate these lessons learned. It will continuously look for innovative ways to deal with the United States and its armed forces.

Adaptability

The OPFOR has developed its doctrine, force structure, and capabilities with an eye toward employing them against both regional and extraregional opponents, if necessary. It has thought about and trained for how to adapt once an extraregional force becomes engaged. It has included this adaptability in its doctrine in the form of general principles, based on its perceptions of the United States and other threats to its goals and aspirations. It will seek to avoid types of operations and environments for which U.S. forces are optimized. During the course of conflict, it will make further adaptations, based on experience and opportunity.

In general, the OPFOR will be less predictable than OPFORs in the past. It will be difficult to template as it adapts and attempts to create opportunity. Its patterns of operation will change as it achieves success or experiences failure. The OPFOR's doctrine might not change, but its way of operating will.

Initiative

Like U.S. Army doctrine, OPFOR doctrine must allow sufficient freedom for bold, creative initiative in any situation. OPFOR doctrine is descriptive, but not prescriptive; authoritative, but not authoritarian; definitive, but not dogmatic. The OPFOR that U.S. units encounter in various training venues will not apply this doctrine blindly or unthinkingly, but will use its experience and assessments to interpolate from this baseline in light of specific situations. Thus, U.S. units can no longer say that the OPFOR has to do certain things and cannot do anything that is not expressly prescribed in established OPFOR doctrine. Doctrine guides OPFOR actions in support of the State's objectives; OPFOR leaders apply it with judgment and initiative.

Terminology

Since OPFOR baseline doctrine is a composite of how various forces worldwide might operate, it uses some terminology that is in common with that of other countries, including the United States. Whenever possible, OPFOR doctrine uses established U.S. military terms—with the same meaning as defined in FM 1-02 and/or JP 1-02. However, the TC 7-100 series also includes some concepts for things the OPFOR does differently from how the U.S. military does them. Even if various real-world foreign countries might use the same concept, or something very close to it, different countries might give it different names. In those cases, the OPFOR TCs either use a term commonly accepted by one or more other countries or create a new, "composite" term that makes sense and is clearly understandable. In any case where an operational or tactical term is not further specifically defined in the TC 7-100 series, it is used in the same sense as in the U.S. definition.

> *Note.* After this introduction, the chapters of this TC address their topics from the OPFOR point of view. So, *friendly* refers to the OPFOR and allied or affiliated forces. Likewise, *enemy* refers to the enemy of the OPFOR, which may be an opponent within its own country or region or an extraregional opponent (normally the United States or a U.S.-led coalition).

THE COE AND OPFOR CONTINUE TO EVOLVE

Taking into consideration adversaries in real-world OEs and desired training outcomes and leader development goals, the authors of TC 7-100 series have developed an OPFOR doctrine and structure that reflect those of forces that could be encountered in actual OEs. While the Army integrates this OPFOR and other operational variables into training scenarios, the authors of the TC 7-100 series are continuing to research real-world OEs and to mature the OPFOR in training in order to provide a richer, appropriately challenging training environment and keep the OPFOR and the COE truly "contemporary."

The nature of real-world OEs and potential OEs is extremely fluid, with rapidly changing regional and global relationships. New actors—both nation-states and non-state actors—are constantly appearing and disappearing from the scene. As the United States and its military forces interact with various OEs worldwide, the OEs change, and so does a composite of those OEs. Therefore, the nature of the COE for training is adaptive and constantly changing. As the Army applies the lessons learned from training, the OPFOR and potential real-world adversaries will also learn and adapt.

The OPFOR tactical doctrine provided in this TC should meet most of the U.S. Army's training needs for the foreseeable future. In the near- and mid-term, almost anyone who fights the United States would probably have to use the same kinds of adaptive action as outlined in this doctrine.

However, as real-world conditions, forces, or capabilities change over time, OPFOR doctrine and its applications will evolve along with them, to continue to provide the Army an appropriate OPFOR. Thus, the OPFOR will remain capable of presenting a challenge that is appropriate to meet evolving training requirements.

This page intentionally left blank.

Chapter 1
Strategic and Operational Framework

This chapter describes the State's national security strategy and how the State designs campaigns and operations to achieve strategic goals outlined in that strategy. This provides the general framework within which the OPFOR plans and executes military actions at the tactical level, which are the focus of the remainder of this TC. See FM 7-100.1 for more detail on OPFOR operations.

Note. The State and its armed forces may act independently, as part of a multinational alliance or coalition, or as part of the Hybrid Threat (HT). When part of the HT, its regular forces will act in concert with irregular forces and/or criminal elements to achieve mutually benefitting effects. Is such cases, the national-level strategy, operational designs, and courses of action of the State may coincide with those of the HT. (See TC 7-100 for HT strategy and operations.)

NATIONAL-LEVEL ORGANIZATION

1-1. The State intends to achieve its strategic goals and objectives through the integrated use of four instruments of national power:
- Diplomatic-political.
- Informational.
- Economic.
- Military.

The four instruments are interrelated and complementary. A clear-cut line of demarcation between military, economic, and political matters does not exist. The informational instrument cuts across the other three. Thus, the State believes that its national security strategy must include all the instruments of national power, not just the military. Power is a combination of many elements, and the State can use them in varying combinations as components of its overall national security strategy.

Note. The term *the State* is simply a generic placeholder until trainers replace it. In specific U.S. Army training environments, the generic name of the State may give way to other (fictitious) country names. (See guidance in AR 350-2.)

NATIONAL COMMAND AUTHORITY

1-2. The National Command Authority (NCA) exercises overall control of the application of all instruments of national power in planning and carrying out the national security strategy. Thus, the NCA includes the cabinet ministers responsible for those instruments of power:
- The Minister of Foreign Affairs.
- The Minister of Public Information.
- The Minister of Finance and Economic Affairs.
- The Minister of the Interior.
- The Minister of Defense.

It may include other members selected by the State's President, who chairs the NCA.

1-3. The President also appoints a Minister of National Security, who heads the Strategic Integration Department (SID) within the NCA. The SID is the overarching agency responsible for integrating all the instruments of national power under one cohesive national security strategy. The SID coordinates the plans and actions of all State ministries, but particularly those associated with the instruments of power. (See figure 1-1.)

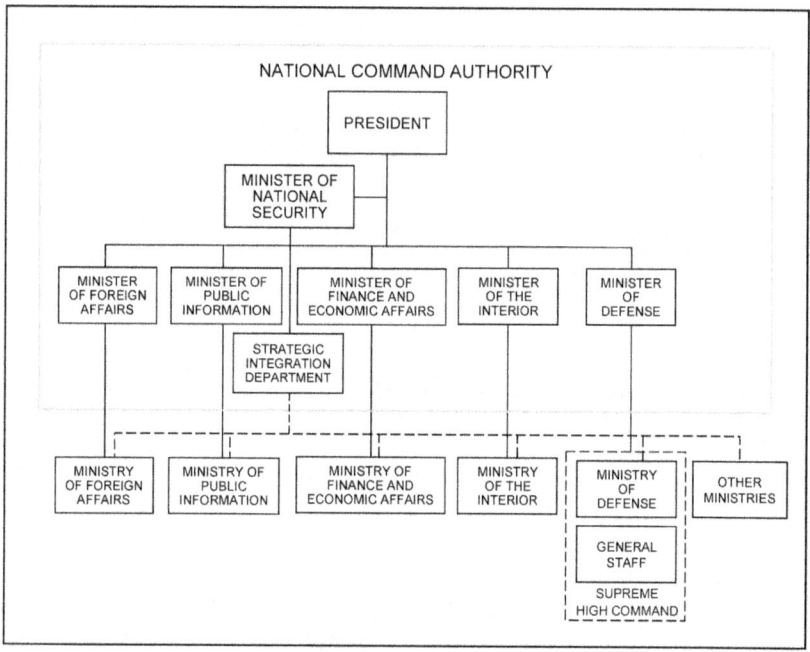

Figure 1-1. National Command Authority

ARMED FORCES

1-4. The NCA exercises command and control (C2) of the State's armed forces (OPFOR) via the Supreme High Command (SHC). The SHC includes the Ministry of Defense (MOD) and a General Staff drawn from all the service components. (See figure 1-2 on page 1-3.) In peacetime, the MOD and General Staff operate closely but separately. The MOD is responsible for policy, acquisitions, and financing the armed forces. The General Staff promulgates policy and supervises the service components. Its functional directorates are responsible for key aspects of defense planning. During wartime, the MOD and General Staff merge to form the SHC, which functions as a unified headquarters.

1-5. The State organizes its armed forces into six service components:
- **Army.** The largest of the six services, although it relies on mobilization of reserve and militia forces to conduct sustained operations.
- **Navy.** Includes naval infantry.
- **Air Force.** Includes national-level Air Defense Forces.
- **Strategic Forces.** With long-range rockets and missiles.
- **Special-Purpose Forces (SPF) Command.** Includes SPF and commando units.

- **Internal Security Forces.** Subordinate to the Ministry of the Interior in peacetime, but can be resubordinated to the SHC as a sixth service in time of war.

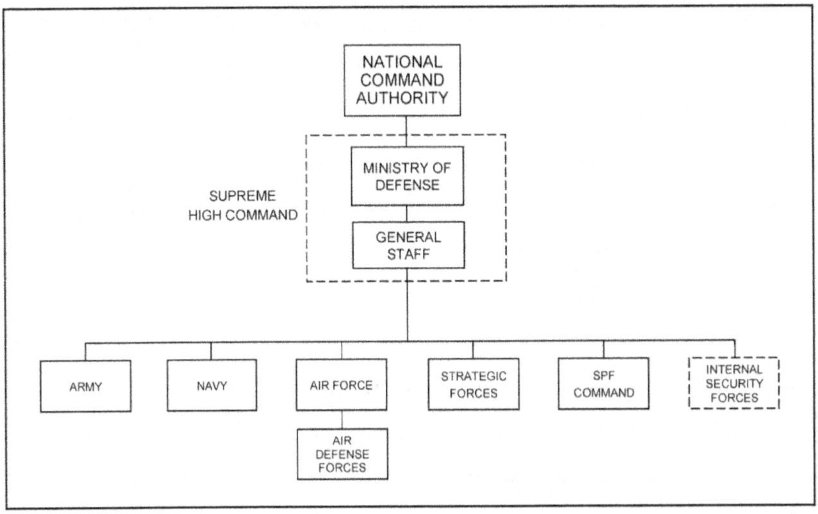

```
                    NATIONAL
                    COMMAND
                    AUTHORITY
                        |
          +-------------+-------------+
          |    MINISTRY OF DEFENSE    |
  SUPREME |                           |
  HIGH COMMAND                        |
          |    GENERAL STAFF          |
          +-------------+-------------+
                        |
  +--------+--------+--------+----------+--------+----------+
  | ARMY   | NAVY   | AIR    | STRATEGIC| SPF    | INTERNAL |
  |        |        | FORCE  | FORCES   |COMMAND | SECURITY |
  |        |        |        |          |        | FORCES   |
  +--------+--------+--------+----------+--------+----------+
                        |
                    AIR DEFENSE
                    FORCES
```

Figure 1-2. The State's armed forces

ADMINISTRATIVE FORCE STRUCTURE

1-6. The OPFOR has an administrative force structure (AFS) that manages its military forces in peacetime. This AFS is the aggregate of various military headquarters, facilities, and installations designed to man, train, and equip the forces. In peacetime, forces are commonly grouped into corps, armies, or army groups for administrative purposes. An army group can consist of several armies, corps, or separate divisions and brigades. In some cases, forces may be grouped administratively under geographical commands designated as military regions or military districts. If the SHC elects to create more than one theater headquarters, it may allocate parts of the AFS to each of the theaters, normally along geographic lines. Normally, these administrative groupings differ from the OPFOR's go-to-war (fighting) force structure. Other parts of the AFS consist of assets centrally controlled at the national level.

1-7. In wartime, the normal role of administrative commands is to serve as force providers during the creation of operational- and tactical-level fighting commands. Typically, an administrative command transfers control of its major fighting forces to one or more task-organized fighting commands. After doing so, the administrative headquarters, facility, or installation may continue to provide depot- and area support-level administrative, supply, and maintenance functions. A geographically based administrative command also provides a framework for the continuing mobilization of reserves to complement or supplement regular forces. In rare cases, an administrative command could function as a fighting command. (See FM 7-100.4 for the basic structures of OPFOR organizations in the AFS and guidance on how they can be task-organized in the fighting force structure.)

Chapter 1

NATIONAL SECURITY STRATEGY

1-8. The *national security strategy* is the State's vision for itself as a nation and the underlying rationale for building and employing its instruments of national power. It outlines how the State plans to use all its instruments of national power to achieve its strategic goals. Despite the term *security*, this strategy defines not just what the State wants to protect or defend, but what it wants to achieve.

NATIONAL STRATEGIC GOALS

1-9. The NCA determines the State's strategic goals. The State's overall goals are to continually expand its influence within its region and possibly to enhance its position within the global community. These are the long-term aims of the State. Supporting the overall, long-term, strategic goals, there may be one or more specific goals, each based on a particular threat or opportunity.

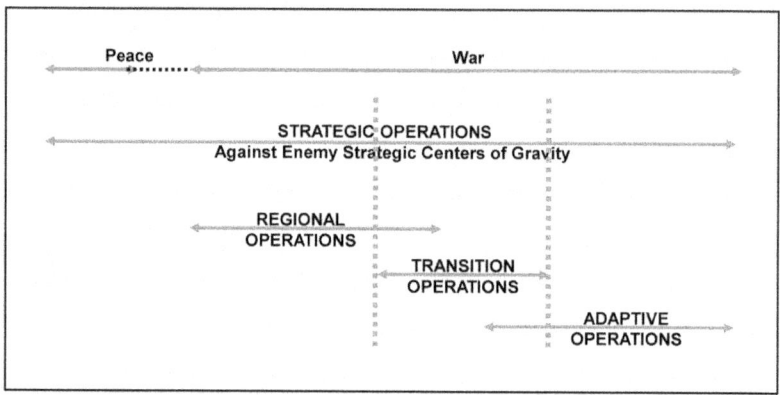

Figure 1-3. Conceptual framework for implementing the State's national security strategy

FRAMEWORK FOR IMPLEMENTING NATIONAL SECURITY STRATEGY

1-10. In pursuit of its national security strategy, the State is prepared to conduct four basic types of strategic-level courses of action (COA). (See figure 1-3.) Each COA involves the use of all four instruments of national power, but to different degrees and in different ways. The State gives the four types the following names:
- **Strategic operations.** A strategic-level COA that uses all instruments of power in peace and war to achieve the goals of the State's national security strategy by attacking the enemy's strategic centers of gravity.
- **Regional operations.** A strategic-level COA (including conventional, force-on-force military operations) against regional adversaries and internal threats.
- **Transition operations.** A strategic-level COA that bridges the gap between regional and adaptive operations and contains some elements of both. The State continues to pursue its regional goals while dealing with the development of outside intervention with the potential for overmatching the State's capabilities.
- **Adaptive operations.** A strategic-level COA to preserve the State's power and apply it in adaptive ways against opponents that may overmatch the State.

1-11. Although the State refers to them as "operations," each of these COAs is actually a subcategory of strategy. Each of these types of "operations" is actually the aggregation of the effects of tactical, operational, and strategic actions. Those actions, in conjunction with the other three instruments of national power, contribute to the accomplishment of strategic goals. The type(s) of operations the State employs at a given time will depend on the types of threats and opportunities present and other conditions in the operational environment (OE).

1-12. Strategic operations are a continuous process not limited to wartime or preparation for war. Once war begins, they continue during regional, transition, and adaptive operations and complement those operations. The latter three types of strategic COAs are also operational designs (see Operational Designs, below). Each of those three occurs only during war and only under certain conditions.

Strategic Operations

1-13. What the State calls "strategic operations" is actually a universal strategic COA it would use to deal with all situations. Strategic operations can occur in peacetime and war—against all kinds of opponents, potential opponents, or neutral parties. The nature of strategic operations at any particular time corresponds to the conditions perceived by the NCA. Depending on the situation, the State may first try to achieve its ends through strategic operations alone, without having to resort to armed conflict. It may be able to achieve the desired goal through pressure applied by other-than-military instruments of power, perhaps with the mere threat of using its military power against a regional opponent. For additional information on strategic operations see FM 7-100.1.

1-14. Once war begins, the State will employ all means available against the enemy's strategic centers of gravity:
- Diplomatic initiatives.
- Information warfare.
- Economic pressure.
- Terror attacks.
- State-sponsored insurgency.
- Direct action by SPF.
- Long-range precision fires.
- Even weapons of mass destruction against selected targets.

These efforts allow the enemy no sanctuary and often place noncombatants at risk.

Regional Operations

1-15. When nonmilitary means are not sufficient or expedient, the State may resort to armed conflict as a means of creating conditions that lead to the desired end state. However, strategic operations continue even if a particular regional threat or opportunity causes the State to undertake "regional operations" that include military means.

1-16. Prior to initiating armed conflict and throughout such conflict with its regional opponent, the State would continue to use strategic operations to preclude intervention by outside actors. Such actors could include other regional neighbors or an extraregional power that could overmatch the State's forces. However, plans for regional operations always include branches and sequels for dealing with the possibility of intervention by an extraregional power.

1-17. At the military level, regional operations may be combined arms, joint, interagency, and/or multinational operations. They are conducted in the State's region and, at least at the outset, against a regional opponent. The State's doctrine, organization, capabilities, and national security strategy allow the OPFOR to deal with regional threats and opportunities primarily through offensive action.

1-18. Regionally focused operations typically involve "conventional" patterns of operation. However, the term *conventional* does not mean that the OPFOR will use only conventional forces and conventional weapons in such a conflict. Nor does it mean that the OPFOR will not use some adaptive approaches. Regional operations may also consist of military and paramilitary forces designed to destabilize the government of the opponent. An example of this might be one or more operational-strategic commands

(OSCs) composed of SPF brigade(s), affiliated guerrilla brigade(s), information warfare battalion(s), and several different affiliated insurgent organizations. The headquarters of an OSC may not necessarily be colocated with its subordinate elements in the neighboring country, but may physically remain within the boundaries of the State.

1-19. Regional operations are not limited to only a single country. Similar operations to those in the example in the paragraph above may be conducted simultaneously in several different countries. A regional "neighbor" is not limited only to those countries sharing a physical international boundary with the State.

Transition Operations

1-20. When unable to limit the conflict to regional operations, the State is prepared to engage extraregional forces through a series of "transition and adaptive operations." Usually, the State does not shift directly from regional to adaptive operations. The transition is incremental and does not occur at a single, easily identifiable point. If the State perceives that intervention is likely, transition operations may begin simultaneously with regional and strategic operations.

1-21. Transition operations allow the State to shift gradually to adaptive operations or back to regional operations. At some point, the State either seizes an opportunity to return to regional operations, or it reaches a point where it must complete the shift to adaptive operations. Even after shifting to adaptive operations, the State tries to set conditions for transitioning back to regional operations. Thus, a period of transition operations overlaps both regional and adaptive operations.

1-22. When an extraregional force starts to deploy into the region, the balance of power may begin to shift away from the State. Although the State may not yet be overmatched, it faces a developing threat it may not be able to handle with normal, "conventional" patterns of operation designed for regional conflict. Therefore, the State must begin to adapt its operations to the changing threat. Transition operations serve as a means for the State to retain the initiative and still pursue its overall strategic goals.

Adaptive Operations

1-23. Once an extraregional force intervenes with sufficient power, the full conventional design used in regionally focused operations may no longer be sufficient to deal with this threat. The State has developed its doctrine, organization, capabilities, and strategy with an eye toward dealing with both regional and extraregional opponents.

1-24. The OPFOR still has the same forces and technology that were available to it for regional operations. However, it must use them in creative and adaptive ways. It has already thought through how it will adapt to this new or changing threat in general terms. It has already developed appropriate branches and sequels to its basic strategic campaign plan (SCP) and does not have to rely on improvisation. During the course of combat, it will make further adaptations, based on experience and opportunity.

1-25. Even with the intervention of an advanced extraregional power, the State will not cede the initiative. It will employ military means so long as this does not either place the regime at risk or risk depriving it of sufficient force to pursue its regional goals after the extraregional intervention is over. The primary objectives are to—
- Preserve combat power.
- Degrade the enemy's will and capability to fight.
- Gain time for aggressive strategic operations to succeed.

1-26. The State believes that adaptive operations can lead to several possible outcomes. If the results do not completely resolve the conflict in the State's favor, they may at least allow the State to return to regional operations. Even a stalemate may be a victory for the State, as long as it preserves enough of its instruments of power to preserve the regime and lives to fight another day.

STRATEGIC CAMPAIGN

1-27. To achieve one or more specific strategic goals, the NCA would develop and implement a specific national strategic campaign. Such a campaign is the aggregate of actions of all the State's instruments of power to achieve a specific set of the State's strategic goals. There would normally be a diplomatic-political campaign, an information campaign, and an economic campaign, as well as a military campaign. All of these must fit into a single, integrated national strategic campaign.

1-28. The campaign could include more than one specific strategic goal. For instance, any strategic campaign designed to deal with an insurgency would include contingencies for dealing with reactions from regional neighbors or an extraregional power that could adversely affect the State and its ability to achieve the selected goal. Likewise, any strategic campaign focused on a goal that involves the State's invasion of a regional neighbor would have to take into consideration possible adverse actions by other regional neighbors, the possibility that insurgents might use this opportunity to take action against the State, and the distinct possibility that the original or expanded regional conflict might lead to extraregional intervention.

1-29. Figure 1-4 shows an example of a single strategic campaign that includes three strategic goals. (The map in this diagram is for illustrative purposes only and does not necessarily reflect the actual size, shape, or physical environment of the State or its neighbors.)

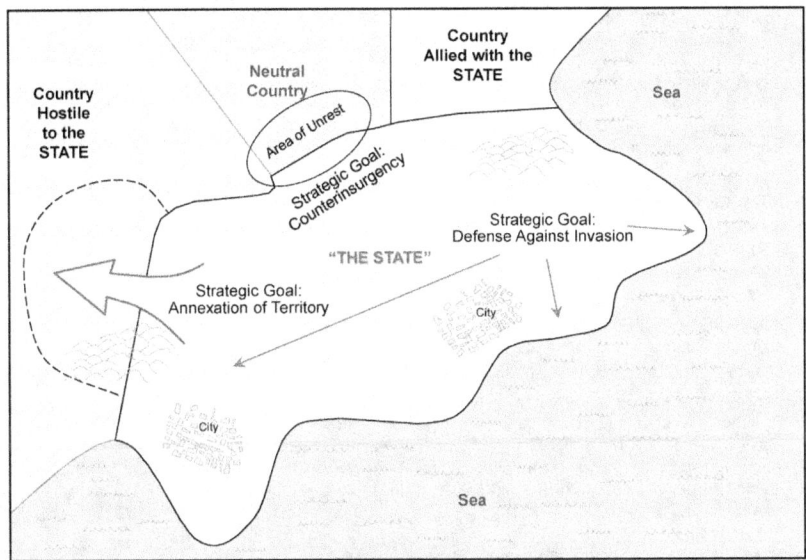

Figure 1-4. Example of a strategic campaign

NATIONAL STRATEGIC CAMPAIGN PLAN

1-30. The national SCP is the plan for integrating the actions of all instruments of national power to set conditions favorable for achieving the central goal(s) identified in the national security strategy. The MOD is only one of several State ministries that provide input and are then responsible for carrying out their respective parts of the consolidated national plan. State ministries responsible for each of the four instruments of power will develop their own campaign plans as part of the unified national SCP.

1-31. A national SCP defines the relationships among all State organizations, military and nonmilitary, for the purposes of executing that SCP. The SCP describes the intended integration, if any, of multinational forces in those instances where the State is acting as part of a coalition. It would also include the State's interactions with irregular forces and/or criminal elements as part of the HT.

MILITARY STRATEGIC CAMPAIGN PLAN

1-32. Within the context of the national strategic campaign, the MOD and General Staff develop and implement a military strategic campaign. During peacetime, the Operations Directorate of the General Staff is responsible for developing, staffing, promulgation, and continuing review of the military strategic campaign plan. It must ensure that the military plan would end in achieving military conditions that would fit with the conditions created by the diplomatic-political, informational, and economic portions of the national plan that are prepared by other State ministries. Therefore, the Operations Directorate assigns liaison officers to other important government ministries.

1-33. Although the State's armed forces (OPFOR) may play a role in strategic operations, the focus of their planning and effort is on the military aspects of regional, transition, and adaptive operations. A military strategic campaign may include several combined arms, joint, and/or interagency operations. If the State succeeds in forming a regional alliance or coalition, these operations may also be multinational.

1-34. The General Staff acts as the executive agency for the NCA. All military forces report through it to the NCA. The Chief of the General Staff (CGS), with NCA approval, defines the theater in which the armed forces will conduct the military campaign and its subordinate operations. He determines the task organization of forces to accomplish the operational-level missions that support the overall campaign plan. He also determines whether it will be necessary to form more than one theater headquarters. For most campaigns, there will be only one theater, and the CGS will serve as theater commander, thus eliminating one echelon of command at the strategic level.

1-35. In wartime, the MOD and the General Staff combine to form the SHC, under the command of the CGS. The Operations Directorate continues to review the military SCP and modify it or develop new plans based on guidance from the CGS. It generates options and contingency plans for various situations that may arise. Once the CGS approves a particular plan for a particular strategic goal, he issues it to the appropriate operational-level commanders.

1-36. The military SCP assigns forces to operational-level commands and designates areas of responsibility (AORs) for those commands. Each command identified in the SCP prepares an operation plan that supports the execution of its role in that SCP.

1-37. From the General Staff down through the operational and tactical levels, the staff of each military headquarters has an operations directorate or section that is responsible for planning. The plan at each level specifies the AOR and task organization of forces allocated to that level of command, in order to accomplish the mission assigned by a higher headquarters. Once the commander at a particular level approves the plan, he issues it to the subordinate commanders who will execute it. Figure 1-5 illustrates the framework for planning from the national level down through military channels to the operational and tactical levels.

OPERATIONAL-LEVEL ORGANIZATION

1-38. In peacetime, tactical-level commands belong to parent organizations in the AFS. In wartime, they typically serve as part of a field group (FG) or an OSC. In rare cases, they might also fight as part of their original parent units from the AFS.

FIELD GROUP

1-39. An FG is the largest operational-level organization, since it has one or more smaller operational-level commands subordinate to it. FGs are always joint and interagency organizations and are often multinational. However, this level of command may or may not be necessary in a particular SCP. The General Staff does not normally form standing FG headquarters, but may organize one or more during full mobilization, if necessary. An FG may be organized when the span of control at theater level exceeds four or five subordinate commands. This can facilitate the theater commander's remaining focused on the theater-strategic level of war and enable him to coordinate effectively the joint forces allocated for his use.

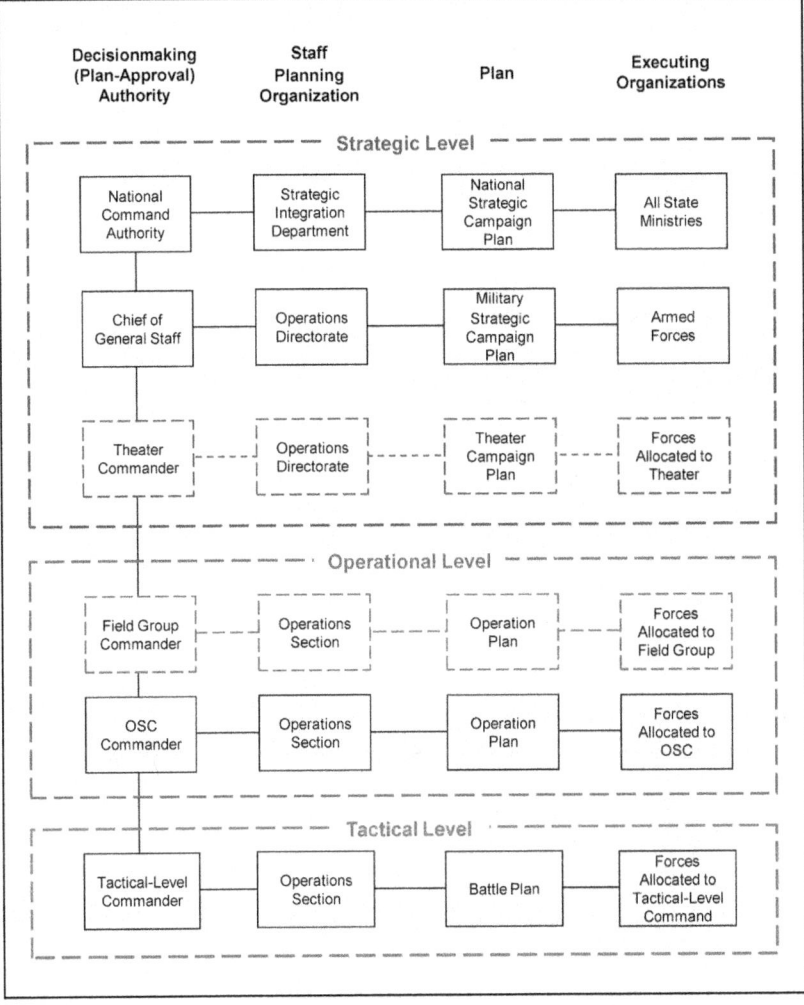

Figure 1-5. State and OPFOR planning framework

1-40. FGs are typically formed for one or more of the following reasons:
- An SCP may require a large number of OSCs and/or operational-level commands from the AFS. When the number of major military efforts in a theater exceeds the theater commander's desired or achievable span of control, he may form one or more FGs.
- In the rare cases when multiple operational-level commands from the AFS become fighting commands, they could come under the command of an FG headquarters.
- Due to modifications to the SCP, a standing operational-level headquarters that was originally designated as an OSC headquarters may receive one or more additional major operational-level commands from the AFS as fighting commands. Then the OSC headquarters would evolve into an FG headquarters.

In the first two cases, a standing FG staff would be formed and identified as having control over two or more OSCs (or operational-level headquarters from the AFS) as part of the same SCP. In the third case, the original OSC headquarters would be redesignated as an FG headquarters. In any case, the FG command group and staff would be structured in the same manner as those of an OSC.

OPERATIONAL-STRATEGIC COMMAND

1-41. The OPFOR's primary operational organization is the OSC. (See FM 7-100.1 for more detail.) Once the General Staff writes a particular SCP, it forms one or more standing OSC headquarters. Each OSC headquarters is capable of controlling whatever combined arms, joint, interagency, or multinational operations are necessary to execute that OSC's part of the SCP. However, the OSC headquarters does not have forces permanently assigned to it.

1-42. When the NCA decides to execute a particular SCP, each OSC participating in that plan receives appropriate units from the AFS, as well as interagency and/or multinational forces. The allocation of organizations to an OSC depends on what is available in the State's AFS and the requirements of other OSCs. Forces subordinated to an OSC may continue to depend on the AFS for support.

1-43. If a particular OSC has contingency plans for participating in more than one SCP, it could receive a different set of forces under each plan. In each case, the forces would be task-organized according to its mission requirements in the given plan. Thus, each OSC consists of those division-, brigade-, and battalion-size organizations allocated to it by the SCP currently in effect. These forces also may be allocated to the OSC for the purpose of training for a particular SCP. When an OSC is neither executing tasks as part of an SCP nor conducting exercises with its identified subordinate forces, it exists as a planning headquarters.

OPERATIONAL DESIGNS

1-44. Of the four types of strategic COA described above, regional, transition, and adaptive operations are also operational designs. While the State and the OPFOR as a whole are in the condition of regional, transition, or adaptive operations, an operational- or tactical-level commander will still receive a mission statement in plans and orders from his higher authority stating the purpose of his actions. To accomplish that purpose and mission, he will use—
- As much as he can of the conventional patterns of operation that were available to him during regional operations.
- As much as he has to of the more adaptive-type approaches dictated by the presence of an extraregional force.

Figure 1-6 illustrates the basic conceptual framework for the three operational designs.

Strategic and Operational Framework

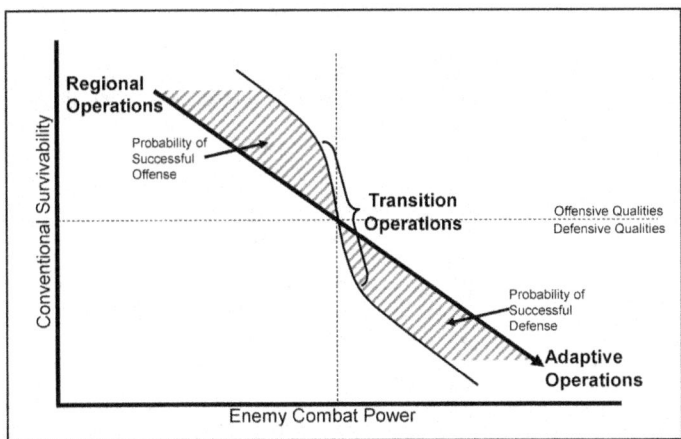

Figure 1-6. Operational designs

REGIONAL OPERATIONS

1-45. Against opponents from within its region, the OPFOR may conduct "regional operations" with a relatively high probability of success in primarily offensive actions. OPFOR offensive operations are characterized by using all available means to saturate the OE with actions designed to disaggregate an opponent's capability, capacity, and will to resist. These actions will not be limited to attacks on military and security forces, but will affect the entire OE. The opponent will be in a fight for survival across many of the variables of the OE: political, military, economic, social, information, and infrastructure.

1-46. The OPFOR may possess an overmatch in some or all elements of combat power against regional opponents. It is able to employ that power in an operational design focused on offensive action. A weaker regional neighbor may not actually represent a threat, but rather an opportunity that the OPFOR can exploit. To seize territory or otherwise expand its influence in the region, the OPFOR must destroy a regional enemy's will and capability to continue the fight.

1-47. During regional operations, the OPFOR relies on the State's continuing strategic operations (see above) to preclude or control outside intervention. It tries to keep foreign perceptions of its actions during a regional conflict below the threshold that will invite intervention by other regional actors or extraregional forces. The OPFOR wants to achieve its objectives in the regional conflict, but has to be careful how it does so. It works to prevent development of international consensus for intervention and to create doubt among possible participants. Still, at the very outset of regional operations, it lays plans and positions forces to conduct access-limitation operations in the event of outside intervention.

1-48. Although the OPFOR would prefer to achieve its objectives through regional operations, it has the flexibility to change and adapt if required. Since the OPFOR assumes the possibility of extraregional intervention, its operation plans will already contain thorough plans for transition operations, as well as adaptive operations, if necessary.

TRANSITION OPERATIONS

1-49. Transition operations serve as a pivotal point between regional and adaptive operations. The transition may go in either direction. The fact that the OPFOR begins transition operations does not necessarily mean that it must complete the transition from regional to adaptive operations (or vice versa).

Chapter 1

As conditions allow or dictate, the "transition" could end with the OPFOR conducting the same type of operations as before the shift to transition operations.

1-50. The OPFOR conducts transition operations when other regional and/or extraregional forces threaten its ability to continue regional operations in a conventional design against the original regional enemy. At the point of shifting to transition operations, the OPFOR may still have the ability to exert its combat power against an overmatched regional enemy. Indeed, it may have already defeated its original adversary. However, its successful actions in regional operations have prompted either other regional actors or an extraregional actor to contemplate intervention. The OPFOR will use all means necessary to preclude or defeat intervention.

1-51. Even extraregional forces may be vulnerable to "conventional" operations during the time they require to build combat power and create support at home for their intervention. Against an extraregional force that either could not fully deploy or has been successfully separated into isolated elements, the OPFOR may still be able to use some of the more conventional patterns of operation.

1-52. As the OPFOR begins transition operations, its immediate goal is preservation of its combat power while seeking to set conditions that will allow it to transition back to regional operations. Transition operations feature a mixture of offensive and defensive actions that help the OPFOR control the tempo while changing the nature of conflict to something for which the intervening force is unprepared. Transition operations can also buy time for the State's strategic operations to succeed.

1-53. There are two possible outcomes to transition operations:
- The extraregional force suffers sufficient losses or for other reasons must withdraw from the region. In this case, the OPFOR's operations may begin to transition back to regional operations, again becoming primarily offensive.
- The extraregional force is not compelled to withdraw and continues to build up power in the region. In this case, the OPFOR's transition operations may begin to gravitate in the other direction, toward adaptive operations.

ADAPTIVE OPERATIONS

1-54. At some point, an extraregional force may intervene with sufficient power to overmatch the OPFOR at least in certain areas. When that occurs, the OPFOR has to adapt its patterns of operation to deal with this threat. The OPFOR will seek to conduct adaptive operations in circumstances and terrain that provide opportunities to optimize its own capabilities and degrade those of the enemy. It will employ a force that is optimized for the terrain or for a specific mission. For example, it will use its antitank capability, tied to obstacles and complex terrain, inside a defensive structure designed to absorb the enemy's momentum and fracture his organizational framework.

1-55. At least at the tactical and operational levels, the types of adaptive actions and methods that characterize adaptive operations can also serve the OPFOR well in regional or transition operations. However, the OPFOR will conduct such adaptive actions more frequently and on a larger scale during adaptive operations against a fully-deployed extraregional force. During the course of operations, the OPFOR will make further adaptations, based on what works or does not work against a particular opponent.

TYPES OF OFFENSIVE AND DEFENSIVE ACTION

1-56. The types of offensive action in OPFOR doctrine are both tactical methods and guides to the design of operational COAs. An OSC offensive operation plan may include subordinate units that are executing different offensive and defensive COAs within the overall offensive mission framework. The OPFOR recognizes three basic types of offensive action at OSC level: attack, limited-objective attack, and strike. (See chapter 3 for discussion of the first two types of action at the tactical group, division, and brigade level.) A strike is an offensive action that rapidly destroys a key enemy organization through a synergistic combination of massed precision fires and maneuver (see FM 7-100.1 for more detail).

1-57. The OPFOR's types of defensive action also are both tactical methods and guides to the design of operational COAs. The two basic types are maneuver defense and area defense. An OSC defensive operation plan may include subordinate units that are executing various combinations of maneuver and area defenses, along with some offensive COAs, within the overall defensive mission framework. (See chapter 4 for discussion of the same types of defensive action at the tactical group, division, and brigade level.)

SYSTEMS WARFARE

1-58. The OPFOR defines a *system* as a set of different elements so connected or related as to perform a unique function not performable by the elements or components alone. The essential ingredients of a system include—

- The components.
- The synergy among components and other systems.
- Some type of functional boundary separating it from other systems.

Therefore, a "system of systems" is a set of different systems so connected or related as to produce results unachievable by the individual systems alone. The OPFOR views the OE, the battlefield, the State's own instruments of power, and an opponent's instruments of power as a collection of complex, dynamic, and integrated systems composed of subsystems and components.

PRINCIPLE

1-59. The primary principle of systems warfare is the identification and isolation of the critical subsystems or components that give the opponent the capability and cohesion to achieve his aims. While the aggregation of these subsystems or components is what makes the overall system work, the interdependence of these subsystems is also a potential vulnerability. The focus is on disaggregating the system by attacking critical subsystems in a way that will degrade or destroy the use, effectiveness, or importance of the overall system. Systems warfare has applicability or impact at all three levels of warfare.

APPLICATION AT THE STRATEGIC LEVEL

1-60. At the strategic level, the instruments of national power and their application are the focus of analysis. National power is a system of systems in which the instruments of national power work together to create a synergistic effect. Each instrument of power (diplomatic-political, informational, economic, and military) is also a collection of complex and interrelated systems.

1-61. The State clearly understands how to analyze and locate the critical components of its own instruments of power. It will aggressively aim to protect its own systems from attack or vulnerabilities. It also understands that an adversary's instruments of power are similar to the State's. Thus, at the strategic level, the State can use the OPFOR and its other instruments of power to counter or target the systems and subsystems that make up an opponent's instruments of power. The primary purpose is to subdue, control, or change the opponent's behavior.

1-62. If an opponent's strength lies in his military power, the State and the OPFOR can attack the other instruments of power as a means of disaggregating or disrupting the enemy's system of national power. Thus, it is possible to render the overall system ineffective without necessarily having to defeat the opponent militarily.

APPLICATION AT THE OPERATIONAL LEVEL

1-63. At the operational level, the application of systems warfare pertains only to the use of armed forces to achieve a result. Therefore, the "system of systems" in question at this level is the combat system of the OPFOR and/or the enemy.

Combat System

1-64. A *combat system* (see figure 1-7 on page 1-14) is the "system of systems" that results from the synergistic combination of four basic subsystems that are integrated to achieve a military function. The subsystems are as follows:

- **Combat forces**—such as main battle tanks, infantry fighting vehicles (IFVs) and/or armored personnel carriers (APCs), or infantry.
- **Combat support forces**—such as artillery, surface-to-surface missiles (SSMs), air defense, engineers, and direct air support.
- **Logistics forces**—such as transportation, ammunition, fuel, rations, maintenance, and medical.
- **C2 and reconnaissance, intelligence, surveillance and target acquisition (RISTA)**—such as headquarters, signal nodes, satellite downlink sites, and reconnaissance sensors.

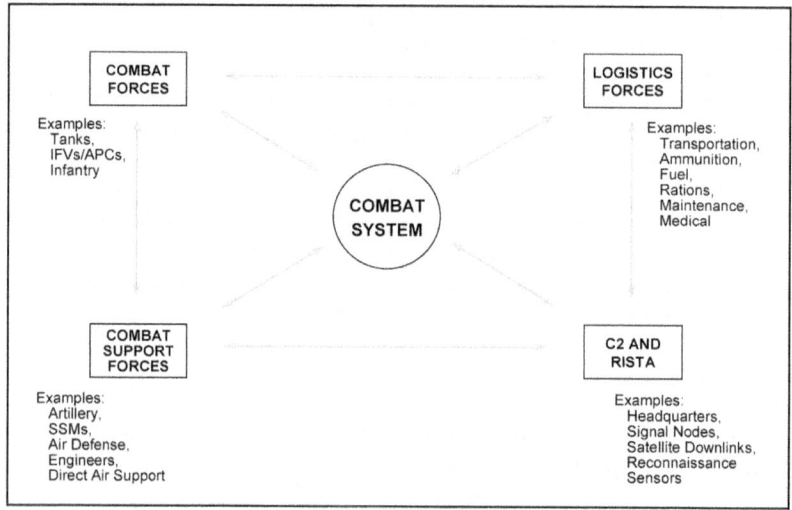

Figure 1-7. Combat system

1-65. The combat system is characterized by interaction and interdependence among its subsystems. Therefore, the OPFOR will seek to identify key subsystems of an enemy combat system and target them and destroy them individually. Against a technologically superior extraregional force, the OPFOR will often use any or all subcomponents of its own combat system to attack the most vulnerable parts of the enemy's combat system rather than the enemy's strengths. For example, attacking the enemy's logistics, C2, and RISTA can undermine the overall effectiveness of the enemy's combat system without having to directly engage his superior combat and combat support forces. Aside from the physical effect, the removal of one or more key subsystems can have a devastating psychological effect, particularly if it occurs in a short span of time.

Planning and Execution

1-66. The systems warfare approach to combat is a means to assist the commander in the decisionmaking process and the planning and execution of his mission. The OPFOR believes that a qualitatively and/or quantitatively weaker force can defeat a superior foe, if the lesser force can dictate the terms of combat. It believes that the systems warfare approach allows it to move away from the traditional attrition-based approach to combat. It is no longer necessary to match an opponent system-for-system or capability-for-capability. Commanders and staffs will locate the critical component(s) of the enemy combat system, patterns of interaction, and opportunities to exploit this connectivity. The OPFOR will seek to disaggregate enemy combat power by destroying or neutralizing single points of failure in the enemy's combat system. Systems warfare has applications in both offensive and defensive contexts.

1-67. The essential step after the identification of the critical subsystems and components of a combat system is the destruction or degradation of the synergy of the system. This may take one of three forms—
- Total destruction of a subsystem or component.
- Degradation of the synergy of components.
- The simple denial of access to critical links between systems or components.

1-68. The destruction of a critical component or link can achieve one or more of the following:
- Create windows of opportunity that can be exploited.
- Set the conditions for offensive action.
- Support a concept of operation that calls for exhausting the enemy on the battlefield.

Once the OPFOR has identified and isolated a critical element of the enemy combat system that is vulnerable to attack, it will select the appropriate method of attack.

1-69. Today's state-of-the-art combat and combat support systems are impressive in their ability to deliver precise attacks at long standoff distances. However, the growing reliance of some extraregional forces on these systems offers opportunity. For example, attacking critical ground-based C2 and RISTA nodes or logistics systems and lines of communication (LOCs) may have a very large payoff for relatively low investment and low risk. Modern logistics systems assume secure LOCs and voice or digital communications. These characteristics make such systems vulnerable. Therefore, the OPFOR can greatly reduce a military force's combat power by attacking a logistics system that depends on "just-in-time delivery."

1-70. For the operational commander, the systems warfare approach to combat is not an end in itself. It is a key component in his planning and sequencing of tactical battles and engagements aimed toward achieving assigned strategic goals. Systems warfare supports his concept; it is not the concept. The ultimate aim is to destroy the enemy's will and ability to fight.

APPLICATION AT THE TACTICAL LEVEL

1-71. It is at the tactical level that systems warfare is executed in attacking the enemy's combat system. While the tactical commander may use systems warfare in the smaller sense to accomplish assigned missions, his attack on systems normally will be in response to missions assigned him by the operational commander.

APPLICATION ACROSS ALL TYPES OF STRATEGIC-LEVEL ACTIONS

1-72. Systems warfare is applicable against all types of opponents in all strategic-level COAs. In regional operations, the OPFOR will seek to render a regional opponent's systems ineffective to support his overall concept of operation. However, this approach is especially conducive to the conduct of transition and adaptive operations. The very nature of this approach lends itself to adaptive and creative options against an adversary's technological overmatch.

RELATIONSHIP TO THE C2 PROCESS

1-73. The systems warfare approach to combat is an important part of OPFOR planning. It serves as a means to analyze the OPFOR's own combat system and how it can use the combined effects of this system to degrade the enemy's combat system. The OPFOR believes that the approach allows its decisionmakers to be anticipatory rather than reactive.

THE ROLE OF PARAMILITARY FORCES IN OPERATIONS

1-74. Paramilitary forces are those organizations that are distinct from the regular armed forces but resemble them in organization, equipment, training, or purpose. Basically, any organization that accomplishes its purpose, even partially, through the force of arms could be considered a paramilitary organization. These organizations can be part of a government infrastructure or operate outside of any government or any institutionalized controlling authority.

1-75. The OPFOR views these organizations as assets that can be used to its advantage in time of war. Within its own structure, the OPFOR has formally established this concept by assigning the Internal Security Forces, part of the Ministry of the Interior in peacetime, to the SHC during wartime. Additionally, the OPFOR cultivates relationships with and covertly supports nongovernment paramilitary organizations to achieve common goals while at peace and to have a high degree of influence on them when at war.

1-76. The primary paramilitary organizations are the Internal Security Forces, irregular forces, and criminal organizations. The degree of control the OPFOR has over these organizations varies from absolute, in the case of the Internal Security Forces, to tenuous when dealing with irregular forces and criminal organizations. In the case of those organizations not formally tied to the OPFOR structure, control can be enhanced through the exploitation of common interests and ensuring that these organizations see personal gain in supporting OPFOR goals. Common interests may result in the State's regular military forces acting as part of the HT, which also includes irregular forces and/or criminal elements.

1-77. The OPFOR views the creative use of these organizations as a means of providing depth and continuity to its operations. A single attack by an irregular force will not in itself win the war. However, the use of paramilitary organizations to carry out a large number of planned actions, in support of strategy and operations, can play an important part in assisting the OPFOR in achieving its goals. These actions, taken in conjunction with other adaptive actions, can also supplement a capability degraded due to enemy superiority.

Chapter 2

Command and Control

This chapter focuses on tactical command and control (C2). It explains how the OPFOR expects to direct the forces and actions described in other chapters of this TC. Most important, it shows how OPFOR commanders and staffs think and work. In modern war, the overriding need for speedy decisions to seize fleeting opportunities drastically reduces the time available for decisionmaking and for issuing and implementing orders. Moreover, the tactical situation is subject to sudden and radical changes, and the results of combat are more likely to be decisive than in the past. OPFOR C2 participants, processes, and systems are designed to operate effectively and efficiently in this environment.

CONCEPT AND PRINCIPLES

2-1. The OPFOR defines *command and control* as the actions of commanders, command groups, and staffs of military headquarters to maintain continual combat readiness and combat efficiency of forces, to plan and prepare for combat operations, and to provide leadership and direction during the execution of assigned missions. It views the C2 process as the means for assuring both *command* (establishing the aim) and *control* (sustaining the aim). The OPFOR's tactical C2 concept is based on the following key principles:

MISSION TACTICS

2-2. OPFOR tactical units focus on the purpose of their tactical missions. They continue to act on that purpose even when the details of an original plan have become irrelevant through enemy action or unforeseen events.

FLEXIBILITY THROUGH BATTLE DRILL

2-3. True flexibility comes from soldiers in tactical units understanding basic battlefield functions to such a degree that they are second nature. Battle drills are not viewed as a restrictive methodology. Only when common battlefield functions can be performed rapidly without further guidance or orders do tactical commanders achieve the flexibility to modify the plan on the move.

ACCOUNTING FOR MISSION DYNAMICS

2-4. The OPFOR recognizes that enemy action and battlefield conditions may make the originally selected mission irrelevant and require an entirely new mission be acted upon without an intermediate planning session. An example would be an OPFOR fixing force that finds itself the target of an enemy fixing action. To continue solely as a fixing force would actually assist the enemy in achieving his mission. In this case, the OPFOR unit might choose to change its task organization on the move and allocate a part of the fixing force to the exploitation force and use a smaller amount of combat power to keep the enemy fixing force from being able to influence the fight. OPFOR tactical headquarters constantly evaluate the situation to determine if the mission being executed is still relevant and, if not, to advise the commander on how best to shift to a relevant course of action. Each situation requires the commander at each level of command to act flexibly, exercising his judgment as to what best meets and sustains the aim of his superior.

COMMAND AND SUPPORT RELATIONSHIPS

2-5. OPFOR units are organized using four command and support relationships, summarized in table 2-1 and described in the following paragraphs. These relationships may shift during the course of an operation in order to best align the force with the tasks required. The general category of *subordinate units* includes both constituent and dedicated relationships; it can also include interagency and multinational (allied) subordinates.

Table 2-1. Command and support relationships

Relationship	Commanded by	Logistics from	Positioned by	Priorities from
Constituent	Gaining	Gaining	Gaining	Gaining
Dedicated	Gaining	Parent	Gaining	Gaining
Supporting	Parent	Parent	Supported	Supported
Affiliated	Self	Self or "Parent"	Self	Mutual Agreement

CONSTITUENT

2-6. *Constituent* units are those forces assigned directly to a unit and forming an integral part of it. They may be organic to the table of organization and equipment (TOE) of the administrative force structure forming the basis of a given unit, assigned at the time the unit was created, or attached to it after its formation.

DEDICATED

2-7. *Dedicated* is a command relationship identical to constituent with the exception that a dedicated unit still receives logistics support from a parent headquarters of similar type. An example of a dedicated unit would be the case where a specialized unit, such as an attack helicopter company, is allocated to a brigade tactical group (BTG). The base brigade does not possess the technical experts or repair facilities for the aviation unit's equipment. However, the dedicated relationship permits the company to execute missions exclusively for the BTG while still receiving its logistics support from its parent organization. In OPFOR plans and orders, the dedicated command and support relationship is indicated by (DED) next to a unit title or symbol.

SUPPORTING

2-8. *Supporting* units continue to be commanded by and receive their logistics from their parent headquarters, but are positioned and given mission priorities by their supported headquarters. This relationship permits supported units the freedom to establish priorities and position supporting units while allowing higher headquarters to rapidly shift support in dynamic situations. An example of a supporting unit would be a multiple rocket launcher battalion supporting a BTG for a particular phase of an operation but ready to rapidly transition to a different support relationship when the BTG becomes the division tactical group (DTG) reserve in a later phase. The supporting unit does not necessarily have to be within the supported unit's area of responsibility (AOR). In OPFOR plans and orders, the supporting command and support relationship is indicated by (SPT) next to a unit title or symbol.

AFFILIATED

2-9. *Affiliated* organizations are those operating in a unit's AOR that the unit may be able to sufficiently influence to act in concert with it for a limited time. No "command relationship" exists between an affiliated organization and the unit in whose AOR it operates. Affiliated organizations are typically nonmilitary or paramilitary groups such as criminal cartels or insurgent organizations. In some cases, affiliated forces may receive support from the DTG or BTG as part of the agreement under which they cooperate. Although there will typically be no formal indication of this relationship in OPFOR plans and orders, in rare cases (AFL) is used next to unit titles or symbols.

Note. In organization charts, the affiliated status is reflected by a dashed (rather than solid) line connecting the affiliated force to the unit with which it is affiliated (see the examples in figures 2-1 and 2-2). This is not to be confused with dashed boxes, which indicate additional units that may or may not be present.

TACTICAL-LEVEL ORGANIZATIONS

2-10. OPFOR tactical organizations fight battles and engagements. They execute the combat actions described in the remainder of this TC.

2-11. In the OPFOR's administrative force structure (AFS), the largest *tactical-level* organizations are divisions and brigades. In peacetime, they are often subordinate to a larger, operational-level administrative command. However, a service of the Armed Forces might also maintain some separate single-service tactical-level commands (divisions, brigades, or battalions) directly under the control of their service headquarters. (See FM 7-100.4.) For example, major tactical-level commands of the Air Force, Navy, Strategic Forces, and the Special-Purpose Forces (SPF) Command often remain under the direct control of their respective service component headquarters. The Army component headquarters may retain centralized control of certain elite elements of the ground forces, including airborne units and Army SPF. This permits flexibility in the employment of these relatively scarce assets in response to national-level requirements.

2-12. For these tactical-level organizations (division and below), the organizational directories of FM 7-100.4 contain standard "TOE" structures of the AFS. However, these administrative groupings normally differ from the OPFOR's go-to-war (fighting) force structure. (See FM 7-100.4 on task-organizing.)

DIVISIONS

2-13. In the OPFOR's AFS, the largest tactical formation is the division. Divisions are designed to be able to—

- Serve as the basis for forming a DTG, if necessary. (See discussion of Tactical Groups, below.)
- With or without becoming a DTG, fight as part of an operational-strategic command (OSC) or an organization from the AFS (such as army or military region) or as a separate unit in a field group (FG).
- Sustain independent combat operations over a period of several days.
- Integrate interagency forces up to brigade or group size.
- Execute all of the actions discussed in this TC.

Integrated Fires Command

2-14. The integrated fires command (IFC) is a combination of a standing C2 structure and task-organizing of constituent and dedicated fire support units. Division or DTG and above have IFCs. Brigades, BTGs, and below do not. All division-level and above OPFOR organizations possess an IFC C2 structure—staff, command post (CP), communications and intelligence architecture, and automated fire control system. The IFC exercises C2 of all constituent and dedicated fire support assets retained by its level of command. This includes army aviation, artillery, and missile units. It also exercises C2 over all reconnaissance, intelligence, surveillance, and target acquisition (RISTA) assets allocated to it. (See chapter 9 for more detail on the IFC.)

Note. Based on mission requirements, the division or DTG (or above) commander may also place maneuver forces under the command of the IFC commander. One possibility would be for the IFC CP to command the disruption force, the exploitation force, or any other functional force whose actions must be closely coordinated with fires delivered by the IFC.

Chapter 2

Integrated Support Command

2-15. The *integrated support command* (ISC) is the aggregate of combat service support units (and perhaps some combat support units) organic to a division and additional assets allocated from the AFS to a DTG. It contains such units that the division or DTG does not suballocate to lower levels of command in a constituent or dedicated relationship. The division or DTG further allocates part of its ISC units as an integrated support group (ISG) to support its IFC, and the remainder supports the rest of the division or DTG, as a second ISG.

2-16. For organizational efficiency, combat service support units may be grouped in this ISC and its ISGs, although they may support only one of the major units of the division or DTG or its IFC. Sometimes, an ISC or ISG might also include units performing combat support tasks (such as chemical warfare, engineer, or law enforcement) that support the division or DTG and its IFC. (See chapter 14 for more detail on the ISC and ISG.)

MANEUVER BRIGADES

2-17. The OPFOR's basic combined arms unit is the maneuver brigade. In the AFS, maneuver brigades are typically constituent to divisions, in which case the OPFOR refers to them as *divisional brigades*. However, some are organized as *separate brigades*, designed to have greater ability to accomplish independent missions without further allocation of forces from higher-level tactical headquarters. In OPFOR plans and orders, the status of separate brigades may be indicated by (Sep) next to a unit title or symbol. Similarly, a brigade that is part of a division may be marked as (Div) in order to distinguish it from a separate brigade.

2-18. Maneuver brigades are designed to be able to—
- Serve as the basis for forming a BTG, if necessary.
- Fight as part of a division or DTG.
- Fight as a separate unit in an OSC, an organization from the AFS (such as army, corps, or military district), or an FG.
- Sustain independent combat operations over a period of 1 to 3 days.
- Integrate interagency forces up to battalion size.
- Execute all of the actions discussed in this TC.

TACTICAL GROUPS

2-19. A *tactical group* is a task-organized division or brigade that has received an allocation of additional land forces in order to accomplish its mission. These additional forces may come from within the Ministry of Defense, from the Ministry of the Interior, or from affiliated forces. Typically, these assets are initially allocated to an OSC or FG, which further allocates them to its tactical subordinates. The purpose of a tactical group is to ensure unity of command for all land forces in a given AOR. Tactical groups formed from divisions are *division tactical groups* (DTGs) and those from brigades are *brigade tactical groups* (BTGs). A DTG may fight as part of an OSC or as a separate unit in an FG. A BTG may fight as part of a division or DTG or as a separate unit in an OSC or FG. Figures 2-1 and 2-2 give examples of the types of units that could comprise possible DTG and BTG organizations.

2-20. In addition to augmentation received from a higher command, a DTG or BTG normally retains the assets that were originally subordinate to the division or brigade that served as the basis for the tactical group. However, it is also possible that the higher command could use units from one division or brigade as part of a tactical group that is based on another division or brigade.

Note. Any division or brigade receiving additional assets from a higher command becomes a DTG or BTG.

Command and Control

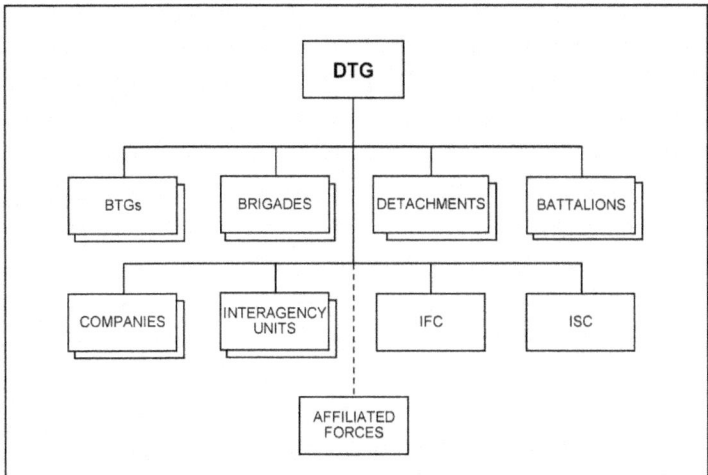

Figure 2-1. Possible DTG organization (example)

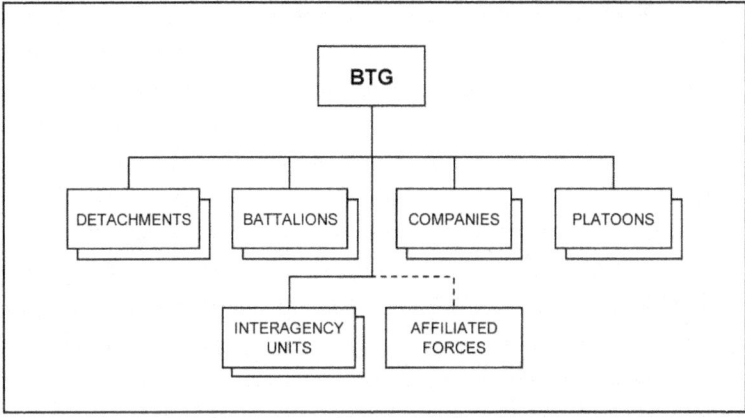

Figure 2-2. Possible BTG organization (example)

2-21. The division that serves as the basis for a DTG may have some of its brigades task-organized as BTGs. However, just the fact that a division becomes a DTG does not necessarily mean that it forms BTGs. A DTG could augment all of its brigades, or one or two brigades, or none of them as BTGs. A division could augment one or more brigades into BTGs, using the division's own constituent assets, without becoming a DTG. If a division receives additional assets and uses them all to create one or more BTGs, it is still designated as a DTG. Within a DTG or BTG, some battalions and companies may become task-organized as detachments, while others retain their original structures. (See discussion of Detachments, below.)

Chapter 2

Note. Unit symbols for all OPFOR units use the diamond-shaped frame. All OPFOR task organizations use the "task force" symbol placed over the "echelon" (unit size) modifier above the diamond frame. When there is a color capability, there are two options for use of red: all parts of the symbol that would otherwise be black can use red, or the diamond can have red fill color with the frame and other parts of the symbol in black. (See figures 2-3 through 2-5 and also figures 2-10 through 2-14 on pages 2-9 through 2-11 for examples.)

2-22. Unit symbols for tactical groups show the unit type and size of the "base" unit (division or brigade) around which the task organization was formed and whose headquarters serves as the headquarters for the tactical group. Figures 2-3 through 2-5 show examples of unit symbols for various types of OPFOR tactical groups.

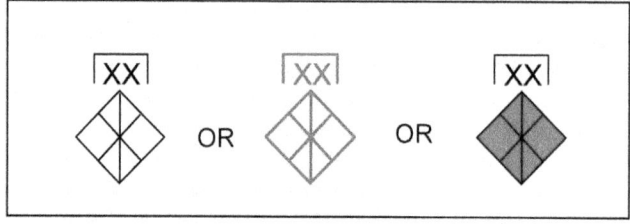

Figure 2-3. Motorized infantry-based DTG symbol

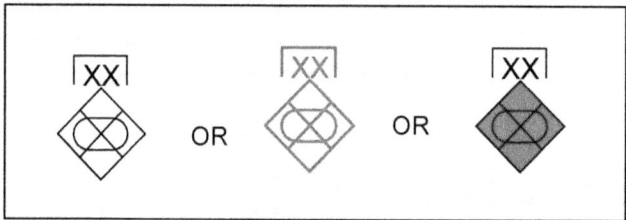

Figure 2-4. Mechanized infantry-based DTG symbol

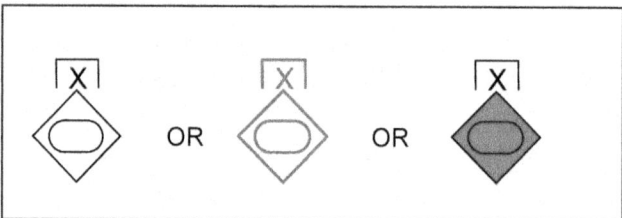

Figure 2-5. Tank-based BTG symbol

BATTALIONS

2-23. In the OPFOR's force structure, the basic unit of action is the *battalion*. (See figure 2-6.) Battalions are designed to be able to—

- Serve as the basis for forming a battalion-size detachment (BDET), if necessary. (See discussion of Detachments below.)
- Fight as part of a brigade, BTG, division, or DTG.
- Execute basic combat missions as part of a larger tactical force.
- Plan for operations expected to occur 6 to 24 hours in the future.
- Execute all of the tactical actions discussed in this TC.

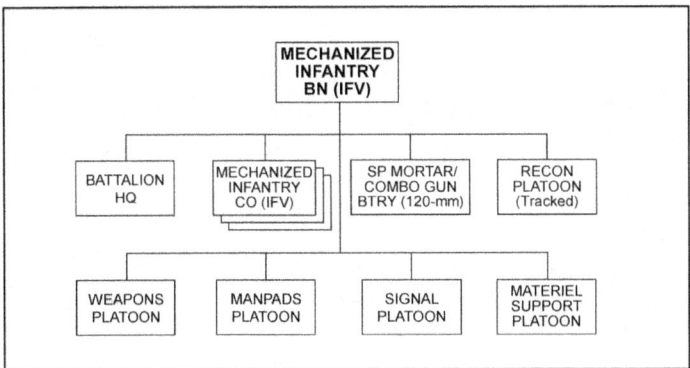

Figure 2-6. Battalion example

COMPANIES

2-24. In the OPFOR's force structure, the largest unit without a staff is the company. In fire support units, this level of command is commonly called a battery. (See figure 2-7.) Companies are designed to be able to—

- Serve as the basis for forming a company-size detachment (CDET), if necessary. (See discussion of Detachments below.)
- Fight as part of a battalion, BDET, brigade, BTG, division, or DTG.
- Execute tactical tasks. (A company will not normally be asked to perform two or more tactical tasks simultaneously.)

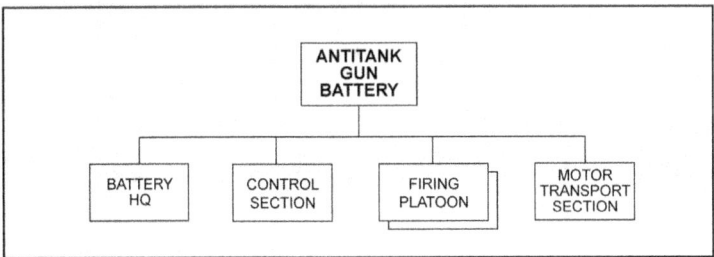

Figure 2-7. Company (battery) example

Chapter 2

DETACHMENTS

2-25. A *detachment* is a battalion or company designated to perform a specific mission and allocated the forces necessary to do so. (See figures 2-8 and 2-9.) Detachments are the smallest combined arms formations and are, by definition, task-organized. To further differentiate, detachments built from battalions can be termed *battalion-size detachments* (BDETs), and those formed from companies can be termed *company-size detachments* (CDETs). The forces allocated to a detachment suit the mission expected of it. They may include—

- Artillery or mortar units.
- Air defense units.
- Engineer units (with obstacle, survivability, or mobility assets).
- Heavy weapons units (including heavy machineguns, automatic grenade launchers, and antitank guided missiles).
- Units with specialty equipment such as flame weapons, specialized reconnaissance assets, or helicopters.
- Interagency forces up to company size for BDETs, or platoon size for CDETs.
- Chemical defense, antitank, medical, logistics, signal, and electronic warfare units.

BDETs can accept dedicated and supporting SPF, aviation (combat helicopter, transport helicopter), and unmanned aerial vehicle units.

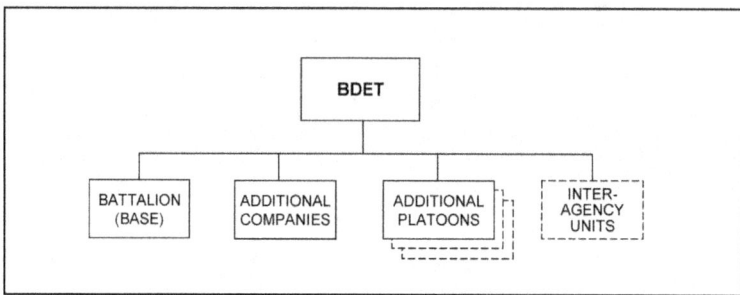

Figure 2-8. Battalion-size detachment (BDET) example

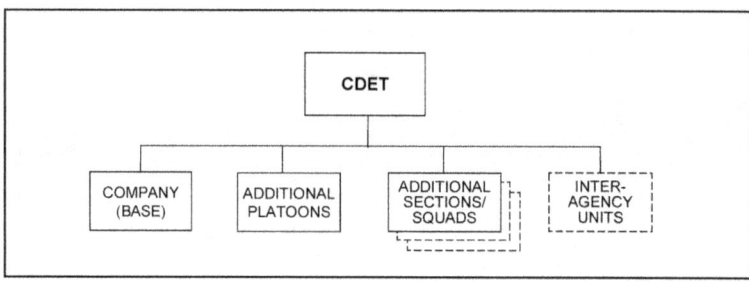

Figure 2-9. Company-size detachment (CDET) example

2-26. The basic type of OPFOR detachment—whether formed from a battalion or a company—is the *independent mission detachment* (IMD). IMDs are formed to execute missions that are separated in space

Command and Control

and/or time from those being conducted by the remainder of the forming unit. IMDs can be used for a variety of missions, some of which are listed here as examples:
- Seizing key terrain.
- Linking up with airborne or heliborne forces.
- Conducting tactical movement on secondary axes.
- Pursuing or enveloping an enemy force.
- Conducting a raid or ambush.

2-27. Other types of detachments and their uses are described in subsequent chapters. These detachments include—
- Counterreconnaissance detachment. (See chapter 5.)
- Urban detachment. (See chapter 5.)
- Security detachment. (See chapter 5.)
- Reconnaissance detachment. (See chapter 7.)
- Movement support detachment. (See chapter 12.)
- Obstacle detachment. (See chapter 12.)

2-28. Unit symbols for detachments show the unit type and size of the "base" unit (battalion or company) around which the task organization was formed and whose headquarters serves as the headquarters for the detachment. Figures 2-10 through 2-12 on pages 2-9 and 2-10 show examples of unit symbols for various types of OPFOR detachments.

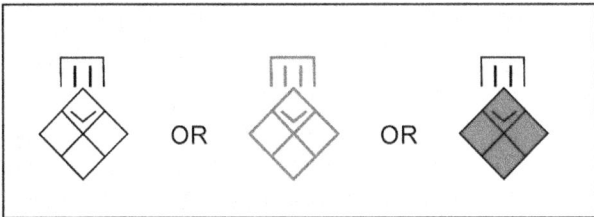

Figure 2-10. Heliborne infantry-based BDET symbol

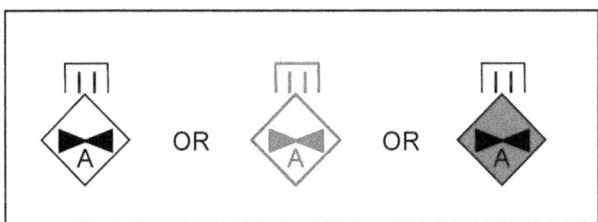

Figure 2-11. Attack helicopter-based BDET symbol

Chapter 2

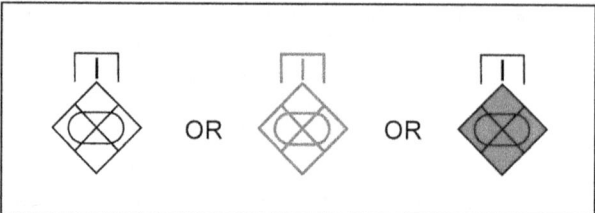

Figure 2-12. Mechanized infantry-based CDET symbol

PLATOONS AND SQUADS

2-29. In the OPFOR's force structure, the smallest unit typically expected to conduct independent fire and maneuver is the platoon. Platoons are designed to be able to—
- Serve as the basis for forming a functional element or patrol.
- Fight as part of a company, battalion, or detachment.
- Execute tactical tasks. (A platoon will not be asked to perform two or more tactical tasks simultaneously.)
- Exert control over a small riot, crowd, or demonstration.

2-30. Platoons and squads within them can be task-organized for specific missions. Figures 2-13 and 2-14 show examples of unit symbols for various types of OPFOR task-organized platoons and squads.

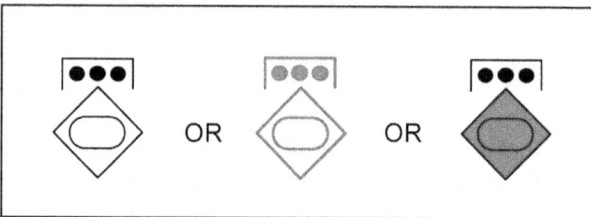

Figure 2-13. Tank-based task-organized platoon symbol

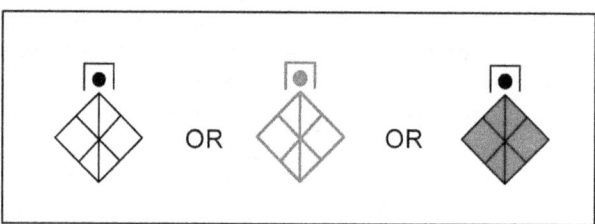

Figure 2-14. Motorized infantry-based task-organized squad symbol

ORGANIZING THE TACTICAL BATTLEFIELD

2-31. The OPFOR organizes the battlefield in such a way that it can rapidly transition between offensive and defensive actions and between linear and nonlinear dispositions. This flexibility can help the OPFOR adapt and change the nature of conflict to something for which the enemy is not prepared.

2-32. In his combat order, the commander specifies the organization of the battlefield from the perspective of his level of command. Within his unit's AOR, as defined by the next-higher commander, he designates specific AORs for his subordinates, along with zones, objectives, and axes related to his own overall mission.

AREAS OF RESPONSIBILITY

2-33. The OPFOR defines an *area of responsibility* (AOR) as the geographical area and associated airspace within which a commander has the authority to plan and conduct combat operations. An AOR is bounded by a *limit of responsibility* (LOR) beyond which the organization may not operate or fire without coordination through the next-higher headquarters. AORs may be linear or nonlinear in nature. Linear AORs may contain subordinate nonlinear AORs and vice versa. (See figures 2-15 through 2-18 on pages 2-12 and 2-13 for examples of tactical-level AORs. See chapters 3 and 4 for additional examples of AORs and zones in offense and defense.)

2-34. A combat order normally defines AORs (and zones within them) by specifying boundary lines in terms of distinct local terrain features through which a line passes. The order specifies whether each of those terrain features is included or excluded from the unit's AOR or zones within it. Normally, a specified terrain feature is included unless the order identifies it as "excluded." For example, the left boundary of the DTG AOR in figure 2-16 on page 2-12 runs from hill 108, to hill 250 (excluded), to junction of highway 52 and road 98, to the well. That example also illustrates that, even in a linear AOR, not all boundaries have to be straight lines.

2-35. It is possible, although not likely, that a higher commander may retain control of airspace over a lower commander's AOR. This would be done through the use of standard airspace management measures.

ZONES

2-36. AORs typically consist of three basic *zones*: *battle zone*, *disruption zone*, and *support zone*. An AOR may also contain one or more *attack zones* and/or *kill zones*. The various zones in an AOR have the same basic purposes within each type of offensive and defensive action. Zones may be linear or nonlinear in nature. The size of these zones depends on the size of the OPFOR units involved, engagement ranges of weapon systems, the terrain, and the nature of the enemy's operation. Within the LOR, the OPFOR normally refers to two types of control lines. The *support line* separates the support zone from the battle zone. The *battle line* separates the battle zone from the disruption zone.

2-37. An AOR is not required to have any or all of these zones in any particular situation. A command might have a battle zone and no disruption zone. It might not have a battle zone, if it is the disruption force of a higher command. If it is able to forage, it might not have a support zone. The intent of this method of organizing the battlefield is to preserve as much flexibility as possible for subordinate units within the parameters that define the aim of the senior commander. An important feature of the basic zones in an AOR is the variations in actions that can occur within them in the course of a specific battle.

Chapter 2

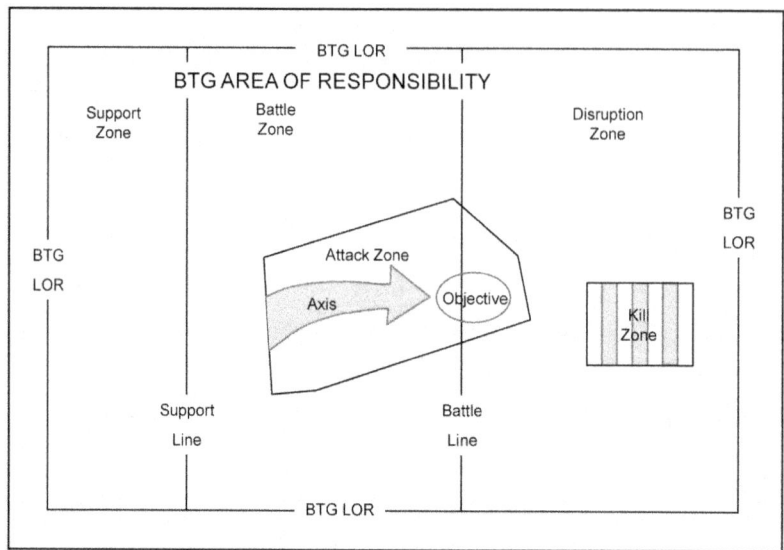

Figure 2-15. Linear AOR (example 1)

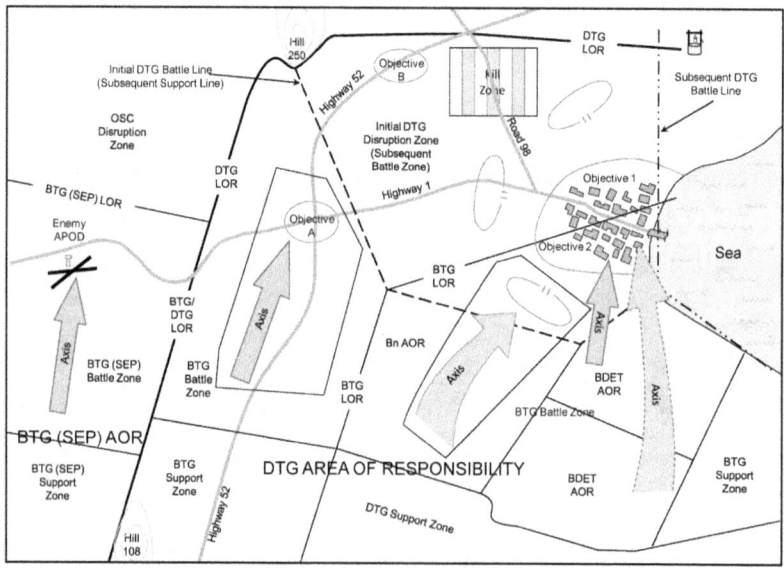

Figure 2-16. Linear AOR (example 2)

Command and Control

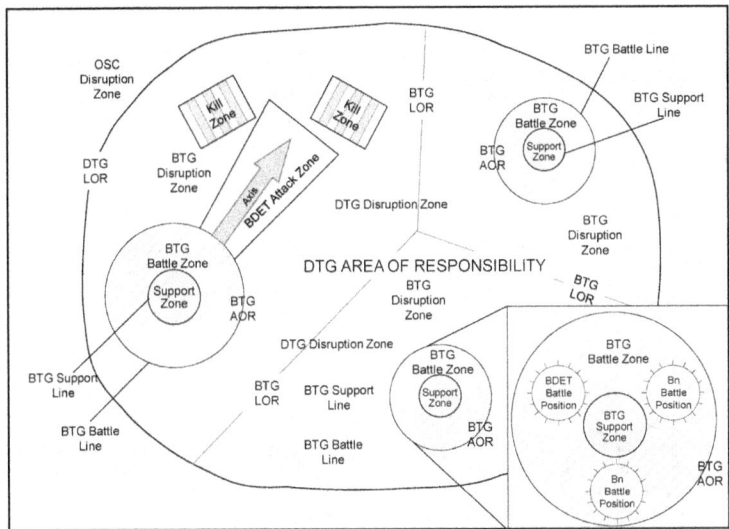

Figure 2-17. Nonlinear AOR (example 1)

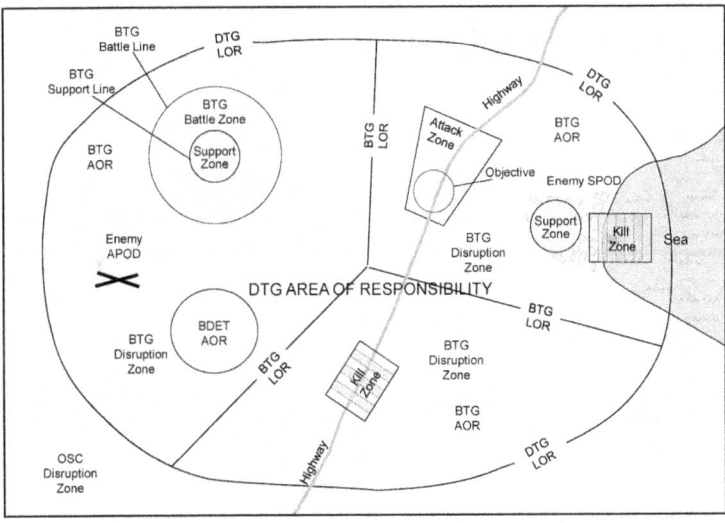

Figure 2-18. Nonlinear AOR (example 2)

Disruption Zone

2-38. The *disruption zone* is the AOR of the disruption force. It is that geographical area and airspace in which the unit's disruption force will conduct disruption tasks. This is where the OPFOR will set the conditions for successful combat actions by fixing enemy forces and placing long-range fires on them. Units in this zone begin the attack on specific components of the enemy's combat system, to begin the disaggregation of that system. Successful actions in the disruption zone will create a window of opportunity that is exploitable in the battle zone.

2-39. Specific actions in the disruption zone can include—
- Attacking the enemy's engineer elements. This can leave his maneuver force unable to continue effective operations in complex terrain—exposing them to destruction by forces in the battle zone.
- Stripping away the enemy's reconnaissance assets while denying him the ability to acquire and engage OPFOR targets with deep fires. This includes an air defense effort to deny aerial attack and reconnaissance platforms from targeting OPFOR forces.
- Forcing the enemy to deploy early or disrupting his offensive preparations.
- Gaining and maintaining reconnaissance contact with key enemy elements.
- Deceiving the enemy as to the disposition of OPFOR units.

2-40. The disruption zone is bounded by the battle line and the LOR of the overall AOR. In linear offensive combat, the higher headquarters may move the battle line and LOR forward as the force continues successful offensive actions. Thus, the boundaries of the disruption zone will also move forward during the course of a battle. (See the example in figure 2-16 on page 2-12.) The higher commander can push the disruption zone forward or outward as forces adopt a defensive posture while consolidating gains at the end of a successful offensive battle and/or prepare for a subsequent offensive battle. Disruption zones may be contiguous or noncontiguous. They can also be "layered," in the sense that one command's disruption zone is part of the disruption zone of the next-higher command. (See an example of this layering in figure 2-17 on page 2-13.)

2-41. Battalions and below do not typically have their own disruption zones. However, they may conduct actions within the disruption zone of a higher command.

Battle Zone

2-42. The *battle zone* is the portion of the AOR where the OPFOR expects to conduct decisive actions. Forces in the battle zone will exploit opportunities created by actions in the disruption zone. Using all elements of combat power, the OPFOR will engage the enemy in close combat to achieve tactical decision in this zone.

2-43. In the battle zone, the OPFOR is typically trying to accomplish one or more of the following:
- Create a penetration in the enemy defense through which exploitation forces can pass.
- Draw enemy attention and resources to the action.
- Seize terrain.
- Inflict casualties on a vulnerable enemy unit.
- Prevent the enemy from moving a part of his force to impact OPFOR actions elsewhere on the battlefield.

2-44. A division or DTG does not always form a division- or DTG-level battle zone per se—that zone may be the aggregate of the battle zones of its subordinate units. In nonlinear situations, there may be multiple, noncontiguous brigade or BTG battle zones, and within each the division or DTG would assign a certain task to the unit charged to operate in that space. The brigade or BTG battle zone provides each of those subordinate unit commanders the space in which to frame his actions. Battalion and below units often have AORs that consist almost entirely of battle zones with a small support zone contained within them.

2-45. The battle zone is separated from the disruption zone by the battle line and from the support zone by the support line. In the offense, the commander may adjust the location of these lines in order to accommodate successful offensive action. In a linear situation, those lines can shift forward during the course of a successful attack. Thus, the battle zone would also shift forward. (For an example of this, see figure 2-16 on page 2-12.)

Support Zone

2-46. The *support zone* is that area of the battlefield designed to be free of significant enemy action and to permit the effective logistics and administrative support of forces. Security forces will operate in the support zone in a combat role to defeat enemy special operations forces. Camouflage, concealment, cover, and deception (C3D) measures will occur throughout the support zone to protect the force from standoff RISTA and precision attack. A division or DTG support zone may be dispersed within the support zones of subordinate brigades or BTGs, or the division or DTG may have its own support zone that is separate from subordinate AORs. If the battle zone moves during the course of a battle, the support zone would move accordingly. The support zone may be in a sanctuary that is noncontiguous with other zones of the AOR.

Attack Zone

2-47. An *attack zone* is given to a subordinate unit with an offensive mission, to delineate clearly where forces will be conducting offensive maneuver. Attack zones are often used to control offensive action by a subordinate unit inside a larger defensive battle or operation.

Kill Zone

2-48. A *kill zone* is a designated area on the battlefield where the OPFOR plans to destroy a key enemy target. A kill zone may be within the disruption zone or the battle zone. In the defense, it could also be in the support zone.

FUNCTIONAL ORGANIZATION OF FORCES AND ELEMENTS

2-49. An OPFOR commander specifies in his combat order the initial organization of forces or elements within his level of command, according to the specific *functions* he intends his various subordinate units to perform. At brigade or BTG and above, the subordinate units performing these functions are referred to as *forces*, while at battalion or BDET and below, they are called *elements*.

> *Note.* This portion of chapter 2 provides a brief overview of functional organization as a key part of the OPFOR C2 process. This provides a common language and a clear understanding of how the commander intends his subordinates to fight functionally. Thus, subordinates that perform common tactical tasks such as disruption, fixing, assault, exploitation, security, deception, or main defense are logically designated as disruption, fixing, assault, exploitation, security, deception, or main defense forces or elements. OPFOR commanders prefer using the clearest and most descriptive term to avoid any confusion. For more detailed discussion and examples of the roles of various functional forces and elements in offense and defense, see chapters 3 and 4, respectively.

2-50. The OPFOR organizes and designates various forces and elements according to their function in the planned offensive or defensive action. A number of different functions must be executed each time an OPFOR unit attempts to accomplish a mission. The functions do not change, regardless of where the force or element might happen to be located on the battlefield. However, the function (and hence the functional designation) of a particular force or element may change during the course of the battle. The use of precise functional designations for every force or element on the battlefield allows for a clearer understanding by subordinate units of the distinctive functions their commander expects them to perform. It also allows each force or element to know exactly what all of the others are doing at any time. This knowledge facilitates the OPFOR's ability to make quick adjustments and to adapt very rapidly to shifting tactical situations. This practice also assists in a more comprehensive planning process by eliminating the likelihood of some confusion (especially on graphics) of who is responsible for what. Omissions and errors are much easier to spot using these functional labels rather than relying on unit designators, numbers, or code words.

> *Note.* A unit or group of units designated as a particular functional force or element may also be called upon to perform other, more specific functions. Therefore, the function of that force or element, or part(s) of it, may be more accurately described by a more specific functional designation. For example, a disruption force generally "disrupts," but also may need to "fix" a part of the enemy forces. In that case, the entire disruption force could become the fixing force, or parts of that force could become fixing elements.

2-51. The various functions required to accomplish any given mission can be quite diverse. However, they can be broken down into two very broad categories: action and enabling.

ACTION FORCES AND ELEMENTS

2-52. One part of the unit or grouping of units conducting a particular offensive or defensive action is normally responsible for performing the primary function or task that accomplishes the overall mission goal or objective of that *action*. In most general terms, therefore, that part can be called the *action force* or *action element*. In most cases, however, the higher unit commander will give the action force or element a more specific designation that identifies the specific function or task it is intended to perform, which equates to achieving the objective of the higher command's mission.

2-53. For example, if the objective of the action at detachment level is to conduct a *raid*, the element designated to complete that action may be called the *raiding element*. In offensive actions at brigade or BTG and higher, a force that completes the primary offensive mission by *exploiting* a window of opportunity created by another force is called the *exploitation force*. In defensive actions, the unit or grouping of units that performs the *main defensive* mission in the battle zone is called the *main defense force* or *main defense element*. However, in a maneuver defense, the main defensive action is executed by a combination of two functional forces: the *contact force* and the *shielding force*.

ENABLING FORCES AND ELEMENTS

2-54. In relation to the action force or element, all other parts of the organization conducting an offensive or defensive action provide *enabling* functions of various kinds. In most general terms, therefore, each of these parts can be called an *enabling force* or *enabling element*. However, each subordinate force or element with an enabling function can be more clearly identified by the specific function or task it performs. For example, a force that enables by *fixing* enemy forces so they cannot interfere with the primary action is a *fixing force*. Likewise, an element that *clears* obstacles to permit an action element to accomplish a detachment's tactical task is a *clearing element*.

2-55. Other types of enabling forces or elements designated by their specific function may include—

- *Disruption force or element.* Operates in the disruption zone; disrupts enemy preparations or actions; destroys or deceives enemy reconnaissance; begins reducing the effectiveness of key components of the enemy's combat system.
- *Fixing force or element.* Fixes the enemy by preventing a part of his force from moving from a specific location for a specific period of time, so it cannot interfere with the primary OPFOR action.
- *Security force or element.* Provides security for other parts of a larger organization, protecting them from observation, destruction, or becoming fixed.
- *Deception force or element.* Conducts a deceptive action (such as a demonstration or feint) that leads the enemy to act in ways prejudicial to enemy interests or favoring the success of an OPFOR action force or element.

- *Support force or element.* Provides support by fire; other combat or combat service support; or C2 functions for other parts of a larger organization.

OTHER FORCES AND ELEMENTS

2-56. In initial orders, some subordinates are held in a status pending determination of their specific function. At the commander's discretion, some forces or elements may be held out of initial action, in *reserve*, so that he may influence unforeseen events or take advantage of developing opportunities. These are designated as *reserves* (*reserve force* or *reserve element*). If and when such units are subsequently assigned a mission to perform a specific function, they receive the appropriate functional force or element designation. For example, a reserve force in a defensive operation might become the counterattack force.

2-57. In defensive actions, there may be a particular unit or grouping of units that the OPFOR commander wants to be *protected* from enemy observation or fire, to ensure that it will be available after the current battle or operation is over. This is designated as the *protected force*.

COMMAND OF FORCES AND ELEMENTS

2-58. Each of the separate functional forces or elements—even when it involves a grouping of multiple units—has an identified commander. This is often the senior commander of the largest subordinate unit assigned to that force or element.

> *Note.* A *patrol* is a platoon- or squad-size grouping task-organized to accomplish a specific reconnaissance and/or security mission. There are two basic types of patrol: fighting patrol and reconnaissance patrol. Both are described in chapter 8.

HEADQUARTERS, COMMAND, AND STAFF

2-59. All OPFOR levels of command share parallel staff organization. However, command and staff elements at various levels are tailored to match differences in scope and span of control.

DTG OR DIVISION COMMAND GROUP AND STAFF

2-60. A DTG or division headquarters includes the command group and the staff. (See figure 2-19 on page 2-18.) These elements perform the functions required to control the activities of forces preparing for and conducting combat.

2-61. The primary functions of headquarters are to—
- Make decisions.
- Plan combat actions that accomplish those decisions.
- Acquire and process the information needed to make and execute effective decisions.
- Support the missions of subordinates.

The commander exercises C2 functions through his command group, staff, and subordinate commanders.

Chapter 2

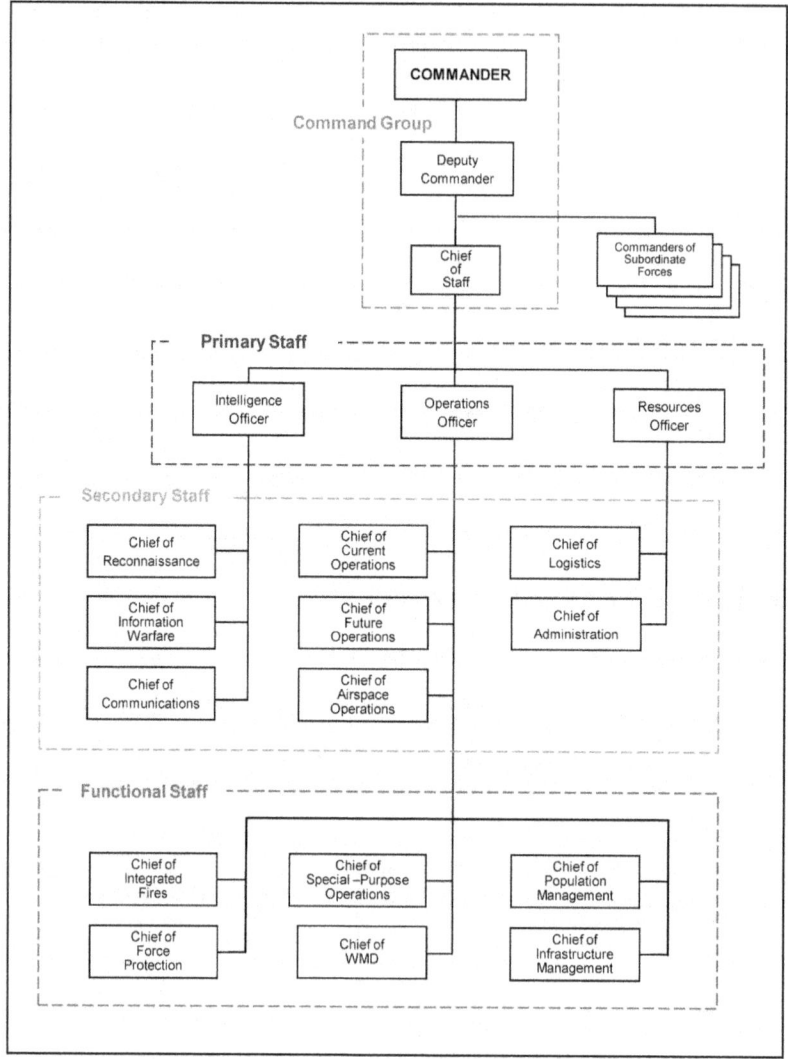

Figure 2-19. DTG or division command group and staff

Command Group

2-62. The command group consists of the commander, deputy commander (DC), and chief of staff (COS). Together, they direct and coordinate the activities of the staff and of subordinate forces.

Commander

2-63. The commander directs subordinate commanders and, through his staff and liaison officers, controls any supporting elements. OPFOR commanders have complete authority over their subordinates and overall responsibility for those subordinates' actions. Under the fluid conditions of modern warfare, even in the course of carefully planned actions, the commander must accomplish assigned missions on his own initiative without constant guidance from above.

2-64. The commander is responsible for—
- The combat capability of subordinate units.
- The organization of combat actions.
- The maintenance of uninterrupted C2.
- The successful conduct of combat missions.

2-65. The commander examines and analyzes the mission he receives (that is, he determines his forces' place in the senior commander's concept of the battle or operation). He may do this alone or jointly with the COS. He then gives instructions to the COS on preparing his forces and staff for combat. He also provides instructions about the timing of preparations. The commander makes his own assessment of intelligence data supplied by the chief of intelligence. Then, with advice from all the primary staff officers, he makes an assessment of his own forces. After discussing his deductions and proposals with the operations officer and his staff, the commander reaches a decision, issues combat missions to subordinates, and gives instructions about planning the battle. He then directs coordination within his organization and with adjacent forces and other elements operating in his AOR.

2-66. During the course of combat, the commander must constantly evaluate the changing situation, predict likely developments, and issue new combat missions in accordance with his vision of the battlefield. He also keeps his superiors informed as to the situation and character of friendly and enemy actions and his current decisions.

Deputy Commander

2-67. In the event the commander is killed or incapacitated, the *deputy commander* (DC) would assume command. Barring that eventuality, the primary responsibility of the DC of a DTG or division is to command the IFC. As IFC commander, he is responsible for executing tactical-level fire support in a manner consistent with the commander's intent.

Chief of Staff

2-68. Preeminent among OPFOR staff officers is the *chief of staff* (COS) position (found at every level from the General Staff down to battalion). He exercises direct control over the primary staff. During combat, he is in charge of the main CP when the commander moves to the forward CP. He has the power to speak in the name of the commander and DC, and he normally countersigns all written orders and combat documents originating from the commander's authority. He alone has the authority to sign orders for the commander or DC and to issue instructions in the commander's name to subordinate units. In emergency situations, he can make changes in the tasks given to subordinate commanders. Thus, it is vital that he understands not merely the commander's specific instructions but also his general concept and train of thought. He controls the battle during the commander's absences.

2-69. The COS is a vital figure in the C2 structure. His role is to serve as the director of staff planning and as coordinator of all staff inputs that assist the commander's decisionmaking. He is the commander's and DC's focal point for knowledge about the friendly and enemy situation. He has overall responsibility for providing the necessary information for the commander to make decisions. Thus, he plays a key role in structuring the overall reconnaissance effort, which is a combined arms task, to meet the commander's information requirements.

Staff

2-70. A staff provides rapid, responsive planning for combat activity, and then coordinates and monitors the execution of the resulting plans on behalf of the commander. Proper use of this staff allows the commander to focus on the most critical issues in a timely manner and to preserve his energies. For additional detail on the organization of the command group and staff, see FM 7-100.4.

2-71. The staff releases the commander from having to solve administrative and technical problems, thereby allowing him to concentrate on the battle. The primary function of the staff is to plan and prepare for combat. Evaluation and knowledge of the operational environment is fundamental to the decisionmaking process and the direction of troops. After the commander makes the decision, the staff must organize, coordinate, disseminate, and support the missions of subordinates. Additionally, it is their responsibility to train and prepare troops for combat, and to monitor the pre-combat and combat situations.

2-72. In the decisionmaking and planning process, the staff—
- Prepares the data and estimates the commander uses to make a decision.
- Plans and implements the basic measures for comprehensive support of a combat action.
- Organizes communications with subordinate and adjacent headquarters and the next-higher staff.
- Monitors the activities of subordinate staffs.
- Coordinates ongoing activity with higher-level and adjacent staffs during a battle or operation.

2-73. The staff consists of three elements: the primary staff, the secondary staff, and the functional staff. Figure 2-19 on page 2-18 depicts the primary, secondary, and functional staff officers of a DTG or division headquarters. (It does not show the liaison teams, which support the primary, secondary, and functional staff.)

Primary and Secondary Staff

2-74. Each member of the primary staff heads a staff section. Within each section are two or three secondary staff officers heading subsections subordinate to that primary staff officer.

2-75. **Operations Officer.** The *operations officer* heads the operations section, and conducts planning and prepares plans and orders. Thus, the operations section is the principal staff section. It includes current operations, future operations, and airspace operations subsections, as well as the functional staff.

2-76. The operations officer also serves as *deputy chief of staff*. He is responsible for—
- Training.
- Formulating and writing plans, combat orders, and important combat reports.
- Monitoring the work of all other staff sections.
- Remaining knowledgeable of the current situation.
- Being ready to present information and recommendations concerning the situation.

In coordination with the intelligence and information section, the operations officer keeps the commander informed on the progress of the battle and the overall operation.

2-77. Specific duties of the operations section include—
- Assisting the commander in the making and execution of combat decisions.
- Collecting information concerning the situation of friendly forces.
- Preparing and disseminating orders, plans and reports, summaries, and situation overlays.
- Providing liaison for the exchange of information within the headquarters and with higher, subordinate, and adjacent units.
- Organizing the main CP.

Command and Control

- Organizing troop movement and traffic control.
- Coordinating the organization of reconnaissance with the intelligence and information section.

2-78. The *chief of current operations* is a secondary staff officer who proactively monitors the course of current operations and coordinates the actions of forces to ensure execution of the commander's intent. He serves as the representative of the commander, COS, and operations officer in their absence and has the authority to control forces in accordance with the battle plan.

2-79. The *chief of future operations* is a secondary staff officer who heads the planning staff and ensures continuous development of future plans and possible branches, sequels, and contingencies. While the commander and the chief of current operations focus on the current battle, the chief of future operations and his subsection monitor the friendly and enemy situations and their implications for future battles. They try to identify any developing situations that require command decisions and/or adaptive measures. They advise the commander on how and when to make adjustments to the battle plan during the fight. Planning for various contingencies and anticipated opportunities can facilitate immediate and flexible response to changes in the situation.

2-80. The *chief of airspace operations* (CAO) is a secondary staff officer who is responsible for the control of the division's or DTG's airspace. See chapters 9 and 10 for further information on his duties.

2-81. **Intelligence Officer.** The *intelligence officer* heads the *intelligence and information section*, which consists of the *reconnaissance subsection*, the *information warfare (INFOWAR) subsection*, and the *communications subsection*. The intelligence officer is responsible for the acquisition, synthesis, analysis, dissemination, and protection of all information and intelligence related to and required by the division's or DTG's combat actions. He ensures the commander's intelligence requirements are met. He provides not only intelligence on the current and future operational environment, but also insight on opportunities for adaptive and creative responses to ongoing operations. The intelligence officer works in close coordination with the chief of future operations to establish feedback and input for future operations and the identification of possible windows of opportunity.

2-82. The intelligence officer also formulates the division's or DTG's INFOWAR plan and must effectively task-organize his staff resources to conduct and execute INFOWAR in a manner that supports the strategic INFOWAR plan. He is responsible for the coordination of all necessary national or theater-level assets in support of the INFOWAR plan and executes staff supervision over the INFOWAR and communications plans. He is supported by three secondary staff officers: the chief of reconnaissance, the chief of INFOWAR, and the chief of communications.

2-83. The *chief of reconnaissance* develops reconnaissance plans, gathers information, and evaluates data on the operational environment. During combat, he supervises the efforts of subordinate reconnaissance units and reconnaissance staff subsections of subordinate units. Specific responsibilities of the reconnaissance subsection include—

- Continuously collecting, analyzing, and disseminating information on the operational environment to the commander and subordinate, higher, and adjacent units.
- Organizing reconnaissance missions, including requests for aerial reconnaissance, in coordination with the operations section and in support of the INFOWAR plan.
- Preparing the reconnaissance plan, in coordination with the operations section.
- Preparing the reconnaissance portion of battle plans and combat orders.
- Preparing intelligence reports.
- Supervising the exploitation of captured enemy documents and materiel.
- Supervising interrogation and debriefing activities throughout the command.
- Providing targeting data for long-range fires.

2-84. The *chief of information warfare* is responsible supervising the execution of the division's or DTG's INFOWAR plan. (See chapter 6 for details and components of the INFOWAR plan.) These responsibilities include—

- Coordinating the employment of INFOWAR assets, both those subordinate to the division or DTG and those available at higher levels.

- Planning for and supervising all information protection and security measures.
- Supervising the implementation of the deception and perception management plans.
- Working with the operations staff to ensure that targets scheduled for destruction support the INFOWAR plan, and if not, resolving conflicts between INFOWAR needs and operational needs.
- Recommending to the intelligence officer any necessary actions required to implement the INFOWAR plan.

2-85. The *chief of communications* develops a communications plan for the command that is approved by the intelligence officer and COS. He organizes communications with subordinate, adjacent, and higher headquarters. To ensure that the commander has continuous and uninterrupted control, the communications subsection plans the use of all forms of communications, to include satellite communications (SATCOM), wire, radio, digital, cellular, and couriers. Specific responsibilities of the communications subsection include—

- Establishing SATCOM and radio nets.
- Establishing call signs and radio procedures.
- Organizing courier and mail service.
- Operating the command's message center.
- Supervising the supply, issue, and maintenance of signal equipment.

An additional and extremely important role of the communications officer is to ensure the thorough integration of interagency, allied, subordinate, supporting, and affiliated forces into the division's or DTG's communications and C2 structure. The division or DTG headquarters is permanently equipped with a full range of C2 systems compatible with each of the services of the State's Armed Forces as well as with other government agencies commonly operating as part of DTGs.

2-86. **Resources Officer.** The *resources officer* is responsible for the requisition, acquisition, distribution, and care of all of the division's or DTG's resources, both human and materiel. He ensures the commander's logistics and administrative requirements are met and executes staff supervision over the command's logistics and administrative procedures. (Logistics procedures are detailed in chapter 14.) He is supported by two secondary staff officers: the chief of logistics and the chief of administration. One additional major task of the resources officer is to free the commander from the need to bring his influence to bear on priority logistics and administrative functions. He is also the officer in charge of the sustainment CP.

2-87. The *chief of logistics* heads the logistics system. He is responsible for managing the order, receipt, and distribution of supplies to sustain the command. He is responsible for the condition and combat readiness of armaments and related combat equipment and instruments. He is also responsible for their supply, proper utilization, repair, and evacuation. He oversees the supply and maintenance of the division's or DTG's combat and technical equipment. These responsibilities encompass the essential wartime tasks of organizing and controlling the division's or DTG's recovery, repair, and replacement system. During combat, he keeps the commander informed on the status of the division's or DTG's equipment.

2-88. The *chief of administration* supervises all personnel actions and transactions in the division or DTG. His subsection—

- Maintains daily strength reports.
- Records changes in TOE of units in the AFS.
- Assigns personnel.
- Requests replacements.
- Records losses.
- Administers awards and decorations.
- Collects, records, and disposes of war booty.

Functional Staff

2-89. The *functional staff* consists of experts in a particular type of military operation or function. (See figure 2-19 on page 2-18.) These experts advise the command group and the primary and secondary staff on issues pertaining to their individual areas of expertise. The functional staff consists of the following elements:

- Integrated fires.
- Force protection.
- Special-purpose operations.
- Weapons of mass destruction (WMD).
- Population management.
- Infrastructure management.

2-90. In peacetime, the functional staff is a cadre with personnel assigned from appropriate branches. It has enough personnel to allow continuous 24-hour capability and the communications and information management tools to allow them to support the commander's decisionmaking process and exercise staff supervision over their functional areas throughout the AOR. In wartime, the functional staff receives liaison teams from subordinate, supporting, allied, and affiliated units that perform tasks in support of those functional areas.

2-91. **Chief of Integrated Fires.** The chief of integrated fires is responsible for integrating C2 and RISTA means with fires and maneuver. He works closely with the division or DTG chief of reconnaissance and the IFC staff. He also coordinates with the chief of INFOWAR to ensure that deception and protection and security measures contribute to the success of fire support of offensive and defensive actions.

2-92. **Chief of Force Protection.** The chief of force protection is responsible for coordinating activities to prevent or mitigate the effects of hostile actions against OPFOR personnel, resources, facilities, and critical information. This protection includes—

- Air, space, and missile defense.
- CBRN protection.
- Defensive INFOWAR.
- Antiterrorism measures.
- Counterreconnaissance.
- Engineer survivability measures.

This subsection works closely with those of the chief of WMD and the chief of INFOWAR. Liaison teams from internal security, air defense, chemical defense, and engineer forces provide advice within their respective areas of protection.

2-93. **Chief of Special-Purpose Operations.** The chief of special-purpose operations is responsible for planning and coordinating the actions of SPF units allocated to a DTG or supporting it from OSC level. When possible, this subsection receives liaison teams from any affiliated forces that act in concert with the SPF.

2-94. **Chief of WMD.** The chief of WMD is responsible for planning the offensive use of WMD. This functional staff element receives liaison teams from any subordinate or supporting units that contain WMD delivery means.

2-95. **Chief of Population Management.** The chief of population management is responsible for coordinating the actions of Internal Security Forces, as well as psychological warfare, perception management, civil affairs, and counterintelligence activities. This subsection works closely with the chief of INFOWAR and receives liaison teams from psychological warfare, civil affairs, counterintelligence, and Internal Security Forces units allocated to the DTG or operating within the division's or DTG's AOR. There is always a representative of the Ministry of the Interior, and frequently one from the Ministry of Public Information.

2-96. **Chief of Infrastructure Management.** The chief of infrastructure management is responsible for establishing and maintaining—
- Roads.
- Airfields.
- Railroads.
- Hardened structures (warehouses and storage facilities).
- Inland waterways.
- Ports.
- Pipelines.

He coordinates with the division or DTG resources officer regarding improvement and maintenance of supply and evacuation routes. He exercises staff supervision or cognizance over the route construction and maintenance functions of both civil and combat engineers operating in the division's or DTG's AOR. He coordinates with civilian agencies and the division or DTG chief of communications to ensure adequate telecommunications support.

Liaison Teams

2-97. Liaison teams support brigade and division staffs (as well as those of tactical groups and detachments) with detailed expertise in the mission areas of their particular branch or service. They also provide direct communications to subordinate and supporting units executing missions in those areas. They are not a permanent part of the staff structure. Liaison team chiefs speak for the commanders of their respective units. All liaison teams are under the direct control of the operations section. The operations officer is responsible for ensuring proper placement and utilization of the teams.

2-98. Liaison teams to DTGs and divisions are generally organized with a liaison team chief, two current operations officers or senior NCOs, and two future operations officers or senior NCOs. This gives liaison teams the ability to conduct continuous operations and simultaneously execute current plans and develop future plans. The staff will also receive liaison teams from multinational and interagency subordinates and from affiliated forces. The number and types of liaison teams is fluid and is determined by many factors. Liaison teams provide their own equipment. A detailed breakout of personnel and equipment of a typical liaison team is available in FM 7-100.4.

BTG OR BRIGADE COMMAND GROUP AND STAFF

2-99. Generally speaking, the command group and staff of a BTG or brigade are smaller versions of those previously described for DTG or division level. The following paragraphs highlight the differences other than size.

Command Group

2-100. The BTG or brigade command group consists of the commander, DC, and COS. The primary difference is that the DC does not serve as IFC commander, since there is no IFC at this level of command.

Commander

2-101. Compared to higher-level commands, much more of the BTG or brigade fight is the direct fire battle. Therefore, BTG or brigade commanders typically spend more time at the forward CP or with forward-deployed subordinate units than do DTG or division commanders.

Deputy Commander

2-102. At the BTG or brigade (and below), IFCs do not exist. Thus, the DC is not also the commander of the IFC.

Command and Control

Chief of Staff

2-103. At BTG or brigade level, the COS position retains all the characteristics of the DTG or division COS position. The COS is in charge of the main CP in the absence of the commander.

Staff

2-104. BTG and brigade staffs are naturally smaller and less capable than DTG and division staffs. In particular, the sections responsible for planning are much reduced, providing the BTG or brigade with the ability to plan combat actions only 24 to 48 hours into the future.

2-105. Another key difference from the DTG or division staff is that the functional staff is organized differently. A BTG or brigade functional staff consists of the following elements:

- Fire support coordination.
- Force protection.
- Special-purpose operations.
- WMD.
- Population management.
- Infrastructure management.

Note. There is no chief of integrated fires or dedicated staff element for integrating fires at the BTG or brigade level, since this level of command has no IFC. However, the staff at that level does include a *chief of fire support coordination* and a fire support coordination staff element within the functional staff of the operations section. This staff performs all of the necessary coordination between constituent, dedicated, and supporting fire support elements. See FM 7-100.4.

2-106. Liaison teams support brigade and BTG staffs with detailed expertise in the mission areas of subordinate and supporting units. A typical liaison team to a BTG or brigade staff consists of four personnel: a team chief, an assistant team chief, a staff officer or NCO, and their driver. A detailed breakout of personnel and equipment of a typical liaison team is available in FM 7-100.4.

BATTALION OR BDET COMMAND SECTION AND STAFF

2-107. The OPFOR battalion or BDET headquarters is function-based and is composed of two sections—the command section and the staff section. They are highly streamlined and do not contain the robust planning and control capabilities necessary for higher staffs. For details on the personnel and equipment in the battalion command and staff sections, see, FM 7-100.4.

Command Section

2-108. The battalion command section consists of the commander, the DC, their vehicle drivers and radio telephone operators (RTOs). The COS is part of both the command section and the staff. On the battlefield, however, he is generally located with the staff section because he exercises direct control over the battalion staff. He is also in direct charge of the main CP in the absence of the commander. In maneuver units where the battalion command section employs combat vehicles, the command section may include vehicle gunners as well. Often a staff officer or NCO from the operations and/or intelligence section accompanies the battalion command section on the battlefield.

2-109. The battalion commander positions himself where he can best influence the critical action on the battlefield. The DC is typically supervising the execution of the battalion's second most critical operation and separate from the battalion commander so that both are not killed by the same engagement. The DC's vehicle may be used as a forward or auxiliary CP or observation post (OP).

Staff

2-110. At battalion level, the staff consists of two elements: the primary staff and the secondary staff. There is no functional staff below brigade or BTG level. Figure 2-20 on page 2-27 depicts the staff of a combat battalion headquarters and the liaison teams that support it.

2-111. The battalion staff consists of the operations officer (who also serves as the deputy COS), the assistant operations officer, the intelligence officer, and the resources officer. At battalion, typically each member of the primary staff heads a small staff section consisting of themselves and two other soldiers of the correct specialty. Two staff NCOs serve as day and night shift leaders in the main CP along with several enlisted soldiers. For more specifics, see FM 7-100.4.

2-112. The battalion staff has the mission to coordinate battlefield functions and to anticipate the battalion's needs 6-12 hours into the future. It supports the commander in the completion of combat orders and ensures his intent is being executed. It is not capable of executing long-range planning or supporting complex, joint, or interagency operations without augmentation.

2-113. The staff section operates the main CP. Staff vehicles are configured to supplement operations or for use as a forward or auxiliary CP or as an OP. In the latter case, they will be manned with appropriate staff personnel. The vehicles in the signal platoon may also be used. The COS may choose to ride in a utility vehicle or other vehicle.

2-114. Liaison teams may or may not be attached to the battalion. They are not a permanent part of the battalion staff structure. They may support the battalion staff with detailed expertise in the mission areas of their own particular branch or service. They provide direct communications to subordinate and supporting units executing missions in those areas. The battalion staff may also receive liaison teams from multinational and interagency subordinates and from affiliated forces. The operations officer is responsible for ensuring proper placement and utilization of the teams. The number and types of liaison teams is fluid and is determined by many variables. A typical liaison team consists of four personnel: a team chief, an assistant team chief, a staff NCO, and a driver/RTO. Liaison teams augmenting the battalion staff provide their own equipment. A detailed breakout of personnel and equipment of a typical liaison team is available in FM 7-100.4.

Staff Command

2-115. At BTG or brigade level and above, the OPFOR does not use the concept of commanders of combat support units acting as senior staff officers. However, in order to streamline staff functions at battalion, the OPFOR relies on commanders of supporting arms units to exercise staff supervision over their areas.

2-116. The leaders of the battalion's specialty platoons serve in a staff command role to coordinate key battlefield functions:

- The commander of the mortar battery also serves as the *chief of fire support coordination* for the battalion. (The assistant operations officer functions as the chief of fire support in those units without a mortar battery but still requiring fire and/or targeting support.)
- The signal platoon leader also serves as the battalion *chief of communications.*
- The platoon leader of the reconnaissance platoon serves as the battalion *chief of reconnaissance.*
- The platoon leader of the materiel support platoon serves as the battalion *chief of logistics.* (The battalion resources officer may also function as chief of administration.)

The chiefs of reconnaissance and communications coordinate with the main and forward CPs, and the chief of logistics operates the battalion trains.

2-117. The OPFOR will only employ staff command functions—

- In secondary staff areas (such as administration, logistics, reconnaissance, and communications).
- With commanders who have sufficient control over the area that the additional staff supervision functions are not a serious burden to their command responsibilities.

Command and Control

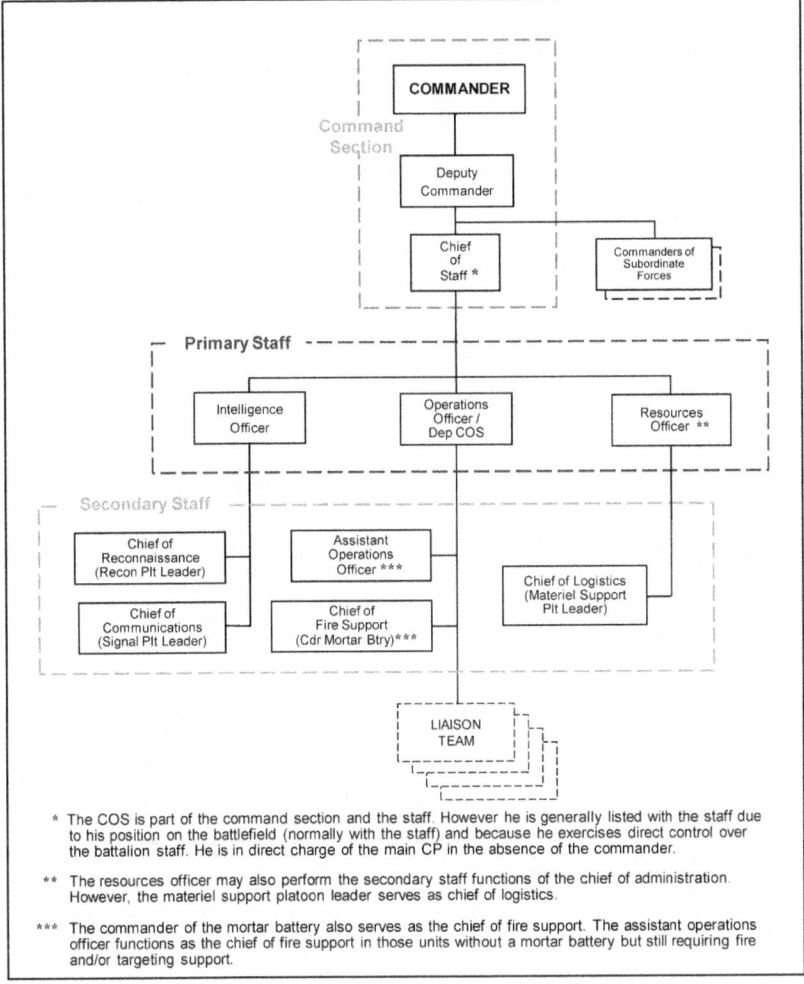

Figure 2-20. Battalion command section and staff

COMPANY OR CDET COMMAND AND STAFF

2-118. The basic structure of OPFOR companies or CDETs includes a headquarters and service section. This section consists of a command team, a support team, and a supply and transport team. Key personnel are—

- **Command Team.** Company commander and staff NCO.
- **Support Team.** DC and first sergeant.
- **Supply and Transport Team.** Supply sergeant.

Chapter 2

COMMAND POSTS

2-119. The OPFOR plans to exercise tactical control over its wartime forces from an integrated system of CPs. It has designed this system to ensure uninterrupted control of forces. All OPFOR levels of command share a parallel CP structure, tailored to match the differences in scope and span of control.

2-120. CPs are typically formed in three parts: a control group, a support group, and a communications group. The control group includes members of the command group or section and staff. The support group consists of the transport and logistics elements. Whenever possible, the communications group is remoted from the control and support groups because of its large number of signal vans, generators, and other special vehicles that would provide a unique signature.

2-121. Because the OPFOR expects its C2 to come under heavy attack in wartime, its military planners have created a CP structure that emphasizes survivability through dispersal, stringent security measures, redundancy, and mobility. They have constructed a CP system that can sustain damage with minimum disruption to the actual C2 process. In the event of disruption, they can quickly reestablish control. This extensive system of CPs extends from the hardened command facilities of the National Command Authority to the specially designed command vehicles from which OPFOR tactical commanders control their units. Tactical CPs and most operational-level CPs have been designed to be very mobile and smaller than comparable enemy CPs. The number, size, and types of CPs depend on the level of command.

COMMAND POST TYPES

2-122. OPFOR ground forces use five basic and three special types of CPs. Not all levels of command use all types at all times. (See table 2-2, where parentheses indicate that a type of CP may or may not be employed at a certain level.) The redundancy provided by multiple CPs helps to ensure that the C2 process remains survivable.

Table 2-2. Command post system

Level of Command	CP Types							
	Main CP	IFC CP	Forward CP	Sustainment CP	Airborne CP	Alternate CP	Auxiliary CP	Deception CP
DTG/Division	X	X	X	X	(X)	(X)	(X)	(X)
BTG/Brigade	X	-	X	X	(X)	(X)	(X)	(X)
BDET/Battalion	X	-	X	X	-	(X)	-	(X)
CDET/Company	-	-	X	X	-	-	-	(X)

2-123. For brevity, OPFOR plans and orders may use acronyms for the various types of CP. Thus, main CP may appear as MCP, integrated fires command CP as IFC CP, forward CP as FCP, sustainment CP as either SCP or SUSCP, airborne CP as AIRCP, alternate CP as ALTCP, auxiliary CP as AUXCP, and deception CP as DCP.

Main Command Post

2-124. The *main CP* generally is located in a battle zone or in a key sanctuary area or fortified position. It contains the bulk of the staff. The COS directs its operation. Its primary purpose is to simultaneously coordinate the activities of subordinate units not yet engaged in combat and plan for subsequent missions. The particular emphasis on planning in the main CP is on the details of transitioning between current and future operations. The main CP is the focus of control. It is less mobile and much larger than the forward CP. It makes use of hardened sites when possible.

2-125. The COS directs the staff in translating the commander's decisions into plans and orders. He also coordinates the movement and deployment of all subordinate units not yet in combat and monitors their progress and combat readiness. In addition to the COS, personnel present at the main CP include the liaison teams from subordinate, supporting, allied, and affiliated units, unless their presence is required in another CP.

IFC Command Post

2-126. The DC of a DTG or division directs the IFC from the *IFC CP*. The IFC CP possesses the communications, airspace control, and automated fire control systems required to integrate RISTA means and execute long-range fires. Each secondary staff subsection and some functional staff subsections have an element dedicated to the IFC CP. The IFC CP includes liaison teams from fire support, army aviation, and long-range reconnaissance elements. The IFC CP is typically separated from the main CP. Also for survivability, the various sections of the IFC headquarters that make up the IFC CP do not necessarily have to be located in one place.

Forward Command Post

2-127. A commander often establishes a *forward CP* with a small group of selected staff members. Its purpose is to provide the commander with information and communications that facilitate his decisions. The forward CP is deployed at a point from which he can more effectively and personally observe and influence the battle.

2-128. The personnel at the forward CP are not permanent. The assignment of officers to accompany the commander is dependent on the mission, situation, and availability of officers, communications, and transport means. Officers who may accompany the commander include the operations officer and the chief of reconnaissance. Other primary and or secondary staff officers may also deploy with the forward CP, depending on the needs of the situation. The secondary staff contains enough personnel to man the forward CP without degrading its ability to man the main or IFC CPs.

2-129. When formed, and when the commander is present, the forward CP is the main focus of *command*, though the COS (remaining in the main CP) has the authority to issue directives in the commander's absence.

Sustainment Command Post

2-130. The resources officer establishes and controls the sustainment CP. This CP is deployed in a position to permit the supervision of execution of sustainment procedures and the movement of support troops, typically in the support zone. It contains staff officers for fuel supply, medical support, combat equipment repair, ammunition supply, clothing supply, food supply, prisoner-of-war, and other services. It interacts closely with the subordinate units to ensure sustained combat capabilities. In nonlinear situations, multiple sustainment CPs may be formed.

Airborne Command Post

2-131. To maintain control in very fluid situations, when subordinates are spread over a wide area, or when the other CPs are moving, a commander may use an *airborne CP*. This is very common in DTG- or division-level commands, typically aboard helicopters.

Alternate Command Post

2-132. The *alternate CP* provides for the assumption of command should the CP containing the commander be incapacitated. The alternate CP is a designation given to an existing CP and is not a separately established element. The commander will establish which CP will act as an alternate CP to take command if the main (or forward) CP is destroyed or disabled. For example, the commander might designate the IFC CP as the alternate CP during a battle where long-range fires are critical to mission success. For situations that require reconstituting, he might designate the sustainment CP instead. Alternate CPs are also formed when fighting in complex terrain, or if the organization is dispersed over a wider area than usual and lateral communication is difficult.

Auxiliary Command Post

2-133. The commander may create an *auxiliary CP* to provide C2 over subordinate units fighting on isolated or remote axes. He may also use it in the event of disrupted control or when he cannot adequately maintain control from the main CP. An officer appointed at the discretion of the commander mans it.

Deception Command Post

2-134. As part of the overall INFOWAR plan, the OPFOR very often employs deception CPs. These are complex, multi-sensor-affecting sites integrated into the overall deception plan to assist in achieving battlefield opportunity by forcing the enemy to focus their command and control warfare efforts against meaningless positions.

COMMAND POST MOVEMENT

2-135. Plans for relocating the CPs are prepared by the operations section. The CPs are deployed and prepared in order to ensure that they are reliably covered from enemy ground and aerial reconnaissance, or from attack by enemy raiding forces.

2-136. Commanders deploy CPs in depth to facilitate control of their AORs. During lengthy moves, CPs may bound forward along parallel routes, preceded by reconnaissance elements that select the new locations. Normally, the main and forward CPs do not move at the same time, with one moving while the other is set up and controlling the battle. During an administrative movement, when there is little or no likelihood of contact with the enemy, a CP may move into a site previously occupied by another CP. However, during a tactical movement or when contact is likely, the OPFOR will not occupy a site twice, because to do so would increase the chances of an enemy locating a CP. While on the move, CPs maintain continuous contact with subordinates, higher headquarters, and flanking organizations. During movement halts, the practice is to disperse the CP in a concealed area, camouflaging it if necessary and locating radio stations and special vehicles some distance from the control and support groups. Because of dispersion in a mobile environment, CPs are often responsible for their own local ground defenses.

2-137. During the movement of a main CP, the OPFOR maintains continuity of control by handing over control to either the forward or airborne CP or, more rarely, to the alternate CP. Key staff members often move to the new location by helicopter to reduce the time spent away from their posts. Before any move, headquarters' troops carefully reconnoiter and mark the new location. Engineer preparation provides protection and concealment.

COMMAND POST LOCATION

2-138. The OPFOR locates CPs in areas affording good concealment, with good road network access being a secondary consideration. It situates CPs so that no single weapon can eliminate more than one. Remoting communications facilities lessens the chance of the enemy's locating the actual command element by radio direction finding.

2-139. During some particularly difficult phases of a battle, where close cooperation between units is essential, the forward CP of one element may be colocated with the forward or main CP of another. Examples are the commitment of an exploitation force or the passing of one organization through another.

COMMAND POST SECURITY

2-140. Security of CPs is important, and the OPFOR takes a number of measures to ensure it. CPs are a high priority for air defense protection. Ideally, main CPs also locate near reserve elements to gain protection from ground attack. Nevertheless, circumstances often dictate that they provide for their own local defense. Engineers normally dig in and camouflage key elements.

2-141. Good camouflage, the remoting of communications facilities, and the deployment of alternate CPs make most of the C2 structure fairly survivable. Nevertheless, one of the most important elements, the forward CP, often remains vulnerable. It forms a distinctive, if small, grouping, usually well within enemy artillery range. The OPFOR will therefore typically provide key CPs with sufficient engineer and combat arms support to protect them from enemy artillery or special operations raids.

COMMAND AND CONTROL SYSTEMS

2-142. The OPFOR commander's C2 requirements are dictated generally by the doctrine, tactics, procedures, and operational responsibilities applicable to commanders at higher echelons. Battlefield dispersion, mobility, and increasing firepower under conventional or WMD conditions require reliable, flexible, and secure C2.

2-143. Expanding C2 requirements include the need for—
- High mobility of combat headquarters and subordinate elements.
- Rapid collection, analysis, and dissemination of information as the basis for planning and decisionmaking.
- Maintaining effective control of forces operating in a hostile INFOWAR environment.

Supporting communications systems, which are the principal means of C2, must have a degree of mobility, reliability, flexibility, security, and survivability comparable to the C2 elements being supported.

2-144. Modern warfare has shifted away from large formations arrayed against one another in a linear fashion, to maneuver warfare conducted across large areas with more lethal, yet smaller, combat forces. C2 must provide the reliable, long-range communications links necessary to control forces deployed over greater distances. In order to move with the maneuver forces, the communications systems must be highly mobile.

COMMUNICATIONS

2-145. The chief characteristics of communications supporting the C2 structure are security, survivability, and flexibility. Redundancy in equipment, communications links, CPs, and other parts of the C2 system is the primary means of ensuring the control structure's security and survivability.

2-146. The organization of communications to meet operational and tactical requirements is the responsibility of the commander at each level. Prior to combat, the chief of communications, under the personal direction of the intelligence officer, prepares the communications plan. After approval by the COS, it becomes an annex to the combat order, for implementation by subordinate signal units. OPFOR communications reflect the concern of commanders to maintain uninterrupted C2, flexibility, and security.

Signal Assets

2-147. Communication systems employed include—
- Man-portable high-frequency (HF) and very-high-frequency (VHF) radios.
- HF radio stations.
- VHF and ultra-high-frequency (UHF) multichannel radio relay.
- Super-high-frequency (SHF) troposcatter systems.
- Satellite communications (SATCOM).
- Wire and cable (landline as far forward as possible).
- Commercial communications networks (including cellular, microwave, radio, wire, digital, and satellite).
- Local area networks and wide area networks.
- Internet and intranet.

2-148. Encrypted communications are common from battalion upward and may extend to the lowest levels in many OPFOR units. Units without encrypted communications mitigate the effects of not having secure communications by their strict adherence to communications and operations security principles, techniques and procedures. Some examples of these are their use of couriers and other messengers, pre-arranged signals and techniques, visual and sound signals, directional antennas, and wire and land lines whenever possible. The OPFOR assumes someone is always listening and takes appropriate countermeasures.

2-149. Headquarters normally task-organize their signal assets to support the formation of forward, main, IFC, and sustainment CPs. The numbers and types of signal units can vary greatly depending on the size and makeup of the force grouping under a particular headquarters. It is possible to extend mobile communications through the integration of wire and wireless systems and by connecting with fixed military and civil communications facilities.

2-150. Signal communications are organized through the communications centers that are established to provide communications for the CPs. All available communications methods are used to integrate the control and support groups of the CPs with the communications centers.

2-151. The OPFOR also stresses the use of *non-electronic* means of communications. While radio must be the principal means of communication in a fluid, mobile battle, the OPFOR is aware of the threat from enemy signals intelligence, direction finding, and communications jamming. Also, wire and cable are often not practical in fast-moving situations.

2-152. During periods of radio silence or disruption of radio communications, the OPFOR employs messengers, liaison officers, and visual and sound signals. Messengers are the preferred method for delivering combat orders at any time. Representatives from the staff may observe and supervise the execution of orders. Whenever possible, the OPFOR prefers personal contact between commanders (or their representatives) and subordinates.

Communications Nets

2-153. C2 of OPFOR organizations relies on extensive and redundant communications. The OPFOR primarily uses—

- UHF and SHF SATCOM.
- Radio relay multichannel.
- HF radio stations.
- HF and VHF single-channel radios.
- Wire or cable.

2-154. Tactical commands operate two *command nets*. The commander normally controls the *primary command net* from the forward CP, while the COS maintains control of the *alternate net* from the main CP. Depending on the distances involved, the primary net may be either HF or VHF. All of the command's constituent and dedicated units monitor the command nets. The DTG or division IFC CP also monitors the command nets.

2-155. The operations officer maintains an *operations net* monitored by the commander, subordinate and supporting units, and any alternate or auxiliary CP created. The resources officer also monitors this net from the sustainment CP.

2-156. The DTG or division DC, as IFC commander, maintains the integrated fires net. This net is monitored by the elements of the IFC.

2-157. The resources officer maintains the *support net*. Materiel support, maintenance, and medical units monitor this net. Subordinate combat arms units may also use this net when requiring additional, immediate assistance that constituent support assets are unable to provide.

2-158. The chief of reconnaissance maintains an *intelligence net*, monitored by reconnaissance units, maneuver units, the commander, DC, COS, and resources officer.

2-159. The CAO maintains the *airspace control net* for the purpose of controlling the command's airspace. Elements on this net include aviation units, air defense assets, and army aviation liaison teams.

2-160. When required, the commander will create a *special mission net*, monitored by the COS, that is employed to control the activities of units conducting a special mission. Examples could be a reconnaissance detachment or an airborne or heliborne landing force deployed behind enemy lines. Specific communications systems employed depend on the depth and type of mission.

2-161. The command maintains an *air defense and CBRN warning communications net*, monitored by all constituent, dedicated, and supporting units. This net is used for passing tactical alerts and CBRN and air warning notices. The COS maintains a watch on the DTG- or division-level warning nets at the main CP; he then disseminates warning where appropriate.

2-162. The command establishes multichannel links between the main and sustainment CPs and the CPs of subordinate units. These links are used for high-capacity voice and data transmissions. The division or DTG also establishes multichannel links between the main and sustainment CPs.

2-163. The primary responsibility for maintaining communications of a tactical command with its parent headquarters rests with the main CP. With the larger staffs and greater communications capabilities of the main CP, the commander is allowed to focus more on the actual conduct of the battle from the forward CP. Obviously, when staff members (such as the CAO or chief of reconnaissance) accompany the commander, they will establish control over their respective nets as required.

2-164. The chief of INFOWAR may also control one or more deception nets designed to mislead enemy signals intelligence analysis. Integrated into the INFOWAR plan are a description of these nets and procedures for their use.

Procedures

2-165. Before making contact with the enemy, most radio and radio-relay systems maintain a listening watch with transmission forbidden or strictly controlled. OPFOR units usually observe radio silence when defending or departing assembly areas. During radio silence, wire and courier are the primary communications means. While moving toward the enemy, units normally limit radio transmissions to various code words informing commanders they have accomplished assigned tasks or have encountered unexpected difficulties. The OPFOR also uses visual signals, such as flags and flares, to a great extent during movement. Usually only the commander and reconnaissance elements have permission to transmit.

2-166. In the offense, OPFOR units maintain radio silence until the outbreak of battle, when those authorized to transmit may do so without restriction. When contact with the enemy occurs, units initiate normal radio procedures. Subordinate commanders inform the commander—usually by code word—when they reach objectives, encounter CBRN contamination, make contact with the enemy, or have important information to report.

COMMAND AND CONTROL SYSTEMS SURVIVABILITY

2-167. Survivability of C2 systems is of great concern, since the C2 elements are typically located within range of enemy standoff systems, with increased potential for disruption or destruction. The OPFOR stresses the need to maintain continuous, reliable control of its forces and has undertaken a number of measures to prevent disruption and enhance survivability, while remaining flexible enough to retain control of units in combat. These include—
- High mobility of C2 systems and facilities.
- Redundancy of the C2 elements and networks.
- Adherence to operations- and information-security measures.
- Deception.

2-168. INFOWAR activities contribute to C2 survivability. The survivability of the headquarters' command group is facilitated by the fact that the commander, DC, and COS can be in separate CPs (forward, IFC, and main CPs, respectively).

Mobility

2-169. C2 elements must be highly mobile, due to the emphasis on maintaining combat at a rapid tempo. Because of their proximity to the enemy, CPs and supporting communications must frequently relocate to avoid detection and subsequent destruction.

2-170. CPs are usually mobile (that is, in vehicles) but may also be fixed. By emphasizing the use of multiple, mobile CPs, planners minimize the disruption of C2 that would occur with the enemy's destruction of these elements of the C2 structure. Highly mobile signal units employing transportable communications equipment support mobile CPs. This gives OPFOR commanders great flexibility in organizing and deploying their C2 elements. Thus, they are able to provide effective control in varied situations.

Redundancy

2-171. The OPFOR has built extensive redundancy into the C2 structure. Multiple CPs are fielded as low as possible. For communications between levels of command, multiple communications types are employed. Providing a variety of single- and multichannel links, these systems operate over a wide frequency spectrum.

Operations and Information Security

2-172. The consistent adherence to operations- and information-security measures is especially critical, given the increased capabilities of enemy reconnaissance, the increased role of surprise, and the proliferation of precision weapons. Given the high priority the enemy places on C2 elements as targets, maintaining operations security is an important requirement for C2 nodes. This is achieved by the stringent adherence to information-security procedures and extensive use of C3D.

Chapter 3
Offense

The offense carries the fight to the enemy. The OPFOR sees this as the decisive form of combat and the ultimate means of imposing its will on the enemy. While conditions at a particular time or place may require the OPFOR to defend, defeating an enemy force ultimately requires shifting to the offense. Even within the context of defense, victory normally requires some sort of offensive action. Therefore, OPFOR commanders at all levels seek to create and exploit opportunities to take offensive action, whenever possible.

The aim of offense at the tactical level is to achieve tactical missions in support of an operation. A tactical command ensures that its subordinate commands thoroughly understand both the overall goals of the operation and the specific purpose of a particular mission they are about to execute. In this way, subordinate commands may continue to execute the mission without direct control by a higher headquarters, if necessary.

PURPOSE OF THE OFFENSE

3-1. All tactical offensive actions are designed to achieve the goals of an operation through active measures. However, the purpose or reason, of any given offensive mission varies with the situation, as determined through the decisionmaking process. The primary distinction among types of offensive missions is their purpose which is defined by what the commander wants to achieve tactically. Thus, the OPFOR recognizes six general purposes of tactical offensive missions:

- Gain freedom of movement.
- Restrict freedom of movement.
- Gain control of key terrain, personnel, or equipment.
- Gain information.
- Dislocate.
- Disrupt.

3-2. These general purposes serve as a guide to understanding the design of an offensive mission and not as a limit placed on a commander as to how he makes his intent and aim clear. These are not the only possible purposes of tactical missions but are the most common.

3-3. These six general purposes are only a few of the many reasons the OPFOR might have for attacking an enemy, a potential enemy, a neighbor, or someone else. The true intent of an attack may reside at the operational or strategic level, but the attack is executed at the tactical level. Therefore, the actual reason for the attack may often be difficult to discern. In addition to those listed above, a few other reasons to attack may be to destroy, deceive, demonstrate dominance, deter (such as to discourage a neighbor from joining a coalition or alliance), or any number of other purposes.

3-4. In each of these general purposes of tactical offensive missions, the enemy may be destroyed or attrited to varying levels. Destruction is an inherent part of any attack. The critical tactical factor to the OPFOR commander initially is not *how* to conduct the offensive mission—but rather *why*. Once the *why* has been decided the method with the best chance of achieving tactical success becomes the *how*.

Chapter 3

ATTACK TO GAIN FREEDOM OF MOVEMENT

3-5. An *attack to gain freedom of movement* creates a situation in an important part of the battlefield where other friendly forces can maneuver in a method of their own choosing with little or no opposition. Such an attack can take many forms, of which the following are some examples:

- Seizing an important mobility corridor to prevent a counterattack into the flank of another moving force.
- Destroying an air defense unit so that a combat helicopter may use an air avenue of approach at lower risk.
- Breaching a complex obstacle to allow an exploitation force to pass through.
- Executing security tasks such as screen, guard, and cover. Such tasks may involve one or more attacks to gain freedom of movement as a component of the scheme of maneuver.

ATTACK TO RESTRICT FREEDOM OF MOVEMENT

3-6. An *attack to restrict freedom of movement* prevents the enemy from maneuvering as he chooses. Restricting attacks can deny key terrain, ambush moving forces, dominate airspace, or fix an enemy formation. Tactical tasks often associated with restricting attacks are ambush, block, canalize, contain, fix, interdict, and isolate. The attrition of combat elements and equipment may also limit the enemy units' ability to move. An example of this may be a preemptive strike on the enemy's water-crossing or mineclearing equipment.

ATTACK TO GAIN CONTROL OF KEY TERRAIN, PERSONNEL, OR EQUIPMENT

3-7. An *attack to gain control of key terrain, personnel, or equipment* is not necessarily terrain focused—a raid with the objective of taking prisoners or key equipment is also an attack to gain control. Besides the classic seizure of key terrain that dominates a battlefield, an attack to control may also target facilities such as economic targets, ports, or airfields. Tactical tasks associated with an attack to control are raid, clear, destroy, occupy, retain, secure, and seize. Some non-traditional attacks to gain control may be information attack, computer warfare, electronic warfare, or other forms of information warfare (INFOWAR).

ATTACK TO GAIN INFORMATION

3-8. An *attack to gain information* is a subset of the reconnaissance attack. (See Reconnaissance Attack later in this chapter.) In this case, the purpose is not to locate to destroy, fix, or occupy but rather to gain information about the enemy. Quite often the OPFOR will have to penetrate or circumvent the enemy's security forces and conduct an attack in order to determine the enemy's location, dispositions, capabilities, and intentions.

ATTACK TO DISLOCATE

3-9. An *attack to dislocate* employs forces to obtain significant positional advantage, rendering the enemy's dispositions less valuable, perhaps even irrelevant. It aims to make the enemy expose forces by reacting to the dislocating action. Dislocation requires enemy commanders to make a choice: accept neutralization of part of their force or risk its destruction while repositioning. Turning movements and envelopments produce dislocation. Artillery or other direct or indirect fires may cause an enemy to either move to a more tenable location or risk severe attrition. Typical tactical tasks associated with dislocation are ambush, interdict, and neutralize.

ATTACK TO DISRUPT

3-10. An *attack to disrupt* is used to prevent the enemy from being able to execute an advantageous course of action (COA) or to degrade his ability to execute that COA. It is also used to create windows of opportunity to be exploited by the OPFOR. It is an intentional interference (disruption) of enemy plans and intentions, causing the enemy confusion and the loss of focus, and throwing his battle synchronization into turmoil. The OPFOR then quickly exploits the result of the attack to disrupt. A spoiling attack is an example of an attack to disrupt.

3-11. The OPFOR will use an attack to disrupt in order to upset an enemy's formation and tempo, interrupt the enemy's timetable, cause the enemy to commitment of forces prematurely, and/or cause him to attack in a piecemeal fashion. The OPFOR will either attack the enemy force with enough combat power to achieve the desired results with one mass attack, or sustain the attack until the desired results are achieved.

Note. Disrupt is not only a purpose of the offense, but also a tactical task.

3-12. Attacks to disrupt typically focus on a key enemy capability, intention, or vulnerability. They are also designed to disrupt enemy plans, tempo, infrastructure, logistics, affiliations, C2, formations, or civil order. However, an attack to disrupt is not limited to any of the above. The OPFOR will use any method necessary to upset the enemy and cause disorder, disarray, and confusion.

3-13. Attacks to disrupt often have a strong INFOWAR component and may disrupt, limit, deny, and/or degrade the enemy's use of the electromagnetic spectrum, especially the enemy's C2. They may also take the form of computer warfare and/or information attack.

3-14. Attacks to disrupt are carried out at all levels and are limited only by time and resources available. The attack to disrupt may not be limited by distance. It may be carried out in proximity to the enemy (as in an ambush) or from an extreme distance (such as computer warfare or information attack from another continent) or both simultaneously. The attack to disrupt may be conducted by a single component (an ambush in contact) or a coordinated attack by several components such as combined arms using armored fighting vehicles, infantry, artillery, and several elements of INFOWAR (for example, electronic warfare, deception, perception management, information attack, and/or computer warfare).

3-15. The OPFOR does not limit its attacks to military targets or enemy combatants. The attack to disrupt may be carried out against noncombatant civilians (even family members of enemy soldiers at home station or in religious services), diplomats, contractors, or whomever and/or whatever the OPFOR commanders believe will enhance their probabilities of mission success.

PLANNING THE OFFENSE

3-16. For the OPFOR, the key elements of planning offensive missions are—
- Determining the objective of the offensive action.
- Defining time available to complete the action.
- Determining the level of planning possible (planned versus situational offense).
- Organizing the battlefield.
- Organizing forces and elements by function, including affiliated forces.
- Organizing INFOWAR activities in support of the offense (see chapter 7).

PLANNED OFFENSE

3-17. A *planned offense* is an offensive mission or action undertaken when there is sufficient time and knowledge of the situation to prepare and rehearse forces for specific tasks. Typically, the enemy is in a defensive position and in a known location. Key considerations in offensive planning are—
- Selecting a clear and appropriate objective.
- Determining which enemy forces (security, reaction, or reserve) must be fixed.
- Developing a reconnaissance plan that locates and tracks all key enemy targets and elements.
- Creating or taking advantage of a window of opportunity to free friendly forces from any enemy advantages in precision standoff and situational awareness.
- Determine which component or components of an enemy's combat system to attack.

Chapter 3

SITUATIONAL OFFENSE

3-18. The OPFOR may also conduct a *situational* offense. It recognizes that the modern battlefield is chaotic. Fleeting opportunities to strike at an enemy weakness will continually present themselves and just as quickly disappear. Although detailed planning and preparation greatly mitigate risk, they are often not achievable if a window of opportunity is to be exploited.

3-19. The following are examples of conditions that might lead to a situational offense:
- A key enemy unit, system, or capability is exposed.
- The OPFOR has an opportunity to conduct a spoiling attack to disrupt enemy defensive preparations.
- An OPFOR unit makes contact on favorable terms for subsequent offensive action.

3-20. In a situational offense, the commander develops his assessment of the conditions rapidly and without a great deal of staff involvement. He provides a basic COA to the staff, which then quickly turns that COA into an executable combat order. Even more than other types of OPFOR offensive action, the situational offense relies on implementation of battle drills by subordinate tactical units (see chapter 5).

3-21. Organization of the battlefield in a situational offense will normally be limited to minor changes to existing control measures. The nature of situational offense is such that it often involves smaller, independent forces accomplishing discrete missions. These missions will typically require the use of task-organized tactical groups and detachments of various types.

> *Note.* Any division or brigade receiving additional assets from a higher command becomes a division tactical group (DTG) or brigade tactical group (BTG). Therefore, references to a tactical group throughout this chapter may also apply to division or brigade, unless specifically stated otherwise.

FUNCTIONAL ORGANIZATION OF FORCES FOR THE OFFENSE—TACTICAL GROUPS, DIVISIONS, AND BRIGADES

3-22. In planning and executing offensive actions, OPFOR commanders at brigade level and above organize and designate various *forces* within his level of command according to their *function*. (See chapter 2.) Thus, subordinate forces understand their roles within the overall battle. However, the organization of forces can shift dramatically during the course of a battle, if part of the plan does not work or works better than anticipated. For example, a unit that started out being part of a fixing force might split off and become an exploitation force, if the opportunity presents itself.

3-23. Each functional force has an identified commander. This is often the senior commander of the largest subordinate unit assigned to that force. For example, if two BTGs are acting as the DTG's fixing force, the senior of the two BTG commanders is the fixing force commander. Since, in this option, each force commander is also a subordinate unit commander, he controls the force from his unit's command post (CP). Another option is to have one of the DTG's CPs be in charge of a functional force. For example, the forward CP could control a disruption force or a fixing force. Another possibility would be for the integrated fires command (IFC) CP to command the disruption force, the exploitation force, or any other force whose actions must be closely coordinated with fires delivered by the IFC.

3-24. In any case, the force commander is responsible to the tactical group commander to ensure that combat preparations are made properly and to take charge of the force during the battle. This frees the tactical group commander from decisions specific to the force's mission. Even when tactical group subordinates have responsibility for parts of the disruption zone, there is still an overall tactical group disruption force commander.

3-25. A battalion or below organization can serve as a functional force (or part of one) for its higher command. At any given time, it can be part of only a single functional force or a reserve. If, for example, a

Offense

BTG needed one part of one of its battalions to serve as an enabling force, but needed another part to join the exploitation force, one of the two battalion subunits would be task organized as a separate detachment.

Enabling Force(s)

3-26. Various types of *enabling forces* are charged with creating the conditions that allow the action force the freedom to operate. In order to create a window of opportunity for the action force to succeed, the enabling force may be required to operate at a high degree of risk and may sustain substantial casualties. However, an enabling force may not even make contact with the enemy, but instead conduct a demonstration.

3-27. Battalions and below serving as an enabling force are often required to conduct breaching or obstacle-clearing tasks. However, it is important to remember that the requirements laid on the enabling force are tied directly to the type and mission of the action force.

Disruption Force

3-28. In the offense, the *disruption force* would typically include the disruption force that already existed in a preceding defensive situation (see chapter 4). It is possible that forces assigned for actions in the disruption zone in the defense might not have sufficient mobility to do the same in the offense or that targets may change and require different or additional assets. Thus, the disruption force might require augmentation. For example, the disruption force for a division or DTG is typically a BTG especially task-organized for that function. Battalions and below can serve as disruption forces for brigades or BTGs. However, this mission typically is complex enough for them to be task-organized as detachments.

Fixing Force

3-29. OPFOR offensive actions are founded on the concept of fixing enemy forces so that they are not free to maneuver. The OPFOR recognizes that units and soldiers can be fixed in a variety of ways. For example—

- They find themselves without effective communication with higher command.
- Their picture of the battlefield is unclear.
- They are (or believe they are) decisively engaged in combat.
- They have lost mobility due to complex terrain, obstacles, or weapons of mass destruction (WMD).

3-30. In the offense, planners will identify which enemy forces need to be fixed and the method by which they will be fixed. They will then assign this responsibility to a force that has the capability to fix the required enemy forces with the correct method. The fixing force may consist of a number of units separated from each other in time and space, particularly if the enemy forces required to be fixed are likewise separated. A fixing force could consist entirely of affiliated irregular forces. It is possible that a discrete attack on logistics, command and control (C2), or other systems could fix an enemy without resorting to deploying large fixing forces.

3-31. Battalions and below often serve as fixing forces for BTGs and are also often capable of performing this mission without significant task organization. This is particularly true in those cases where simple suppressive fires are sufficient to fix enemy forces.

Assault Force

3-32. At BTG level, the commander may employ one or more *assault forces*. This means that one or more subordinate detachments would conduct an assault to destroy an enemy force or seize a position. However, the purpose of such an assault is to create or help create the opportunity for the action force to accomplish the BTG's overall mission. (See the section on Assault below, under Types of Offensive Action—Detachment, Battalion, and Company.)

Chapter 3

Security Force

3-33. The *security force* conducts activities to prevent or mitigate the effects of hostile actions against the overall tactical-level command and/or its key components. If the commander chooses, he may charge this security force with providing force protection for the entire area of responsibility (AOR), including the rest of the functional forces; logistics and administrative elements in the support zone; and other key installations, facilities, and resources. The security force may include various types of units—such as infantry, special-purpose forces (SPF), counterreconnaissance, and signals reconnaissance assets—to focus on enemy special operations and long-range reconnaissance forces operating throughout the AOR. It can also include Internal Security Forces units allocated to the tactical-level command, with the mission of protecting the overall command from attack by hostile insurgents, terrorists, and special operations forces. The security force may also be charged with mitigating the effects of WMD.

Deception Force

3-34. When the INFOWAR plan requires combat forces to take some action (such as a demonstration or feint), these forces will be designated as deception forces in close-hold executive summaries of the plan. Wide-distribution copies of the plan will refer to these forces according to the designation given them in the deception story.

Support Force

3-35. A support force provides support by fire; other combat or combat service support; or C2 functions for other parts of the tactical group. (When fire support units in the figures in this chapter are not identified as performing another function, they are probably acting as a support force.)

Action Force(s)

3-36. One part of the tactical group conducting a particular offensive action is normally responsible for performing the primary function or task that accomplishes the overall goal or objective of that *action*. In most general terms, therefore, that part can be called the *action force*. In most cases, however, the tactical group commander will give the action force a more specific designation that identifies the specific function it is intended to perform, which equates to achieving the objective of the tactical group's mission.

3-37. There are three basic types of action forces: exploitation force, strike force, and mission force. In some cases there may be more than one such force.

Exploitation Force

3-38. In most types of offensive action at tactical group level, an *exploitation force* is assigned the task of achieving the objective of the mission. It typically exploits a window of opportunity created by an enabling force. In some situations, the exploitation force could engage the ultimate objective with fires only.

Strike Force

3-39. A strike is an offensive COA that rapidly destroys a key enemy organization through a synergistic combination of massed precision fires and maneuver. The primary objective of a strike is the enemy's will and ability to fight. A strike is typically planned and coordinated at the operational level. However, it is often executed by a tactical-level force. The force that actually accomplishes the final destruction of the targeted enemy force is called the *strike force*. (See FM 7-100.1 for more detailed discussion of a strike.)

Mission Force

3-40. In those non-strike offensive actions where the mission can be accomplished without the creation of a specific window of opportunity, the set of capabilities that accomplish the mission are collectively known as a *mission force*. However, the tactical group commander may give a mission force a more specific designation that identifies its specific function.

Offense

Reserves

3-41. OPFOR offensive reserve formations will be given priorities in terms of whether the staff thinks it most likely that they will act as a particular type of enabling or action force. The size and composition of an offensive reserve are entirely situation-dependent.

FUNCTIONAL ORGANIZATION OF ELEMENTS FOR THE OFFENSE—DETACHMENTS, BATTALIONS, AND BELOW

3-42. An OPFOR detachment is a battalion or company designated to perform a specific mission and task-organized to do so. Commanders of detachments, battalions, and companies organize their subordinate units according to the specific functions they intend each subordinate to perform. They use a methodology of "functional organization" similar to that used by used by brigades and above (see chapter 2). However, one difference is that commanders at brigade and higher use the term *forces* when designating functions within their organization. Commanders at detachment, battalion, and below use the term *element*. Elements can be broken down into two very broad categories: action and enabling. However, commanders normally designate functional elements more specifically, identifying the specific action or the specific means of accomplishing the function during a particular mission. Commanders may also organize various types of specialist elements. Depending on the mission and conditions, there may be more than one of some of these specific element types.

> *Note.* A detachment as a whole can receive a specific mission assigned by a higher commander. That commander gives the detachment a functional designation based on the role it will play in his overall mission or the specific function it will perform for him. For example, a detachment assigned to conduct a raid may be called the *raiding detachment* or the *raiding force* of the higher command.

3-43. The number of functional elements is unlimited and is determined by any number of variables, such as the size of the overall organization, its mission, and its target. Quite often the distinction between exactly which element is an action element or an enabling element is blurred because, as the mission progresses, conditions change or evolve and require adaptation.

Action Elements

3-44. The *action element* is the element conducting the primary action of the overall organization's mission. However, the commander normally gives this element a functional designation that more specifically describes exactly what activity the element is performing on the battlefield at that particular time. For example, the action element in a raid may be called the *raiding element*. If an element accomplishes the objective of the mission by exploiting an opportunity created by another element, is may be called the *exploitation element*. Throughout this TC, the clearest and most descriptive term will be used to avoid any confusion.

3-45. At a different time or place on the battlefield, the element performing the "primary action" might have a completely different function and would be labeled accordingly. These are not permanent labels and differ with specific functions performed.

Enabling Elements

3-46. Enabling elements can enable the primary action in various ways. The most common types are security elements and support elements. The *security element* provides local tactical security for the overall organization and prevents the enemy from influencing mission accomplishment. (A security element providing front, flank, or rear security may be identified more specifically as the "front security element," "flank security element," or "rear security element.") The *support element* provides combat and combat service support and C2 for the larger organization. Due to such considerations as multiple avenues of approach, a commander may organize one or more of each of these elements in specific cases.

Specialist Elements

3-47. In certain situations, a detachment may organize one or more specialist elements. Specialist elements are typically formed around a unit with a specific capability, such as an obstacle-clearing, reconnaissance, or deception. Detachments formed around such specialist elements may or may not have a security or support element depending on their specialty, their location on the battlefield, and the support received from other units. For example, a movement support detachment (MSD) typically has a reconnaissance and obstacle-clearing element, plus one or two road and bridge construction and repair elements. If an MSD is receiving both security and other support from the infantry or mechanized units preparing to move through the cleared and prepared area, it probably will not have its own support and security elements. In this case, all of the elements will be dedicated to various types of engineer mobility functions.

PREPARING FOR THE OFFENSE

3-48. In the preparation phase, the OPFOR focuses on ways of applying all available resources and the full range of actions to place the enemy in the weakest condition and position possible. Commanders prepare their organizations for all subsequent phases of the offense. They organize the battlefield and their forces and elements with an eye toward capitalizing on conditions created by successful attacks.

ESTABLISH CONTACT

3-49. The number one priority for all offensive actions is to gain and maintain contact with key enemy forces. As part of the decisionmaking process, the commander and staff identify which forces must be kept under watch at all times. The OPFOR will employ whatever technical sensors it has at its disposal to locate and track enemy forces, but the method of choice is ground reconnaissance. It may also receive information on the enemy from the civilian populace, local police, or affiliated irregular forces.

MAKE THOROUGH LOGISTICS ARRANGEMENTS

3-50. The OPFOR understands that there is as much chance of an offense being brought to culmination by a lack of sufficient logistics support as by enemy action. Careful consideration will be given to carried days of supply and advanced caches to obviate the need for easily disrupted lines of communications (LOCs).

MODIFY THE PLAN WHEN NECESSARY

3-51. The OPFOR takes into account that, while it might consider itself to be in the preparation phase for one battle, it is continuously in the execution phase. Plans are never considered final. Plans are checked throughout the course of their development to ensure they are still valid in light of battlefield events.

REHEARSE CRITICAL ACTIONS IN PRIORITY

3-52. The commander establishes the priority for the critical actions expected to take place during the battle. The force rehearses those actions in as realistic a manner as possible for the remainder of the preparation time.

EXECUTING THE OFFENSE

3-53. The degree of preparation often determines the nature of the offense in the execution phase. Successful execution depends on forces that understand their roles in the battle and can swiftly follow preparatory actions with the maximum possible shock and violence and deny the enemy any opportunity to recover. A successful execution phase often ends with transition to the defense in order to consolidate gains, defeat enemy counterattacks, or avoid culmination. In some cases, the execution phase is followed by continued offensive action to exploit opportunities created by the battle just completed.

MAINTAIN CONTACT

3-54. The OPFOR will go to great lengths to ensure that its forces maintain contact with key elements of the enemy force throughout the battle. This includes rapid reconstitution of reconnaissance assets and units and the use of whatever combat power is necessary to ensure success.

IMPLEMENT BATTLE DRILLS

3-55. The OPFOR derives great flexibility from battle drills. (See chapter 5 for more detail.) Contrary to the U.S. view that battle drills, especially at higher levels, reduce flexibility, the OPFOR uses minor, simple, and clear modifications to thoroughly understood and practiced battle drills to adapt to ever-shifting conditions. It does not write standard procedures into its combat orders and does not write new orders when a simple shift from current formations and organization will do. OPFOR offensive battle drills will include, but not be limited to, the following:
- React to all seven forms of contact—direct fire, indirect fire, visual, obstacle, CBRN, electronic warfare, and air attack.
- Fire and maneuver.
- Fixing enemy forces.
- Situational breaching.

MODIFY THE PLAN WHEN NECESSARY

3-56. The OPFOR is sensitive to the effects of mission dynamics and realizes that the enemy's actions may well make an OPFOR unit's original mission achievable, but completely irrelevant. As an example, a unit of the fixing force in an attack may be keeping its portion of the enemy force tied down while another portion of the enemy force is maneuvering nearby to stop the exploitation force. In this case, the OPFOR unit in question must be ready to transition to a new mission quickly and break contact to fix the maneuvering enemy force.

SEIZE OPPORTUNITIES

3-57. The OPFOR places maximum emphasis on decentralized execution, initiative, and adaptation. Subordinate units are expected to take advantage of fleeting opportunities so long as their actions are in concert with the goals of the higher command.

DOMINATE THE TEMPO OF COMBAT

3-58. Through all actions possible, the OPFOR plans to control the tempo of combat. It will use continuous attack, INFOWAR, and shifting targets, objectives, and axes to ensure that tactical events are taking place at the pace it desires.

TYPES OF OFFENSIVE ACTION—TACTICAL GROUPS, DIVISIONS, AND BRIGADES

3-59. The types of offensive action in OPFOR doctrine are both tactical methods and guides to the design of COAs. An offensive mission may include subordinate units that are executing different offensive and defensive COAs within the overall offensive mission framework.

Note. Any division or brigade receiving additional assets from a higher command becomes a division tactical group (DTG) or brigade tactical group (BTG). Therefore, references to a tactical group throughout this chapter may also apply to division or brigade, unless specifically stated otherwise.

Chapter 3

ATTACK

3-60. An *attack* is an offensive operation that destroys or defeats enemy forces, seizes and secures terrain, or both. It seeks to achieve tactical decision through primarily military means by defeating the enemy's military power. This defeat does not come through the destruction of armored weapons systems but through the disruption, dislocation, and subsequent paralysis that occurs when combat forces are rendered irrelevant by the loss of the capability or will to continue the fight. Attack is the method of choice for OPFOR offensive action. There are two types of attack: *integrated attack* and *dispersed attack*.

3-61. The OPFOR does not have a separate design for "exploitation" as a distinct offensive COA. Exploitation is considered a central part of all integrated and dispersed attacks.

3-62. The OPFOR does not have a separate design for "pursuit" as a distinct offensive COA. A pursuit is conducted using the same basic COA framework as any other integrated or dispersed attack. The fixing force gains contact with the fleeing enemy force and slows it or forces it to stop while the assault and exploitation forces create the conditions for and complete the destruction of the enemy's C2 and logistics structure or other systems.

3-63. The OPFOR recognizes that moving forces that make contact must rapidly choose and implement an offensive or defensive COA. The OPFOR methodology for accomplishing this is discussed in chapter 5.

Integrated Attack

3-64. *Integrated attack* is an offensive action where the OPFOR seeks military decision by destroying the enemy's will and/or ability to continue fighting through the application of combined arms effects. Integrated attack is often employed when the OPFOR enjoys overmatch with respect to its opponent and is able to bring all elements of offensive combat power to bear. It may also be employed against a more sophisticated and capable opponent, if the appropriate window of opportunity is created or available. See figures 3-1 through 3-3, on pages 3-10 through 3-12 for examples of integrated attacks.

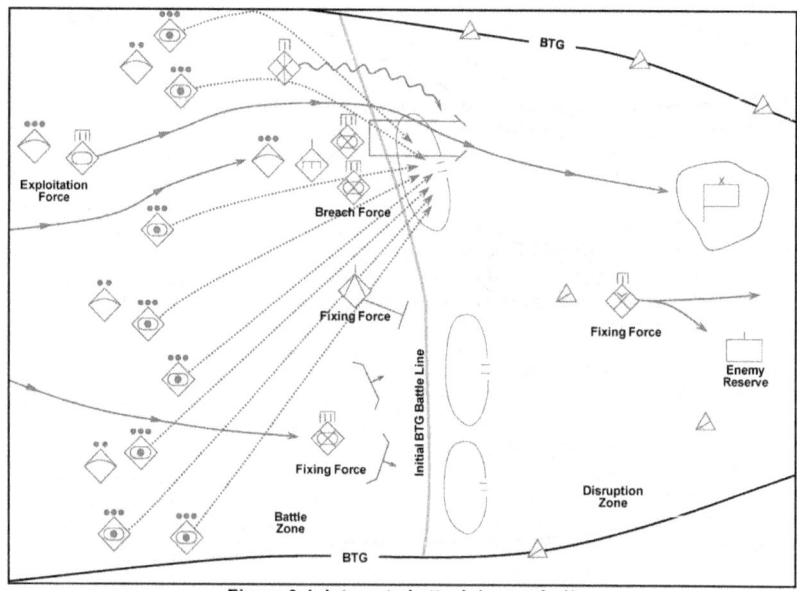

Figure 3-1. Integrated attack (example 1)

Offense

3-65. The primary objective of an integrated attack is destroying the enemy's will and ability to fight. The OPFOR recognizes that modern militaries cannot continue without adequate logistics support and no military, modern or otherwise, can function without effective command and control.

3-66. Integrated attacks are characterized by—
- Not being focused solely on destruction of ground combat power but often on C2 and logistics.
- Fixing the majority of the enemy's force in place with the minimum force necessary.
- Isolating the targeted subcomponent(s) of the enemy's combat system from his main combat power.
- Using complex terrain to force the enemy to fight at a disadvantage.
- Using deception and other components of INFOWAR to degrade the enemy's situational understanding and ability to target OPFOR formations.
- Using flank attack and envelopment, particularly of enemy forces that have been fixed.

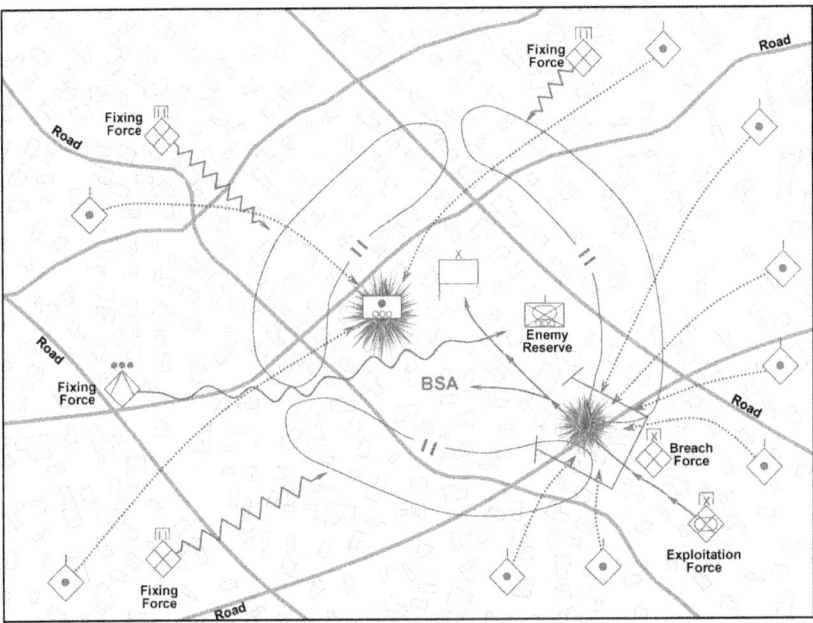

Figure 3-2. Integrated attack (example 2)

3-67. The OPFOR prefers to conduct integrated attacks when most or all of the following conditions exist:
- The OPFOR possesses significant overmatch in combat power over enemy forces.
- It possesses at least air parity over the critical portions of the battlefield.
- It is sufficiently free of enemy standoff reconnaissance and attack systems to be able to operate without accepting high levels of risk.

Chapter 3

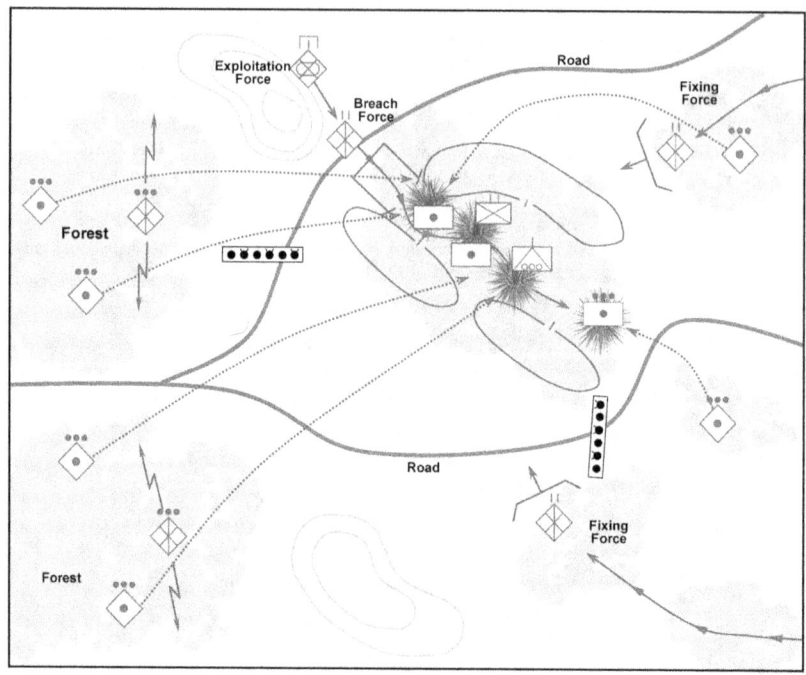

Figure 3-3. Integrated attack (example 3)

Functional Organization for an Integrated Attack

3-68. An integrated attack employs various types of functional forces. The tactical group commander assigns subordinate units functional designations that correspond to their intended roles in the attack.

Enabling Forces

3-69. An integrated attack often employs fixing, assault, and support forces. A disruption force exists, but is not created specifically for this type of offensive action.

3-70. **Fixing Force.** The fixing force in an integrated attack is required to prevent enemy defending forces, reserves, and quick-response forces from interfering with the actions of the assault and exploitation forces. The battle will develop rapidly, and enemy forces not in the attack zone cannot be allowed to reposition to influence the assault and exploitation forces. Maneuver forces, precision fires, air defense units, long-range antiarmor systems, situational obstacles, chemical weapons, and electronic warfare are well suited to fix defending forces.

3-71. **Assault Force.** At BTG level, the commander may employ one or more *assault forces*. The assault force in an integrated attack is charged with destroying a particular part of the enemy force or seizing key positions. Such an assault can create a window of opportunity for an exploitation force. An assault force may successfully employ infiltration of infantry to carefully pre-selected points to assist the exploitation force in its penetration of enemy defenses.

3-72. **Support Force.** A support force provides support to the attack by fire; other combat or combat service support; or C2 functions. Smoke and suppressive artillery and rocket fires, combat engineer units, and air-delivered weapons are well suited to this function.

Action Force

3-73. The most common type of action force in an integrated attack is the *exploitation force*. Such a force must be capable of penetrating or avoiding enemy defensive forces and attacking and destroying the enemy's support infrastructure before he has time to react. An exploitation force ideally possesses a combination of mobility, protection, and firepower that permits it to reach the target with sufficient combat power to accomplish the mission.

Dispersed Attack

3-74. *Dispersed attack* is the primary manner in which the OPFOR conducts offensive action when threatened by a superior enemy and/or when unable to mass or provide integrated C2 to an attack. This is not to say that the dispersed attack cannot or should not be used against peer forces, but as a rule integrated attack will more completely attain objectives in such situations. Dispersed attack relies on INFOWAR and dispersion of forces to permit the OPFOR to conduct tactical offensive actions while overmatched by precision standoff weapons and imagery and signals sensors. The dispersed attack is continuous and comes from multiple directions. It employs multiple means working together in a very interdependent way. The attack can be dispersed in time as well as space. See figures 3-4 through 3-6 on pages, 3-14 through 3-16 for examples of dispersed attacks.

3-75. The primary objective of dispersed attack is to take advantage of a window of opportunity to bring enough combined arms force to bear to destroy the enemy's will and/or capability to continue fighting. To achieve this, the OPFOR does not necessarily have to destroy the entire enemy force, but often just destroy or degrade a key component of the enemy's combat system.

3-76. Selecting the appropriate component of the enemy's combat system to destroy or degrade is the first step in planning the dispersed attack. This component is chosen because of its importance to the enemy and varies depending on the force involved and the current military situation. For example, an enemy force dependent on one geographical point for all of its logistics support and reinforcement would be most vulnerable at that point. Disrupting this activity at the right time and to the right extent may bring about tactical decision on the current battlefield, or it may open further windows of opportunity to attack the enemy's weakened forces at little cost to the OPFOR. In another example, an enemy force preparing to attack may be disrupted by an OPFOR attack whose purpose is to destroy long-range missile artillery, creating the opportunity for the OPFOR to achieve standoff with its own missile weapons. In a final example, the key component chosen may be the personnel of the enemy force. Attacking and causing mass casualties among infantrymen may delay an enemy offensive in complex terrain while also being politically unacceptable for the enemy command structure.

3-77. Dispersed attacks are characterized by—

- Not being focused on complete destruction of ground combat power but rather on destroying or degrading a key component of the enemy's combat system (often targeting enemy C2 and logistics).
- Fixing and isolating enemy combat power.
- Using smaller, independent subordinate elements.
- Conducting rapid moves from dispersed locations.
- Massing at the last possible moment.
- Conducting simultaneous attack at multiple, dispersed locations.
- Using deception and other elements of INFOWAR to degrade the enemy's situational understanding and ability to target OPFOR formations.

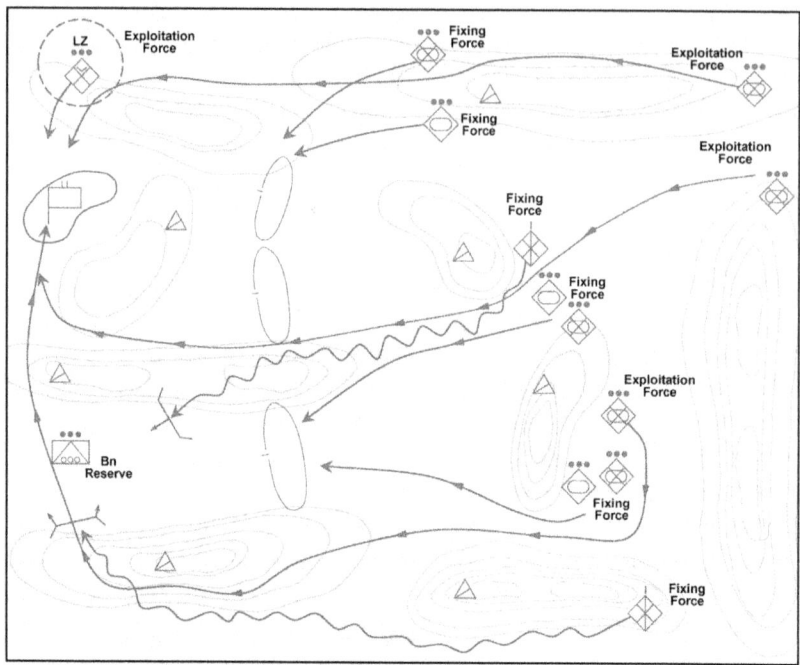

Figure 3-4. Dispersed attack (example 1)

3-78. The window of opportunity needed to establish conditions favorable to the execution of a dispersed attack may be one created by the OPFOR or one that develops due to external factors in the operational environment. When this window must be created, the OPFOR keys on several tasks that must be accomplished:

- Destroy enemy ground reconnaissance.
- Deceive enemy imagery and signals sensors.
- Create an uncertain air defense environment.
- Selectively deny situational awareness.
- Maximize use of complex terrain.

Offense

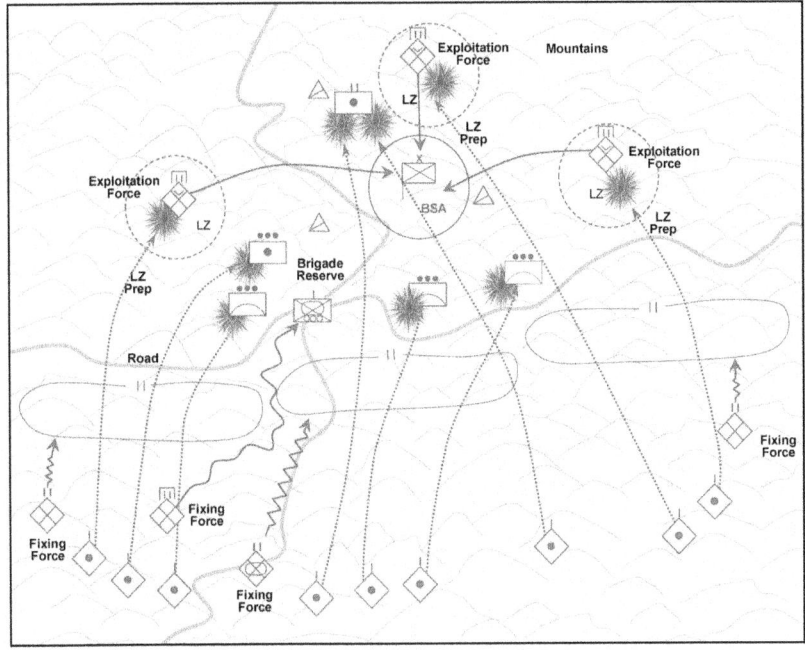

Figure 3-5. Dispersed attack (example 2)

Functional Organization for a Dispersed Attack

3-79. A dispersed attack employs various types of functional forces. The tactical group commander assigns subordinate units functional designations that correspond to their intended roles in the attack.

Enabling Forces

3-80. A dispersed attack often employs fixing, assault, and support forces. A disruption force exists, but is not created specifically for this type of offensive action. Deception forces can also play an important role in a dispersed attack.

3-81. **Fixing Force.** The fixing force in a dispersed attack is primarily focused on fixing enemy response forces. Enemy reserves, quick response forces, and precision fire systems that can reorient rapidly will be those elements most capable of disrupting a dispersed attack. Maneuver forces, precision fires, air defense and antiarmor ambushes, situational obstacles, chemical weapons, and electronic warfare are well suited to fix these kinds of units and systems. Dispersed attacks often make use of multiple fixing forces separated in time and/or space.

3-82. **Assault Force.** At BTG level, the commander may employ one or more *assault forces*. The assault force in an integrated attack is charged with destroying a particular part of the enemy force or seizing key positions. Such an assault can create favorable conditions for the exploitation force to rapidly move from dispersed locations and penetrate or infiltrate enemy defenses. An assault force may successfully employ infiltration of infantry to carefully pre-selected points to assist the exploitation force in its penetration. Dispersed attacks often make use of multiple assault forces separated in time and/or space.

Chapter 3

3-83. **Support Force.** A support force provides support to the attack by fire; other combat or combat service support; or C2 functions. Smoke and suppressive artillery and rocket fires, combat engineer units, and air-delivered weapons are well suited to this function.

Action Forces

3-84. The most common type of action force in an integrated attack is the *exploitation force*. Such a force must be capable, through inherent capabilities or positioning relative to the enemy, of destroying the target of the attack. In one set of circumstances, a tank brigade may be the unit of choice to maneuver throughout the battlefield as single platoons in order to have one company reach a vulnerable troop concentration or soft C2 node. Alternatively, the exploitation force may be a widely dispersed group of SPF teams set to attack simultaneously at exposed logistics targets. Dispersed attacks often make use of multiple exploitation forces separated in time and/or space, but often oriented on the same objective or objectives.

Note. The example of a dispersed attack by a BTG in figure 3-6 would unfold in the following sequence. The disruption force—consisting primarily of SPF teams and a UAV company (shown at its launch and recovery position)—locates and destroys enemy reconnaissance. Fixing forces 1, 2, and 3—consisting primarily of mechanized battalion-size detachments (BDETs)—ensure that the enemy combat battalions and brigade reserve do not have freedom to maneuver. Support forces 1 and 2—consisting primarily of artillery and antitank weapons—provide direct and indirect fires that assist the fixing forces. With all enemy combat forces fixed, exploitation forces 1, 2, and 3—consisting of guerillas, insurgents, INFOWAR, and combined arms forces—destroy the enemy C2 and logistics structure.

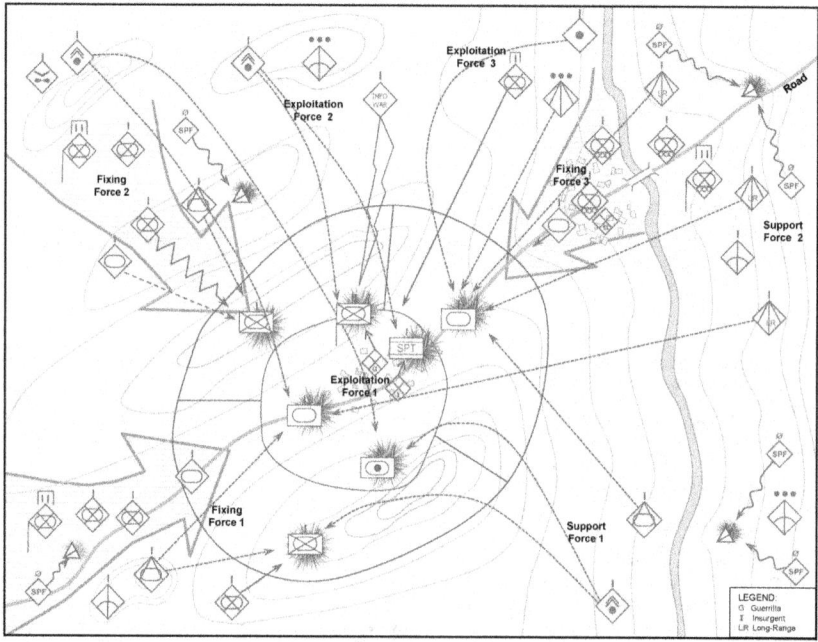

Figure 3-6. Dispersed attack (example 3)

Limited-Objective Attack

3-85. A *limited-objective attack* seeks to achieve results critical to the battle plan or even the operation plan by destroying or denying the enemy key capabilities through primarily military means. The results of a limited-objective attack typically fall short of tactical or operational decision on the day of battle, but may be vital to the overall success of the battle or operation. Limited-objective attacks are common while fighting a stronger enemy with the objective of preserving forces and wearing down the enemy, rather than achieving decision.

3-86. The primary objective of a limited-objective attack is a particular enemy capability. This may or may not be a particular man-made system or group of systems, but may also be the capability to take action at the enemy's chosen tempo.

3-87. Limited-objective attacks are characterized by—

- Not being focused solely on destruction of ground combat power but often on C2 and logistics.
- Denying the enemy the capability he most needs to execute his plans.
- Maximal use of the systems warfare approach to combat (see chapter 1).
- Significant reliance on a planned or seized window of opportunity.

3-88. Quite often, the limited-objective attack develops as a situational offense. This occurs when an unclear picture of enemy dispositions suddenly clarifies to some extent and the commander wishes to take advantage of the knowledge he has gained to disrupt enemy timing. This means that limited-objective attacks are often conducted by reserve or response forces that can rapidly shift from their current posture to strike at the enemy.

3-89. There are two types of tactical limited-objective attack: spoiling attack and counterattack. These share some common characteristics, but differ in purpose.

Spoiling Attack

3-90. The purpose of a spoiling attack is to preempt or seriously impair an enemy attack while the enemy is in the process of planning, forming, assembling, or preparing to attack. It is designed to control the tempo of combat by disrupting the timing of enemy operations. The spoiling attack is a type of attack to disrupt.

3-91. Spoiling attacks do not have to accomplish a great deal to be successful. Conversely, planners must focus carefully on what effect the attack is trying to achieve and how the attack will achieve that effect. In some cases, the purpose of the attack will be to remove a key component of the enemy's combat system. A successful spoiling attack can make this key component unavailable for the planned attack and therefore reduces the enemy's overall chances of success. More typically, the attack is designed to slow the development of conditions favorable to the enemy's planned attack. See figure 3-7 on page 3-18 for an example of a spoiling attack.

Chapter 3

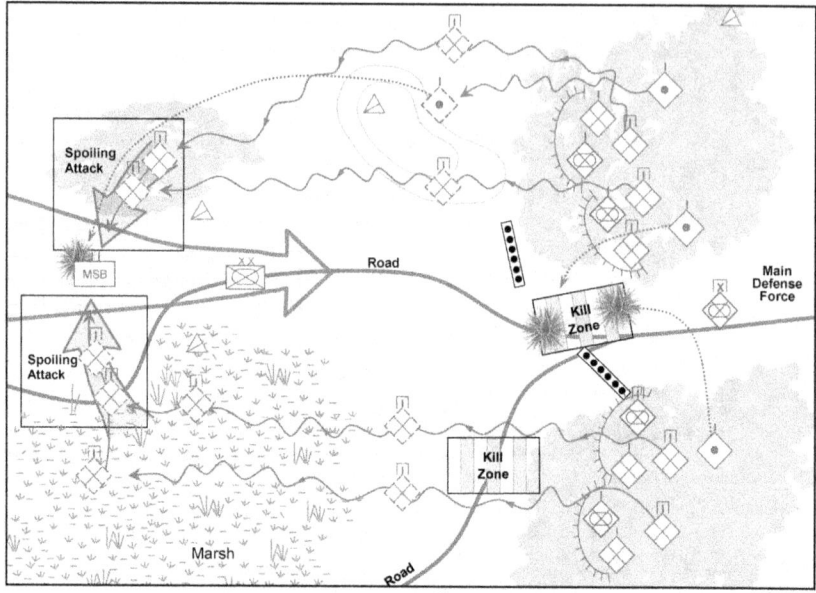

Figure 3-7. Spoiling attack (example)

3-92. Spoiling attacks are characterized by—
- A requirement to have a clear picture of enemy preparations and dispositions.
- Independent, small unit action.
- Highly focused objectives.
- The possibility that a spoiling attack may open a window of opportunity for other combat actions.

3-93. The OPFOR seeks to have the following conditions met in order to conduct a spoiling attack:
- Reconnaissance, intelligence, surveillance, and target acquisition (RISTA) establishes a picture of enemy attack preparations.
- Enemy security, reserve, and response forces are located and tracked.
- Enemy ground reconnaissance in the attack zone is destroyed or rendered ineffective.

Functional Organization for a Spoiling Attack

3-94. A spoiling attack may or may not involve the use of enabling forces. If enabling forces are required, the part of the tactical group that actually executes the spoiling attack would be an *exploitation force*. Otherwise, it may be called the *mission force*. The exploitation or mission force will come from a part of the tactical group that is capable of acting quickly and independently to take advantage of a fleeting opportunity. Since the spoiling attack is a type of attack to disrupt, the exploitation or mission force might come from the disruption force. If more combat power is required, it might come from the main defense force, which would not yet be engaged. A third possibility is that it could come from the tactical group reserve.

Offense

Counterattack

3-95. A *counterattack* is a form of attack by part or all of a defending force against an enemy attacking force with the general objective of denying the enemy his goal in attacking. It is designed to cause an enemy offensive operation to culminate and allow the OPFOR to return to the offense. A counterattack is designed to control the tempo of combat by returning the initiative to the OPFOR. See figure 3-8 for an example of a counterattack.

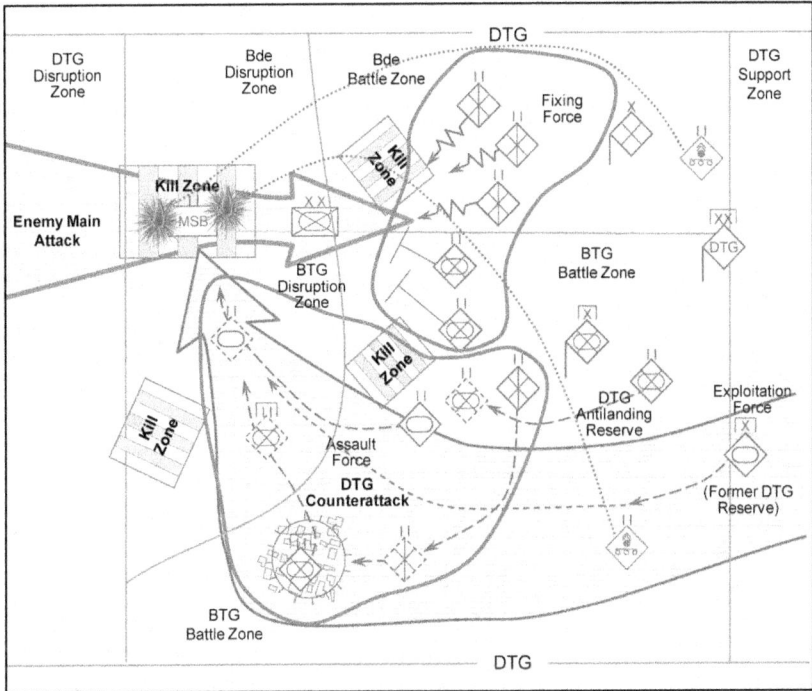

Figure 3-8. Counterattack (example)

3-96. Counterattacks are characterized by—
- A shifting in command and support relationships to assume an offensive posture for the counterattacking force.
- A proper identification that the enemy is at or near culmination.
- The planned rapid transition of the remainder of the force to the offense.
- The possibility that a counterattack may open a window of opportunity for other combat actions.

3-97. The OPFOR seeks to set the following conditions for a counterattack:
- Locate and track enemy reserve forces and cause them to be committed.
- Destroy enemy reconnaissance forces that could observe counterattack preparations.

Chapter 3

Functional Organization for a Counterattack

3-98. Functional organization for a counterattack involves many of the same types of forces as for an integrated or dispersed attack. However, the exact nature of their functions may differ due to the fact that the counterattack comes out of a defensive posture. (To a large degree, functional organization of forces is the same for spoiling attacks and counterattacks. That is because both types of limited-objective attack are carried out under similar conditions.)

Enabling Forces

3-99. A counterattack often employs fixing, assault, and support forces. The disruption force was generally part of a previous OPFOR defensive posture.

3-100. **Fixing Force.** The fixing force in a counterattack is that part of the force engaged in defensive action with the enemy. These forces continue to fight from their current positions and seek to account for the key parts of the enemy array and ensure they are not able to break contact and reposition. Additionally, the fixing force has the mission of making contact with and destroying enemy reconnaissance forces and any combat forces that may have penetrated the OPFOR defense.

3-101. **Assault Force.** In a counterattack, the assault force (if one is used) is often assigned the mission of forcing the enemy to commit his reserve so that the enemy commander has no further mobile forces with which to react. If the fixing force has already forced this commitment, the counterattack design may forego the creation of an assault force.

3-102. **Support Force.** A support force can provide support to a counterattack by fire; other combat or combat service support; or C2 functions. Due to the time-sensitive and highly mobile nature of the counterattack, fire support may be the most critical.

Action Forces

3-103. The most common type of action force in a counterattack is an exploitation force. A larger-scale counterattack may involve more than one exploitation force.

3-104. The exploitation force in a counterattack bypasses engaged enemy forces to attack and destroy the enemy's support infrastructure before he has time or freedom to react. An exploitation force ideally possesses a combination of mobility, protection, and firepower that permits it to reach the target with sufficient combat power to accomplish the mission.

TACTICAL OFFENSIVE ACTIONS—DETACHMENTS, BATTALIONS, AND BELOW

3-105. OPFOR commanders of detachments, battalions, and below select the offensive action best suited to accomplishing their mission. Units at this level typically are called upon to execute one combat mission at a time. Therefore, it would be rare for such a unit to employ more than one type of offensive action simultaneously. At the tactical level, all OPFOR units, organizations, elements, and even plans are dynamic and adapt very quickly to the situation.

> *Note.* Any battalion or company receiving additional assets from a higher command becomes a battalion-size detachment (BDET) or company-size detachment (CDET). Therefore, references to a detachment throughout this chapter may also apply to battalion or company, unless specifically stated otherwise.

ASSAULT

3-106. An assault is an attack that destroys an enemy force through firepower and the physical occupation and/or destruction of his position. An assault is the basic form of OPFOR tactical offensive combat. Therefore, other types of offensive action may include an element that conducts an assault to complete the mission; however, that element will typically be given a designation that corresponds to the specific mission accomplished. (For example, an element that conducts an assault in the completion of an ambush would be called the ambush element.)

3-107. The OPFOR does not have a separate design for "mounted" and "dismounted" assaults, since the same basic principles apply to any assault action. An assault may have to make use of whatever units can take advantage of a window of opportunity, but the OPFOR views all assaults as combined arms actions. See figures 3-9 through 3-11 on pages 3-21 through 3-23 for examples of assault.

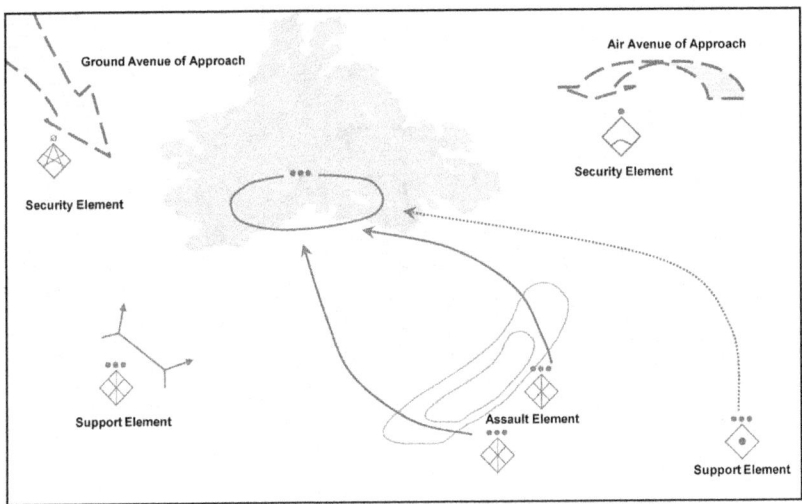

Figure 3-9. Assault (example 1)

Functional Organization for an Assault

3-108. A detachment conducting an ambush typically is organized into three elements: the *assault element*, the *security element*, and the *support element*.

Assault Element.

3-109. The *assault element* is the action element. It maneuvers to and seizes the enemy position, destroying any forces there.

Security Element

3-110. The *security element* provides early warning of approaching enemy forces and prevents them from reinforcing the assaulted unit. Security elements often make use of terrain choke points, obstacles, ambushes and other techniques to resist larger forces for the duration of the assault. The commander may (or may be forced to) accept risk and employ a security element that can only provide early warning but is not strong enough to halt or repel enemy reinforcements. This decision is based on the specific situation.

Chapter 3

Support Element

3-111. The *support element* provides the assaulting detachment with one or more of the following:
- C2.
- Combat service support (CSS).
- Supporting direct fire (such as small arms, grenade launchers, or infantry antitank weapons).
- Supporting indirect fire (such as mortars or artillery).
- Mobility support.

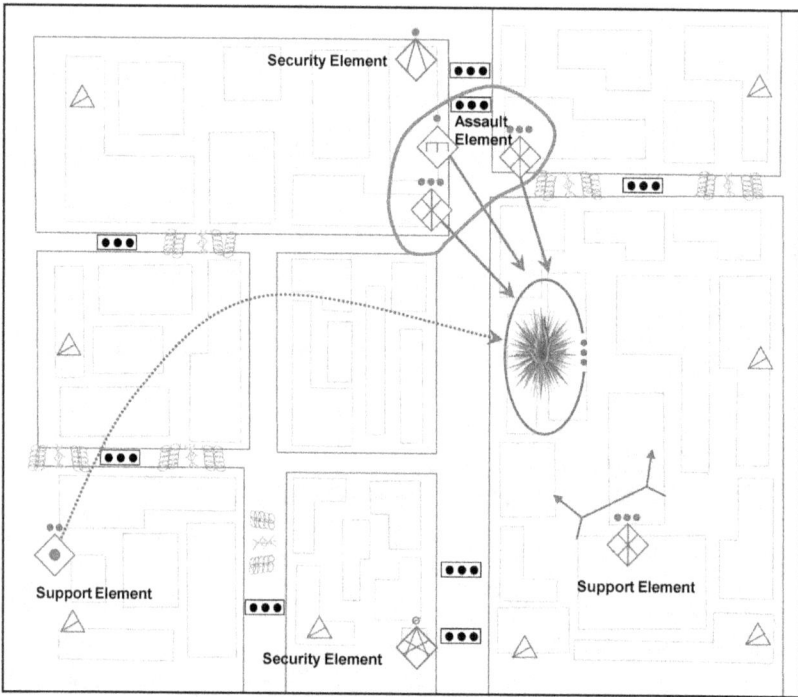

Figure 3-10. Assault (example 2)

Organizing the Battlefield for an Assault

3-112. The detachment conducting an assault is given an AOR in which to operate. A key decision with respect to the AOR will be whether or not a higher headquarters is controlling the airspace associated with the assault.

3-113. The combat order, which assigns the AOR, will often identify the enemy position being assaulted as the primary objective, with associated attack routes and/or axes. Support by fire positions will typically be assigned for use by the support element. The security element will have battle positions that overwatch key enemy air and ground avenues of approach with covered and concealed routes to and from those positions.

Offense

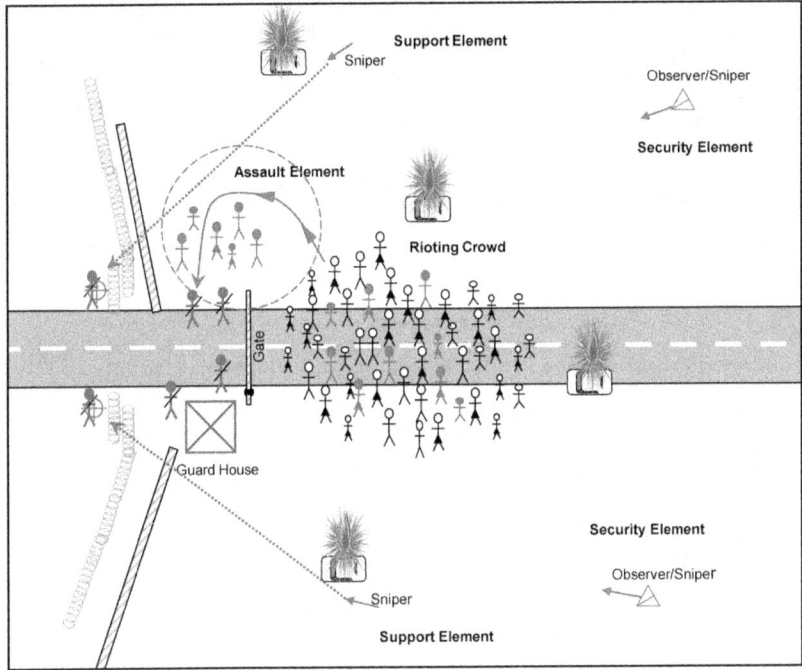

Figure 3-11. Assault (example 3)

Executing an Assault

3-114. An assault is the most violent COA a military force can undertake. The nature of an assault demands an integrated combined-arms approach. Indeed, a simple direct assault has a very low chance of success without some significant mitigating factors. Decisive OPFOR assaults are characterized by—

- Isolation of the objective (enemy position) so that it cannot be reinforced during the battle.
- Effective tactical security.
- Effective suppression of the enemy force to permit the assault element to move against the enemy position without receiving destructive fire.
- Violent fire and maneuver against the enemy.

Assault Element

3-115. The assault element must be able to maneuver from its assault position to the objective and destroy the enemy located there. It can conduct attack by fire, but this is not an optimal methodology and should only be used when necessary. Typical tactical tasks expected of the assault element are—

- Clear.
- Destroy.
- Occupy.
- Secure.
- Seize.

3-116. Speed of execution is critical to an assault. At a minimum, the assault element must move with all practical speed once it has left its attack position. However, the OPFOR goal in an assault is for all the elements to execute their tasks with as much speed as can be achieved. For example, the longer the security element takes to move to its positions and isolate the objective, the more time the enemy has to react even before the assault element has begun maneuvering. Therefore, the OPFOR prefers as much of the action of the three elements of an assault to be simultaneous as possible. OPFOR small units practice the assault continually and have clear battle drills for all of the key tasks required in an assault.

3-117. In addition to speed, the assault element will use surprise; limited visibility; complex terrain; and/or camouflage, concealment, cover, and deception (C3D). These can allow the assault element to achieve the mission while remaining combat effective. The OPFOR intent is to maintain lethal velocity in the assault by using the synergistic effects of any or all methods and/or options available.

Security Element

3-118. The security element is typically the first element to act in an assault. The security element moves to a position (or positions) where it can deny the enemy freedom of movement along any ground or air avenues of approach that can reinforce the objective or interfere with the mission of the assault element. The security element is equipped and organized such that it can detect enemy forces and prevent them from contacting the rest of the detachment. The security element normally is assigned a screen, guard, or cover task, but may also be called upon to perform other tactical tasks in support of its purpose:

- Ambush.
- Block.
- Canalize.
- Contain.
- Delay.
- Destroy.
- Disrupt.
- Fix.
- Interdict.
- Isolate.

Support Element

3-119. The support element can have a wide range of functions in an assault. Typically the detachment commander exercises C2 from within a part of the support element, unless his analysis deems success requires he leads the assault element personally. The support element controls all combat support (CS) and CSS functions as well as any supporting fires. The support element typically does not become decisively engaged but parts of it may employ direct suppressive fires. Tasks typically expected of support elements in the assault are—

- Attack by fire.
- Disrupt.
- Fix.
- Neutralize.
- Support by fire.
- Canalize.
- Contain.

Command and Control of an Assault

3-120. Typically, the commander positions himself with the support element and the deputy commander moves with the assault element, although this may be reversed. The primary function of control of the assault is to arrange units and tasks in time and space so that the assault element begins movement with all

Offense

capabilities of the support element brought to bear, the security element providing the detachment's freedom to operate and the objective isolated.

Assaults in Complex Terrain

3-121. Fighting in complex terrain slows the rate of advance, requiring a high consumption of manpower and materiel, especially ammunition. In the offense, combat in cities is avoided whenever possible, either by bypassing defended localities or by seizing towns from the march before the enemy can erect defenses. When there is no alternative, units reorganize their combat formations to attack an urban area by assault. The attackers can exploit undefended towns by using them as avenues of approach or assembly areas. For fighting in urban areas, the OPFOR prefers to establish urban detachments. See chapter 6 for additional information on urban detachments, their composition, and how they fight.

> *Note. Complex terrain* is a topographical area consisting of an urban center larger than a village and/or of two or more types of restrictive terrain or environmental conditions occupying the same space. (Restrictive terrain or environmental conditions include, but are not limited to, slope, high altitude, forestation, severe weather, and urbanization.)

3-122. The burden of combat in cities usually falls on infantry troops, supported by other arms. Tank units can be used to seal off pockets of resistance en route to the town or city, to envelop and cut off the city, or to provide reinforcements or mobile reserves for mechanized infantry units.

3-123. Complex terrain has both advantages and disadvantages for assaulting troops. Due to its unique combination of restrictive terrain and environmental conditions, complex terrain imposes significant limitations on observation, maneuver, fires, and intelligence collection. It reduces engagement ranges, thereby easing the task of keeping the assault element protected during its approach to the objective. However, it also provides the enemy cover and concealment on the objective as well as natural obstacles to movement and good ambush positions along that same approach. This is especially the case when the enemy is defending in an urban area.

3-124. Assaults in close, mixed or open terrain always face the possibility of obstacles restricting movement to the objective. Obstacles in complex terrain—man-made or not—are virtually a certainty. Typically then, assault detachments include a specialist element designed to execute mobility and/or shaping tasks in support of the assault element. This element, made up of sappers and other supporting arms, may be known as an obstacle-clearing element. Its functions would be similar to those of reconnaissance and obstacle-clearing element in a movement support detachment (see chapter 12).

Support of the Assault

3-125. An assault typically requires several types of support. These can include reconnaissance, fire support, air defense, and INFOWAR.

Reconnaissance

3-126. Reconnaissance effort in support of an assault can begin long before the assault is executed. A higher level of command may control this effort, or it may allocate additional reconnaissance assets to a detachment. However, the unit conducting an assault often has to rely on its own reconnaissance effort to a large degree in order to take advantage of a window of opportunity to execute an assault. Reconnaissance patrols in support of an assault are typically given the following missions:

- Determine and observe enemy reinforcement and counterattack routes.
- Determine composition and disposition of the forces on the objective.
- Locate and mark enemy countermobility and survivability effort.
- Locate and track enemy response forces.
- Defeat enemy C3D effort.

Fire Support

3-127. The primary mission of fire support in an assault is to suppress the objective and protect the advance of the assault element. Precision munitions may be used to destroy key systems that threaten the assault force and prevent effective reinforcement of the objective. Fire support assets committed to the assault are typically part of the support element.

Air Defense

3-128. The typical purpose of air defense support to an assault is to prevent enemy air power from influencing the action of the assault element. It is possible to find air defense systems and measures in all three elements of an assault.

3-129. Air defense systems in the security element provide early warning and defeat enemy aerial response to the assault. Such systems also target enemy aerial reconnaissance such as unmanned aerial vehicles (UAVs) to prevent the enemy from having a clear picture of the assault action. Air defense systems in the support element provide overwatch of the assault element and the objective.

3-130. Air defense systems are least likely to be found in the assault element. However, such situations as long attack axes, which require the assault force to operate out of the range of systems in the support element for any length of time, may dictate this disposition of air defense systems.

3-131. Some air defense systems may prove useful in close combat such as urban areas. Air defense weapons usually have a very high angle of fire allowing them to target the upper stories of buildings. High-explosive rounds allow the weapons to shoot through the bottom floor of the top story, successfully engaging enemy troops and/or equipment located on roof tops. The accuracy and lethality of air defense weapons also facilitates their role as a devastating ground weapon when used against personnel, equipment, buildings, and lightly armored vehicles.

INFOWAR

3-132. INFOWAR supports the assault primarily by helping isolate the objective. This is often done by—
- Deceiving forces at the objective as to the timing, location, and/or intent of the assault.
- Conducting deception operations to fix response forces.
- Isolating the objective with electronic warfare.

Note: A simple, effective, and successful technique often employed by the OPFOR is to distract and then flank with multiple coordinated assaults.

AMBUSH

3-133. An ambush is a surprise attack from a concealed position, used against moving or temporarily halted targets. Such targets could include truck convoys, railway trains, boats, individual vehicles, or dismounted troops. In an ambush, enemy action determines the time, and the OPFOR sets the place. Ambushes may be conducted to—
- Destroy or capture personnel and supplies.
- Harass and demoralize the enemy.
- Delay or block movement of personnel and supplies.
- Canalize enemy movement by making certain routes useless for traffic.

3-134. The OPFOR also uses ambush as a primary psychological warfare tool. The psychological effect is magnified by the OPFOR use of multi-tiered ambushes. A common tactic is to spring an ambush and set other ambushes along the relief or reaction force's likely avenues of approach. (For an example, see figure 3-17 on page 3-34.) Another tactic is to attack enemy medical evacuation assets, especially if the number of these assets is limited. The destruction of means to evacuate wounded instills a sense of tentativeness in the enemy soldiers because they realize that, should they become wounded or injured, medical help will not be forthcoming.

Note. Prior to springing an ambush, the OPFOR may prefer to wait until enemy sweep operations end. As enemy units return to garrison, tired, low on fuel, and lax after searching for an extended time and not engaging, the OPFOR strikes hard.

3-135. Cutting LOCs is a basic OPFOR tactic. The OPFOR attempts to avoid enemy maneuver units and concentrate on ambushes along supporting LOCs. Attacking LOCs damages an enemy's economic infrastructure, isolates military units, and attrits the enemy with little risk to the ambush force. Successful ambushes usually result in concentrating the majority of movements to principal roads, railroads, or waterways where targets are more vulnerable to attack by other forces.

3-136. Key factors in an ambush are—
- Surprise.
- Control.
- Coordinated fires and shock (timing).
- Simplicity.
- Discipline.
- Security (and enemy secondary reaction).
- Withdrawal.

Functional Organization for an Ambush

3-137. Similar to an assault, a detachment conducting an ambush is typically organized into three elements: the *ambush element*, the *security element*, and the *support element*. There may be more than one of each of these types of element.

Ambush Element

3-138. The *ambush element* of an ambush has the mission of attacking and destroying enemy elements in the kill zone(s). The ambush element conducts the main attack against the ambush target that includes halting the column, killing or capturing personnel, recovering supplies and equipment, and destroying unwanted vehicles or supplies that cannot be moved.

Security Element

3-139. The *security element* of an ambush has the mission to prevent enemy elements from responding to the ambush before the main action is concluded. Failing that, it prevents the ambush element from becoming decisively engaged. This is often accomplished simply by providing early warning.

3-140. Security elements are placed on roads and trails leading to the ambush site to warn the ambush element of the enemy approach. These elements isolate the ambush site using roadblocks, other ambushes, and outposts. They also assist in covering the withdrawal of the ambush element from the ambush site. The distance between the security element and the ambush element is dictated by terrain. In many instances, it may be necessary to organize secondary ambushes and roadblocks to intercept and delay enemy reinforcements.

Support Element

3-141. The *support element* of an ambush has the same basic functions as that of an assault. It is quite often involved in supporting the ambush element with direct and/or indirect fires.

Planning and Preparation for an Ambush

3-142. The planning and preparation of an ambush is similar that for a raid, except that selecting an ambush site is an additional consideration. A detachment is typically assigned a battle zone in which to execute an ambush, since such attacks typically do not require control of airspace at the detachment level. The area where the enemy force is to be destroyed is delineated by one or more kill zones.

3-143. The mission may be a single ambush against one target or a series of ambushes against targets on one or more LOCs. The probable size, strength, and composition of the enemy force that is to be ambushed, formations the enemy is likely to use, and enemy reinforcement capabilities are considered. Favorable terrain for an ambush, providing unobserved routes for approach and withdrawal, must be selected.

3-144. The time of the ambush should coincide with periods of limited visibility, offering a wider choice of positions and better opportunities to surprise and confuse the enemy. However, movement and control are more difficult during a night ambush. Night ambushes are more suitable when the mission can be accomplished during, or immediately following, the initial burst of fire. They require a maximum number of automatic weapons to be used at close range. Night ambushes can hinder the enemy's use of LOCs at night, while friendly aircraft can attack the same routes during the day (if the enemy does not have air superiority). Daylight ambushes facilitate control and permit offensive action for a longer period of time. Daylight ambushes also provide an opportunity for more effective fire from such weapons as rocket launchers and antiarmor weapons.

Site Selection

3-145. In selecting the ambush site, the basic consideration is favorable terrain, although limitations such as deficiencies in firepower and lack of resupply during actions may govern the choice of the ambush site. The site should have firing positions offering concealment and favorable fields of fire. Whenever possible, firing should be conducted from the screen of foliage. The terrain at the site should serve to canalize the enemy into a kill zone. The entire kill zone is covered by fire so that there is no dead space that would allow the enemy to organize resistance. The ambush force should take advantage of natural obstacles such as defiles, swamps, and cliffs to restrict enemy maneuvers against the force. When natural obstacles do not exist, mines, demolitions, barbed or concertina wire, and other concealed obstacles are employed to canalize the enemy.

Movement

3-146. The ambush force moves over a pre-selected route or routes to the ambush site. One or more mission support sites (MSSs) or rendezvous points usually are necessary along the route to the ambush site. Last-minute intelligence is provided by reconnaissance elements, and final coordination for the ambush is made at the MSS or the assembly area.

Action at the Ambush Site

3-147. The ambush force moves to an assembly area near the ambush site. Security elements take up their positions first and then the ambush elements move into place. As the approaching enemy force is detected, or at a predesignated time, the ambush force commander decides whether or not to execute the ambush. This decision is based on the size of the enemy force, enemy guard and security measures, and the estimated value of the target in light of the probable cost to the ambush force.

Executing an Ambush

3-148. If the decision is made to execute the ambush, enemy security elements or advance guards are allowed to pass through the main ambush position. When the head of the enemy main force reaches a predetermined point, then fire, demolitions, or obstacles halt it. At this signal, the entire ambush element opens fire. Designated security elements engage the enemy's advance and rear security elements to prevent reinforcement of the main force. The volume of fire is violent, rapid, directed at enemy personnel exiting from vehicles, and concentrated on vehicles mounting automatic weapons. Antiarmor weapons (such as recoilless rifles and rocket launchers) are used against armored vehicles. Machineguns lay bands of fixed fire across escape routes, as well as in the kill zone. Mortar projectiles, hand grenades, and rifle grenades, and directional antipersonnel mines are fired into the kill zone.

3-149. If the commander decides to assault the enemy force, the assault is initiated with a prearranged signal. After the enemy has been rendered combat ineffective, designated ambush element personnel move into the kill zone to recover supplies, equipment, and ammunition.

Types of Ambush

3-150. There are three types of OPFOR ambush—annihilation, harassment, or containment—based on the desired effects and the resources available. Ambushes are frequently employed because they have a great chance of success and provide force protection. The OPFOR conducts ambushes to kill or capture personnel, destroy or capture equipment, restrict enemy freedom of movement, and collect information and supplies.

Annihilation Ambush

3-151. The purpose of an *annihilation ambush* is to destroy the enemy force. These are violent attacks designed to ensure the enemy's return fire, if any, is ineffective. Generally, this type of ambush uses the terrain to the attacker's advantage and employs mines and other obstacles to halt the enemy in the kill zone. The goal of the obstacles is to keep the enemy in the kill zone throughout the action. Using direct, or indirect, fire systems, the support element destroys or suppresses all enemy forces in the kill zone. It remains in a concealed location and may have special weapons, such as antitank weapons.

3-152. The support and ambush elements kill enemy personnel and destroy equipment within the kill zone by concentrated fires. The ambush element remains in covered and concealed positions until the enemy is rendered combat ineffective. Once that occurs, the ambush element secures the kill zone and eliminates any remaining enemy personnel that pose a threat. The Ambush element remains in the kill zone to thoroughly search for any usable information and equipment, which it takes or destroys.

3-153. The security element positions itself to ensure early warning and to prevent the enemy from escaping the kill zone. Following the initiation of the ambush, the security element seals the kill zone and does not allow any enemy forces in or out. The ambush force withdraws in sequence; the ambush element withdraws first, then the support element, and lastly the security element. The entire ambush force reassembles at a predetermined location and time. For examples of annihilation ambushes see figures 3-12 through 3-14 on pages 3-30 and 3-31.

Chapter 3

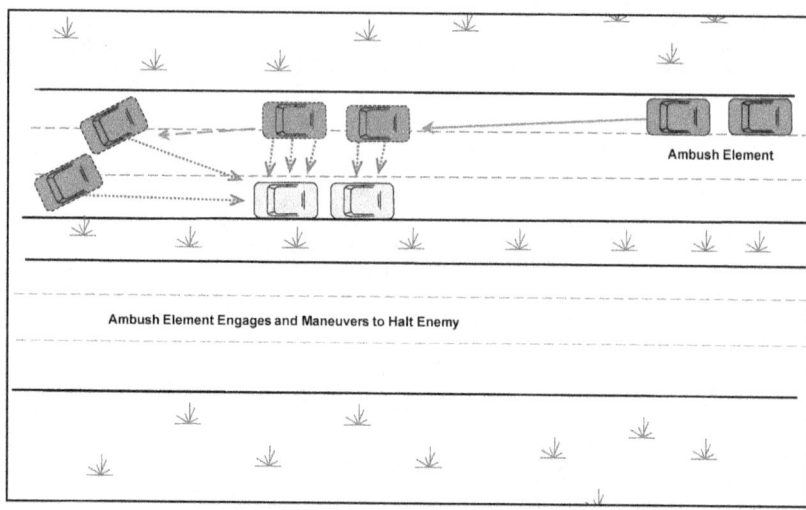

Figure 3-12. Annihilation ambush (example 1)

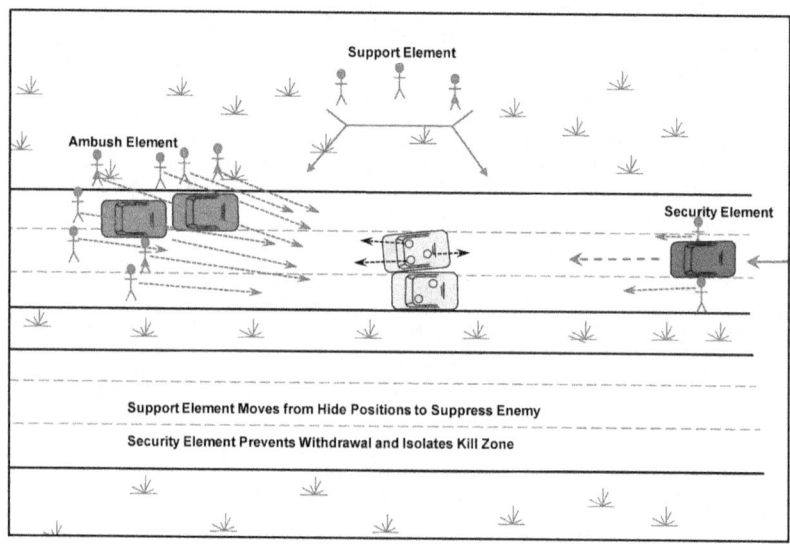

Figure 3-13. Annihilation ambush (example 2)

Offense

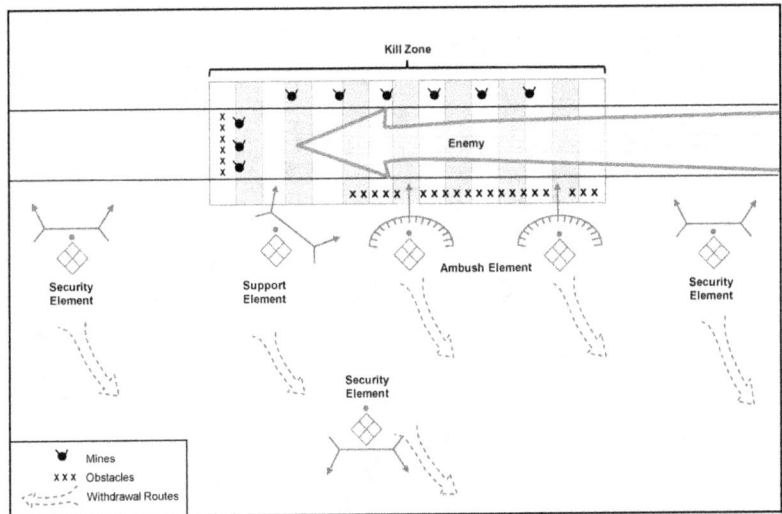

Figure 3-14. Annihilation ambush (example 3)

3-154. Annihilation ambushes in complex terrain, including urban environments, often involve task organizations that the OPFOR calls "hunter-killer (HK) teams." The HK team structure is extremely lethal and is especially effective for close fighting in such environments. Although other companies may be used as well, generally, infantry companies are augmented and task-organized into these HK teams. When task-organized to ambush armored vehicles, they may be called "antiarmor HK teams" or "HK infantry antiarmor teams."

3-155. HK teams primarily fight from ground level. In urban environments, however, they prefer to attack multi-dimensionally, from basements or sewers and from upper stories and on the tops of buildings. The targets are engaged simultaneously to maximize effectiveness and confusion.

3-156. At a minimum, each HK infantry antiarmor team is composed of gunners of infantry antiarmor weapons, a machinegunner, a sniper, and one or more riflemen. Multiple HK teams simultaneously attack a single armored vehicle. Kill shots are generally made against the top, rear, and sides of vehicles from multiple dimensions. The teams prefer to trap vehicle columns in city streets where destruction of the first and last vehicles will trap the column and allow its total destruction. Single vehicles are easily defeated.

3-157. The HK teams use command detonated, controllable, and side-attack mines (antitank, anti-vehicle, and antipersonnel) in conjunction with predetermined artillery and mortar fires. Side-attack (off-route) mines may be placed out of sight, such as inside windows and alleys. Figure 3-15 on page 3-32 is an example an annihilation ambush conducted by HK teams in a complex, urban environment.

Chapter 3

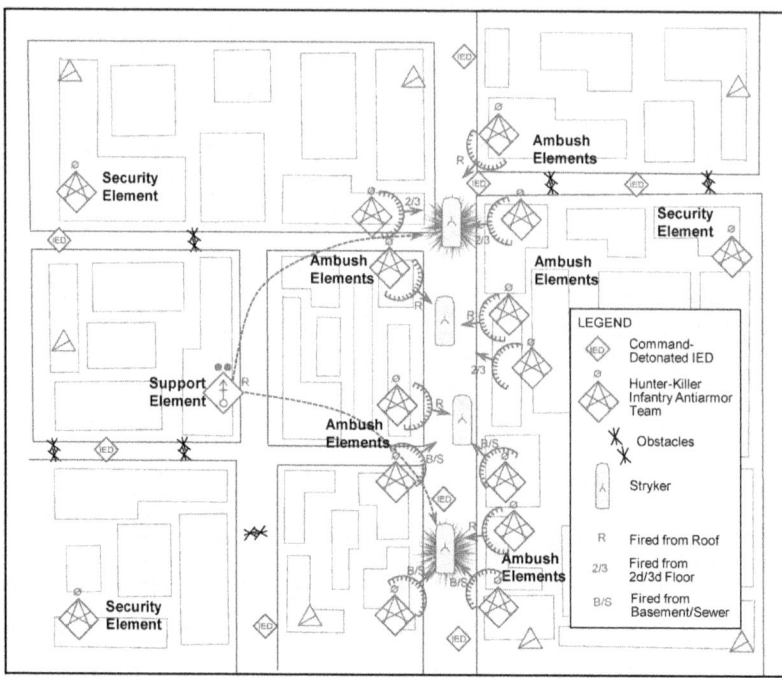

Figure 3-15. Annihilation ambush using infantry antiarmor (hunter-killer) teams (example 4)

Harassment Ambush

3-158. A *harassment ambush* interferes with routine enemy activities, impedes the enemy's freedom of movement, and has a psychological impact on enemy personnel. The OPFOR may choose to conduct a harassment ambush if the enemy has superior combat power. This type of ambush does not require the use of obstacles to keep the enemy in the kill zone. Compared to an annihilation ambush, the detachment typically conducts a harassment ambush at a greater distance from the enemy, up to the maximum effective range of its weapons. See the example of a harassment ambush at figure 3-16.

Offense

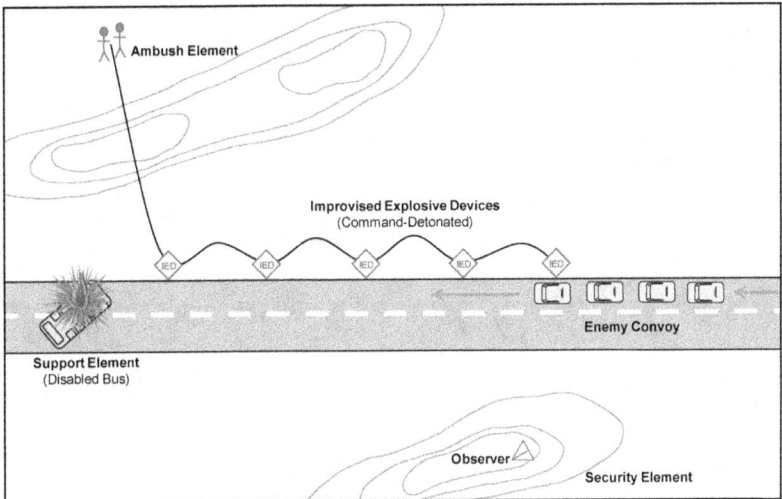

Figure 3-16. Harassment ambush (example)

3-159. Many times, the ambush and support elements are combined to provide better control of fires throughout the kill zone, which may be quite wide. The ambush and support elements concentrate massive direct and indirect fires in the kill zone. The security element provides early warning.

3-160. Harassment ambushes may be accomplished with very little in the way of resources. The ambush and security elements may be as little as two to four personnel combined. In cases where the primary weapon system is one or more explosive devices, the ambush may be conducted by one or two personnel or even by a time-delayed fusing system.

3-161. While the ambush and support elements withdraw, the security element may remain to provide warning and to delay enemy forces if necessary. As in all ambushes, the detachment may emplace mines and plan for indirect fires to cover withdrawal routes.

3-162. Harassment ambushes can affect enemy convoy activity. The convoy's security element may be selected as the specific target of the ambush, and the fire of the ambush element is directed against it. Repeated ambushes against the enemy security element can—

- Cause the use of disproportionately strong forces in the enemy security element. This may leave other portions of the enemy main force vulnerable or require the diversion of additional troops to convoy duty.
- Create an adverse psychological effect upon enemy troops, and the continued casualties suffered by the enemy security element may make such duty unpopular.

Chapter 3

Containment Ambush

3-163. A *containment ambush* is a security measure that is usually part of a larger action. It is used to prevent the enemy from using an avenue of approach or interdicting another action, such as a raid or another ambush. The ambush element may secure the kill zone, as described in the annihilation ambush, although this is not required for success. The support and security elements perform the same functions as those described in the annihilation ambush.

3-164. Obstacles may be an integral part of a successful containment ambush. They serve two functions: to prevent the enemy from using the avenue of approach and to hold the enemy in the kill zone. Within time constraints, the ambush force may erect multiple, mutually supporting obstacles covered by direct and indirect fires. See figures 3-17 and 3-18 on pages 3-34 and 3-35 for examples of containment ambushes.

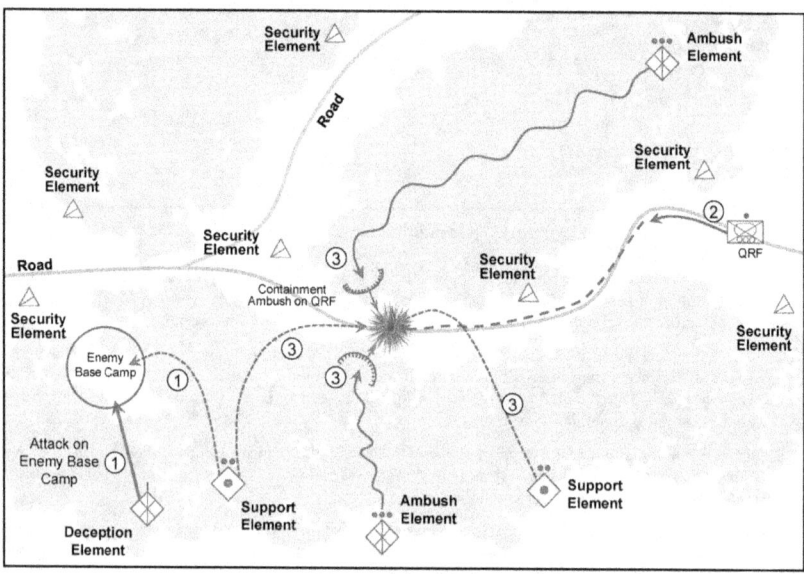

Figure 3-17. Containment ambush (example 1)

Offense

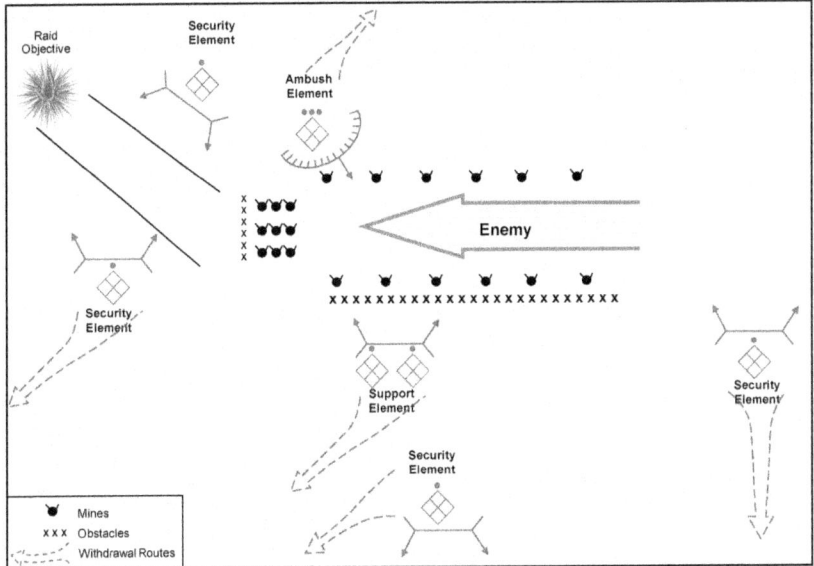

Figure 3-18. Containment ambush (example 2)

Command and Control of an Ambush

3-165. Typically, the commander positions himself with the support element and the deputy moves with the ambush element, although this may be reversed. The primary function of control of the ambush is to arrange units and tasks in time and space so that the ambush element initiates the ambush when the target is most vulnerable while ensuring the detachment is able to break contact when the action is complete.

Support of an Ambush

3-166. An ambush typically requires several types of support. These can include reconnaissance, fire support, air defense, engineer, logistics, and INFOWAR.

Reconnaissance

3-167. Reconnaissance is critical to a successful ambush. Reconnaissance establishes the time the enemy unit will be in the kill zone, determines the best terrain on which to attack and locates response forces and provides early warning.

Fire Support

3-168. Fire support units are almost always in the support element. They provide fires into the kill zone, illumination over it, or smoke to permit withdrawal.

Air Defense

3-169. In the event of an aerial ambush target, air defense units may make up the balance of the ambush element. When the ambush target is a ground unit, air defense is most likely found in the security element(s) where it can provide early warning and fires against aerial response forces.

Chapter 3

Engineer

3-170. The primary task of engineers in an ambush is countermobility—both in support of the security element's mission of isolating the ambush area and to hinder enemy exiting the kill zone.

Logistics

3-171. Ambushes are typically not multi-day battles. The detachment will move from a secured location with everything it needs to complete the mission. In those rare instances where the situation supports a multi-day hide prior to executing the ambush, the detachment will be required to move with its own extra life support. Resupply of an ambush detachment would significantly increase the chances of its detection and defeat its purpose.

INFOWAR

3-172. INFOWAR primarily supports ambushes by concealing the action through deception and information protection. Successful ambushes may be used by an INFOWAR campaign as tools to show the failure of enemy force protection efforts.

Ambushes in Complex Terrain

3-173. Complex terrain is ideal for ambushes due to the cover and concealment it provides. It provides cover and concealment to the ambush detachment, canalizes enemy forces into the kill zone(s) and permits easy withdrawal. Ambushes require that the ambush element remain concealed until the initiation of the attack. Complex generally provides the OPFOR the capability to conduct a 360-degree multi-dimensional ambush.

RAID

3-174. A raid is an attack against a stationary target for the purposes of its capture or destruction that culminates in the withdrawal of the raiding force to safe territory. Raids can also be used to secure information and to confuse or deceive the enemy. The keys to the successful accomplishment of any are raid surprise, firepower, and violence. The raid ends with a planned withdrawal upon completion of the assigned mission. See figures 3-19 and 3-20 for examples of raids.

3-175. Raids are characterized by—

- Destroying or damage key systems or facilities (such as CPs, communication facilities, supply depots, radar sites), providing or denying critical information, or securing hostages or prisoners.
- Destroy, damage, or capture supplies or LOCs.
- Support the INFOWAR plan. Raids can distract attention from other OPFOR actions, to keep the enemy off balance, and to cause the enemy to deploy additional units to protect critical sites.
- OPFOR sensor(s) with capability and mission to find and track the target. Sensors are often ground reconnaissance, but may include UAVs or satellites.
- A C2 method to link raiding force and sensors.
- Supporting operation(s)—usually primarily INFOWAR—to create window of opportunity for raiding force to operate.

Offense

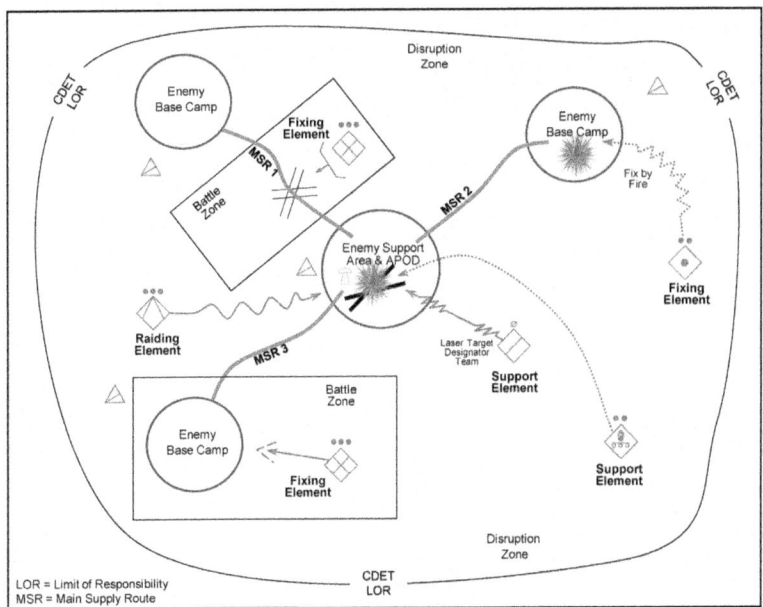

Figure 3-19. Raid (example 1)

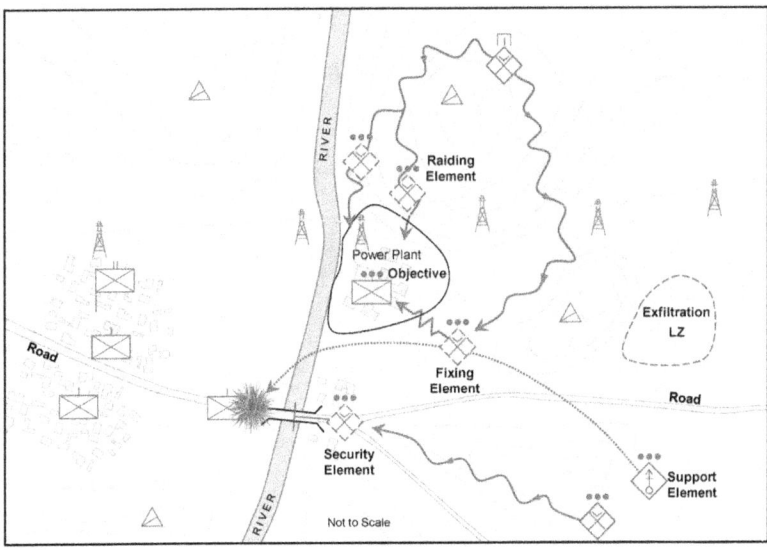

Figure 3-20. Raid (example 2)

Chapter 3

Functional Organization for a Raid

3-176. The size of the raiding force depends upon its mission, the nature and location of the target, and the enemy situation. The raiding force may vary from a detachment attacking a large supply depot to an SPF team attacking a checkpoint or a portion of unprotected railroad track. Regardless of size, the raiding force typically consists of three elements: raiding, security, and support. It may involve other functional elements, such as a fixing element.

> *Note.* The functional designations of elements in figure 3-20 on page 3-37 provide additional examples of the possibilities for identifying functions more specifically. (See Functional Organization of Elements for the Offense—Detachments, Battalions, and Companies earlier in this chapter.) In this case, the air assault infantry platoon labeled "security element" (lower center of figure 3-20) might have been called "blocking element" or "security element/blocking" because that is its specific role in this mission. The light mortar platoon, labeled "support element" in the lower right, might have been labeled "fires element," "fire support element," or "support element/fires." It could have also been labeled as a "blocking element/fires" because its purpose is to prevent the enemy infantry at the left from coming across the river bridge to reinforce or protect the power plant on the right side of the river, which is the objective of the OPFOR raid.

Raiding Element(s)

3-177. The raiding element executes the major task ensuring the success of the raid. It is charged with the actual destruction or seizure of the target of the raid. This element must accomplish this through inherent capabilities and/or positioning relative to the enemy. SPF and infantry trained in night infiltration techniques are all examples of potential components of a raiding element.

3-178. If, for example, the raid mission is to destroy a critical installation such as a railroad bridge or tunnel, the raiding element emplaces and detonates the demolition charges. If the target, such as enemy personnel, is to be neutralized by fire, the raiding element conducts its attack with a high proportion of automatic weapons. In some instances, the raiding element moves physically onto or into the target; in other instances, it is able to accomplish its task from a distance. The other elements of the raid are designed to allow the raiding element access to the target for the time required to accomplish the raid mission.

3-179. The raiding element (and possibly the support element) in a raid may also be required to expose the target to attack, if necessary. It may be, however, that effective INFOWAR, a mismatch in system capabilities, use of geography, or even the enemy's own dispositions create a situation wherein the target is already sufficiently exposed.

Security Element(s)

3-180. The primary threat to all elements of a raiding force is being discovered and defeated by enemy security forces prior to execution of the raid. The security element in a raid is primarily focused on fixing enemy security and response forces or the enemy's escape from the objective area. The security element is equipped and organized such that it can detect enemy forces and prevent them from contacting the rest of the detachment. The security element also covers the withdrawal of the raiding element and act as a rear guard for the raiding force. The size of the security element depends upon the size of the enemy's capability to intervene and disrupt the raid.

3-181. The task of a security element in a raid is to occupy enemy security and response forces and force the enemy to focus on parts of the battlefield away from the raid. Security elements deploy to locations where they can deny the enemy freedom of movement along any ground or air avenues of approach that can reinforce the objective or interfere with the mission of the raiding element. The security element normally gets a screen, guard or cover overall mission, but may also be called upon to perform other tactical tasks in support of its purpose:

- Ambush.
- Block.
- Canalize.
- Contain.
- Destroy.
- Delay.
- Disrupt.
- Fix.
- Interdict.
- Isolate.

Support Element(s)

3-182. The support element serves as an enabling function and assists in setting the conditions for the success of the raid. This support may take several forms. The support element provides fire support, logistics support, reinforcements, to the raiding and security elements. The detachment commander normally controls the raid from within the support element.

3-183. If needed, support elements may assist the raiding element(s) in reaching the target. They can also execute one or more complementary tasks, such as—

- Eliminating guards.
- Breaching and removing obstacles to the objective.
- Conducting diversionary or holding actions.
- Canalizing enemy forces.
- Providing fire support.

Command and Control of a Raid

3-184. A raid is conducted by elements that are autonomous on the battlefield, but linked by C2 and purpose. Although sometimes supported with operational assets, raids are primarily conducted by tactical-level units. They can often involve elements from affiliated forces. A raid is not necessarily tied to scheme of maneuver, in that the larger part of the overall force may be involved in an operation not directly related to the raid.

Support of a Raid

3-185. A raid typically requires several types of support. These can include reconnaissance, armor, fire support, air defense, engineer, logistics, and INFOWAR.

Reconnaissance

3-186. The primary task of reconnaissance in a raid is to locate the target of the raid and track it accurately until the raiding element is in contact. Additionally, reconnaissance determines locations and avenues of approach of response forces.

Chapter 3

Armor

3-187. Armored vehicles, with their advantages of speed and firepower, can be used quite effectively in raids. The challenge in their use is concealing both their movement to the attack and their withdrawal upon completion of the raid.

Fire Support

3-188. Fire support units (and combat helicopters acting in this role) may be the raiding element of a raid with or without additional ground forces. Fire support units support raids in a number of ways:
- Suppression of enemy air defenses (SEAD) to support raiding aviation elements.
- Suppression of response forces.
- Smoke to permit withdrawal.

Air Defense

3-189. In a raid, air defense assets are most likely found in the security element(s). In that role, they can provide early warning and fires against enemy aerial response forces.

Engineer

3-190. Engineers support raids primarily by executing mobility tasks to—
- Permit access to the objective.
- Facilitate withdrawal of the raiding force.

Logistics

3-191. Raids are typically not multi-day battles. The detachment will move from a secured location with everything it needs to complete the mission.

INFOWAR

3-192. INFOWAR primarily supports raids by concealing the action through deception and information protection. Successful raids may be used by an INFOWAR campaign as tools to show the failure of enemy force protection efforts.

RECONNAISSANCE ATTACK

3-193. A *reconnaissance attack* is a tactical offensive action that locates moving, dispersed, or concealed enemy elements and either fixes or destroys them. It may also be used by the commander to gain information about the enemy's location, dispositions, military capabilities, and quite possibly his intentions. The OPFOR recognizes that an enemy will take significant measures to prevent the OPFOR from gaining critical intelligence. Therefore, quite often the OPFOR will have to fight for information, using an offensive operation to penetrate or circumvent the enemy's security forces to determine who and/or what is located where or doing what.

3-194. The reconnaissance attack is the most ambitious—and least preferred—method to gain information. When other means of gaining information have failed, a detachment can undertake a reconnaissance attack.

3-195. Key factors in the reconnaissance attack are—
- Situational awareness.
- Contact conditions.
- Tempo.

Offense

Functional Organization for a Reconnaissance Attack

3-196. Depending on the situation, the detachment commander organizing a reconnaissance attack may designate reconnaissance, security, and/or action elements. There may be more than one of each type. The commander may also form various types of support elements.

Reconnaissance Element(s)

3-197. If the purpose of the reconnaissance attack is merely to gain information, the commander may organize several *reconnaissance elements*. Their role is to locate enemy elements operating in the detachment's AOR. If the purpose is to also be able to fix and/or destroy the located enemy elements, the reconnaissance elements would provide reconnaissance support to the elements that carry out those functions. (If specialized reconnaissance elements are not formed, this role may be performed by security elements.)

Security Element(s)

3-198. If the commander believes he has sufficient combat power to engage enemy elements that may be located, he may also organize one or more *security elements*. Security elements can either work in conjunction with reconnaissance elements (if they exist) or perform the reconnaissance role on their own. Upon locating an enemy element, a security element should be capable of isolating the enemy by blocking avenues of withdrawal or reinforcement of the enemy unit and/or fixing the enemy. (When it performs those functions, it may be called a *blocking element* or *fixing element*.)

Action Element(s)

3-199. If the detachment commander believes he has sufficient combat power to defeat enemy element(s) that are located, he may also organize one or more *action elements*. (These elements may receive a functional designation that more specifically describes the nature of the action by which they accomplish the mission.) Once an enemy element is located and/or fixed, the action element(s) will attack it with the goal of destroying it. There may or may not be one action element for each security element.

Support Element(s)

3-200. One or more *support elements* can perform various supporting tasks. (See Support of a Reconnaissance Attack, below.)

Organizing the Battlefield for a Reconnaissance Attack

3-201. Multiple attack routes often characterize reconnaissance attacks. There may also be objective rally points and orientation objectives.

Executing a Reconnaissance Attack

3-202. Multiple elements normally infiltrate or maneuver separately to find and then fix or destroy enemy elements. A reconnaissance attack is initiated by multiple security elements and/or reconnaissance elements moving through and to likely points of contact with enemy elements that need to be destroyed or fixed. Success relies on the ability of the security elements (and reconnaissance elements, if separate) to operate independently. See figure 3-21 on page 3-42 for an example of a reconnaissance attack.

Chapter 3

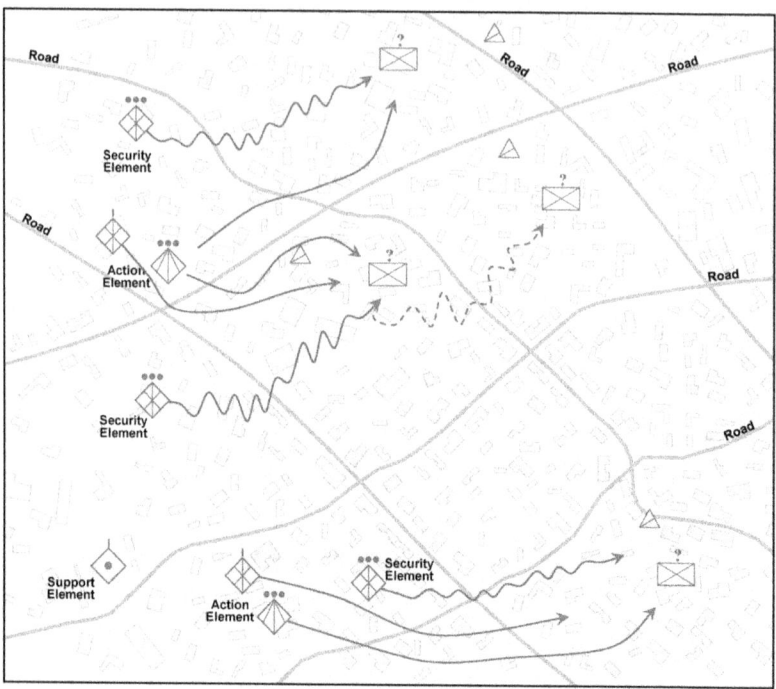

Figure 3-21. Reconnaissance attack (example)

Support of a Reconnaissance Attack

3-203. A reconnaissance attack typically requires several types of support. These can include reconnaissance, armor, fire support, aviation, air defense, engineer, logistics, and INFOWAR.

Reconnaissance

3-204. There are two basic methods for conducting reconnaissance in a reconnaissance attack. The first is for each security element to perform its own reconnaissance tasks, therefore being responsible for both finding an enemy force and fixing it. The second method is for the detachment to organize reconnaissance elements that have the mission to find the enemy forces in the detachment AOR and use security elements solely for the task of fixing those forces.

Armor

3-205. Depending on the situation, armored vehicles may be used in any of the elements of a reconnaissance attack. In a reconnaissance or security element, thermal imagers and other electro-optical aids on armored vehicles can assist in the detection and identification of enemy forces.

3-206. Armored vehicles could provide the security element in a reconnaissance attack with a combination of firepower and protection that significantly enhances the mission of fixing the enemy. The mobility and speed of these forces permit them to serve in the action element of a reconnaissance attack, rapidly orienting on fixed enemy forces and moving to a position of advantage to destroy them.

Offense

Fire Support

3-207. Fire support in a reconnaissance attack focuses on—
- Responsive fires in support of security elements in contact.
- Support of the maneuver of the action element(s).
- Destruction of a fixed enemy, using precision munitions.

If necessary, a fire support element can also cover the withdrawal of reconnaissance, security, action, or other support elements after completing the mission.

Aviation

3-208. Attack helicopters (or their reconnaissance variants) may be allocated to reconnaissance and/or security elements, using armed reconnaissance techniques. They may also be a part (or the entirely) of a highly mobile action element. CS (and sometimes CSS) helicopters can be used to transport infantry and commandos to landing zones where they can attack forces fixed by the security element(s).

Air Defense

3-209. In a reconnaissance attack, air defense prevents enemy response forces and reinforcements from influencing the locating, fixing, and destruction of enemy elements in the AOR. If necessary, aid defense assets can also cover the withdrawal of reconnaissance, security, action, or other support elements after completing the mission.

Engineer

3-210. Engineer support to a reconnaissance attack focuses on mobility, permitting security and action elements freedom of maneuver. Engineer units also conduct countermobility tasks to fix located enemy forces.

Logistics

3-211. A reconnaissance attack can be marked by widely dispersed elements operating over extended time periods and distances. Elements will attempt to carry sufficient resupply with them. The OPFOR will also make use of caches, resupply on the move, dedicated logistics elements and other techniques to sustain the battle without a drop in tempo or loss of contact.

INFOWAR

3-212. INFOWAR activities in a reconnaissance attack are primarily executed to—
- Protect elements of the detachment from being detected.
- Encourage enemy elements to reveal themselves or even surrender.
- Fix enemy elements.

This page intentionally left blank.

Chapter 4
Defense

While the OPFOR sees the offense as the decisive form of military action, it recognizes defense as the stronger form of military action, particularly when faced with a superior foe. Defensive operations can lead to strategic victory if they force a stronger invading enemy to abandon his mission. It may be sufficient for the OPFOR simply not to lose. Even when an operational-level command as a whole is conducting an offensive operation, it is likely that one or more tactical-level subordinate units may be executing defensive missions to preserve offensive combat power in other areas, to protect an important formation or resource, or to deny access to key facilities or geographic areas. The same is true of subordinate units within a tactical-level command.

OPFOR defenses can be characterized as a "shield of blows." Each force and zone of the defense plays an important role in the attack of the enemy's combat system. A tactical-level defense is structured around the concept that disaggregating and destroying the synergy of the enemy's combat system will make enemy forces vulnerable to attack and destruction. Commanders and staffs do not approach the defense with preconceived templates. The tactical situation may cause the commander to vary his defensive methods and techniques. Nevertheless, there are basic characteristics of defensive battles (purposes and types of action) that have applications in all situations.

PURPOSE OF THE DEFENSE

4-1. Defensive battles are designed to achieve the goals of the battle or operation plan through active measures while preserving combat power. A tactical command ensures that its subordinate commands thoroughly understand both the overall goals of the battle plan and the specific purpose of a particular battle they are about to fight. In this way, subordinate commands can continue to fight the battle without direct control by a higher headquarters. The purpose of any given defensive battle depends on the situation, resources, and mission—as determined through the decisionmaking process. The OPFOR recognizes four general purposes of tactical defensive missions:
- Protect personnel and equipment.
- Restrict freedom of movement.
- Control key terrain.
- Gain time.

4-2. These general purposes serve as a guide to understanding the design of a defensive mission and not as a limit placed on a commander as to how he makes his intent and aim clear. These are not the only possible purposes of tactical missions but are the most common.

PROTECT PERSONNEL AND EQUIPMENT

4-3. A *defense to protect key personnel and equipment* creates one or more locations on the battlefield where forces critical to the OPFOR effort are protected from enemy reconnaissance acquisition and destructive action. This can be because these elements are important to the OPFOR effort at an

Chapter 4

operational or even strategic scale or because the OPFOR needs time to reconstitute these elements for future offensive operations.

4-4. Such a defense typically, but not always relies heavily on camouflage, concealment, cover, and deception (C3D) and information warfare (INFOWAR) measures. However, enemy rules of engagement, limited access areas such as nonbelligerent countries, adverse weather conditions, and other such factors may be employed to provide protection to OPFOR forces.

RESTRICT FREEDOM OF MOVEMENT

4-5. A *defense to restrict freedom of movement* prevents the enemy from maneuvering as he chooses. Such defenses can deny key terrain, ambush moving forces, dominate airspace, or fix an enemy formation. Tactical tasks often associated with restricting freedom of movement are ambush, block, canalize, contain, fix, interdict, and isolate.

CONTROL KEY TERRAIN

4-6. A *defense to control key terrain* prevents enemy seizure of geographic features or facilities. Terrain to be protected and controlled can include not only key terrain that dominates a battlefield, but also facilities such as economic targets, ports, or airfields.

GAIN TIME

4-7. A *defense to gain time* prevents the enemy from successfully concluding his scheme of maneuver before a certain point in time or prior to a given event taking place. A defense to gain time is not oriented on either a protected force or a geographic location—it is oriented on the enemy's perceived scheme of maneuver. Disruption, delays, ambushes, and spoiling attacks are often parts of a defense to gain time.

PLANNING THE DEFENSE

4-8. For the OPFOR, the key elements of planning defensive missions are—
- Determining the objective of the defensive action
- Determining the level of planning possible (planned versus situational defense).
- Organizing the battlefield.
- Organizing forces and elements by function.
- Organizing INFOWAR activities in support of the defense (see chapter 7).

4-9. Defensive actions are not limited solely to attrition-based tactics. Some actions against a superior and/or equal force will typically include the increased use of—
- Infiltration to conduct spoiling attacks and ambushes.
- Mitigation of enemy capabilities using INFOWAR, especially perception management and computer attack (see chapter 7), in support of defensive operations.
- Use of affiliated forces for reconnaissance, counterreconnaissance, security, and attacks against key enemy systems and forces.

PLANNED DEFENSE

4-10. A *planned* defense is a defensive mission or action undertaken when there is sufficient time and knowledge of the situation to prepare and rehearse forces for specific tasks. Typically, the enemy is in a staging or assembly area and in a known location and status. Key considerations in defensive planning are—
- Determining which enemy forces will attack, when, and how.
- Determining enemy weakness and how to create and/or exploit them.
- Determining key elements of the enemy's combat system and interdict them, thereby mitigating overall enemy capability.

- Determining defensive characteristics of the terrain. Selecting key positions in complex terrain from which to dominate surrounding avenues of approach.
- Determining the method that will deny the enemy his tactical objectives.
- Developing a plan for reconnaissance, intelligence, surveillance, and target acquisition (RISTA) that locates and tracks major enemy formations, and determines enemy patterns of operations, intentions, timeframes, and probable objectives.
- Creating or taking advantage of a window of opportunity that frees friendly forces from any enemy advantages in precision standoff and situational awareness.
- Planning all aspects of an integrated counterattack making use of all means available, including INFOWAR, unmanned aerial vehicles, special-purpose forces (SPF), and/or affiliated irregular forces.

SITUATIONAL DEFENSE

4-11. The OPFOR may also conduct a *situational* defense. It recognizes that the modern battlefield is chaotic. Circumstances will often change so that the OPFOR is not afforded the opportunity to conduct offensive action, as originally planned, thus forcing it to adopt a defensive posture. If the OPFOR determines that a fleeting, situational window of opportunity is closing, it may assume a situational defense. Although detailed planning and preparation greatly mitigate risk, they are often not achievable if enemy action has taken away the initiative.

4-12. The following are examples of conditions that might lead to a situational defense:
- The enemy is unexpectedly striking an exposed key OPFOR unit, system, or capability.
- The enemy is conducting a spoiling attack to disrupt OPFOR offensive preparations.
- An OPFOR unit makes contact on unfavorable terms for subsequent offensive action.
- The enemy gains or regains air superiority sooner than anticipated.
- An enemy counterattack was not effectively fixed

4-13. In a situational defense, the commander develops his assessment of the conditions rapidly and without a great deal of staff involvement. He provides a basic course of action (COA) to the staff, which then quickly turns that COA into an executable combat order. Even more than other types of OPFOR defensive action, the situational defense relies on implementation of battle drills by subordinate tactical units.

Note. Any division or brigade receiving additional assets from a higher command becomes a division tactical group (DTG) or brigade tactical group (BTG). Therefore, references to a tactical group, DTG, or BTG throughout this chapter may also apply to division or brigade, unless specifically stated otherwise.

FUNCTIONAL ORGANIZATION OF FORCES FOR THE DEFENSE—TACTICAL GROUPS, DIVISIONS, AND BRIGADES

4-14. In his combat order, the commander of a division, DTG, brigade, or BTG also specifies the initial functional organization of the forces within his level of command. However, the organization of forces can shift dramatically during the course of a battle. For example, a unit that initially was part of a disruption force may eventually occupy a battle position within the battle zone and become part of the main defense force or act as a reserve.

4-15. Each of the separate functional forces has an identified commander. This is often the senior commander of the largest subordinate unit assigned to that force. For example, if two BTGs and an independent mission detachment (IMD) are acting as the DTG's main defense force, the senior of the two BTG commanders is the main defense force commander. During dispersed and decentralized operations, even when the force consists of like units of the same command level, control can be delegated to the

Chapter 4

senior commander of that force's like units. Since, in this option, each force commander is also a subordinate unit commander, he controls the force from his unit's command post (CP).

4-16. Another option is to have one of the higher unit's CPs command and control a functional force. Particularly during dispersed defensive operations, functional forces that contain units of the same command level might be controlled from the forward, auxiliary, or airborne CP of the tactical group. For example, the forward CP could control a disruption force. Another possibility would be for the integrated fires command (IFC) CP to command the disruption force or any other force whose actions must be closely coordinated with fires delivered by the IFC.

4-17. In any case, the force commander is responsible to the division, brigade, or tactical group commander to ensure that combat preparations are made properly and to take charge of the force during the operation. This frees the higher-level commander from decisions specific to the force's mission. Even when subordinates of a tactical group have responsibility for parts of the tactical group disruption zone, there is still an overall tactical group disruption force commander.

Disruption Force

4-18. The OPFOR commander may create a single cohesive disruption force with a single overall commander or he may create multiple (probably dispersed) forces operating in the disruption zone with numerous commanders. Activities in the disruption zone may be independent of each other, integrated, continuous, or sporadic.

4-19. The size and composition of forces in the disruption zone depends on the level of command involved, the commander's concept of the battle, and the circumstances in which the unit adopts the defense. The function of the disruption force is to prevent the enemy from conducting an effective attack. Therefore, the size of the disruption force is not linked to any specific echelon, but rather to the function. A tactical commander will always make maximum use of stay-behind forces and affiliated forces existing within his AOR. Subordinate commanders can employ forces in a higher command's disruption zone with tactical group approval.

4-20. While a DTG disruption force is typically a BTG, a BTG disruption force is typically an IMD. However, a disruption force has no set order of battle and will be whatever available unit(s) best fit the commanders needs. The disruption force may contain—

- Ambush teams (ground and air defense).
- Long-range reconnaissance patrols and/or SPF teams.
- RISTA assets and forces.
- Counterreconnaissance detachments.
- Artillery systems.
- Target designation teams.
- Elements of affiliated forces (such as guerrillas, terrorists, insurgents, or criminals).
- Antilanding reserves.

4-21. The purpose of the disruption force is to prevent the enemy from conducting an effective attack. The disruption force does this by initiating the attack on key components of the enemy's combat system. Successful attack of designated components or subsystems begins the disaggregation of the enemy's combat system and creates vulnerabilities for exploitation in the battle zone. Skillfully conducted disruption operations will effectively deny the enemy the synergy of effects of his combat system.

4-22. For example, the tactical group commander may determine that destruction of the enemy's mobility assets will create an opportunity to destroy maneuver units in the battle zone. The disruption force would be given the mission of seeking out and destroying enemy mobility assets while avoiding engagement with maneuver forces.

4-23. The disruption force may also have a counterreconnaissance mission (see chapter 6). It may selectively destroy or render irrelevant the enemy's RISTA forces and deny him the ability to acquire and engage OPFOR targets with deep fires. It employs OPFOR RISTA assets to locate and track enemy

RISTA forces and then directs killing systems to destroy them. For this purpose, the disruption force may include operational-level RISTA assets, SPF, and helicopters. There will be times, however, when the OPFOR wants enemy reconnaissance to detect something that is part of the deception plan. In those cases, the disruption force will not seek to destroy all of the enemy's RISTA assets.

4-24. The disruption force may deceive the enemy as to the location and configuration of the main defense in the battle zone, while forcing him to show his intent and deploy early. Some other results of actions in the disruption zone can include delaying the enemy to allow time for preparation of the defense or a counterattack, canalizing the enemy onto unfavorable axes, or ambushing key systems and vulnerable troop concentrations.

Main Defense Force

4-25. The *main defense* force is the functional force charged with execution of the primary defensive mission. It operates in the battle zone to accomplish the purpose of the defense. (During a maneuver defense, the main defense force is typically broken down into a contact force and a shielding force.)

Protected Force

4-26. The *protected force* is the force being kept from detection or destruction by the enemy. It may be in the battle zone or the support zone.

Security Force

4-27. The *security force* conducts activities to prevent or mitigate the effects of hostile actions against the overall command and/or its key components. If the commander chooses, he may charge this security force with providing force protection for the entire AOR, including the rest of the functional forces; logistics and administrative elements in the support zone; and other key installations, facilities, and resources.

4-28. The security force may include various types of units—such as infantry, SPF, counterreconnaissance, and signals reconnaissance assets—to focus on enemy special operations and long-range reconnaissance forces operating throughout the AOR. It can also include Internal Security Forces with the mission of protecting the overall command from attack by hostile insurgents, terrorists, and special operations forces. The security force may also be charged with mitigating the effects of weapons of mass destruction (WMD). The security force commander can be given control over one or more reserve formations, such as the antilanding reserve. (See also Tactical Security in chapter 6.)

Counterattack Forces

4-29. A defensive battle may include a planned counterattack scheme. This is typical of a maneuver defense, but could also take place within an area defense. In these cases, the tactical commander will designate one or more *counterattack forces*. He will also shift his task organization to create a counterattack force when a window of opportunity opens that leaves the enemy vulnerable to such an action. The counterattack force can have within it fixing, mission, and exploitation forces (as outlined in chapter 3). It will have the mission of causing the enemy's offensive operation to culminate. The tactical group commander uses counterattack forces to complete the defensive mission and regain the initiative for the offense.

Types of Reserves

4-30. At the commander's discretion, forces may be held out of initial action so that he may influence unforeseen events or take advantage of developing opportunities. He may employ a number of different types of reserve forces of varying strengths, depending on the situation.

Chapter 4

Maneuver Reserve

4-31. The size and composition of a reserve force is entirely situation-dependent. However, the reserve is normally a force strong enough to respond to unforeseen opportunities and contingencies at the tactical level. A reserve may assume the role of counterattack force to deliver the final blow that ensures the enemy can no longer conduct his preferred COA. Reserves are almost always combined arms forces.

4-32. A reserve force will be given a list of possible missions for rehearsal and planning purposes. The staff assigns to each of these missions a priority, based on likelihood that the reserve will be called upon to execute that mission. Some missions given to the reserve may include—

- Conducting a counterattack. (The counterattack goal is not limited to destroying enemy forces, but may also include recovering lost positions or capturing positions advantageous for subsequent combat actions.)
- Conducting counterpenetration (blocking or destroying enemy penetrations).
- Conducting antilanding missions (eliminating vertical envelopments).
- Assisting forces heavily engaged on a defended line to break contact and withdraw.
- Acting as a deception force.

Antitank Reserve

4-33. OPFOR commanders faced with significant armored threats may keep an antitank reserve (ATR). It is generally an antitank unit and often operates in conjunction with an obstacle detachment (OD). Based on the availability of antitank and engineer assets, a division- or brigade-size unit may form more than one ATR.

Antilanding Reserve

4-34. Because of the potential threat from enemy airborne or heliborne troops, a commander may designate an antilanding reserve (ALR). While other reserves can perform this mission, the commander may create a dedicated ALR to prevent destabilization of the defense by vertical envelopment of OPFOR units or seizure of key terrain. ALRs will be resourced for rapid movement to potential drop zones (DZs) and landing zones (LZs). The ALR commander will have immediate access to the operational and tactical intelligence system for early warning of potential enemy landing operations. ALRs typically include maneuver, air defense, and engineer units, but may be allocated any unit capable of disrupting or defeating an airborne or heliborne landing, such as smoke or INFOWAR. ALRs assume positions prepared to engage the enemy primary DZ or LZ as a kill zone. They rehearse and plan for rapid redeployment to other suspected DZs or LZs.

Special Reserves

4-35. In addition to their obstacle detachments (ODs), units may form an *engineer reserve* of earthmoving and obstacle-creating equipment. A commander can deploy this reserve to strengthen defenses on a particularly threatened axis during the course of the battle. A unit threatened by enemy use of weapons of mass destruction (WMD) may also form a *chemical defense reserve*.

Deception Force

4-36. When the INFOWAR plan requires the creation of nonexistent or partially existing formations, these forces will be designated *deception forces* in close-hold executive summaries of the battle plan. Wide-distribution copies of the plan will make reference to these forces according to the designation given them in the deception story. The deception force in the defense is typically given its own command structure, both to replicate the organization(s) necessary to the deception story and to execute the multidiscipline deception required to replicate an actual military organization. Tactical group commanders can use deception forces to replicate subordinate tactical group and detachment command structures, in order to deny enemy forces information on battle plans for the defense.

Defense

Note. Any battalion or company receiving additional assets from a higher command becomes a battalion-size detachment (BDET) or company-size detachment (CDET). Therefore, references to a detachment, BDET, or CDET throughout this chapter may also apply to battalion or company, unless specifically stated otherwise.

FUNCTIONAL ORGANIZATION OF ELEMENTS FOR THE DEFENSE—DETACHMENTS, BATTALIONS, AND BELOW

4-37. Detachments, battalions, and companies employ a similar but different scheme for organizing functional elements than the functional force methodology used by tactical groups. This is because the OPFOR tends to use detachments to accomplish a single tactical task rather than a multi-task mission.

4-38. The standard functional organization of a detachment for defense is into four parts: the *disruption element*, the *main defense element*, the *support element*, and the *reserve element*. There may also be *specialist elements*. Due to such considerations as multiple avenues of approach, a detachment may organize one or more of each of these elements in specific cases.

4-39. The *disruption element* of a detachment can provide security for the detachment, prevents the enemy from influencing mission accomplishment, and prevents the enemy from conducting an effective attack by targeting key systems and subcomponents of the enemy's combat system in the disruption zone. The *main defense element* accomplishes the detachment's tactical task. The *support element* provides combat and combat service support and C2 for the detachment. The *reserve element* provides the defender with the tactical flexibility to influence unforeseen events or to take advantage of developing opportunities.

4-40. In certain situations, a detachment may organize one or more *specialist elements*. Specialist elements are typically formed around a unit with a specific capability such as an obstacle-clearing element, reconnaissance element, or deception element.

4-41. At any given time, a detachment will only be associated with a single functional force (disruption, main defense, security, counterattack, or reserve force) of a higher command. If a higher command needs to divide a detachment to accomplish other tasks, it will require a change in task-organizing. For example, if a BTG needed one part of one of its battalions or BDETs to serve as the main defense force, but needed another part to join the reserve, the two parts would be task-organized as separate detachments and assigned different functional element designations.

4-42. Detachments may be assigned one of several tasks while conducting a defense. Some examples are—

- Defend a simple battle position.
- Defend a complex battle position.
- Act as counterreconnaissance detachment.
- Act as deception force.
- Act as security force.
- Act as counterattack force.
- Act as reserve.

PREPARING FOR THE DEFENSE

4-43. In the preparation phase, the OPFOR focuses on ways of applying all available resources and the full range of actions to conduct the defense in the strongest condition and strongest positions possible. Commanders organize the battlefield and their functional forces or elements with an eye toward capitalizing on conditions created by successful defensive actions, and seizing opportunities for offensive actions wherever possible.

4-44. The defensive dispositions are based on the application of the systems warfare approach to combat, as described in chapter 1. OPFOR defensive actions focus on attacking components or subsystems of the enemy's combat system to disaggregate the "system of systems." By denying the enemy the synergy created by an integrated, aggregated combat system, vulnerabilities are created that defensive forces can exploit.

DENY ENEMY INFORMATION

4-45. Tactical commanders realize that enemy operations hinge on awareness and understanding of the situation. Defensive preparations will focus on destruction and deception of enemy sensors in order to limit the ability of enemy forces to understand the OPFOR defensive plan. A high priority for all defensive preparations is to deny the enemy the ability to maintain reconnaissance contact on the ground. The OPFOR recognizes that, when conducting operations against a stronger enemy, it will often be impossible to destroy the ability of the enemy's standoff RISTA means to observe its defensive preparations. However, the OPFOR also recognizes the reluctance of enemy military commanders to operate without human confirmation of intelligence, as well as the relative ease with which imagery and signals sensors may be deceived. OPFOR tactical commanders consider ground reconnaissance by enemy special operations forces as a significant threat in the enemy RISTA suite and will focus significant effort to ensure its removal. While the OPFOR will execute missions to destroy standoff RISTA means, C3D will be the method of choice for degrading the capability of such systems.

MAKE THOROUGH COUNTERMOBILITY AND SURVIVABILITY PREPARATIONS

4-46. The more time available, the greater the preparation of a battle position, zone, or area of responsibility (AOR). This is a reflection of engineer effort and time to devote to that effort. The OPFOR employs every method to maximize the time available to prepare for the defense.

4-47. Tactical commanders realize that engineer works are vital to the stability of the defense. They will use engineer assets to improve the advantages of complex terrain in protecting friendly forces and exposing enemy forces to engagement. Engineer efforts can contribute to creating windows of opportunity by degrading the ability of the enemy's combat system to integrate the effects of its subsystems. Of course, such work is not just an engineer responsibility; it is a combined arms task.

4-48. Engineer units specializing in rapid obstacle construction and minelaying form mission-specific units known as obstacle detachments (ODs). These ODs normally deploy in conjunction with reserves to block enemy penetrations or to protect the flanks of counterattack forces. In the initial stages of the defense, engineer assets concentrate on creating obstacles in the disruption zone, in gaps in the combat formation, and to the flanks, and preparing lines for counterpenetration and counterattack and routes to such lines. The obstacle plan ensures that the effort is coordinated with fires and maneuver to produce the desired effects. In conjunction with other tasks, engineers support the INFOWAR plan through activities such as constructing false defensive positions and preparing false routes. See chapter 12 for more information on countermobility and survivability planning.

MAKE USE OF COMPLEX TERRAIN

4-49. The OPFOR will make maximum use of complex terrain in all defensive actions. Complex terrain provides cover from fires, concealment from standoff RISTA assets, and intelligence and logistics support from the population of urban areas. It plays into the strength of OPFOR resolve to win through any means and through protracted conflict if necessary.

MAKE THOROUGH LOGISTICS ARRANGEMENTS

4-50. The overwhelming ability of a powerful, modern enemy to strike exposed logistics elements makes it difficult to resupply forces. The OPFOR understands that there is as much chance of a defensive action being brought to culmination by a lack of sufficient logistics support as there is by enemy action. Careful consideration will be given to carried days of supply and advanced caches to obviate the need for easily disrupted lines of communication (LOCs).

Defense

MODIFY THE PLAN WHEN NECESSARY

4-51. The OPFOR takes into account that, while it might consider itself to be in the preparation phase for one battle, it is continuously in the execution phase. Plans are never considered final and are continually checked throughout the course of their development to ensure they are still valid in light of battlefield events.

REHEARSE EVERYTHING POSSIBLE, IN PRIORITY

4-52. The commander establishes the priority for critical parts of the battle. Then he rehearses those actions with his subordinates in as realistic a manner as possible for the remainder of the preparation time. Typical actions to be rehearsed in preparation for a defense include—
- Counterreconnaissance plan.
- Commitment of reserve.
- Initiation of a counterattack.
- Execution of the fire support plan.
- Integration of the INFOWAR plan.

EXECUTING THE DEFENSE

4-53. Successful execution depends on forces and elements that understand their roles in the battle and can swiftly follow preparatory actions with implementation of the battle plan or rapid modifications to the plan, as the situation requires. A successful execution phase results in the culmination of the enemy's offensive action. It ideally ends with transition to the offense in order to keep the enemy under pressure and destroy him completely. Against a superior enemy force, however, a successful defense may end in a stalemate.

4-54. A successful defense sets the military conditions for a return to the offense or a favorable political resolution of the conflict. The OPFOR may have to surrender territory to preserve forces. Territory can always be recaptured, but the destruction of OPFOR major combat formations threatens the survival of the State. Destruction of the protected force is unacceptable.

4-55. Success criteria for a tactical commander conducting an area or maneuver defense may include—
- Major combat formations remain intact.
- The enemy is forced to withdraw or, at a minimum, forego offensive operations due to losses.
- A stalemate allows operational-, theater-, and national-level assets time to conduct attacks against enemy strategic centers of gravity.

MAINTAIN CONTACT

4-56. OPFOR commanders will go to great lengths to maintain contact with key elements of the enemy force throughout the battle. This includes rapid reconstitution of reconnaissance assets and units and the use of whatever combat power is necessary to ensure success.

IMPLEMENT BATTLE DRILLS

4-57. The OPFOR derives great flexibility from battle drills. Unlike the U.S. view that battle drill, especially at higher levels, reduces flexibility, the OPFOR uses minor, simple, and clear modifications to thoroughly understood and practiced battle drills to adapt to ever-shifting conditions. It does not write standard procedures into its combat orders and does not write new orders when a simple shift from current formations and organization will do.

4-58. Battle drills are slightly less important in defensive situations, but the standardized battle drills for reacting to all seven forms of contact (direct fire, indirect fire, visual, obstacle, CBRN, electronic warfare, and air attack) will have defensive counterparts.

MODIFY THE PLAN WHEN NECESSARY

4-59. The OPFOR is sensitive to the effects of mission dynamics and realizes that the enemy's actions may well make the original mission of an OPFOR unit achievable, but completely irrelevant. For example, a disruption force may be capable of fixing a key element of the enemy's attack because the enemy is using a small force to attack the OPFOR in one area while attacking strongly in another. In this case, the OPFOR unit in question must be ready to transition to a new mission quickly and break contact with sufficient combat power to fix the maneuvering enemy force. Then the original disruption force, or part of it, may be redesignated as a fixing force.

SEIZE OPPORTUNITIES

4-60. The OPFOR places maximum emphasis on decentralized execution, initiative, and adaptation. Subordinate units are expected to take advantage of fleeting opportunities so long as their actions are in concert with the purpose of the combat order or battle plan.

TYPES OF DEFENSIVE ACTION—TACTICAL GROUPS, DIVISIONS, AND BRIGADES

4-61. The types of defensive action in OPFOR doctrine are both tactical methods and guides to the design of operational COAs. The two basic types are maneuver and area defense. A tactical group commander may use both forms of defense simultaneously across his AOR. A defensive battle plan may include subordinate units that are executing various combinations of maneuver and area defenses, along with some offensive actions, within the overall defensive mission framework.

MANEUVER DEFENSE

4-62. In situations where the OPFOR is not completely overmatched, it may conduct a tactical *maneuver defense*. This type of defense is designed to achieve tactical decision by skillfully using fires and maneuver to destroy key elements of the enemy's combat system and deny enemy forces their objective, while preserving the friendly force. Maneuver defenses cause the enemy to continually lose effectiveness until he can no longer achieve his objectives. They can also economize force in less important areas while the OPFOR moves additional forces onto the most threatened axes.

4-63. Even within a maneuver defense, the tactical group commander may use area defense on some enemy attack axes, especially on those where he can least afford to lose ground. Conversely, he may employ maneuver defense techniques to conduct actions in the disruption zone if it enhances the attack on the enemy's combat system and an area defense in the battle zone.

Method

4-64. Maneuver defense inflicts losses on the enemy, gains time, and protects friendly forces. It allows the defender to choose the place and time for engagements. Each portion of a maneuver defense allows a continuing attack on the enemy's combat system. As the system begins to disaggregate, more elements are vulnerable to destruction. The maneuver defense accomplishes this through a succession of defensive battles in conjunction with short, violent counterattacks and fires. It allows abandoning some areas of terrain when responding to an unexpected enemy attack or when conducting the battle in the disruption zone. In the course of a maneuver defense, the tactical commander tries to force the enemy into a situation that exposes enemy formations to destruction. See figure 4-1 for an example of maneuver defense.

Defense

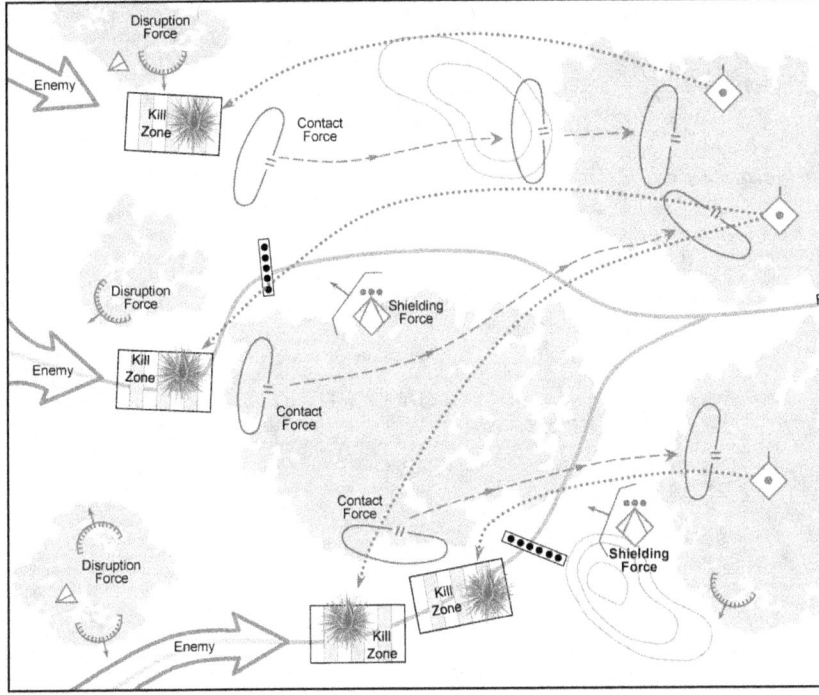

Figure 4-1. Maneuver defense (example 1)

4-65. A maneuver defense trades terrain for the opportunity to destroy portions of the enemy formation and render the enemy's combat system ineffective. The OPFOR might use a maneuver defense when—
- It can afford to surrender territory.
- It possesses a mobility advantage over enemy forces.
- Conditions are suitable for canalizing the enemy into areas where the OPFOR can destroy him by fire or deliver decisive counterattacks.

4-66. Compared to area defense, the maneuver defense involves a higher degree of risk for the OPFOR, because it does not always rely on the inherent advantages of complex battle positions. Units conducting a maneuver defense typically place smaller forces or elements forward in defensive positions and retain much larger reserves than in an area defense.

Defensive Arrays

4-67. The basis of maneuver defense is for units to conduct maneuver from position to position through a succession of defensive arrays. A defensive array is a group of positions in which one or more subordinate units have orders to defend for a certain time within a higher unit's AOR. The OPFOR can accept large intervals between defensive positions in such an array. Part of the array may consist of natural or manmade obstacles or of deception defensive positions.

4-68. Defensive arrays are generally integrated into the terrain. In the spaces between arrays, the defenders typically execute disruption. Thus, it is difficult for the enemy to predict where he will encounter resistance.

4-69. The number of arrays and duration of defense on each array depend on the nature of the enemy's actions, the terrain, and the condition of the defending units. Arrays are selected based on the availability of obstacles and complex terrain.

Defensive Maneuver

4-70. *Defensive maneuver* consists of movement by bounds and the maintenance of continuous fires on enemy forces. A disruption force and/or a main defense force (or part of it) can perform defensive maneuver. In either case, the force must divide its combat power into two smaller forces: a contact force and a shielding force. The *contact force* is the force occupying the defensive array in current or imminent contact with the enemy. The *shielding force* is the force occupying a defensive array that permits the contact force to reposition to a subsequent array.

4-71. The contact force ideally forces the enemy to deploy his maneuver units and perhaps begin his fires in preparation for the attack. Then, before the contact force becomes decisively engaged, it maneuvers to its next preplanned array, protected by the array occupied by the shielding force. While the original contact force is moving, the shielding force is able to keep the enemy under continuous observation, fires, and attack. When the original contact force assumes positions in its subsequent defensive array, it becomes the shielding force for the new contact force. In this manner, units continue to move by bounds to successive arrays, preserving their own forces while delaying and destroying the enemy.

4-72. Subsequent arrays are far enough apart to permit defensive maneuver by friendly units. The distance should also preclude the enemy from engaging two arrays simultaneously without displacing his indirect fire weapons. This means that the enemy, having seized a position in one array, must change the majority of his firing positions and organize his attack all over again in order to get to a position in the next array. However, the arrays are close enough to allow the defending units to maintain coordinated, continuous engagement of the enemy while moving from one to the other. It is possible that not all of the forces executing contact and shielding functions have the same number of arrays.

4-73. OPFOR commanders may require a unit occupying an array to continue defending, even if this means the unit becomes decisively engaged or enveloped. This may be necessary in order to allow time for the construction of defenses farther from contact with the enemy. This may be the case when a unit is conducting maneuver defense in the disruption zone while the main defense force is preparing for an area defense in the battle zone. At some point, a unit conducting maneuver defense as part of the main defense force may be ordered to continue to defend an array, if conditions are favorable for defeating the enemy or repelling the attack at that array.

4-74. The example of maneuver defense in figure 4-2 shows that the shielding force does not necessarily have to remain in place to do its job. It can go out to meet the enemy (perhaps in an ambush) and then maneuver into another array. This type of maneuver can force the enemy into a nonlinear fight.

Defense

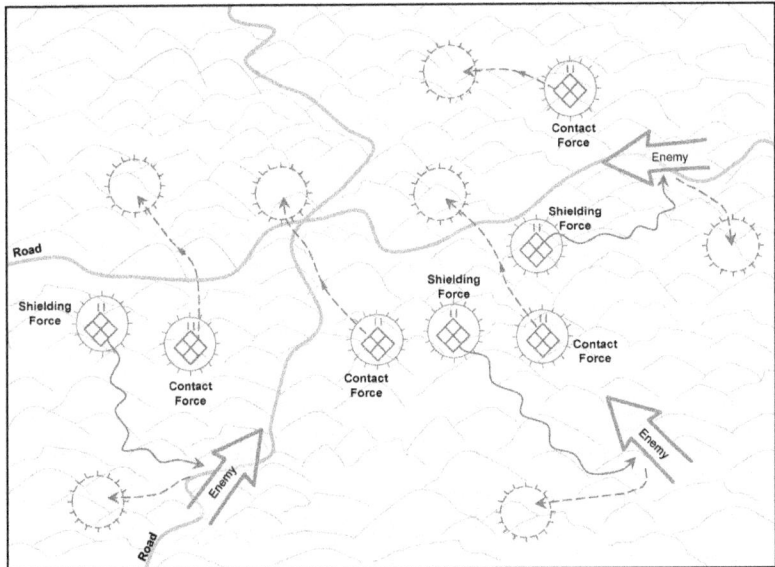

Figure 4-2. Maneuver defense (example 2)

Disruption Force

4-75. The disruption force initiates the attack on the enemy's combat system by targeting and destroying subsystems that are critical to the enemy. If successful, the disruption force can cause culmination of the enemy attack before the enemy enters the battle zone. In the worst case, the enemy would enter the battle zone unable to benefit from an integrated combat system and vulnerable to defeat by the main defense force.

4-76. In a maneuver defense, the disruption force often occupies battle positions in the disruption zone and seeks to force the enemy to fight on disadvantageous ground and at a tempo of the OPFOR's choosing. A maneuver defense disruption force also can set the conditions for a spoiling attack or counterattack (see chapter 3). The disruption force mission includes disaggregating the enemy attack and, if possible, destroying the enemy force.

4-77. Maneuver units conduct the defense from successive battle positions. Intervals between these positions provide space for deployment of mobile attack forces, precision fire systems, and reserves.

4-78. The distance between successive positions in the disruption zone is such that the enemy is forced to displace the majority of his supporting weapons to continue the attack on the subsequent positions. This aids the force in breaking contact and permits time to occupy subsequent positions. Long-range fires, ODs, and ambushes to delay pursuing enemy units can assist units in breaking contact and withdrawing.

4-79. If the disruption force has not succeeded in destroying or halting the attacking enemy, but is not under too great a pressure from a pursuing enemy, it may occupy prepared battle positions in the battle zone and assist in the remainder of the defensive mission as part of the main defense force. A disruption force may have taken losses and might not be at full capability; a heavily damaged disruption force may pass into hide positions. In that case, main defense or reserve forces occupy positions to cover the disruption force's disengagement.

Main Defense Force

4-80. The mission of the main defense force is to complete the defeat of the enemy by engaging portions of the force exposed by actions of the disruption force and by enemy reactions to contact. This may involve resubordination of units and in some cases attacks by fire or maneuver forces across unit limits of responsibility.

4-81. The main defense force in a maneuver defense divides its combat power into contact and shielding forces. These forces move in bounds to successive arrays of defensive positions.

4-82. The basic elements of the battle zone are battle positions, firing lines, and repositioning routes. Battle positions use the terrain to protect forces while providing advantage in engagements.

4-83. The commander may order a particular unit to stand and fight long enough to repel an attack. He may order this if circumstances are favorable for defeating the enemy at that point. The unit also might have to remain in that position because the next position is still being prepared or a vertical envelopment threatens the next position or the route to it.

Reserves

4-84. A commander in the maneuver defense can employ a number of reserve forces of varying types and strengths. The maneuver reserve is a force strong enough to defeat the enemy's exploiting force. The commander positions this reserve in an assembly area using C3D to protect it from observation and attack. From this position, it can transition to a situational defense or conduct a counterattack. The reserve must have sufficient air defense coverage and mobility assets to allow maneuver. If the commander does not commit the reserve from its original assembly area, it maneuvers to another assembly area, possibly on a different axis, where it prepares for other contingencies. (See the Types of Reserves section earlier in this chapter for discussion of other types of reserves.)

AREA DEFENSE

4-85. In situations where the OPFOR must deny key areas (or the access to them) or where it is overmatched, it may conduct a tactical area defense. Area defense is designed to achieve a decision in one of two ways:

- By forcing the enemy's offensive operations to culminate before he can achieve his objectives.
- By denying the enemy his objectives while preserving combat power until decision can be achieved through strategic operations or operational mission accomplishment.

4-86. The area defense does not surrender the initiative to the attacking forces, but takes action to create windows of opportunity that permit forces to attack key components of the enemy's combat system and cause unacceptable casualties. Area defense can set the conditions for destroying a key enemy force. Extended windows of opportunity permit the action of maneuver forces to prevent destruction of key positions and facilitate transition to a larger offensive action. INFOWAR is particularly important to the execution of the area defense. Deception is critical to the creation of complex battle positions, and effective perception management is vital to the creation of the windows of opportunity needed to execute maneuver and fires.

Method

4-87. Area defense inflicts losses on the enemy, retains ground, and protects friendly forces. It does so by occupying complex battle positions and dominating the surrounding area with reconnaissance fire (see chapter 9). These fires attack designated elements of the enemy's combat system to destroy components and subsystems that create an advantage for the enemy. The intent is to begin disaggregating the enemy combat system in the disruption zone. When enemy forces enter the battle zone, they should be incapable of synchronizing combat operations. See figures 4-3 through 4-5 on pages 4-15 through 4-17 for examples of area defense.

Defense

4-88. Area defense creates windows of opportunity in which to conduct spoiling attacks or counterattacks and destroy key enemy systems. In the course of an area defense, the tactical commander uses terrain that exposes the enemy to continuing attack.

4-89. An area defense trades time for the opportunity to attack enemy forces when and where they are vulnerable. The OPFOR might use an area defense when—
- It is conducting access-control operations.
- Enemy forces enjoy a significant RISTA and precision standoff advantage.
- Conditions are suitable for canalizing the enemy into areas where the OPFOR can destroy him by fire and/or maneuver.

4-90. A skillfully conducted area defense can allow a significantly weaker force to defeat a stronger enemy force. However, the area defense relies to a significant degree on the availability of complex terrain and decentralized logistics. Units conducting an area defense typically execute ambushes and raids in complex terrain throughout the AOR to force the enemy into continuous operations and steadily drain his combat power and resolve.

4-91. Within an overall area defense, the OPFOR might use maneuver defense on some portions of the AOR, especially on those where it can afford to lose ground. This occurs most often as OPFOR forces and elements are initially occupying the complex terrain positions necessary for the execution of the area defense.

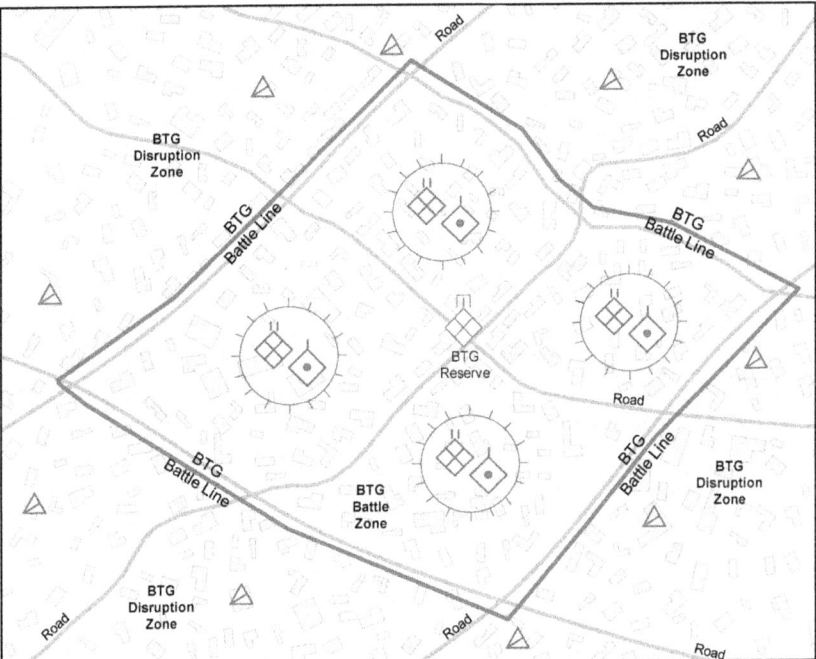

Figure 4-3. Area defense (example 1)

Chapter 4

Disruption Force

4-92. In an area defense, the disruption zone is the area surrounding its battle zone(s) where the OPFOR may cause continuing harm to the enemy without significantly exposing itself. For example, counterreconnaissance activity may draw the attention of enemy forces and cause them to enter the kill zone of an ambush using long-range precision fires. RISTA assets and counterreconnaissance forces occupy the disruption zone, along with affiliated forces. Paramilitary forces may assist other disruption force elements by providing force protection, controlling the civilian population, and executing deception operations as directed. See figure 4-4 for an example of a disruption force in an area defense.

4-93. The disruption zone of an area defense is designed to be an area of uninterrupted battle. OPFOR RISTA elements contact with enemy forces, and other parts of the disruption force attack them incessantly with ambush and precision fires.

4-94. The disruption force has many missions. The most important mission at the tactical level is destruction of appropriate elements of the enemy's combat system, to begin its disaggregation. The following list provides examples of other tasks a disruption force may perform:

- Detect the enemy's main groupings.
- Force the enemy to reveal his intentions.
- Deceive the enemy as to the location and configuration of battle positions.
- Delay the enemy, allowing time for preparation of defenses and counterattacks.
- Force the enemy into premature deployment.
- Attack lucrative targets (key systems, vulnerable troops).
- Canalize the enemy into situations unfavorable to him.

The disruption force mission also includes maintaining contact with the enemy and setting the conditions for successful reconnaissance fire and counterattacks.

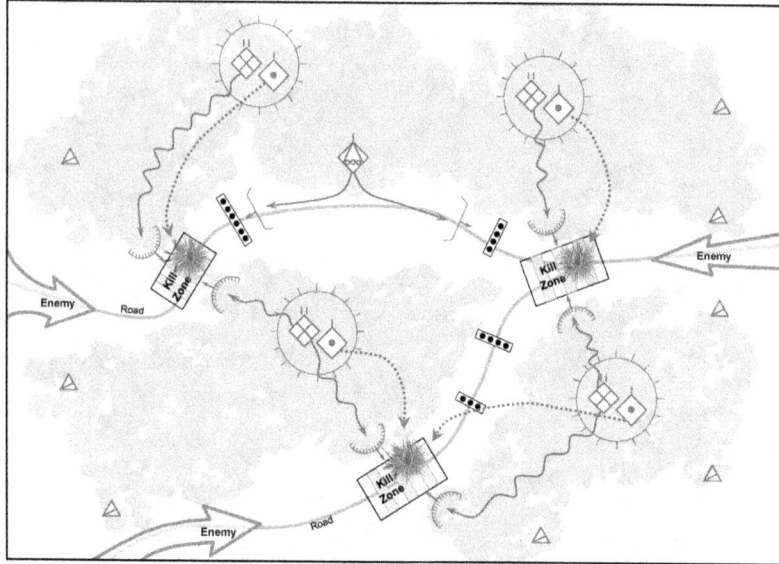

Figure 4-4. Area defense (example 2)

4-95. In an area defense, the disruption force often occupies and operates out of battle positions in the disruption zone and seeks to inflict maximum harm on selected enemy units and destroy key enemy systems operating throughout the AOR. An area defense disruption force permits the enemy no safe haven and continues to inflict damage at all hours and in all weather conditions.

4-96. Disruption force units break contact after conducting ambushes and return to battle positions for refit and resupply. Long-range fires, ODs, and ambushes to delay pursuing enemy units can assist units in breaking contact and withdrawing.

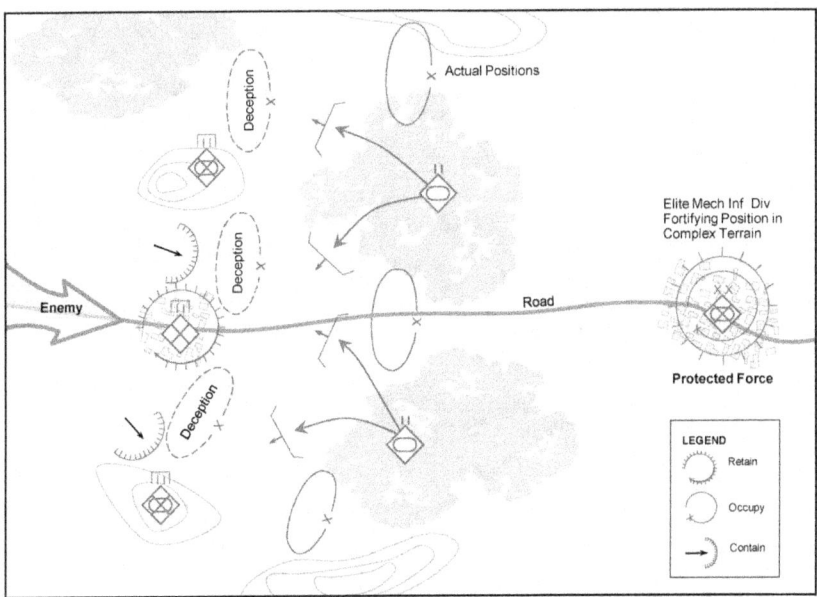

Figure 4-5. Area defense (example 3)

4-97. Even within the overall context of an area defense, the disruption force might employ a maneuver defense. In this case, the distance between positions in the disruption zone is such that the enemy will be forced to displace the majority of his supporting weapons to continue the attack on the subsequent positions. This aids the force in breaking contact and permits time to occupy subsequent positions.

4-98. The disruption zone will often include a significant obstacle effort. Engineer effort in the disruption zone also provides mobility support to portions of the disruption force requiring maneuver to conduct attacks or ambushes. Especially when overmatched by enemy forces, the OPFOR may use booby traps and other types of improvised obstacles.

4-99. Within the overall structure of the area defense, the disruption force seeks to conduct highly damaging local attacks. Units selected for missions in the disruption zone deploy on likely enemy avenues of approach. They choose the best terrain to inflict maximum damage on the attacking enemy and use obstacles and barriers extensively. They defend aggressively by fire and maneuver. When enemy pressure grows too strong, these forces can conduct a maneuver defense, withdrawing from one position to another in order to avoid envelopment or decisive engagement.

4-100. Since a part of the disruption force mission is to attack the enemy's combat system, typical targets for attack by forces in the disruption zone are—

Chapter 4

- C2 systems.
- RISTA assets.
- Precision fire systems.
- Aviation assets in the air and on the ground—at attack helicopter forward arming and refueling points (FARPs) and airfields. (Air defense ambushes are particularly effective in the disruption zone. See chapter 11).
- Logistics support areas.
- LOCs.
- Mobility and countermobility assets.
- Casualty evacuation routes and means.

4-101. In some cases, the disruption force can have a single mission of detecting and destroying a particular set of enemy capabilities. This does not mean that no other targets will be engaged. It simply means that, given a choice between targets, the disruption force will engage the targets that are the most damaging to the enemy's combat system.

Main Defense Force

4-102. The units of the main defense force conducting an area defense occupy complex battle positions (CBPs) within the battle zone. The complex terrain is reinforced by engineer effort and C3D measures. These CBPs are designed to prevent enemy forces from being able to employ precision standoff attack means and force the enemy to choose costly methods in order to affect forces in those positions. They are also arranged in such a manner as to deny the enemy the ability to operate in covered and concealed areas himself.

4-103. The main defense force in an area defense conducts attacks and employs reconnaissance fire against enemy forces in the disruption zone. Disruption zone forces may also use the CBPs occupied by the main defense force as refit and rearm points.

Reserves

4-104. A commander in an area defense can employ a number of reserve forces of varying types and strengths. In addition to its other functions, the maneuver reserve in an area defense may have the mission of winning time for the preparation of positions. This reserve is a unit strong enough to defeat the enemy's exploitation force in a maneuver battle during a counterattack. The commander positions its reserve in an assembly area within one or more of the battle positions, based on his concept for the battle. (See the Types of Reserves section earlier in this chapter for discussion of other types of reserves.)

TACTICAL DEFENSIVE ACTIONS—DETACHMENTS, BATTALIONS, AND BELOW

4-105. OPFOR detachments, battalions, and companies generally participate as part of a maneuver or area defense organized by a higher command, as opposed to conducting one independently. Commanders of OPFOR detachments, battalions, or companies select the defensive action they deem to be best suited to accomplishing their mission. OPFOR detachments and below are typically called upon to execute one combat mission at a time. Therefore, it would be rare for such a unit to employ more than one of these methods simultaneously. As part of either an area defense or maneuver defense, such units often conduct tactical defensive actions employing simple battle positions (SBPs). Alternatively, as part of an area defense, they may employ complex battle positions (CBPs).

BATTLE POSITIONS

4-106. A *battle position* (BP) is a defensive location oriented on a likely enemy avenue of approach. A BP is designed to maximize the occupying unit's ability to accomplish its mission. A BP is selected such

Defense

that the terrain in and around it is complementary to the occupying unit's capabilities and its tactical task. There are two kinds of BPs: simple and complex (see figure 4-6).

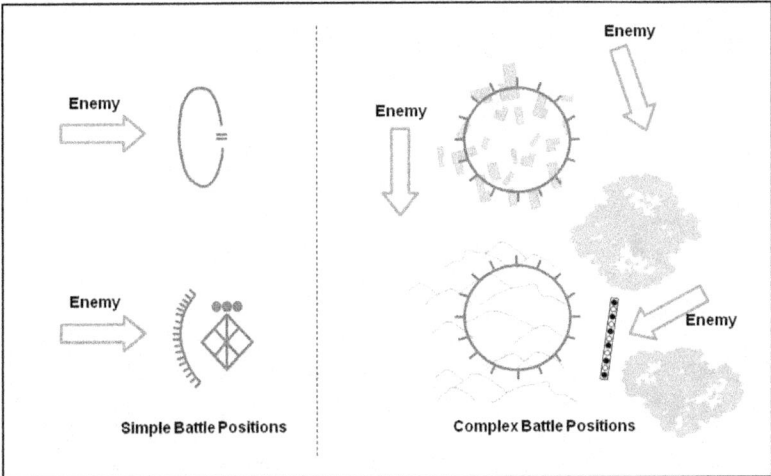

Figure 4-6. Simple and complex battle positions

Note. Of the two types of symbol for an SBP, this TC typically uses the "goose egg" type to show an SBP occupied by a generic unit of a particular size, and the "arc" type to show an SBP of a specifically identified unit (usually a detachment or below). Sometimes graphics show a large unit, such as a BTG or even a DTG inside a symbol for a CBP. This actually means that such a unit's subordinates occupy a series of CBPs within that area.

Simple Battle Position

4-107. A *simple battle position* (SBP) is a defensive location oriented on the most likely enemy avenue of approach. SBPs are not necessarily tied to complex terrain. However, they often employ as much engineer effort and/or camouflage, concealment, cover, and deception (C3D) measures as time allows.

Complex Battle Position

4-108. A *complex battle position* (CBP) is a defensive location designed to employ a combination of complex terrain, C3D, and engineer effort to protect the unit(s) within them from detection and attack while denying their seizure and occupation by the enemy. CBPs typically have the following characteristics that distinguish them from SBPs:

- Limited avenues of approach. (CBPs are not necessarily tied to an avenue of approach.)
- Any existing avenues of approach are easily observable by the defender.
- 360-degree fire coverage and protection from attack. (This may be due to the nature of surrounding terrain or engineer activity such as tunneling.)
- Engineer effort prioritizing C3D measures; limited countermobility effort that might reveal the CBP location.
- Large logistics caches.
- Sanctuary from which to launch local attacks.

DEFENSE OF A SIMPLE BATTLE POSITION

4-109. An SBP is typically oriented on the most likely enemy avenue of approach. SBPs may or may not be tied to restrictive terrain but will employ as much engineer effort as possible to restrict enemy maneuver. Defenders of SBPs will take all actions necessary to prevent enemy penetration of their position, or defeat a penetration once it has occurred.

Functional Organization of Elements to Defend an SBP

4-110. The commander of a detachment, battalion, or company defending an SBP designates his subordinate units as functional elements. The name of the element describes its function within the defensive action.

Disruption Element

4-111. Unit(s) assigned to the disruption element have the mission of defeating enemy reconnaissance efforts; determining the location, disposition, and composition of attacking forces; and in some cases they will also target designated subsystems of the attacking enemy's combat system. To accomplish these tasks, the disruption element may form combat security outposts (CSOPs) and ambush teams.

4-112. **CSOPs.** CSOPs prevent enemy reconnaissance or small groups from penetrating friendly positions and force the enemy to prematurely deploy and lose his momentum in the attack. CSOPs are generally composed of task-organized platoon- or squad-size elements. In a battalion or BDET, the platoon or squad(s) forming the CSOP is generally drawn from the battalion reserve element. Companies or CDETs may also form their own CSOPs. CSOPs are positioned forward of the battle zone on key terrain or along key avenues of approach. They typically will not be positioned directly astride avenues of approach into kill zones, but may cover them with fire. If decisively overmatched by enemy combat power, CSOPs may withdraw to the battle zone. An OPFOR battalion or BDET may employ more than one CSOP. During the counterreconnaissance battle, other forces may augment CSOPs, covering those avenues of approach that the CSOPs do not cover. CSOPs are typically assigned one or more of the following tactical tasks:

- **Ambush.** A CSOP with this task generally will avoid contact with superior enemy forces and only engage key enemy targets. When assigning this task, the OPFOR commander must also describe desired effects on the enemy (such as destroy, fix, or suppress).
- **Attack by fire.** A CSOP with this task is normally attempting to shape the battlefield in some fashion, either by turning an attacking enemy force into a kill zone or by denying the enemy a key piece of terrain. A CSOP with this task may also be required to target a key element of the enemy force.
- **Delay.** A CSOP with this task will attempt to buy time for the OPFOR to accomplish some other task such as defensive preparations, launch a counterattack, or complete a withdrawal. Normally, the CSOP will withdraw (remaining in the disruption zone, or moving to the battle or support zone) after engaging for a set amount of time.
- **Disrupt.** A CSOP with this task will attempt to weaken an enemy attack by using fires to cause premature commitment of the enemy, break apart his formation, and desynchronize his plan.
- **Fix.** A CSOP with this task will use fires to prevent a key element of the enemy force from moving from a specific place or halt them for a specific amount of time.

See figure 4-7 for an example of a CSOP.

Defense

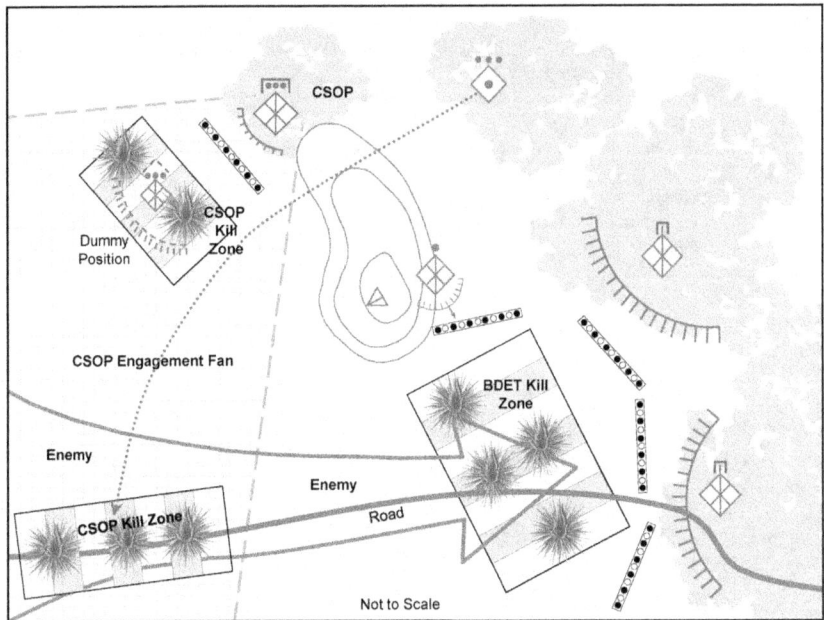

Figure 4-7. Combat security outpost (example)

4-113. **Ambush Teams.** Ambush teams (independent from CSOPs) remain concealed forward of the battle zone, and may allow some enemy forces to bypass their position. Once they identify key enemy targets, they will engage them by employing flanking or surprise close-range fire.

Main Defense Element

4-114. The main defense element of an SBP is responsible for defeating an attacking force, and for maneuvering to defeat the penetration or seizure of other SBPs. For examples of a CDET in an SBP, see figures 4-8 and 4-9 on pages 4-22 and 4-23.

Reserve Element

4-115. The reserve element of an SBP exists to provide the OPFOR commander with tactical flexibility. During the counterreconnaissance battle, the reserve may augment forces in the disruption zone, in order to provide additional security to the main defense element. During this time, the reserve element will also rehearse potential counterattack routes, although to avoid detection it will rarely do so en masse. Once a significant attacking force is detected, the reserve element will withdraw to a covered and concealed position, conduct resupply, and prepare for additional tasks. Some typical additional tasks given to the reserve may include—

- Conducting a counterattack.
- Conducting counterpenetration (blocking or destroying enemy penetration of the SBP).
- Conducting antilanding defense.
- Assisting engaged forces in breaking contact.
- Acting as a deception element.

Chapter 4

Support Element

4-116. The support element of an SBP has the mission of providing one or more of the following to the defending force:
- Combat service support (CSS).
- C2.
- Supporting direct fire (such as heavy machinegun, antitank guided missile [ATGM], recoilless rifle, or automatic grenade launcher).
- Supporting indirect fire (mortar or artillery).
- Supporting nonlethal actions (for example, jamming, psychological warfare, or broadcasts).
- Engineer support.

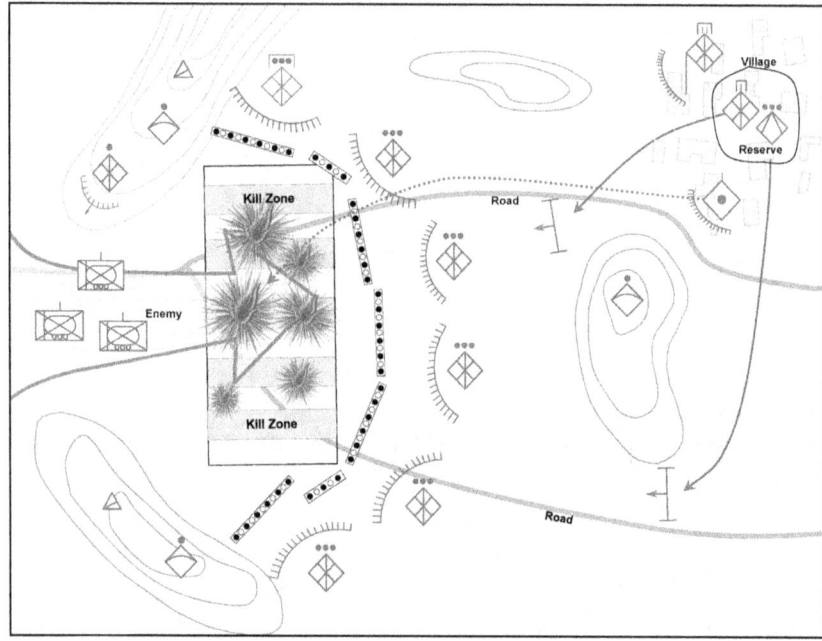

Figure 4-8. CDET in an SBP (example 1)

Organizing the Battlefield for an SBP

4-117. A detachment, battalion or company commander specifies the organization of the battlefield from the perspective of his level of command. As at higher levels, this normally consists of a battle zone and a support zone. It may also include a disruption zone.

Disruption Zone

4-118. The disruption zone is the area forward of the battle zone where the defenders will seek to defeat enemy reconnaissance efforts, detect attacking forces, disrupt and delay an attackers approach, and destroy key attacking elements prior to engagement in the battle zone. A defense of an SBP may or may not include a disruption zone.

Defense

Battle Zone

4-119. The battle zone is the area where the defending commander commits the preponderance of his force to the task of defeating attacking enemy forces. Generally, an SBP will have its battle zone fires integrated with those of any adjacent SBPs. Fires will orient to form kill zones where the OPFOR plans to destroy key enemy targets. When possible, kill zones will be placed on the reverse slope of intervisibility lines within the battle zone.

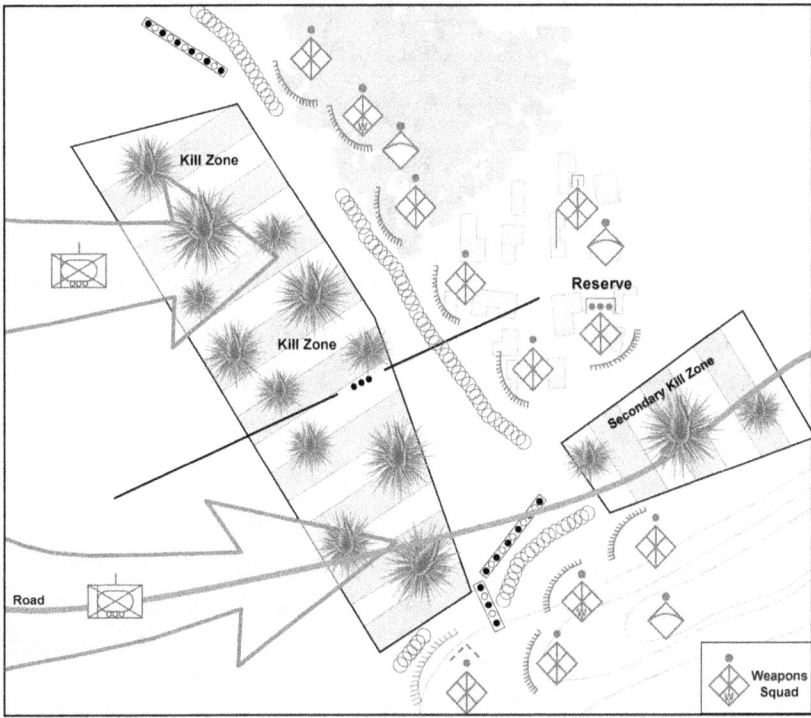

Figure 4-9. CDET in an SBP (example 2)

4-120. **Reverse Slope Defense.** The OPFOR commander will seek a defensive position behind a terrain feature(s) that, in addition to providing an intervisibility line, canalizes attackers into narrow attack frontages that lead into the kill zone. A reverse slope defense is positioned behind an intervisibility line so that is masked from enemy observation and direct fire. The defense is based upon employing the intervisibility line to protect friendly forces and isolate portions of the attacking force as they cross the crest. Although the OPFOR may not occupy the crest in strength, it will control it by fire. OPFOR commanders prefer a reverse slope defensive position because it confers the following advantages:
- It hinders or prevents enemy observation of the defensive position.
- Attacking forces are not able to receive direct fire support from follow-on forces.
- It can negate an enemy stand-off fire advantage
- Attacking enemy forces are silhouetted while crossing the crest of the intervisibility line.
- Engineers can conduct their work out of direct fire and observation from the enemy.

4-121. In some cases, the adoption of a reverse slope defense can prevent the defender's weapon systems from exploiting their maximum range. However, skilful OPFOR commanders will select defensive terrain that allows them to maximize their weapons stand-off range. They do so by emplacing their systems at their maximum effective range behind the crest of the intervisibility line that supports their kill zone. This may mean placing a weapon system on the counterslope behind the terrain forming the intervisibility line.

4-122. Maintaining observation of the enemy while on the forward slope of an intervisibility line can be difficult. To alleviate this disadvantage, OPFOR commanders will employ reconnaissance assets to observe forward of the reverse slope defensive position.

4-123. **Fire Planning.** Fire is the basic means of destroying the enemy in the defense. To perform this task, the OPFOR will employ lethal and nonlethal weaponry in a unified manner, often directed into a kill zone. The normal basis of a battalion's or BDET's system of fire is the antitank (AT) fire of its companies (and any additional units task-organized into the BDET) and supporting artillery. In areas that are not accessible to vehicles, the basis of fire will primarily be machinegun, grenade launcher, mortar, and artillery fires. In this case, where possible, AT systems will be employed in an antipersonnel role.

4-124. During the OPFOR fire planning process, the commander and staff delineate key enemy targets. The planners then appoint reconnaissance elements to identify targets and weapons systems to engage them. The OPFOR battalion's or BDET's fire planning includes sectors of concentrated fire and barrier fire lines of artillery and mortars in the disruption zone, on flanks, and throughout the depth of the battle zone. Subordinate units and weapons are expected to coordinate with each other as well as flanks units in the coverage of kill zones.

4-125. Kill zones will be covered by frontal and flanking or cross fires of the OPFOR battalion's or BDET's and other supporting weapons systems. The OPFOR will employ obstacles and fire concentrations to halt and hold the enemy within kill zones. Terrain considerations and available weaponry will dictate the size of the kill zone and the width of the OPFOR defense.

Support Zone

4-126. The support zone may contain C2, CSS, indirect and direct support fire assets, and the reserve, as well as other supporting assets. The support zone will normally be located in the SBP. Support zones are not typically found below the company level.

Executing Defense of an SBP

4-127. SBP defenders will conduct aggressive counterreconnaissance throughout their occupation of the battle position. Such counterreconnaissance will occur primarily in the disruption zone, but measures will also be taken in the battle and support zones. OPFOR electronic warfare assets will attempt to detect the presence and location of enemy reconnaissance elements. The reserve element may act as a quick-response force to destroy any enemy reconnaissance assets discovered in the battle or support zones. Once a significant attacking force is detected, the OPFOR will employ fires (direct or indirect) to delay and attrit attackers in the disruption zone.

Battle Zone

4-128. Defenders in the battle zone will attempt to defeat attacking forces. Should the enemy penetrate the main defenses or capture a position, defenders will take measures to defeat the penetration or recapture the position, to include the commitment of reserves and repositioning forces from other areas within the SBP.

Support Zone

4-129. Defenders in the support zone will provide support to defenders in the disruption and battle zones as required. In the event of the defeat or penetration of the SBP, they will maneuver as needed to avoid destruction or to support counterattacks.

Defense

Deception

4-130. To keep the enemy from discovering the nature of the OPFOR defenses and to draw fire away from actual units, defenders will establish dummy firing positions and battle positions. In addition to enhancing force protection, the OPFOR will employ deception positions as an economy-of-force measure to portray strength. These measures will include the creation of false entrenchments, heat signatures, and dummy vehicles.

Command and Control of an SBP

4-131. To maintain security during defensive preparations, defenders will make all possible use of secure communications, such as couriers and wire. However, once the main battle is joined, communications measures will tend to be those that support maneuver, such as radio and cellular technology.

Support of an SBP

4-132. Depending on the situation, the SBP will require support. This support may include combat support (CS) and/or CSS or a mixture of both. While some of this support will be provided from within the parent organization, other support may be from other organizations.

Reconnaissance

4-133. SBP defenders will perform aggressive counterreconnaissance activities to prevent the enemy from remaining in reconnaissance contact with the SBP. The OPFOR will observe avenues of approach to provide early warning; determine location, composition, and disposition of attackers; and direct fires against key enemy systems or components of systems. Figure 4-10 is an example of reconnaissance in support of an SBP. In this example, each of the three squads of the BDET's reconnaissance platoon serves as an observation post (OP).

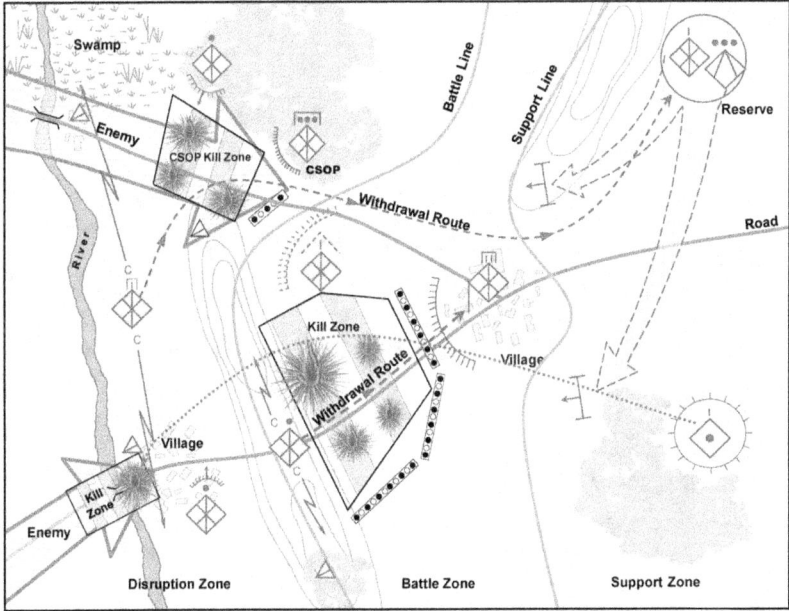

Figure 4-10. Reconnaissance support to an SBP (example)

Chapter 4

Armored Fighting Vehicles

4-134. When employed within an SBP, armored fighting vehicles will typically serve an anti-armor role, but can also serve as in an anti-infantry role. They may also be massed as a counterattack reserve. Defending armored vehicles will be in two-tier (turret defilade) vehicle fighting positions to provide maximum cover and concealment, or will fight above ground to take maximum advantage of maneuver capabilities. Armored vehicles defending SBPs do not prefer single-tier (hull defilade) vehicle fighting positions, since they provide insufficient cover and concealment against precision munitions and restrict vehicular mobility.

Fire Support

4-135. SBPs may receive fire support both from constituent assets and from higher echelon supporting forces. Fire support is integrated with other adjacent units to ensure appropriate coverage. Defenders will employ fires to—

- Attrit attackers along the avenues of approach and in LZs.
- Defeat attackers in the battle zone.
- Defeat penetrations of battle positions.
- Support counterattacking forces.

Air Defense

4-136. SBPs employ both active and passive air defense measures to protect the defender from air threats. Antiaircraft guns and shoulder-fired surface-to-air missiles (SAMs) may be found interspersed throughout the SBP, including antilanding ambushes. Electronically integrated air defense systems may be present when allocated to the defending force from higher-echelon supporting units.

Engineer

4-137. When available, engineers support the SBP initially by preparing survivability positions and countermobility works that support the disruption and battle zones. Once these preparations are complete, engineer support will shift to mobility support for the reserve force to ensure that it maintains freedom of maneuver.

4-138. Engineer tasks are a shared responsibility throughout the OPFOR. Although engineers have the bulk of specialized equipment for constructing fortified positions, this work exceeds the capability of organic constituent engineers and even those likely allocated from higher command. Therefore, the OPFOR uses all available personnel and equipment.

4-139. SBP obstacles are normally employed to shape the battlefield by disrupting the enemy's approach march, blocking avenues of approach, and turning the enemy into and fixing him in kill zones. Should the OPFOR have a remotely delivered mine capability, it will be used to reinforce pre-existing obstacles, block avenues of approach, or to re-seed breached obstacles.

4-140. Tables 4-1 through 4-3 on pages 4-27, 4-28, and 4-29 show preparation tasks for a battalion or BDET battle position (either simple or complex). Table 4-1 shows tasks that are the first priority in the sequence of position preparation. Combat arms unit personnel clear fields of fire and view. Then they emplace barbed wire, mines, and other obstacles in front of each fighting vehicle, crew-served weapon, and individual infantryman. Personnel use open slit trenches. Using covered slit trenches, engineers dig in headquarters and medical points. Camouflage measures are also performed. If the situation permits, engineers will employ excavating and earthmoving equipment.

Table 4-1. First-priority preparation tasks for a battalion or BDET battle position

Tasks of Combat Troops and Engineers

- Clear fields of observation and fire.
- Emplace obstacles integrated with CSOPs and platoon positions.
- Dig one- or two-man foxholes for riflemen, machinegun crews, snipers, and operators of grenade launchers, man-portable ATGMs, and shoulder-fired SAMs.
- Connect foxholes into a squad trench (open slit trench).
- Prepare a continuous trench in platoon and company positions.
- Prepare emplacements at primary firing positions for IFVs/APCs, tanks, ATGM launchers, and other weapons in the platoon or company position.
- Build basic positions (covered slit trenches) for platoon, company, and battalion or BDET CPs.
- Build basic positions (covered slit trenches) for battalion or company medical points.
- Dig and prepare covered slit trenches for each squad, crew, or team.
- Camouflage positions, weapons, and vehicles against reconnaissance and for protection against enemy precision weapons.

Tasks of Engineers

- Emplace additional obstacles on the most likely axes of enemy attack, in gaps between units, on their flanks, and in the depth of the BP.
- Deepen sections of trenches and communication trenches, and provide covered shelters for equipment on terrain that provides concealment from enemy observation and fire and permits the use of engineer mechanized equipment.
- Prepare lines of firing positions for reserve counterattack forces and prepare forward movement routes to these lines and to lines of deployment for counterattacks.
- Prepare routes for movement to the lines of deployment for the counterattack, lines of deployment of reserves, and firing positions.
- Set up water supply or distribution points.

4-141. Table 4-2 shows tasks that are typically the second priority in the sequence of position preparation. This includes improving positions, creating alternate and temporary positions, and connecting positions with communication trenches.

Table 4-2. Second-priority preparation tasks for a battalion or BDET battle position

Tasks of Combat Troops and Engineers
• Improve company and platoon positions, adding overhead cover if possible.
• Finish building or improve CPs and medical points.
• Dig emplacements at alternate and temporary firing positions of IFVs/APCs, tanks, and other weapons.
• Dig emplacements at firing lines and assembly areas for IFVs/APCs, tanks, and other weapons.
• Dig communication trenches to primary and alternate firing positions for IFVs/APCs, tanks, and other weapons; to shelters; to CPs; and to the rear.
• Prepare dugouts on the basis of one per platoon and one for each company, battalion, or BDET medical point.
• When possible, make covered slit trenches or dugout shelters for each squad, weapon crew, or team.
• Create and upgrade the system of trenches and communication trenches from a combat and housekeeping standpoint. Housekeeping and sanitary preparation or trenches includes making niches for storing food, water, and equipment and making latrines, sumps, soakage pits, and drainage ditches.
Tasks of Engineers
• Connect individual emplacements into emplacements for squads with sections of trench dug with mechanized equipment.
• Prepare a continuous trench in the battalion or BDET BP.
• Make bunkers for each company/battery and at battalion or BDET CPs.
• Make shelters for vehicles, weapons, equipment, missiles, ammunition, and other supplies.
• Prepare main dummy objects in the company position or battalion or BDET BP.
• Prepare for demolition of roads, bridges, overpasses, and other important objectives in the depth of the defense.
• Prepare routes for maneuver, resupply, and evacuation.

4-142. Table 4-3 shows tasks that typically are the third priority in the sequence of position preparation. In addition to the addition and improvement of existing positions and obstacles, engineers connect squad trenches until they run continuously across the entire platoon, company, and battalion or BDET frontage.

Table 4-3. Third-priority preparation tasks for a battalion or BDET battle position

Tasks of Combat Troops and Engineers
• Finish building or improving communication trenches and preparing positions.
• Improve engineer preparation of company positions and the battalion or BDET BP.
• Improve the platoon positions and squad and weapon positions in a tactical and housekeeping respect.
• Connect squad trenches in the platoon and company positions with one another, if this has not already been done.
• Build a system of engineer obstacles.
• Develop a system of trenches and communication trenches in the company position or battalion or BDET BP.
• Establish shelters for personnel and continue building shelters for equipment and deepening trenches and communication trenches.
• Adapt the communication trenches for conducting fire.
• Cover some parts of the trenches.
• Prepare dugout shelters at platoon CPs.
• Set up shelters (one per company and per battalion or BDET CP).
• Dig communication trenches to the rear (first with a depth of 0.6 m and then 1.1 m).
• Equip the trenches and communication trenches with alternate (lateral and forward) foxholes and emplacements for firing machineguns and grenade launchers and with embrasures, overhead protection, and niches or recesses for ammunition.
• Prepare dummy firing positions and BPs.
Tasks of Engineers
• Develop or improve a network of routes for unit maneuver, supply, and evacuation.
• Expand the system of obstacles.
• Improve fighting positions, firing lines, lines of deployment for counterattack, lines of deployment of reserves, CPs, assembly areas of reserves, and logistics elements.

Logistics

4-143. When present, logistics units will normally be found with the support element, to the rear of the SBP. Units in the disruption zone and battle zone will locally stockpile supplies, including multiple basic loads of ammunition, to ensure that they remain self-sufficient during the battle.

INFOWAR

4-144. The SBP is supported by INFOWAR, primarily by deceiving the enemy as to the defenders' actual location. The OPFOR will conduct deception operations that portray inaccurate defender locations and strengths. Such measures will attempt to convince the attacker to strike areas where he will inflict minimal damage to the defenders, or maneuver himself to a position of disadvantage, such as the center of a kill zone.

Chapter 4

DEFENSE OF A COMPLEX BATTLE POSITION

4-145. CBPs are designed to protect the units within them from detection and attack while denying their seizure and occupation by the enemy. Commanders occupying CBPs intend to preserve their combat power until conditions permit offensive action. In the case of an attack, CBP defenders will engage only as long as they perceive an ability to defeat aggressors. Should the defending commander feel that his forces are decisively overmatched, he will attempt a withdrawal in order to preserve combat power. See figure 4-11 for an example of defense in a CBP.

4-146. Units defending in CBPs will use restrictive terrain and engineer countermobility efforts to deny the enemy the ability to approach, seize, and occupy the position. They will also make maximum use of C3D and cultural standoff to deny the enemy the ability to detect and attack the position.

4-147. C3D measures are critical to the success of a CBP, since the defender generally wants to avoid enemy contact. Additionally, forces within a CBP will remain dispersed to negate the effects of precision ordinance strikes. Generally, once the defense is established, non-combat vehicles will be moved away from troop concentrations to reduce their signature on the battlefield.

4-148. To reduce exposure to enemy standoff fires and RISTA, cultural standoff can be used in conjunction with CBPs. Cultural standoff is the fact that protection from enemy weapon systems can be gained through actions that make use of cultural differences to prevent or degrade engagement. Examples of cultural standoff are—

- Using a religious or medical facility as a base of fire.
- Firing from within a crowd of noncombatants.
- Tying prisoners in front of BPs and onto combat vehicles.

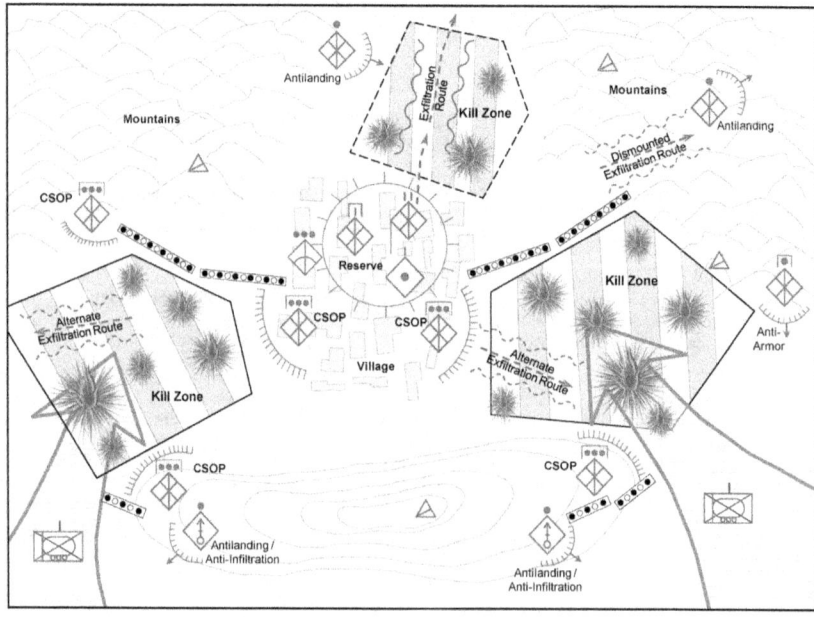

Figure 4-11. Defense of a CBP (example)

Defense

Functional Organization of Elements to Defend a CBP

4-149. The commander of a detachment, battalion, or company defending a CBP designates his subordinate units as functional elements. The name of the element describes its function within the defensive action.

Disruption Element

4-150. The disruption element of a CBP is primarily concerned detecting attackers and providing early warning to the defending force. To accomplish these tasks, the disruption element may form CSOPs and ambush teams. In addition to observation posts and ground ambushes, the disruption element can establish antilanding ambushes and antilanding reserves. When the CBP is attacked, disruption elements will remain in position to provide the OPFOR commander with a reconnaissance capability. The disruption element may also include indirect fire assets, such as mortars, to provide immediate, directly observed, harassing fires.

Main Defense Element

4-151. The main defense element of a CBP is responsible for defeating an attacking force. It can also cover the withdrawal of the support element in the case of an evacuation of the CBP.

Reserve Element

4-152. The reserve element of a CBP exists to provide the OPFOR commander with tactical flexibility. During the counterreconnaissance battle, the reserve may augment disruption elements, in order to provide additional security to the main defense element. However, the reserve will rarely do so if such action would reveal the location of the CBP to the enemy. Some typical additional tasks given to the CBP reserve may include—

- Conducting a counterattack.
- Conducting counterpenetration (blocking or destroying enemy penetration of the CBP).
- Conducting antilanding defense.
- Assisting engaged forces in breaking contact.
- Acting as a deception element.

Support Element

4-153. The support element of a CBP has the mission of providing one or more of the following to the defending force:

- CSS.
- C2.
- Supporting direct fire (such as heavy machinegun, ATGM, recoilless rifle, or automatic grenade launcher).
- Supporting indirect fire (mortar or artillery).
- Supporting nonlethal actions (for example, jamming, psychological warfare, or broadcasts).
- Engineer support.

Organizing the Battlefield for a CBP

4-154. A detachment, battalion, or company commander specifies the organization of the battlefield from the perspective of his level of command. As at higher levels, this normally consists of a battle zone and a support zone. It may also include a disruption zone.

Chapter 4

Disruption Zone

4-155. The battalion, company, or detachment defending in the CBP may send out CSOPs and/or ambush teams into the disruption zone. (See figure 4-12 and figure 4-13 on page 4-33 for examples of CSOPs in the disruption zone supporting a CBP.)

Battle Zone

4-156. The battle zone is the area where the defending commander commits a major part of his force to the task of defeating attacking enemy forces, or delaying them while the defenders withdraw. In the defense of a CBP, the battle zone is typically the area in and surrounding the CBP that the defending force can influence with its direct fires. It may be larger depending on the scheme for maneuver and indirect fires the defending commander wishes to employ.

Support Zone

4-157. The support zone may contain C2, CSS, indirect and direct support fire assets, the reserve, and other supporting assets. The support zone is located within the CBP.

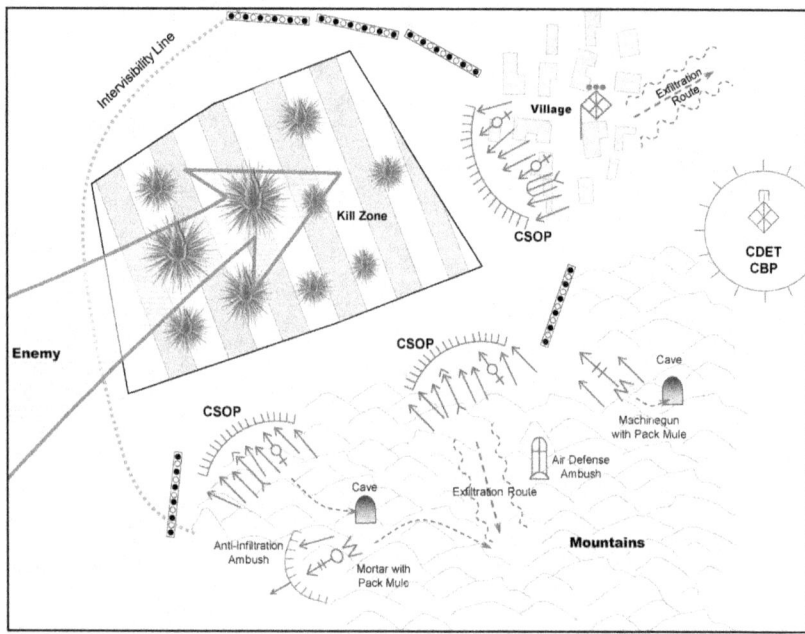

Figure 4-12. CSOPs in the disruption zone supporting a CBP (example 1)

Executing Defense of a CBP

4-158. Most security and/or counterreconnaissance will be passive measures unless attack is imminent. In the event of an attack, fires (direct and/or indirect) will delay and attrit attackers.

4-159. Defenders in the battle zone will attempt to defeat attacking forces. Should the defending commander determine that he lacks the capacity to defeat attackers, defenders in the battle zone will cover

Defense

the withdrawal of the rest of the unit before retiring themselves. Should the enemy penetrate the main defenses, it is likely the defensive commander will determine further resistance to be useless. In this case, he may commit reserves to delay further penetration while the remainder of the defending unit(s) withdraws.

4-160. Defenders in the support zone will provide support to defenders in the disruption and battle zones as required. In the event of a withdrawal, they will be some of the first elements to withdraw via exfiltration routes.

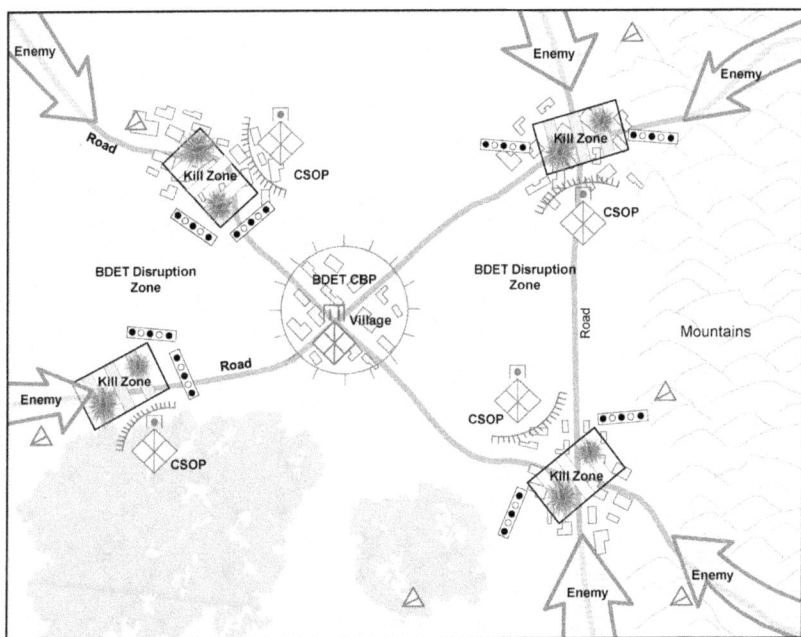

Figure 4-13. CSOPs in the disruption zone supporting a CBP (example 2)

Command and Control of a CBP

4-161. C2 of a CBP is generally more difficult than that of an SBP because the defenders may be more dispersed. To maintain security and avoid detection, defenders will make all possible use of secure communications, such as couriers and wire.

Support of a CBP

4-162. Depending on the situation, the CBP will require support. This support may include CS and/or CSS or a mixture of both. While some of this support will be provided from within the parent organization, other support may be from other organizations.

Reconnaissance

4-163. OPFOR reconnaissance assets will observe avenues of approach key to providing early warning and allow the commander to make "fight or flee" determination. The OPFOR is less likely to engage in counterreconnaissance activities if such actions would reveal CBP location. In order to passively gather information, personnel will imbed themselves within local populations. See figure 4-14 for an example of reconnaissance support to a CBP.

Armor

4-164. Due to the larger signature of armored vehicles (and their tendency to draw precision munitions fire), elements defending a CBP are less likely to employ significant armored assets. When armored vehicles are employed, they will generally remain concealed in reserve and emerge only when needed to defeat attacking enemy forces or to cover a withdrawal.

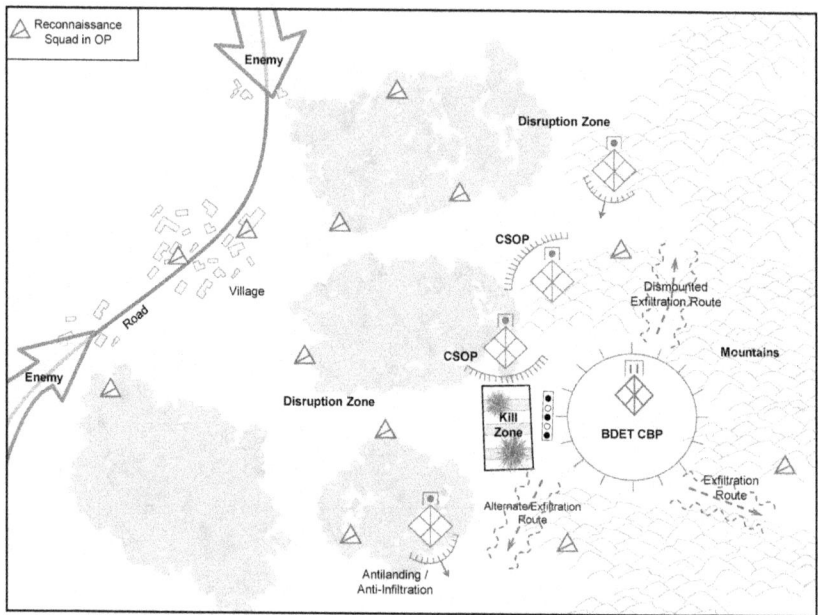

Figure 4-14. Reconnaissance support to a CBP (example)

Fire Support

4-165. Since CBPs are generally independent and self-supporting in their nature, all fire support will come from within the CBP itself. Defenders will employ fires to—
- Attrit attackers along avenues of approach and in LZs.
- Defeat attackers in the battle zone.
- Cover the withdrawal of defenders from the CBP.

Air Defense

4-166. Unlike SBPs, passive air defense is most common to CBPs, and active air defense will generally involve systems that do not emit an electromagnetic signature. Antiaircraft guns and shoulder-fired SAMs may be found interspersed throughout the CBP, including antilanding ambushes.

Defense

Engineer

4-167. Within the CBP, engineer activity will generally be of a non-signature, or low signature-producing variety. Engineers will conceal survivability positions (such as entrenchments, fortifications, improved caves, tunnels, or hardened buildings). Countermobility efforts, such as antipersonnel and/or AT mines and booby traps will likewise be hidden from observation. Wire obstacles, AT ditches, and vehicular survivability positions will be less common due to the difficulty in concealing such works.

4-168. While obstacles may be used in the development of kill zones for a CBP, they are generally more protective in nature than those in a SBP. For example, they may be employed to turn an attacker away from a vulnerable flank, or to protect an exfiltration route by blocking an avenue of approach into it.

4-169. Engineer preparation priorities will otherwise be of a similar nature to those of the SBP. (See SBP preparation tasks in tables 4-1 through 4-3 on pages 4-28, 4-29, and 4-30).

Logistics

4-170. Within a CBP, logistic operations are generally of a self-sustaining nature. Large supply caches will be common.

INFOWAR

4-171. Elements from the CBP may attempt to integrate within any local communities for the purpose of gathering and disseminating information. Generally, the CBP will not have easily detectable INFOWAR activities, since it is attempting to maintain a low profile. INFOWAR may focus on downplaying the existence or significance of the CBP itself. If the CBP cannot be hidden, INFOWAR may attempt to convince enemy forces that the defenders are friendly to them. In some cases, senior OPFOR leaders may conduct INFOWAR from a CBP to convince followers in other locations that they are still alive and leading their organizations in the struggle against the enemy.

This page intentionally left blank.

Chapter 5
Battle Drills

The OPFOR derives great flexibility from battle drills. Unlike the U.S. view that battle drill, especially at higher levels, reduces flexibility, the OPFOR uses minor, simple, and clear modifications to thoroughly understood and practiced battle drills to adapt to ever-shifting conditions. It does not write standard procedures into its combat orders and does not write new orders when a simple shift from current formations and organization will do.

PURPOSE OF BATTLE DRILLS

5-1. The purpose of battle drills is to achieve advantage in controlling the tempo of combat. They allow OPFOR units to perform basic combat functions without hesitation or need for further coordination, assistance, or delay. Battle drills are intended to be the baseline of tactical competence for the OPFOR. Once able to execute all battle drills, units can be directed to act with concise and rapidly formulated combat orders.

5-2. The OPFOR uses battle drills to make the execution of basic tactical tasks that are standard throughout the OPFOR. Battle drills are not designed for a specific unit type, but rather represent a common methodology for executing common, recurring tasks at the tactical level. They are conducted in both offensive and defensive operations.

5-3. Battle drills are detachment-level tactical tasks carried out by functionally organized elements performing various subtasks. The composition of such elements will vary depending on the type of force and the operational environment. However, the subtask(s) each element performs in a given battle drill will be the same for any tactical unit. Most battle drills focus on enabling functions that facilitate the primary action of a larger tactical mission.

Note. Any battalion or company receiving additional assets from a higher command becomes a battalion-size detachment (BDET) or company-size detachment (CDET). Therefore, references to a detachment throughout this chapter may also apply to battalion or company, unless specifically stated otherwise.

ACTIONS ON CONTACT

5-4. When an OPFOR detachment makes contact with the enemy, either expected or unexpected, it executes the actions on contact battle drill. This battle drill is designed to ensure OPFOR units retain the initiative and fight under circumstances of their choosing.

FORMS OF CONTACT

5-5. The OPFOR recognizes seven forms of contact:
- Direct fire.
- Indirect fire.
- Obstacle.
- Air.
- Chemical, biological, radiological, and nuclear (CBRN).

Chapter 5

- Electronic warfare (EW).
- Sensor.

5-6. The actions on contact battle drill is primarily for use by a force making sensor and/or direct fire contact with an enemy force. When making undesired contact (indirect fire, air, CBRN, EW, or ground contact made by a noncombat unit), the break contact battle drill is employed instead. When making contact with an isolated obstacle, the situational breach battle drill may be selected.

CONDITIONS

5-7. The commander will take action after determining the type of contact made, which may be—
- Expected contact in his course of action.
- Unexpected contact regarding time.
- Unexpected contact regarding location.
- Unexpected contact regarding the enemy.
- Unexpected contact regarding the any combination of the above.

5-8. The OPFOR considers it highly unlikely that contact will be made in the expected location at the expected time with the expected enemy force. Battle drill actions on contact are designed to provide the commander with the flexibility to either continue with the planned course of action or rapidly adopt a new course of action more suited to the new circumstances.

5-9. This flexibility is achieved by—
- Ensuring that contact is made with one or more security elements before the remainder of the force becomes engaged.
- Employing one or more security elements to shape the engagement area by either fixing or isolating the enemy to avoid additionally committing the action element.
- Providing the commander with the ability to make his own decisions if communication with higher authority is impractical.
- Using cover camouflage, concealment, cover, and deception (C3D) to prevent unwanted engagements.

EXECUTION

5-10. Execution of actions on contact varies depending on the situation and the commander's battle plan. The actions on contact battle drill is accomplished by performing one or a combination of the five subtasks below. Figure 5-1 shows an example of actions on contact involving some of these subtasks.

Fix

5-11. The security element making contact fixes the enemy. This security element is then known as the fixing element. It continues to provide early warning of approaching enemy forces and prevents them from gaining further information on the rest of the OPFOR force. Fixing elements often make use of terrain choke points, obstacles, ambushes, and other techniques to fix a larger force.

Note. When an element that is not a security element makes contact with the enemy, the commander will designate that element as the fixing element.

Assess and Report

5-12. Based on reports he receives from element(s) in contact, the detachment commander must make an assessment of the tactical situation that determines whether or not making contact in this manner and with this enemy constitutes a change in his course of action. This determination is the most vital step in successful execution of actions on contact because if it is performed incorrectly, the unit will subsequently be executing a course of action inappropriate to the mission and situation. Concurrent with his assessment, the commander reports to the chain of command what contact has been made with the enemy force, critical details of its composition, and his assessment.

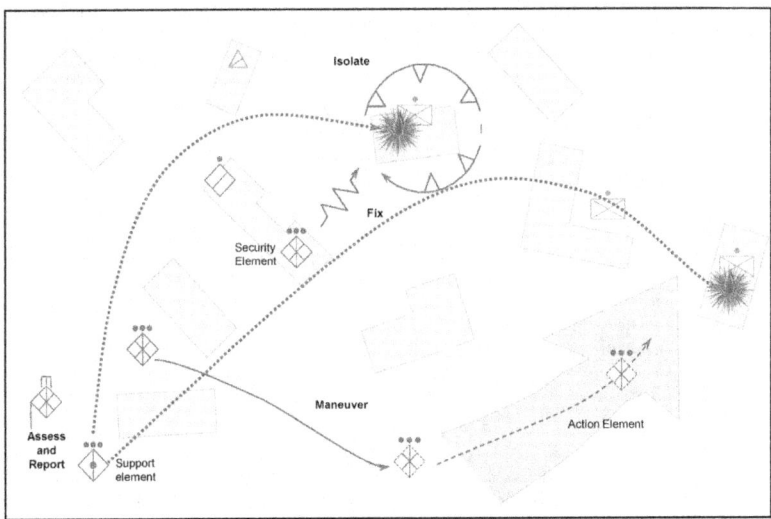

Figure 5-1. Actions on contact (example)

Isolate

5-13. The detachment making contact maneuvers and deploys security elements to ensure additional enemy forces do not join the battle unexpectedly. Indirect fire and close air support can be used either individually or combined with other means to achieve the same effect.

Maintain Freedom to Maneuver

5-14. The commander of the contacting unit ensures he makes contact with the minimum part of his force necessary to fix the enemy. He makes use of C3D and the break contact battle drill to prevent his force from becoming decisively engaged. Security elements determine safe maneuver avenues for him to employ. Freedom to maneuver is also maintained by—
- Dominating avenues of approach into the engagement area.
- Determining location of enemy flanks or exposed areas of weakness.

Execute Course of Action

5-15. The contacting unit either continues with its original course of action if deemed appropriate or executes a new one that suits the situation. A new course of action could be one given to the unit based on the assessment it provide to its higher command or one chosen by the commander in absence of time or guidance. The unit making contact ensures follow-on units are aware of the contact and deconflict positioning, typically through the use of a standard marking system.

Chapter 5

BREAKING CONTACT

5-16. The primary objective in breaking contact is to remove the enemy's ability to place destructive or suppressive fires on the greater portion of the OPFOR force. This is accomplished by fixing the enemy; regaining freedom to maneuver; and employing fires, C3D, and countermobility. The OPFOR will routinely break contact in order to maneuver into predesignated defensive positions or to draw the enemy force into an ambush. In other cases, the OPFOR breaks contact when faced with no other tactical option.

CONDITIONS

5-17. The commander will break contact under the following conditions:

- **Included in the battle plan.** The OPFOR may include breaking contact with the enemy as part of the scheme of maneuver for its battle plan.
- **Loss of time is especially critical.** If the OPFOR expects to engage the enemy for an overly extended period, it will break contact in order to exploit an alternative avenue of approach.
- **Loss of terrain is not critical.** If the location in which the OPFOR engages the enemy is not suited to its posture or force structure, it will break contact in order to either bypass the enemy or to engage him at a more favorable location.
- **Enemy is too strong to engage with the force on hand.** If the enemy force is overwhelming and or the OPFOR has sustained excessive damage to its force, the OPFOR will break contact in order to recover from the engagement.

EXECUTION

5-18. Execution of the breaking contact battle drill varies depending on the situation and the commander's battle plan. The breaking contact battle drill is accomplished by performing the following subtasks. In most cases, all subtasks are part of the breaking contact battle drill. However, the first three subtasks may be executed in a variety of ways. Figure 5-2 shows an example of breaking contact.

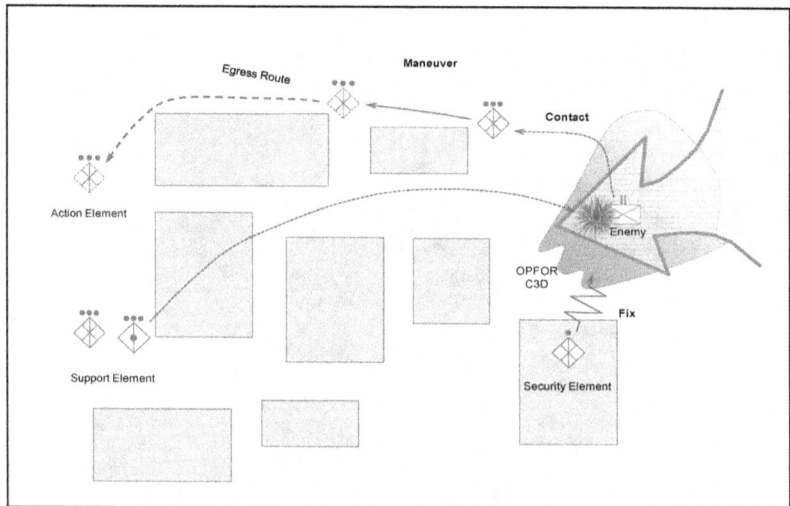

Figure 5-2. Breaking contact (example)

Protect

5-19. The detachment commander takes immediate steps, using a variety of means, to protect his force while it maneuvers to a position out of contact. The security element fixes the enemy. It prevents the enemy force from maneuvering in order to remain in contact with the rest of the OPFOR force. It may employ INFOWAR to appear to be larger than it is or even to appear to be the entire OPFOR detachment.

5-20. The detachment commander employs fires as part of the break contact battle drill to suppress the enemy and prevent him from returning fire effectively and to fix him and restrict his maneuver. If available, he may use indirect fires, close air support, EW, and/or CBRN means to fix the enemy. C3D is employed to limit or remove the enemy's ability to maintain situational awareness of the OPFOR force. This may be as simple as placing obscuring smoke between the enemy and the detachment or as complex as a sophisticated deception plan making use of decoys and mock-ups. Another way to protect the force is to use alternate positions or assembly areas. Countermobility actions, such as the emplacement of dynamic obstacles or the destruction of man-made structures can also restrict the enemy's ability to maneuver and maintain contact with the detachment.

Retain Freedom to Maneuver

5-21. The commander reduces his elements in contact to only security element(s). For any other element(s) that originally made contact, he identifies egress routes. He selects one or more routes from his current location that enable his detachment to remain out of contact while permitting him to maneuver in support of his mission.

5-22. Once the rest of the force has maneuvered out of contact, the security element(s) that performed a fixing function can rejoin the rest of the force. Separating these fixing elements from the enemy may require further use of C3D, fires, and countermobility measures.

Assess and Report

5-23. The commander receives reports from the subordinate element(s) that first made contact and/or the fixing element(s) that remain in contact. Based on those reports, he must make an assessment of the tactical situation. Concurrent with his assessment, the commander reports to the chain of command what form of contact has been made with the enemy force, critical details of the enemy force's composition, and his assessment of the situation.

Continue or Change Course of Action

5-24. Once freedom to maneuver has been retained or regained, the OPFOR force executes the basic course of action. The course of action is usually the primary action of the unit's original tactical mission. However, the detachment commander makes an assessment of the tactical situation to determine whether or not making contact in this manner and with this enemy force dictates a change in the course of action.

SITUATIONAL BREACH

5-25. A situational breach is the reduction of and passage through an obstacle encountered in the due course of executing another tactical task. The unit conducting a situational breach may have expected an obstacle or not, but in either case conducts a situational breach with the resources at hand and does not wait for specialized equipment and other support. This allows the unit to maintain momentum rather than being stopped or impeded by the obstacle. The decision to attempt the situational breach is based on the OPFOR commander's knowledge of the enemy forces in the area and the expected tactical advantage in terms of key terrain and time.

Conditions

5-26. The commander will order a situational breach under the following conditions:
- **Included in the battle plan.** The OPFOR expects to breach enemy obstacles.
- **Time constraints.** The OPFOR commander assesses that by breaching an obstacle he will save more time than if he bypasses it.
- **Terrain is crucial.** The OPFOR commander decides that key terrain can be seized by breaching the obstacle.
- **Exposes enemy weaknesses.** The OPFOR commander decides that by doing so he can engage the enemy decisively and he has a clear advantage.

Execution

5-27. In order to execute the situational breach effectively, the OPFOR must be prepared to provide the necessary security to allow movement through the obstacle. This is accomplished by isolating the potential enemy avenues of approach while reducing the obstacle for the rest of the unit to pass through. See figure 5-3 for an example of a situational breach.

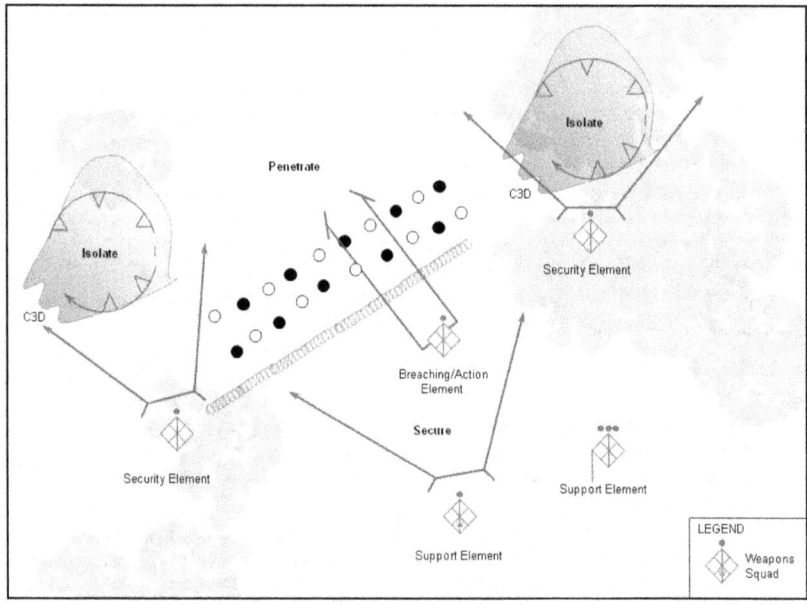

Figure 5-3. Situational breach (example)

Isolate

5-28. The security element takes action to ensure enemy elements cannot reinforce those defending the obstacle. It might accomplish this through C3D measures, countermobility tasks, direct or indirect fire engagements, or a variety of other means.

Battle Drills

Secure

5-29. A support element establishes a support-by-fire position and takes action to ensure enemy elements defending the obstacle are neutralized. It also supports movement through the obstacle.

Penetrate

5-30. The breaching element reduces the obstacle such that it can complete its mission (as the action element) and/or enable a follow-on force to do so. All OPFOR organizations carry sufficient equipment, whether field expedient or constituent, to penetrate basic enemy obstacle systems and urban construction and debris. Precise descriptions of OPFOR obstacle reduction techniques can be found in chapter 12.

Execute Course of Action

5-31. Once the obstacle has been penetrated and the lanes isolated and secured, the action element and/or a follow on force continues the mission, if that is deemed appropriate. However, based on the commander's assessment of the situation and/or guidance from a higher command, the unit may adopt a new course of action.

FIRE AND MANEUVER

5-32. Fire and maneuver is the way in which OPFOR units move while in contact with the enemy. When required to move under such conditions, the OPFOR commander selects part of his force to be the *firing element* and part to be the *moving element*. The firing element fires from a position of concealment or cover in order to support the moving element. This is the most basic of all OPFOR battle drills.

CONDITIONS

5-33. The commander will employ fire and maneuver under the following conditions:
- **Included in the battle plan.** The OPFOR plans for a movement to contact and expects to fire and maneuver on its way to the objective.
- **Time constraints.** Time is not a critical factor, since the fire and maneuver battle drill will slow progress along the OPFOR avenues of approach.
- **Exposes enemy weaknesses.** If the OPFOR realizes a clear advantage by maintaining contact with the enemy, it may use fire and maneuver to lure him into an ambush or attrit his forces to the point where he has to withdrawal from the engagement.

EXECUTION

5-34. The critical aspect of executing fire and maneuver is the commander's selection of the right amount of combat power and resources to assign to each of the elements of his force. If the firing element does not have the ability to significantly reduce the effectiveness of the enemy, the moving element will be destroyed. If the moving element does not have the combat power to take the objective or assume its new role as firing element, the mission will fail.

5-35. The part of the force initially designated as the firing element directs suppressing fire against any enemy that has the ability to influence the movement of the moving element. The moving element then moves to the next firing line. Once the moving element reaches that new position, it becomes the new firing element, and the former firing element becomes the new moving element. This continues until a moving element reaches the objective. See figure 5-4 on page 5-8 for an example of fire and maneuver.

Make Contact

5-36. Normally, a security element makes first contact with the enemy. It observes the enemy force and reports on its activity. Security element(s) continue to provide early warning of approaching enemy forces and prevent them from gaining further information on the rest of the OPFOR unit. If the enemy force

Chapter 5

attempts to move in a direction that could influence the movement of the OPFOR unit, the security element becomes a fixing element.

Fix

5-37. The security element making contact fixes the enemy. Once the firing element moves into a suitable position, it can also fix the enemy, often by delivering suppressing fires against an enemy force that has the ability to influence the movement of the moving element. (While performing this function, the firing element could be called a fixing element.)

Isolate

5-38. Security elements ensure additional enemy forces do not join the battle unexpectedly. Indirect fire and close air support can be used either individually or combined with other means to achieve the same effect.

Maneuver

5-39. The moving element maneuvers to a new position of advantage with respect to the enemy. On order, the moving element assumes the role of the new firing element. If further maneuver is required, the moving and firing elements continue alternation of fixing the enemy and maneuvering against the enemy.

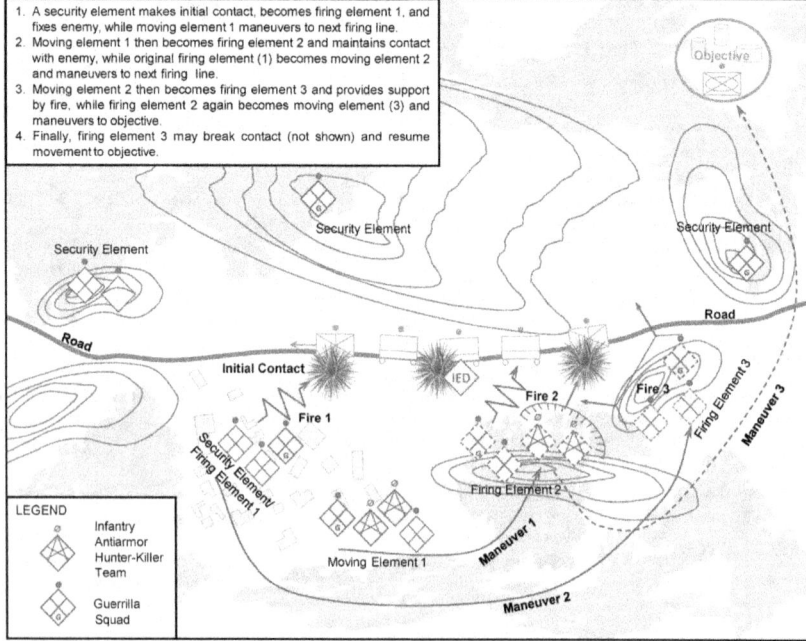

Figure 5-4. Fire and maneuver (example)

FIXING

5-40. Fixing is a tactical task intended to prevent the enemy from moving any part of his force from a specific location for a period of time. It is one of the most critical battle drills OPFOR units execute, and it is often a subtask in other battle drills. The ability to fix the enemy at crucial points is the fundamental way by which units maintain the freedom to maneuver and retain the initiative. An enemy becomes fixed in one of three basic ways:

- He cannot physically move.
- He does not want to move.
- He does not think he can move.

5-41. An enemy that cannot physically move is constrained in some real way. Fixing an enemy by physically preventing him from moving is the most difficult and resource-intensive method. An enemy does not want to move when he feels that in doing so he takes great risk to life and material. Suppressive fires are the primary method by which an enemy is fixed in this way. Suppressive fires are simple to employ and are the least difficult and resource intensive means. However, they are also the means that places the OPFOR at the greatest risk—the soldiers and systems providing the suppressive fires are vulnerable to detection and return fire. The use of snipers who target individual soldiers and materiel and/or the deployment of scatterable mines can fix the enemy by halting their movement and disrupting their operations for medical evacuation and repairs. Information warfare (INFOWAR) actions such as deception can also achieve the effects of physically fixing the enemy when feasible.

CONDITIONS

5-42. The OPFOR will employ fixing under the following conditions:

- **Included in the battle plan.** The OPFOR expects to have the enemy fixed at a designated time and location as part of its battle plan.
- **Time is required for follow-on forces.** Fixing can allow an action element or other follow-on force to maneuver into place or allow reconnaissance elements to help assess the situation.
- **Enemy is located on a preplanned target.** The OPFOR fixes the enemy in a predesignated kill zone in order to mass fires.
- **INFOWAR assets achieve desired effects.** The OPFOR fixes the enemy through the use of INFOWAR. Because of lack of training of the enemy forces in countering INFOWAR effects, the enemy is removed from the fight without committing other OPFOR maneuver forces.

EXECUTION

5-43. The OPFOR will fix the enemy using the method most likely to achieve the results with the minimum risk to its forces. (See figure 5-5 on page 5-10 for an example of fixing.) The following are the primary methods.

Fires

5-44. Fires fix the enemy by killing enemy soldiers or wounding them enough to prevent relocation (destructive fires) or by making it too dangerous for them to reposition (suppressive fires). Indirect fires and/or close air support are also employed to fix the enemy in situations where distance and terrain make it difficult to achieve the effect through direct fire alone. Fires are the main method for decisively engaging the enemy.

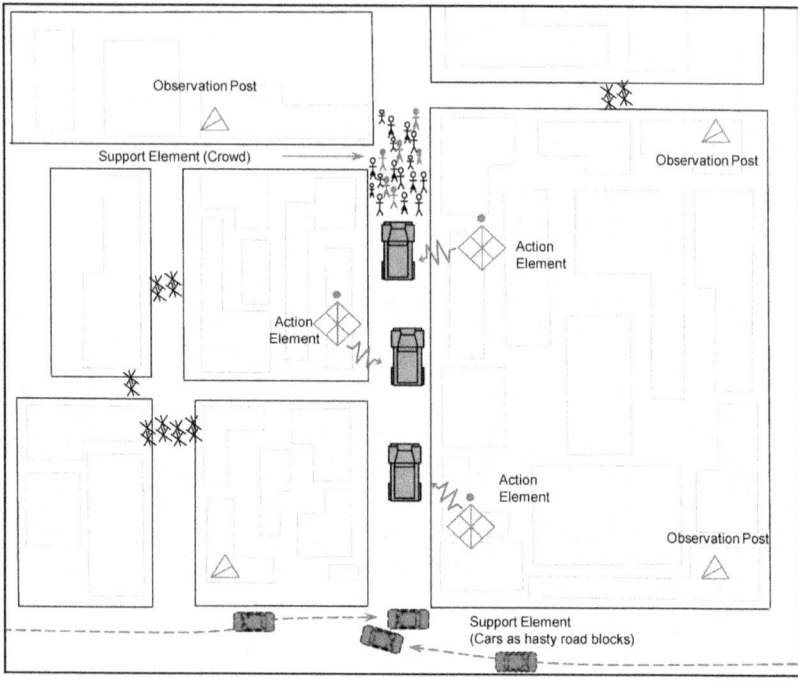

Figure 5-5. Fixing (example)

INFOWAR

5-45. INFOWAR fixes the enemy by convincing the enemy he does not want to move or by making him think he cannot move. Some examples of INFOWAR used to fix the enemy are—

- Propaganda claiming the enemy will be destroyed if he moves in the open. Effectively employing snipers in the area will reinforce this claim and cause trepidation among the enemy troops.
- Deception that simulates the enemy higher commander ordering the enemy unit to remain in place.
- Information attack on enemy sensors to register that the fixing element is stronger than it is, or at least capable of destroying the enemy force if it relocates.

Countermobility

5-46. Countermobility actions fix the enemy primarily by physically restraining his movement. In actuality, there is no obstacle that cannot be breached with effort. This fact typically makes countermobility actions time-sensitive. The more time spent and resources gathered by the enemy, the less effective countermobility actions would be in fixing him.

Chapter 6

Other Combined Arms Actions

The OPFOR's basic combined arms unit is the maneuver brigade. Brigades, divisions, and tactical groups conduct or organize combined arms actions other than the basic types of offensive and defensive action outlined in chapters 3 and 4. The tactics described in this chapter are employed in combat actions that could be either offensive or defensive in nature.

ACTIONS OF THE DISRUPTION FORCE

6-1. The purpose of the disruption force is to significantly degrade the enemy's combat capability and to prevent the enemy from conducting an effective operation. The primary task of the disruption force is to initiate the attack against one or more key components of the enemy's combat system. (See Systems Warfare in chapter 1.) Successful attack of designated components or subsystems begins the disaggregation of the enemy's combat system and creates vulnerabilities for exploitation in the battle zone. Skillfully conducted disruption operations will effectively deny the enemy the synergy of effects of his combat system. In addition, the disruption force—

- Destroys enemy reconnaissance.
- Forces the enemy to deploy early or disrupts his offensive preparations.
- Gains and maintains reconnaissance contact with key enemy elements.
- Deceives the enemy as to the disposition of OPFOR units.

6-2. The disruption force may be given any offensive, defensive, or security mission that best suits the disruption of the particular enemy force. To accomplish these missions, the disruption force executes a combination of tactical tasks designed to set the conditions for OPFOR success. These tasks include one or more of the following:

- Cover.
- Delay.
- Disrupt.
- Fix.
- Ambush.
- Contain.
- Canalize.
- Isolate.
- Neutralize.
- Interdict.

ORGANIZING THE DISRUPTION ZONE

6-3. The disruption zone is essentially the area of responsibility (AOR) of the disruption force. It may contain subordinate unit battle positions, kill zones, axes, objectives, and attack zones based on the disruption force commander's intent.

Chapter 6

ORGANIZING THE DISRUPTION FORCE

6-4. The size and composition of the disruption force depends on the level of command involved, the commander's concept of the battle, and terrain and enemy involved. A commander will also always make maximum use of stay-behind forces and affiliated forces existing within his AOR. A disruption force has no set order of battle. It may contain—

- Ambush teams (ground and air defense).
- Long-range reconnaissance patrols and/or special-purpose forces (SPF) teams.
- Reconnaissance, intelligence, surveillance, and target acquisition (RISTA) assets and forces.
- Counterreconnaissance detachments.
- Artillery systems.
- Target designation teams.
- Elements of affiliated forces (such as insurgents, guerrillas, or criminals) or local sympathizers.
- Antilanding reserves.

PLANNING DISRUPTION

6-5. The disruption force headquarters plans disruption. Key planning considerations for disruption are—

- Identifying components or subsystems of the enemy's combat system that are priority for attack.
- Identifying priority intelligence tasks to be accomplished by the disruption force.
- Determining the disruption force role in the overall information warfare plan.
- Determining critical OPFOR elements that must be protected from enemy reconnaissance efforts.

EXECUTING DISRUPTION

6-6. The disruption force fixes enemy forces and places long-range fire on key enemy units. It also strips away the enemy's reconnaissance assets while denying him the ability to acquire and engage OPFOR targets with deep fires. This includes an air defense effort to deny aerial attack and reconnaissance platforms from targeting OPFOR elements. The disruption force seeks to conduct highly damaging local attacks.

6-7. Typical systems, units, or facilities to be attacked by the disruption force are—

- Command and control (C2) systems.
- RISTA assets.
- Attack helicopter forward arming and refueling points.
- Airfields.
- Precision fire systems.
- Logistics support areas.
- Lines of communication.
- Mobility and countermobility assets.
- Casualty evacuation routes and means.

COUNTERRECONNAISSANCE

6-8. The OPFOR defines counterreconnaissance (CR) as a continuous combined arms action to locate, track, and destroy all enemy reconnaissance operating in a given AOR. The OPFOR conducts CR at all times and during all types of operations. The OPFOR understands the role of situational awareness in battle and will spare no effort or resource to hunt down and eliminate enemy reconnaissance troops and systems.

Other Combined Arms Actions

ORGANIZING THE BATTLEFIELD FOR COUNTERRECONNAISSANCE

6-9. Control measures key to CR action are those that assist in locating, tracking, and destroying enemy reconnaissance elements. These involve counterreconnaissance zones, reference zones, predicted enemy locations, and kill zones.

Counterreconnaissance Zones

6-10. The AOR will be divided into one or more counterreconnaissance zones (CRZs). There are many ways to do this depending upon the situation. For example, a division tactical group (DTG) could execute CR at its level and make the entire AOR the only CRZ. It could instead give the disruption force the CR mission in the disruption zone and each brigade or brigade tactical group (BTG) responsibility for its own CRZ. A CRZ is the AOR for one counterreconnaissance detachment (CRD, see below). See figure 6-2 on page 6-6 for an example in which a BTG's CRZ equates to the BGT's AOR.

Reference Zones

6-11. Reference zones (RZs) are subdivisions of the CRZ that assist in rapid orientation on the ground and direction of killing forces or systems to enemy reconnaissance elements. RZs may take the form of a grid pattern with individual grids given code names, letters or numbers. RZs may also include target reference points whether for orientation purposes only or also as artillery targets.

Predicted Enemy Locations

6-12. Predicted enemy locations (PELs) are those areas in the AOR where enemy activity, troops, or systems are anticipated. Although not limited to CR, PELs identify specific locations where enemy reconnaissance is expected. PELs are determined by using a wide range of reconnaissance, logic, intelligence, and analytical tools. Whenever possible, all PEL are corroborated by several sources of data.

Kill Zones

6-13. Kill zones are discussed in chapter 2. However, the counterreconnaissance detachment (CRD) commander will often identify his own set of kill zones associated with where he intends to kill enemy reconnaissance on the ground.

ORGANIZING FORCES FOR COUNTERRECONNAISSANCE

6-14. Counterreconnaissance is a combined arms task. Commanders will select the units best suited to locate, track, and kill enemy reconnaissance given the nature of the overall mission and the AOR.

Counterreconnaissance Detachment

6-15. A CRD is a detachment (see chapter 2) task-organized to be able to locate, track, and destroy enemy reconnaissance throughout its CRZ. Each CRZ is the responsibility of one CRD.

Command and Support Relationships

6-16. As a detachment, the CRD is primarily composed of constituent and dedicated units. However, the supporting command and support relationship may be necessary to bring specialized capabilities to bear for limited periods of time. For example, the CRD might receive a precision-capable artillery unit in support in order to destroy enemy mounted reconnaissance targets. In another case, the CRD might receive a night-capable helicopter unit for use during a period of limited visibility.

CRD and the Security Force

6-17. If created, the security force is charged with the force protection of the unit from all threats. A CRD is specifically designed to locate and destroy enemy reconnaissance and intelligence collecting elements. When both exist, the commander has two basic options regarding their relationship. One is that the CRD

Chapter 6

may be a component of the security force. Alternately, the commander may give the CRD additional security responsibilities and resources and charge the CRD with performing the security mission in its AOR, leaving the security force to execute missions in other parts of the higher unit's AOR.

CRD Components

6-18. The CRD is a task organization created specifically for the CR mission. It is a combined arms organization, with various combinations of the following components. See figure 6-1 for an example of a CRD organization with some of these components.

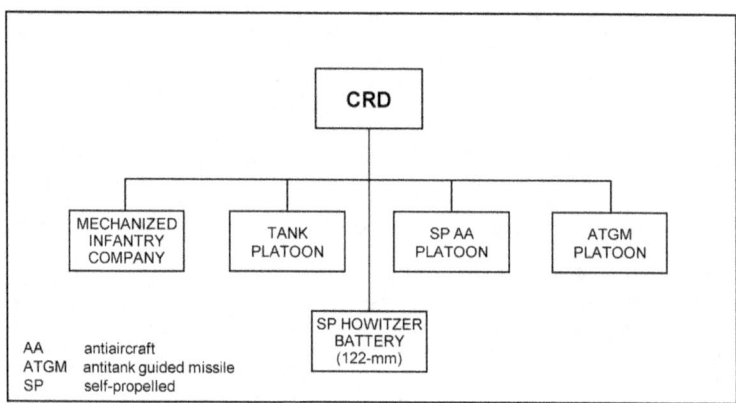

Figure 6-1. Counterreconnaissance detachment (example)

Reconnaissance

6-19. Perhaps the most essential component of a CRD is its reconnaissance elements. If the CRD cannot locate and track enemy reconnaissance elements, it cannot perform its mission. CRD reconnaissance elements take many forms: long-range reconnaissance units, mounted and dismounted (combat) reconnaissance units, signals reconnaissance, aerial reconnaissance, or SPF.

Air Defense

6-20. Air defense assets in the CRD might be used defensively to protect elements of the CRD. They might also be used offensively to destroy enemy aerial reconnaissance systems—unmanned aerial vehicles (UAVs), reconnaissance aircraft, or reconnaissance helicopters.

Aviation

6-21. Aviation assets can play a number of roles in a CRD. They transport infantry rapidly to already located enemy reconnaissance targets. They perform armed and unarmed reconnaissance to locate enemy reconnaissance. Since the CRZ is generally a large area for the forces in the corresponding CRD to cover, aviation assets may be used to resupply dispersed elements of the CRD.

Artillery

6-22. Artillery and other indirect fire systems provide an excellent means of killing enemy reconnaissance without involving direct fire engagements. The challenge is to employ artillery against enemy targets within the battle and support zones without endangering other OPFOR forces. Precision systems are

Other Combined Arms Actions

uniquely suited to the CR task, and the OPFOR considers the expenditure of limited precision resources against enemy reconnaissance targets to be well worth it.

Infantry

6-23. Enemy reconnaissance units often seek concealment in complex terrain. Infantry units in the CRD attack and destroy such targets. Normally, other elements of the CRD locate these targets, but infantry may also be called upon to conduct reconnaissance missions in complex terrain in support of the CR effort.

Engineers

6-24. Typically a CRD does not have significant engineer resources. When present, they execute standard engineer tasks when and where necessary. Combat engineer units also accompany infantry in actions on complex terrain.

Signal

6-25. CRDs often operate over relatively large geographic areas. The CRD will be organized with appropriate signal assets to allow it to transmit and manipulate information securely over large distances.

Electronic Warfare

6-26. CRDs may contain electronic warfare assets. Such assets permit them to block enemy reconnaissance elements from communicating their observations to their higher headquarters and other enemy units.

Armor

6-27. If the enemy has a strong mounted reconnaissance capability, or when terrain conditions are favorable, the CRD may contain armor elements. Armored units can move rapidly to engage and destroy located reconnaissance targets.

PLANNING COUNTERRECONNAISSANCE

6-28. The CR plan is written by the staff of the unit forming the basis of the CRD with guidance from the higher command's staff. CR is treated as an ongoing offensive action no matter what type action is being undertaken by the higher unit. The CR plan is the battle plan of the CRD.

EXECUTING COUNTERRECONNAISSANCE

6-29. The CRD headquarters interacts with the reconnaissance section of the DTG or BTG staff to maintain a clear picture of enemy locations with an emphasis on his reconnaissance systems. As enemy reconnaissance assets are identified, they are tracked by the CRD headquarters, and this information is provided to the CRD elements given the mission to destroy those assets. If other OPFOR combat units in the AOR are closer and/or better suited to find or destroy critical enemy reconnaissance elements, the CRD commander will recommend this action to the higher commander and, if approved, coordinate this effort. See figure 6-2 for an example of CR execution within a BTG's AOR.

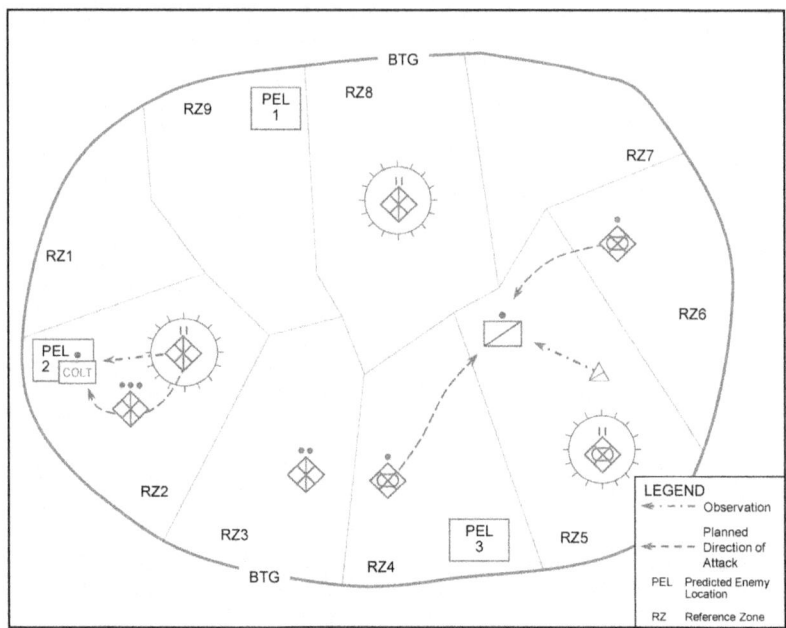

Figure 6-2. Execution of counterreconnaissance (example)

6-30. The CRD commander determines which enemy reconnaissance assets are to be destroyed in accordance with the higher commander's guidance. He assigns missions to attack and destroy those assets to appropriate subordinate elements. The CRD headquarters guides the action element in on the enemy reconnaissance asset until contact is made.

ANTILANDING ACTIONS

6-31. The OPFOR prefers to prevent landings by enemy airborne or heliborne troops through the destruction of the troop transport aircraft in flight. Failing that, it will take significant actions to destroy landing forces on the ground as soon after landing as possible. Antilanding actions can and will be executed by any force or element with the capability to affect the aircraft or the landing forces, but an antilanding action is a combined arms action that primarily falls to the antilanding reserve (ALR) for execution.

ORGANIZING THE BATTLEFIELD FOR ANTILANDING ACTIONS

6-32. Antilanding forces or elements are given their own attack zone to control their actions against landing forces. Such an attack zone may only be activated for the duration of an antilanding action or may be assigned to the ALR permanently. Kill zones are used to control both ground and air defense engagements. Anticipated enemy landing or drop zones (LZs or DZs) are included in the listing of PELs.

ORGANIZING FORCES FOR ANTILANDING ACTIONS

6-33. Commanders form one or more antilanding reserves to conduct antilanding actions during or after an enemy landing operation. ALRs can consist of any units the commander and staff's forces analysis

determines necessary to destroy an enemy airborne or heliborne landing. Typical ALRs may include as subelements—
- Gun and missile air defense units.
- Infantry with antitank weapons.
- Armor.
- Smoke units.
- Engineers.
- Aviation.
- Artillery.

6-34. ALRs are typically detachments. In that case, the detachment commander can organize his force into—
- Disruption element(s) to disrupt the enemy and prevent detection of action element(s).
- Security element(s) to maneuver and fire to ensure the decisive point is isolated and that additional enemy forces do not join the battle unexpectedly.
- Support element(s) to conduct actions to set conditions for action elements' success.
- Action element(s) to destroy the enemy landing force.

However, an ALR for an anticipated major enemy landing operation may be a BTG or even a DTG, should the situation warrant. In that case, the ALR would consist of functional forces rather than elements.

PLANNING ANTILANDING ACTIONS

6-35. The ALR plans actions to attack enemy transport aircraft en route to and in the vicinity of the LZ or DZ. This may require the assistance of other air defense units not in the ALR. The force protection subsection of the BTG or DTG staff performs this coordination.

6-36. The ALR plans and rehearses actions in the vicinity of the LZs or DZs. It also plans and rehearses movement between assembly areas, hide positions, and attack positions and between LZs or DZs.

EXECUTING ANTILANDING ACTIONS

6-37. Early warning is transmitted from the DTG or BTG main command post to the ALR. The ALR moves to positions in the attack zone from which it can engage transport aircraft and destroy landing forces on the ground.

URBAN COMBAT

6-38. The OPFOR sees urban combat as a vital subcomponent of its tactical actions. Complex urban terrain provides significant advantages to the side that is ready to make use of them. OPFOR units train extensively in urban combat and expect to make maximum use of complex urban terrain and to act to deny such use to the enemy.

6-39. Fighting in towns and cities slows the rate of advance, requiring a high consumption of manpower and materiel. In the offense, the OPFOR may prefer to avoid combat in cities, either by bypassing defended localities or by seizing towns from the march before the enemy can erect defenses. When there is no alternative, units reorganize their combat formations to attack a city by assault. The attackers can exploit undefended towns by using them as avenues of approach or assembly areas.

ORGANIZING THE BATTLEFIELD FOR URBAN COMBAT

6-40. AORs in urban combat use the same types of zones as in other actions. However, the OPFOR will place great emphasis in defining and using the third dimension that urban areas create. See figure 6-3 for an example of a multidimensional battlefield.

6-41. All zones will be defined both in terms of horizontal dimensions as well as the vertical. It may well be for example, that the upper floors of a building are a kill zone while the lower floors are still in the

Chapter 6

battle or disruption zone and contain friendly forces. Urban detachments (UDs) are often given an attack zone in which to operate. Support zones are often located in sanctuary areas inside battle positions.

6-42. The OPFOR prefers to attack multidimensionally—from basements or sewers to upper stories or on the tops of buildings. The targets are engaged simultaneously to maximize effectiveness and confusion.

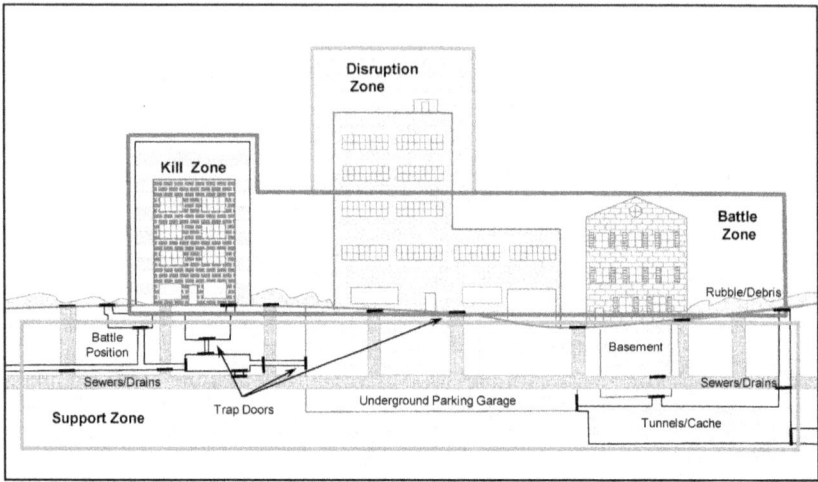

Figure 6-3. Multidimensional battlefield

ORGANIZING FORCES FOR URBAN COMBAT

6-43. A brigade or BTG operating in an urban area would typically organize one or more urban detachments (UDs). The UD is the primary organization used by the OPFOR for urban combat.

Urban Detachments

6-44. A UD is not a specific, standing organization. It is a task-organized battalion or company given the mission to attack and seize selected portions of an urban area. The composition of UDs is mission dependent. The UD attacking along one street may be similar to the detachment on the next street, or it may differ significantly in the number and types of augmenting units. The UD is dissolved when the mission is complete.

6-45. Urban combat consumes ammunition at a much higher rate than some other types of combat. Therefore, detachment personnel are issued increased quantities of certain weapons, especially grenades and flame weapons. Ammunition, specialized equipment, such as grapples, ropes, and ladders, is provided or acquired locally.

Functional Elements

6-46. UDs typically consist of the following functional elements:
- Security element.
- Clearing element.
- Action element.
- Support element.

Other Combined Arms Actions

6-47. The *security element* of a UD provides local tactical security for the detachment and prevents the enemy from influencing mission accomplishment. The *clearing element* ensures the *action element* has an avenue of approach that is clear of obstacles, debris, and rubble that would disrupt its movement. The *action element* moves from a covered and concealed position and accomplishes the UD's tactical task. The *support element* provides combat and combat service support and C2 for the detachment.

Augmentation

6-48. UDs are task-organized for specific missions and conditions. Therefore, there is no fixed template for a UD organization. A typical UD may be augmented with the following:

Tanks

6-49. Tank units are often employed in a decentralized manner, with one or two tanks or a tank platoon allocated to an infantry platoon, or individual tanks allocated to infantry squads. Depending on the circumstances, the tanks may be retained as a unit to serve as a security or support element.

Antitank and Antiarmor

6-50. The UD may receive additional (constituent or dedicated) antitank and/or antiarmor assets. The UD may also be organized into infantry antiarmor hunter-killer teams.

Artillery

6-51. The UD may gain constituent indirect fire support (possibly in the form of 120-mm mortars or 122-mm howitzers). It may also be provided dedicated artillery support.

Air Defense

6-52. The typical purpose of air defense support to urban combat is to prevent enemy air power from influencing the action of the action element. Air defense systems in the security element provide early warning and defeat enemy aerial response to the mission. Such systems also target enemy aerial reconnaissance such as UAVs to prevent the enemy from having a clear picture of the OPFOR action.

6-53. Air defense systems in the support element provide overwatch of the action element and the objective. Some air defense systems may prove useful in close combat in urban areas. Air defense guns usually have a very high angle of fire, allowing them to target the upper stories of buildings. Their high-explosive rounds allow the weapons to shoot through the bottom floor of the top story, successfully engaging enemy troops and/or equipment located on rooftops. The accuracy and lethality of air defense weapons also facilitates their role as a devastating ground weapon when used against personnel, equipment, buildings, and lightly armored vehicles.

Engineers

6-54. Combat in urban terrain always faces the possibility of obstacles restricting movement to the objective. Obstacles in urban terrain—manmade or not—are virtually a certainty. Typically then, UDs include a specialist element made up of sappers and other supporting arms, known as a clearing element, designed to execute mobility and/or shaping, tasks in support of the action element.

6-55. Depending on the mission the engineers may be equipped with obstacle-crossing equipment. They may also have additional flame weapons and CBRN specialists.

PLANNING URBAN COMBAT

6-56. The OPFOR sees certain aspects of urban combat as critical to success and addresses them in the plan for every action on complex urban terrain. The populace of a given urban area may be a key consideration: the side that manages it best has a distinct advantage. This is especially true if large segments of the populace remain in place in the urban area. The OPFOR can use the population to provide camouflage, concealment, cover, and deception (C3D) for its operations, enhancing its mobility in proximity to enemy positions. The OPFOR can take advantage of enemy moral responsibilities and attempt to make the civil population a burden on enemy forces' logistics and force protection resources. It may herd refugees into enemy-controlled sectors, steal from local nationals, and hide among civilians during enemy offensive operations.

6-57. The civil population may also serve as a key intelligence source for the OPFOR. Local hires serving among enemy soldiers, civilians with access to enemy-controlled areas, and refugees moving through enemy-controlled sectors can all be manipulated by the OPFOR to provide information on enemy dispositions, readiness, and intent. Also, OPFOR SPF and reconnaissance assets may infiltrate and move among civilian groups.

6-58. OPFOR planning for urban combat will also include INFOWAR. This may be as important as directly opposing enemy action or perhaps more so. Portable video cameras, Internet access, commercial radios, and cellular phones are all tools that permit the OPFOR to tell its story. This can influence the local population and/or affect the national wills of countries other than the State. The OPFOR may stage and broadcast enemy "atrocities." It may use electronic mail to influence sympathetic groups or undermine enemy resolve. OPFOR-sponsored hackers may gain access to enemy web sites to manipulate information to the OPFOR's advantage.

6-59. The OPFOR plan for urban combat will always address the need for continuous combat. The plan includes a methodology for cycling soldiers out of positions in contact to reduce the effects of combat stress.

EXECUTING URBAN COMBAT

6-60. The OPFOR will identify and quickly seize control of critical components of the urban environment to help shape the battlefield to its own ends. Phone exchanges provide simple and reliable communications that can be easily secured with off-the-shelf technologies. Sewage treatment plants and flood control machinery can be used to implement weapons of mass destruction strategies or to make sections of the urban area uninhabitable. Gaining control of media stations can significantly improve the OPFOR's information warfare capabilities. Power generation and transmission sites provide means to control significant aspects of civilian society over a large area.

6-61. The OPFOR will think of an urban environment in terms of three dimensions. Upper floors and roofs provide the OPFOR excellent observation points and battle positions above many weapons' maximum elevations. Shots from upper floors can strike armored vehicles in vulnerable points. Basements also provide firing points below many weapons' minimum depressions and allow strike at weaker armor. Sewers and subways provide covered and concealed access throughout the AOR. Conventional lateral unit boundaries will often not apply where the OPFOR controls some stories of a building while enemy forces control others. See figure 6-4 for an example of a UD attacking an enemy-controlled building complex.

Other Combined Arms Actions

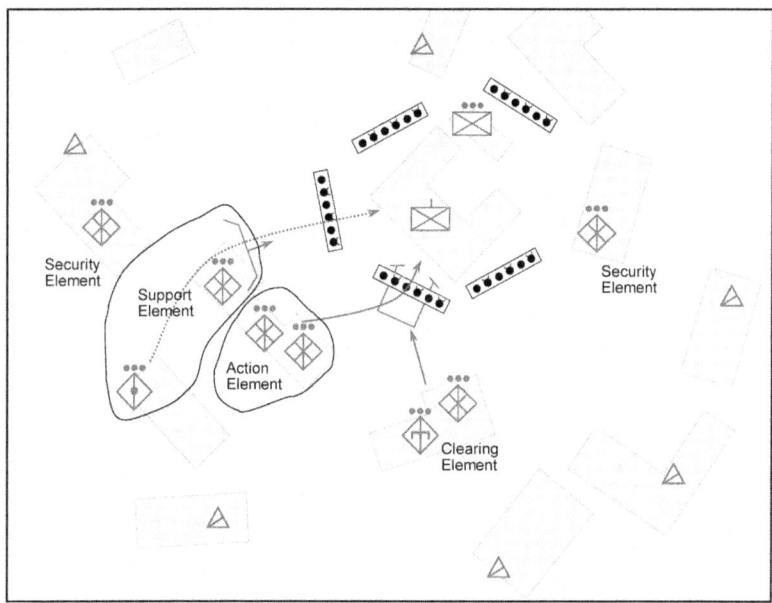

Figure 6-4. Urban detachment attacking enemy-controlled building complex (example)

6-62. Whether they are purpose-built or adapted, many weapons will have greater than normal utility in an urban environment, while others will have significant disadvantages. The following are examples of weapons favored by the OPFOR in urban combat:

- Weapons with no minimum depression or maximum elevation.
- Grenade launchers (automatic and rifle-mounted).
- Antitank grenade launchers and shoulder-fired antitank guided missiles. Some warheads can be removed and dropped from the tops of buildings and overpasses.
- Shoulder-fired "bunker-busters."
- Weapons with reduced backblast (such as gas metered or soft launch).
- Mortars.
- Sniper rifles.
- Machineguns.
- Grenades (including antitank, smoke, and incendiary grenades).
- Flame and incendiary weapons.
- Riot control and tranquilizer gases.
- Mines and booby traps.
- Minefield breaching explosive line charge systems (especially effective when used in narrow streets and/or alleys with high buildings on both sides).
- Artillery pieces used in direct fire mode.

6-63. The OPFOR will "hug" high-technology conventional enemy forces in an urban environment to avoid the effects of high-firepower standoff weapon systems. Additionally, it will attempt to keep all or significant portions of enemy forces engaged in continuous operations to increase their susceptibility to stress-induced illnesses. Urban combat, by its nature, produces an inordinate amount of combat-stress casualties, and continuous operations exacerbate this problem. The OPFOR will maintain a large reserve to minimize the impact of this on its own forces.

6-64. The OPFOR will prey on soldiers untrained in basic infantry skills. Ambushes will focus on such soldiers conducting resupply operations or moving in poorly guarded convoys. Urban combat is characterized by the isolation of small groups and navigational challenges. The OPFOR can use the separation this creates to inflict maximum casualties even when there is no other direct military benefit from the action.

Chapter 7
Information Warfare

The OPFOR is constantly increasing the levels of technology used in its communications, automation, reconnaissance, and target acquisition systems. In order to ensure the successful use of information technologies and to deny the enemy the advantage afforded by such systems, the OPFOR has continued to refine its doctrine and capabilities for information warfare (INFOWAR). The OPFOR knows it cannot maintain continuous information dominance, particularly against peer or more powerful opponents. Therefore, it selects for disruption only those targets most critical to ensuring the successful achievement of its objectives. It attempts to gain an information advantage only at critical times and places on the battlefield. This chapter focuses on INFOWAR activities at the tactical level.

TACTICAL-LEVEL INFOWAR

7-1. The OPFOR defines *information warfare* as specifically planned and integrated actions taken to achieve an information advantage at critical points and times. The primary goals of INFOWAR are to—

- Influence an enemy's decisionmaking through his collected and available information, information systems, and information-based processes.
- Retain the ability to employ friendly information and information-based processes and systems.

7-2. Information and its management, dissemination, and control have always been critical to the successful conduct of tactical missions. Given today's advancements in information and information systems technology, this importance is growing in scope, impact, and sophistication. The OPFOR recognizes the unique opportunities that INFOWAR gives tactical commanders, and it continuously strives to incorporate INFOWAR activities in all tactical missions and battles.

7-3. INFOWAR may help degrade or deny effective enemy communications and blur or manipulate the battlefield picture. In addition, INFOWAR helps the OPFOR achieve the goal of dominating the tempo of combat. Using a combination of perception management activities, deception techniques, and electronic warfare (EW), the OPFOR can effectively slow or control the pace of battle. For example, the OPFOR may select to destroy lucrative enemy targets. It may also orchestrate and execute a perception management activity that weakens the enemy's international and domestic support, causing hesitation or actual failure of the operation. It executes deception plans to confuse the enemy and conceal true OPFOR intentions. More traditional EW activities also contribute to the successful application of INFOWAR at the tactical level by challenging the enemy's quest for information dominance.

7-4. INFOWAR also supports the critical mission of counterreconnaissance at the tactical level. The OPFOR constantly seeks ways to attack, degrade, or manipulate the enemy's reconnaissance, intelligence, surveillance, and target acquisition (RISTA) capabilities. All enemy target acquisition systems and sensors are potential targets.

Chapter 7

ASSOCIATED TACTICAL TASKS

7-5. The effects of INFOWAR can be multidimensional and at times hard to pinpoint. However, the OPFOR highlights the following tasks and associated effects as critical to the application of INFOWAR at the tactical level:

- **Destroy.** Destruction tasks physically render an enemy's information systems ineffective. Destruction is most effective when timed to occur before the enemy executes a command and control (C2) function or when focused on a resource-intensive target that is hard to reconstitute. Neutralizing or destroying the opponent's information capability can be brought about by physical destruction of critical communications nodes and links.
- **Degrade.** Degradation attempts to reduce the effectiveness of the enemy's information infrastructure, information systems, and information collection means.
- **Disrupt.** Disruption activities focus on the disrupting enemy observation and sensor capabilities at critical times and locations. Disruption impedes the enemy's ability to observe and collect information and obtain or maintain information dominance.
- **Deny.** Denial activities attempt to limit the enemy's ability to collect or disseminate information on the OPFOR or deny his collection efforts.
- **Deceive.** Deception activities strive to mislead the enemy's decisionmakers and manipulate his overall understanding of OPFOR activities. Deception manipulates perception and causes disorientation among decisionmakers within their decision cycle.
- **Exploit.** Exploitation activities attempt to use the enemy's C2 or RISTA capabilities to the advantage of the OPFOR. The OPFOR also uses its various INFOWAR capabilities to exploit any enemy vulnerability.
- **Influence.** Influencing information affects an enemy's beliefs, motives, perspectives, and reasoning capabilities, in order to support OPFOR objectives. This may be done through misinformation or by manipulating or "spinning" information.

SYSTEMS WARFARE

7-6. In the *systems warfare* approach to combat (see chapter 1), the OPFOR often focuses on attacking the C2, RISTA, and/or logistics elements that are critical components of the enemy's combat system. It is often more feasible to attack such targets, rather than directly engaging the enemy's combat or combat support forces. Tactical-level INFOWAR can be a primary means of attacking these assets, either on its own or in conjunction with other components of the OPFOR's own combat system.

WINDOWS OF OPPORTUNITY

7-7. To conduct successful action against a more powerful force enjoying a technological overmatch, the OPFOR must exploit windows of opportunity. Sometimes these windows occur naturally, as a result of favorable conditions in the operational environment. Most often, however, the OPFOR will have to create its own opportunities. INFOWAR can help create the necessary windows of opportunity for any type of offensive or defensive action by executing effective deception techniques, EW, and physical destruction.

7-8. When the OPFOR must create a window of opportunity, INFOWAR activities can contribute to this by—

- Destroying or disrupting enemy C2 and RISTA assets.
- Deceiving enemy imagery and signals sensors.
- Selectively denying situational awareness.
- Slowing the tempo of enemy operations.
- Isolating key elements of the enemy force.

ELEMENTS OF INFOWAR

7-9. Integrated within INFOWAR doctrine are the following seven elements:
- Electronic warfare (EW).
- Deception.
- Physical destruction.
- Protection and security measures.
- Perception management.
- Information attack (IA).
- Computer warfare.

7-10. The seven elements of INFOWAR do not exist in isolation from one another and are not mutually exclusive. The overlapping of functions, means, and targets requires that they all be integrated into a single, integrated INFOWAR plan. However, effective execution of INFOWAR does not necessary involve the use of all elements concurrently. In some cases, one element may be all that is required to successfully execute a tactical INFOWAR action. Nevertheless, using one element or subelement, such as camouflage, does not by itself necessarily constitute an application of INFOWAR.

7-11. The use of each element or a combination of elements is determined by the tactical situation and support to the overall operational objective. The size and sophistication of an enemy force also determines the extent to which the OPFOR employs the various elements of INFOWAR. The commander has the freedom to mix and match elements to best suit his tactical needs, within the bounds of guidance from higher authority.

7-12. Tools for waging INFOWAR can include, but are not limited to—
- Conventional physical and electronic destruction means.
- Malicious software.
- Denial-of-service attacks.
- The Internet.
- The media.
- International public opinion.
- Communication networks.
- Various types of reconnaissance, espionage, and eavesdropping technologies.

The OPFOR can employ INFOWAR tools from both civilian and military sources and from assets of third-party actors.

7-13. The OPFOR sees the targets of INFOWAR as an opponent's—
- Decisionmakers.
- Weapons and hardware.
- Critical information infrastructure.
- C2 system.
- Information and telecommunications systems.
- C2 centers and nodes.

Information links, such as transmitters, communication devices, and protocols, will be targeted. The OPFOR is extremely adaptive and will employ the best option available to degrade, manipulate, influence, use, or destroy an information link. See table 7-1 on page 7-4 for typical examples of INFOWAR objectives and targets.

Table 7-1. INFOWAR elements, objectives, and targets

INFOWAR Element	Objectives	Targets
Electronic Warfare	Exploit, disrupt, deny, and degrade the enemy's use of the electromagnetic spectrum.	C2 and RISTA assets and networks.
Deception	Mislead enemy decisionmakers. Cause confusion and delays in the decisionmaking process. Persuade the local population and/or international community to support OPFOR objectives.	Key military decisionmakers. General population and international media sources and Internet sites.
Physical Destruction	Destroy the enemy's information infrastructures.	C2 nodes and links, RISTA assets, telecommunications, and power sources.
Protection and Security Measures	Protect critical assets.	Enemy RISTA assets.
Perception Management	Distort reality or manipulate information to support OPFOR goals.	Enemy RISTA assets. Local populace and leaders. Media sources (international and domestic).
Information Attack	Alter or deny key information.	Decisionmakers and other users of information. Systems reliant on accurate information.
Computer Warfare	Disrupt, deny, or degrade the enemy's computer networks and information flow.	C2 and RISTA assets and networks.

ELECTRONIC WARFARE

7-14. *Electronic warfare* is activity conducted to control or deny the enemy's use of the electromagnetic spectrum, while ensuring its use by the OPFOR. EW capabilities allow an actor to exploit, deceive, degrade, disrupt, damage, or destroy sensors, processors, and C2 nodes. At a minimum, the goal of EW is to control the use of the electromagnetic spectrum at critical locations and times or to attack a specific system. The OPFOR realizes that it cannot completely deny the enemy's use of the spectrum. Thus, the goal of OPFOR EW is to control (limit or disrupt) his use or selectively deny it at specific locations and times, at the OPFOR's choosing. In this way, the OPFOR intends to challenge the enemy's goal of information dominance.

7-15. The OPFOR employs both nonlethal and lethal means for EW. *Nonlethal* means range from signals reconnaissance and electronic jamming to the deployment of corner reflectors, protective countermeasures, and deception jammers. The OPFOR can employ low-cost GPS jammers to disrupt enemy precision munitions targeting, sensor-to-shooter links, and navigation. *Lethal* EW activities include the physical destruction of high-priority targets supporting the enemy's decisionmaking process—such as reconnaissance sensors, command posts (CPs), and communications systems. They also include activities such as lethal air defense suppression measures. If available, precision munitions can degrade or eliminate high-technology C2 assets and associated links.

7-16. EW activities often focus on the enemy's advanced C2 systems developed to provide real-time force synchronization and shared situational awareness. The enemy relies on the availability of friendly and enemy force composition and locations, digital mapping displays, and automated targeting data. By targeting vulnerable communications links, the OPFOR can disrupt the enemy's ability to digitally transfer and share such information. The OPFOR enhances its own survivability through disrupting the enemy's ability to mass fires with dispersed forces, while increasing enemy crew and staff workloads and disrupting his fratricide-prevention measures.

7-17. EW is a perfect example of the integrated nature of OPFOR INFOWAR elements. It overlaps significantly with protection and security measures, deception, and physical destruction. Reconnaissance, aviation, air defense, artillery, and engineer support may all contribute to successful EW for INFOWAR purposes.

Signals Reconnaissance

7-18. *Signals reconnaissance* is action taken to detect, identify, locate, and track high-value targets (HVTs) through the use of the electromagnetic spectrum. It includes both intercept and direction finding, which may enable a near-real-time attack on the target. OPFOR commanders determine the priorities for signals reconnaissance by determining which HVTs must be found in order to have the best chance for success of their plan. If the collected intelligence value is of higher significance than the destruction of the target, the commander determines the best tactical course of action. He may decide either to destroy the target, to jam it, or to continue to exploit the collected information.

7-19. Signals reconnaissance targets must be detectable in some manner in the electromagnetic spectrum. The OPFOR must have some system(s) available that can perform this detection. HVTs that do not generate an electromagnetic signature of some sort must be detected by some means other than signals reconnaissance.

7-20. HVTs sought by signals reconnaissance efforts are specific to the battle, the OPFOR plan and capabilities, and the enemy's plan and capabilities. However, there are some typical targets of signals reconnaissance efforts:

- Maneuver unit CPs.
- Forward air controllers (FACs).
- Logistic CPs.
- Fire support and tactical aviation networks.
- Target acquisition systems.
- Reconnaissance and sensors networks.
- Battlefield surveillance radars.

7-21. Signals reconnaissance information gained from electronic means is fused with information obtained from other sources. For example, the OPFOR can use trained reconnaissance teams or elements to—

- "Put eyes on" targets and objectives.
- Collect required information.
- Provide early warning.
- Monitor lines of communication and movement corridors in a target area.

Such reconnaissance could possibly include a signals reconnaissance capability.

Chapter 7

Note. Successful EW operations are not reliant on high-technology equipment and huge amounts of resourcing. While state actors tend to have higher EW capabilities, nonstate actors (or affiliated forces) can present a challenge to their opponents. For example, Hezbollah effectively used what the OPFOR would call signals reconnaissance and protection and security measures against Israeli forces in the summer of 2006. It successfully monitored Israeli cellular phone communications and was able to evade Israeli jamming devices by using fiber-optic lines instead of wireless signals. Commercial off-the shelf (COTS) equipment is commonly used to provide actors with the means to conduct EW. For example, the Viet Cong (Vietnamese National Liberation Front) used readily available COTS equipment to conduct extremely successful tactical signals reconnaissance operations against U.S. forces. Another nonstate actor, the Fuerzas Armadas Revolucionarias de Colombia (FARC), has utilized ground assets (man-portable radio equipment and other COTS technology) to conduct signals reconnaissance against Colombian and U.S. forces. As these examples prove, an effective EW threat can come from actors with high- and/or low-technology assets. Thus, a sophisticated military, in the Western sense, is not required for a successful signals reconnaissance exploitation and subsequent electronic attack.

Electronic Attack

7-22. Electronic attack (EA) supports the disaggregation of enemy forces. The primary form of EA is jamming—interference with an enemy signals link in order to prevent its proper use. Jamming priorities are similar to those for signals reconnaissance. Maneuver units are jammed in order to disrupt coordination between and within units, especially when enemy units are achieving varying degrees of success. Reporting links between reconnaissance and engineer elements and the supported maneuver units are attacked, since they attempt to exploit OPFOR weaknesses the enemy may have found.

Targets

7-23. The OPFOR can and will conduct EA on virtually any system connected by signals transmitted in the electromagnetic spectrum. This includes communications and non-communications signals and data. As with signals reconnaissance, the choices of which links to disrupt varies with the scheme of maneuver, the impact of the disruption, the enemy's sophistication, and the availability of OPFOR EA assets. A limited but representative list of example targets includes—

- C2 links between a key unit and its higher command.
- Link between a GPS satellite and a receiver.
- Link between a firing system and its fire direction center (FDC).
- Link between a missile and/or munition and its targeting system.
- Computer data links of all types.

Distributive Jamming

7-24. Instead of wideband barrage jamming using large semi-fixed jammers, the OPFOR often fields small distributive jammers. These may be either dispersed throughout the battle area or focused on one or more select targets. These jammers may be both fixed and mobile. Mobility may be by ground vehicle or unmanned aerial vehicle. They can be controlled though civilian cellular phone networks and/or controlled by local forces. Along with known military frequencies, the OPFOR can target civilian radios and/or cellular phones—

- Of a regional neighbor.
- Of nongovernmental organizations (NGOs).
- Of other civilians from outside the region).

7-25. Distributive jamming can cause—
- Loss of GPS, communications, and non-communications data links (such as Blue Force and personal or unit communication).
- Degradation of situational awareness and common operational picture.
- Disruption of tempo.
- Reduction of intelligence feeds to and from CPs.
- Opportunities for ambush, with the resulting ambush videoed and used for perception management operations.
- Enemy units forced to use alternative, less secure communications.

Expendable Jammers

7-26. The OPFOR can take advantage of the time prior to an enemy attack to emplace expendable jammers (EXJAMs). These jammers can disrupt enemy communications nets. When used in conjunction with terrain (such as at natural choke points, mountain passes, or valleys), they can achieve significant results despite their short range and low power. The OPFOR can also use them to support a deception plan, without risking expensive vehicle-based systems. While limited in number, artillery-delivered EXJAMs may be employed. These jammers are especially useful in those areas where support is not available from more powerful vehicle-mounted jammers.

Proximity Fuze Jammers

7-27. Proximity fuzes used on some artillery projectiles rely on return of a radio signal reflected from the target to detonate the round within lethal range of the target. Proximity fuze jammers cause the round to explode at a safe distance. The OPFOR can deploy such jammers to protect high-value assets that are within indirect fire range from enemy artillery.

DECEPTION

7-28. The OPFOR integrates deception into every tactical action. It does not plan deception measures and activities in an ad hoc manner. A deception plan is always a major portion of the overall INFOWAR plan. The extent and complexity of the deception depends on the amount of time available for planning and preparation. The OPFOR formulates its plan of action, overall INFOWAR plan, and deception plan concurrently.

7-29. The OPFOR attempts to deceive the enemy concerning the exact strength and composition of its forces, their deployment and orientation, and their intended manner of employment. When successfully conducted, deception activities ensure that the OPFOR achieves tactical surprise, while enhancing force survivability. All deception measures and activities are continuously coordinated with deception plans and operations at higher levels. Affiliated forces may assist in executing deception activities.

7-30. The OPFOR employs all forms of deception, ranging from physical decoys and electronic devices to tactical activities and behaviors. The key to all types of deception activities is that they must be both realistic and fit the deception story. Due to the sophistication and variety of sensors available to the enemy, successfully deceiving him requires a multispectral effort. The OPFOR must provide false or misleading thermal, visual, acoustic, and electronic signatures.

7-31. While creating the picture of the battlefield the OPFOR wants the enemy to perceive, deception planners have two primary objectives. The first is to cause the enemy to commit his forces and act in a manner that favors the OPFOR's plan. The second objective, and the focus of deception activities when time is limited, is to minimize friendly force signatures. This limits detection and destruction by enemy attack.

7-32. Integral to the planning of deception activities is the OPFOR's identification of the deception target. This target is that individual, organization, or group that has the necessary decisionmaking authority to take action (or to neglect to do so) in line with the OPFOR's deception objective. On the tactical battlefield, this target is typically the enemy commander, although the OPFOR recognizes the importance of focusing actions to affect specific staff elements.

7-33. Successful deception activities depend on the identification and exploitation of enemy information systems and networks, as well as other conduits for introducing deceptive information. Knowing how the conduits receive, process, analyze, and distribute information allows for the provision of specific signatures that meet the conduits' requirements. On the tactical battlefield, the enemy reconnaissance system is the primary information conduit and therefore receives the most attention from OPFOR deception planners. The international media and Internet sites may also be a target for deceptive information at the tactical level. The OPFOR can feed them false stories and video that portray tactical-level actions with the goal of influencing operational or even strategic decisions.

Deception Forces and Elements

7-34. The battle plan and/or INFOWAR plan may call for the creation of one or more deception forces or elements. This means that nonexistent or partially existing formations attempt to present the illusion of real or larger units. When the INFOWAR plan requires forces to take some action (such as a feint or demonstration), these forces are designated as deception forces or elements in close-hold executive summaries of the plan. Wide-distribution copies of the plan make reference to these forces or elements according to the functional designation given them in the deception story.

7-35. The deception force or element is typically given its own command structure. The purpose of this is both to replicate the organization(s) necessary to the deception story and to execute the multidiscipline deception required to replicate an actual or larger military organization. The headquarters of a unit that has lost all of its original subordinates to task organization is an excellent candidate for use as a deception force or element.

Deception Activities

7-36. Deception forces or elements may use a series of feints, demonstrations, ruses, or decoys. All activities must fit the overall deception story and provide a consistent, believable, and multidiscipline representation. Basic tactical camouflage, concealment, cover, and deception (C3D) techniques are used to support all types of deception.

7-37. The OPFOR conducts deception activities to confuse the enemy to the extent that he is unable to distinguish between legitimate and false targets, units, activities, and future intentions. Inserting false or misleading information at any point in the enemy decisionmaking process can lead to increased OPFOR survivability and the inability to respond appropriately to OPFOR tactical actions. Manipulation of the electromagnetic spectrum is often critical to successful deception activities, as the OPFOR responds to the challenge posed by advances in enemy C2 systems and sensors.

7-38. Some example deception activities may include—
- Executing feints and demonstrations to provide a false picture of where the main effort will be.
- Creating the false picture of a major offensive effort.
- Maximizing protection and security measures to conceal movement.
- Creating false high-value assets.

Feints

7-39. Feints are offensive in nature and *require engagement with the enemy* in order to show the appearance of an attack. The goal is to support the mission and ultimately mislead the enemy. Feints can be used to force the enemy to—
- Employ his forces improperly. A feint may cause these forces to move away from the main attack toward the feint, or a feint may be used to fix the enemy's follow-on forces.
- Shift his supporting fires from the main effort.
- Reveal his defensive fires. A feint may cause premature firing, which reveals enemy locations.

Demonstrations

7-40. Demonstrations are a show of force on a portion of the battlefield where no decision is sought, for the purpose of deceiving the enemy. They are *similar to feints, but contact with enemy is not required.* Advantages of demonstrations include—
- Absence of contact with enemy.
- Possibility of using simulation devices in place of real items to deceive the enemy's reconnaissance capabilities.
- Use when a full force is not necessary because of lack of contact with the enemy.

Ruses

7-41. Ruses are tricks designed to deceive the enemy in order to obtain a tactical advantage. They are characterized by deliberately exposing false information to enemy collection means. Information attacks, perception management actions, and basic C3D measures all support this type of deception.

Decoys

7-42. Decoys represent physical imitations of OPFOR systems or deception positions to enemy RISTA assets in order to confuse the enemy. The goal is to divert enemy resources into reporting or engaging false targets. It is not necessary to have specially manufactured equipment for this type of visual deception. Decoys are used to attract an enemy's attention for a variety of tactical purposes. Their main use is to draw enemy fire away from high-value assets. Decoys are generally expendable, and they can be—
- Elaborate or simple. Their design depends on several factors, such as the target to be decoyed, a unit's tactical situation, available resources, and the time available.
- Preconstructed or made from field-expedient materials. Except for selected types, preconstructed decoys are not widely available. A typical unit can construct effective, realistic decoys to replicate its key equipment and features through imaginative planning and a working knowledge of the electromagnetic signatures emitted by the unit.

The two most important factors regarding decoy employment are location and realism.

7-43. Logically placing decoys can greatly enhance their plausibility. Decoys are usually placed near enough to the real target to convince an enemy that he has found the target. However, a decoy must be far enough away to prevent collateral damage to the real target when the decoy draws enemy fire. Proper spacing between a decoy and a target depends on the size of the target, the expected enemy target acquisition sensors, and the type of munitions likely to be directed against the target.

7-44. Decoys must include target features that an enemy will recognize. The most effective decoys are those that closely resemble the real target in terms of electromagnetic signatures. Completely replicating the signatures of some targets, particularly large and complex targets, can be very difficult. Therefore, decoy construction should address the electromagnetic spectral region in which the real target is most vulnerable.

7-45. **Smart Decoys.** Smart decoys are designed to present a high-fidelity simulation of a real vehicle or other system. They may present heat, electromagnetic, electro-optical, audio, and/or visual signatures. They are distributed, controlled decoys. Computerized controls turn on decoy signatures to present a much more valid signature than previous-generation "rubber duck" decoys. Smart decoys can be emplaced close to prohibited targets (such as churches, mosques, schools, or hospitals) and civilian populations. If the enemy engages these decoys, the OPFOR can exploit resulting civilian damage in follow-on perception management activities. Smart decoys cause—
- Loss of situational awareness.
- Flood of fake targets, bogging down the enemy's targeting process.
- Expenditure of limited munitions on non-targets.
- Negation of multispectral RISTA assets (such as night vision goggles, infrared scopes, and other electro-optical devices).
- Negation of critical targeting planning and allocation of assets.

Chapter 7

7-46. **Deception CPs.** The INFOWAR plan may also call for employing deception CPs. These are complex, multi-sensor-affecting sites integrated into the overall deception plan. They can assist in achieving battlefield opportunity by forcing the enemy to expend his command and control warfare effort against meaningless positions.

7-47. **False Deployment.** The OPFOR attempts to deny the enemy the ability to accurately identify its force dispositions and intentions. Knowing it cannot totally hide its forces, it tries to blur the boundaries and compositions of forces, while providing indications of deception units and false targets.

7-48. Specific OPFOR actions taken to hide the exact composition and deployment of forces may include—

- Establishing deception assembly areas or defensive positions supported by decoy vehicles.
- Establishing disruption zones to conceal the actual battle line of friendly defensive positions.
- Concealing unit and personnel movement.
- Creating the perception of false units and their associated activity.
- Creating false high-value assets.

7-49. By providing the appearance of units in false locations, the OPFOR attempts to induce the enemy to attack into areas most advantageous to the OPFOR. When the deception is successful, the enemy attacks where the OPFOR can take maximum advantage of terrain. False thermal and acoustic signatures, decoy and actual vehicles, and corner reflectors, supported by false radio traffic, all contribute to the appearance of a force or element where in fact none exists.

7-50. **Signature Reduction.** The reduction of electromagnetic signatures of OPFOR units and personnel is critical to the success of any deception plan. Minimizing the thermal, radar, acoustic, and electronic signatures of people, vehicles, and supporting systems is critical to ensuring deception of the enemy and enhancing survivability. The OPFOR extensively uses a variety of signature-reduction materials, procedures, and improvised methods that provide protection from sensors and target acquisition systems operating across the electromagnetic spectrum.

Electronic Deception

7-51. Electronic deception is used to manipulate, falsify, and distort signatures received by enemy sensors. It must be conducted in such a manner that realistic signatures are replicated. Electronic deception takes the forms of manipulative, simulative, imitative, and often non-communications deception. The OPFOR may use one or all of these types of electronic deception.

Manipulative Electronic Deception

7-52. Manipulative electronic deception (MED) seeks to counter enemy jamming, signals intelligence (SIGINT), and target acquisition efforts by altering the electromagnetic profile of friendly forces. Specialists modify the technical characteristics and profiles of emitters that could provide an accurate picture of OPFOR intentions. The objective is to have enemy analysts accept the profile or information as valid and therefore arrive at an erroneous conclusion concerning OPFOR activities and intentions.

7-53. MED uses communication or noncommunication signals to convey indicators that mislead the enemy. For example, an OPFOR unit might transmit false fire support plans and requests for ammunition to indicate that the unit is going to attack when it is actually going to withdraw.

7-54. MED can cause the enemy to fragment his intelligence and EW efforts to the point that they lose effectiveness. It can cause the enemy to misdirect his assets and therefore cause fewer problems for OPFOR communications.

Simulative Electronic Deception

7-55. Simulative electronic deception (SED) seeks to mislead the enemy as to the actual composition, deployment, and capabilities of the friendly force. The OPFOR may use controlled breaches of security to add credence to its SED activities. There are a number of techniques the OPFOR uses:

- With *unit simulation*, the OPFOR establishes a network of radio and radar emitters to emulate those emitters and activities found in the specific type unit or activity. The OPFOR may reference the false unit designator in communications traffic and may use false unit call signs.
- With *capability or system simulation*, the OPFOR projects an electronic signature of new or differing equipment to mislead the enemy into believing that a new capability is in use on the battlefield. To add realism and improve the effectiveness of the deception, the OPFOR may make references to "new" equipment designators on related communications nets.
- To provide a *false unit location*, the OPFOR projects an electronic signature of a unit from a false location while suppressing the signature from the actual location. Radio operators may make references to false map locations near the false unit location, such as hill numbers, a road junction, or a river. This would be in accordance with a script as part of the deception plan.

Imitative Electronic Deception

7-56. Imitative electronic deception (IED) injects false or misleading information into enemy communications and radar networks. The communications imitator gains entry as a bona fide member of the enemy communications system and maintains that role until he passes the desired false information to the enemy.

7-57. In IED, the OPFOR imitates the enemy's electromagnetic emissions in order to mislead the enemy. Examples include entering the enemy communication nets by using his call signs and radio procedures, and then giving enemy commanders instructions to initiate actions. Targets for IED include any enemy receiver and can range from cryptographic systems to very simple, plain-language tactical nets. Among other things, IED can cause an enemy unit to be in the wrong place at the right time, to place ordnance on the wrong target, or to delay attack plans. Imitative deception efforts are intended to cause decisions based on false information that appears to the enemy to have come from his own side.

Non-Communications Deception

7-58. The OPFOR continues to develop and field dedicated tactical non-communications means of electronic deception. It can simulate troop movements by such means as use of civilian vehicles to portray to radar the movement of military vehicles, and marching refugees to portray movement of marching troops. Simple, inexpensive radar corner reflectors provide masking by approximating the radar cross sections of military targets such as bridges, tanks, aircraft, and even navigational reference points. Corner reflectors can be quite effective when used in conjunction with other EW systems, such as ground-based air defense jammers.

PHYSICAL DESTRUCTION

7-59. Another method for disrupting enemy control is physical destruction of the target. The OPFOR integrates all types of conventional and precision weapon systems to conduct the destructive fires, to include—
- Fixed- and rotary-wing aviation.
- Cannon artillery.
- Multiple rocket launchers.
- Surface-to-surface missiles.

In some cases, the destruction may be accomplished by ground attack. The OPFOR can also utilize other means of destruction, such as explosives delivered by special-purpose forces or affiliated irregular forces.

7-60. Physical destruction measures focus on destroying critical components of the enemy force. Enemy C2 nodes and target acquisition sensors are a major part of the OPFOR fire support plan during physical destruction actions. Priority targets typically include—
- Battalion, brigade, and division CPs.
- Area distribution system communications centers and nodes.
- Artillery FDCs.

- FACs.
- Weapon system-related target acquisition sensors.
- Jammers and SIGINT systems.

7-61. The OPFOR may integrate all forms of destructive fires, especially artillery and aviation, with other INFOWAR activities. Physical destruction activities are integrated with jamming to maximize their effects. Specific missions are carefully timed and coordinated with the INFOWAR plan and the actions of the supported units.

7-62. Special emphasis is given to destruction of RISTA capabilities prior to an attack on OPFOR defensive positions. Once the attack begins, the OPFOR heavily targets enemy C2 nodes responsible for the planning and conduct of the attack, along with supporting communications. Typically, destruction of C2 nodes prior to the attack may allow the enemy time to reconstitute his control. However, targeting them once forces are committed to the attack can cause a far greater disruptive effect.

7-63. The accuracy of modern precision weapons allows the OPFOR to strike at specific INFOWAR-related targets with deadly accuracy and timing. Due to the mobility and fleeting nature of many INFOWAR targets, precision weapons often deliver the munitions of choice against many high-priority targets.

7-64. The OPFOR continues to research and develop directed energy weapons, to include radio frequency weapons and high-power lasers. While the OPFOR has fielded no dedicated weapon systems, it may employ low-power laser rangefinders and laser target designators in a sensor-blinding role.

PROTECTION AND SECURITY MEASURES

7-65. *Protection and security measures* encompass a wide range of activities, incorporating the elements of deception and EW. Successfully conducted protection and security measures significantly enhance tactical survivability and preserve combat power. The OPFOR would attempt to exploit the large number, and apparently superior technology, of the enemy's sensors. For example, it employs software at the tactical level that allows it to analyze the enemy's satellite intelligence collection capabilities and warn friendly forces of the risk of detection. The use of signature-reducing and -altering devices, along with diligent application of operations security measures, supports deception activities in addition to denying information.

7-66. At the tactical level, protection and security measures focus primarily on—
- Counterreconnaissance.
- C3D.
- Information and operations security.

These and other protection and security measures may overlap into the realms of EW or deception.

Counterreconnaissance

7-67. Winning the counterreconnaissance battle is very important, since it can limit what information the enemy is able to collect and use in the planning and execution of his operations. Tactical commanders realize that enemy operations hinge on situational awareness. Therefore, counterreconnaissance efforts focus on destruction and deception of enemy sensors in order limit the ability of enemy forces to understand the OPFOR battle plan. A high priority for all defensive preparations is to deny the enemy the ability to maintain reconnaissance contact on the ground.

7-68. The OPFOR recognizes that, when conducting operations against a powerful opponent, it will often be impossible to destroy the ability of the enemy's standoff RISTA means to observe its forces. However, the OPFOR also recognizes the reluctance of enemy military commanders to operate without human confirmation of intelligence, as well as the relative ease with which imagery and signals sensors may be deceived. OPFOR tactical commanders consider ground reconnaissance by enemy special operations forces as a significant threat in the enemy RISTA suite and focus significant effort to ensure its removal.

Information Warfare

While the OPFOR may execute missions to destroy standoff RISTA means, C3D is the method of choice for degrading the capability of such systems.

Camouflage, Concealment, Cover, and Deception

7-69. The OPFOR gives particular attention to protective measures aimed at reducing the enemy's ability to target and engage OPFOR systems with precision munitions. Knowing that the enemy cannot attack what his RISTA systems do not find, the OPFOR employs a variety of C3D techniques throughout the disruption, battle, and support zones. These range from the simplest and least expensive methods of hiding from observation to the most modern multispectral signature-reducing technology.

7-70. The OPFOR dedicates extensive effort to employing C3D to protect its defensive positions and high-value assets. All units are responsible for providing protective measures for themselves with their own assets, with possible support from engineer units. The OPFOR employs a variety of signature-reducing or -altering materials and systems, to include infrared- and radar-absorbing camouflage nets and paints.

Information and Operations Security

7-71. Information and operations security can protect the physical and intellectual assets used to facilitate C2. Security must function continuously to be effective. It must conceal not only the commander's intentions and current locations, configurations, and actions of tactical units but also the tactics, techniques, and procedures for employment and operation of information systems.

7-72. The OPFOR clearly understands the importance of information and operations security. Commanders understand their vulnerabilities to being attacked through their own information systems and develop means to protect these systems. In addition, the OPFOR must be capable of isolating attacks on its information systems while maintaining the ability to execute. In order to reduce the vulnerability, the OPFOR emphasizes strong communications, computer, and transmission security.

PERCEPTION MANAGEMENT

7-73. *Perception management* involves measures aimed at creating a perception of truth that best suits OPFOR objectives. It integrates a number of widely differing activities that use a combination of true, false, misleading, or manipulated information. Targeted audiences range from enemy forces, to the local populace, to world popular opinion. At the tactical level, the OPFOR seeks to undermine an enemy's ability to conduct combat operations through psychological warfare (PSYWAR) and other perception management activities aimed at deterring, inhibiting, and demoralizing the enemy and influencing civilian populations.

7-74. The various perception management activities include efforts conducted as part of—
- PSYWAR.
- Direct action.
- Public affairs.
- Media manipulation and censorship.
- Statecraft.
- Public diplomacy.
- Regional or international recruitment and/or fundraising for affiliated irregular forces.

The last three components, while not conducted at the tactical level, can certainly have a great impact on how and where the OPFOR conducts tactical-level perception management activities. Perception management activities conducted at the tactical level must be consistent with, and contribute to, the State's operational and strategic goals.

Psychological Warfare

7-75. PSYWAR is a major contributor to perception management during combat. Targeting the military forces of the enemy, PSYWAR attempts to influence the attitudes, emotions, motivations, aggressiveness, tenacity, and reasoning of enemy personnel. Specialists plan PSYWAR activities at all levels of command.

7-76. In addition to the enemy's military forces, the specialists also concentrate on manipulating the local population and international media in favor of the OPFOR, turning opinion against the enemy's objectives. Planners focus special emphasis on highlighting enemy casualties and lack of success. They also highlight enemy mistakes, especially those that cause civilian casualties. The enemy nation's population is a major target of these activities, due to the criticality of public support for enemy military activities.

7-77. The OPFOR skillfully employs media and other neutral players, such as NGOs, to further influence public and private perceptions. However, if the OPFOR perceives the presence of NGOs to be detrimental to its objectives, it can be extremely effective in hindering their efforts to provide humanitarian assistance to the populace, thus discrediting them.

Public Affairs

7-78. The OPFOR can conduct public affairs actions aimed at winning the favor and/or support of the local leadership and populace—either within the State or in territory it has invaded. This civil support from the OPFOR takes many forms, such as public information and community relations. It can involve providing money, schools, medical support or hospitals, religious facilities, security, other basic services, or just hope. The OPFOR accompanies these support activities with the message or impression that, if the OPFOR loses or leaves, the local population will lose these benefits.

Media Manipulation

7-79. Perception management targeting the media is aimed at influencing domestic and international public opinion. The purpose is to build public and international support for the OPFORs military actions and to dissuade an adversary from pursuing policies perceived to be adverse to the State's interests. The OPFOR exploits the international media's willingness to report information without independent and timely confirmation. While most aspects of media manipulation are applicable to levels well above the tactical, the trickle-down effect can have a major effect on the tactical fight.

7-80. The willingness of the local population to either support or to oppose the OPFOR effort can be critical to OPFOR success. If, for example, media reports convince the populace that the enemy is on a religious vendetta, the local population may decide to join the OPFOR in a fight to the death against the enemy. The OPFOR understands that perception management is not about right and wrong, it is about what people believe is right and wrong. For most people, their perception *is* their reality.

Note. The State employs media censorship to control its own population's access to information and perception of reality. Successful preparation of the population significantly enhances public support for the OPFOR's military actions. As part of this, the State prepares its forces and population for enemy INFOWAR.

Target Audiences

7-81. OPFOR perception management techniques seek to define events in the minds of decisionmakers and populations in terms of the OPFOR's choosing. Successful perception management consists of two key factors: speed and connection. Speed means reaching the target audience before enemy-provided information can alter the perception of events. Connection means having the right media to provide the story to the target audience in a way they will find credible and memorable. World opinion is a primary target of perception management, either to gain support for the OPFOR cause or to turn world opinion and support against the enemy. Reinforcement of its message (preferably by different sources) is also a powerful tool the OPFOR uses to convince the target audience of the OPFOR position.

INFORMATION ATTACK

7-82. *Information attack* (IA) focuses on the intentional disruption or distortion of information in a manner that supports accomplishment of the OPFOR mission. Unlike computer warfare attacks that target the information systems, IAs target the information itself. Attacks on the commercial Internet by civilian hackers have demonstrated the vulnerability of cyber and information systems to innovative and flexible penetration, disruption, or distortion techniques. OPFOR information attackers (cyber attackers) learn from and expand upon these methods. The OPFOR recognizes the increasing dependence of modern armies on tactical information systems. Therefore, the OPFOR attempts to preserve the advantages of such systems for its own use, while exploiting the enemy's reliance on such systems.

7-83. IA is a critical element of INFOWAR, offering a powerful tool for the OPFOR. For example, an information attacker may target an information system for electronic sabotage or manipulate and exploit information. This may involve altering data, stealing data, or forcing a system to perform a function for which it was not intended, such as creating false information in a targeting or airspace control system.

7-84. Data manipulation is potentially one of the most dangerous techniques available to the OPFOR. Data manipulation involves covertly gaining access to an enemy information system and altering key data items without detection. The possibilities are endless with this technique. Some examples are—

- **Navigation.** Altering position data for enemy units, soldiers, and systems, making them think they are in the right place when they are not.
- **Blue Force Tracking.** Altering position data of enemy units, soldiers, and systems to make other units, soldiers and systems believe them to be in one place where they are not or to lose track of them altogether. Alternatively, data manipulation can make OPFOR units appear as enemy or vice versa.
- **Battlefield information systems.** Enhancing OPFOR success by the ability to mitigate and/or influence enemy activities controlled via battlefield information systems.
- **Survey and gun or mortar alignment.** Causing enemy weapons to fire on the wrong target location.
- **Targeting and sensors.** Misdirecting sensors to have false reads, locate false targets, or identify the enemy's own units as OPFOR targets.
- **Weapon guidance.** Sending weapons to the wrong location or wrong target.
- **Timing.** Changing internal clocks, thereby disrupting synchronization.
- **Logistics tracking.** Sending logistics packages to the wrong place or delaying their arrival. This can be done by altering bar codes on equipment or by hacking and altering logistics (delivery or request) data.
- **Aviation operations.** Changing altimeter readings, position location data, or identification, friend or foe codes.

7-85. The OPFOR attempts to inject disinformation through trusted networks. It tries to make the enemy distrust his RISTA and situational awareness assets by injecting incorrect information. Attacks could take the form of icon shifting (blue to red) or moving the icon's location. Fire missions and unit control would require significant human interaction, thus slowing the enemy's target engagement cycle time.

7-86. Likely targets for an IA are information residing in the critical tactical systems of the enemy. Such targets include—

- Telecommunications links and switches.
- Fire control.
- Logistics automation.
- RISTA downlinks.
- Situational awareness networks.
- C2 systems.

Chapter 7

COMPUTER WARFARE

7-87. *Computer warfare* consists of attacks that focus specifically on the computer systems, networks, and/or nodes. This includes a wide variety of activities, including—
- Unauthorized access (hacking) of information systems for intelligence-collection purposes.
- Insertion of malicious software (viruses, worms, logic bombs, or Trojan horses).

Such attacks concentrate on the denial of service and/or disruption or manipulation of the integrity of the information infrastructure. The OPFOR may attempt to accomplish these activities through the use of agents or third-party individuals with direct access to enemy information systems. It can also continually access and attack systems at great distances via communications links such as the Internet.

7-88. Distributed denial of service attacks use a network of slave computers to overwhelm target computers with packets of data and deny them outgoing access to networks. Such attacks could disrupt logistics, communications, intelligence, and other functions.

7-89. The OPFOR can employ various types of malicious software or "malware" on enemy computers to slow operations, extract data, or inject data. Poor operational procedures can enable this type of attack, with significant loss of capability and/or spillage of data to the OPFOR. These attacks also cause the enemy to waste data time and cycles in prevention and remediation. Malware could affect internal clocks (creating positional errors and communications difficulties) and slow the functional speed of computing. Any Internet-capable or networkable system is at potential risk.

7-90. OPFOR computer warfare activities may be conducted prior to or during a military action. For example, by damaging or destroying networks related to an enemy's projected force deployments and troop movements, the OPFOR can effectively disrupt planning and misdirect movement, producing substantial confusion and delays. As modern armies increasingly rely on "just-in-time" logistics support, targeting logistics-related computers and databases can produce delays in the arrival of critical materiel such as ammunition, fuel, and spare parts during critical phases of a conflict.

7-91. The OPFOR can successfully conduct invasive computer warfare activities from the safety of its own territory. It has the distributed ability to reach targeted computers anywhere in the world (as long as they are connected to the Internet). The OPFOR can continuously exploit the highly integrated information systems of an adversary.

Chapter 8

Reconnaissance

To the OPFOR, the single most important component of military action is reconnaissance. Reconnaissance represents all measures associated with organizing, collecting, and studying information on the operational environment associated with the area of upcoming battles. Aggressive, continuous reconnaissance allows the timely accomplishment of combat missions with minimum losses. Poor reconnaissance can lead directly to failure. The OPFOR commits significant resources to any reconnaissance mission.

Note: Reconnaissance is part of the OPFOR military function called *reconnaissance, intelligence, surveillance, and target acquisition* (RISTA). RISTA is the combination of capabilities, operations, and activities using all available means to obtain information concerning foreign nations; areas of actual or potential operations; and/or the strength, capabilities, location, status, nature of operations, and intentions of hostile or potentially hostile forces or elements. It includes production of intelligence resulting from the collection, evaluation, analysis, and interpretation of such information. It also includes detection, identification, and location of a target in sufficient detail to permit the effective employment of weapons.

COMBINED ARMS MISSION

8-1. Reconnaissance is a combined arms mission, not solely the business of reconnaissance troops. It involves the integrated efforts of troops from several branches. OPFOR reconnaissance actions often include the use of affiliated irregular forces and/or friendly civilians. Reconnaissance elements that are defeated before or during the accomplishment of their mission are reconstituted from any appropriate source.

Note. The term *reconnaissance unit* refers to a unit composed of specialized reconnaissance troops. In contrast, the term *reconnaissance element* refers to any unit or task organization given a specific reconnaissance mission, regardless of the type(s) of troops involved. The latter is a functional designation that describes the function an organization is performing.

8-2. This chapter focuses primarily on the reconnaissance activities of ground maneuver forces and specialized ground reconnaissance troops. However, there are also specialized reconnaissance assets in other arms, which other chapters in this manual discuss in more detail:
- Signals reconnaissance (chapter 7).
- Artillery target acquisition (chapter 9).
- Aerial reconnaissance (chapter 10).
- Air defense reconnaissance, early warning, and target acquisition (chapter 11).
- Engineer reconnaissance (chapter 12).
- Chemical, biological, radiological, and nuclear (CBRN) reconnaissance (chapter 13).
- Special reconnaissance (chapter 15).

Chapter 8

The integrated efforts of any or all of these reconnaissance means may be necessary to support specific missions. Efficient and accurate reconnaissance is also crucial to ensuring the success of information warfare (INFOWAR) activities.

CONCEPT

8-3. For the OPFOR, reconnaissance is a critical element of combat support. In modern combat, the battlefield develops unevenly. Therefore, units cannot rely on the security of their flanks or rear—in fact there may not even be "flanks" or "rear." Friendly and enemy forces can become intermingled, with the combat situation developing and changing quickly. Reconnaissance elements must warn commanders of developing threats and identify enemy strengths and vulnerabilities. The OPFOR organizes reconnaissance to acquire continuous, timely, and accurate information on the operational environment. This includes information about—

- The enemy's CBRN and precision weapons, force disposition, and intentions.
- Terrain and weather.

This information is vital to the OPFOR decisionmaking and planning process. Reconnaissance can decisively influence the outcome of a battle.

8-4. The OPFOR treats reconnaissance as an offensive action, since the enemy typically defends vital information with security actions and camouflage, concealment, cover, and deception (C3D) measures. Thus, reconnaissance plans must always have a provision for defeating the enemy's efforts to protect himself.

PRINCIPLES

8-5. The speed and potential nonlinearity of modern combat have increased the importance of reconnaissance. Without decisive actions of reconnaissance elements and assets, it is impossible to preempt the enemy, seize the initiative, and conduct a successful battle. The OPFOR uses the following set of interrelated principles to guide its reconnaissance activities:

Focus

8-6. Reconnaissance action must serve the commander's needs and focus on elements and objectives critical to the execution of combat missions. Each unit develops a comprehensive reconnaissance plan in accordance with the organization's mission. This plan must coordinate the integration of all available assets.

8-7. The OPFOR understands that information is not the same thing as knowledge. With the number of sensors available to the tactical commander, the danger exists that analysts and decisionmakers could become overwhelmed with raw data. Therefore, all reconnaissance activities should focus on answering specific information requirements.

Continuity

8-8. The modern, fluid battlefield demands continuous reconnaissance to provide an uninterrupted flow of information under all conditions. Reconnaissance provides constant coverage of the enemy situation, using a wide variety of redundant assets. Not only must reconnaissance units answer specific requests for information; they also must continuously collect information on all aspects of the operational environment to fully meet future requirements. The variety of overlapping assets ensures greater validity of collected information. Continuous reconnaissance decreases the likelihood that the enemy could carry out successful deception.

8-9. Reconnaissance units attempt to maintain contact with the enemy at all times. They conduct reconnaissance in all directions, in order to prevent surprise. They collect information during all battle phases, 24 hours a day, in all weather conditions.

8-10. To ensure this continuity, units conducting reconnaissance must maintain a high state of combat readiness. They must be able to sustain themselves wherever they deploy, without relying on others for transport or subsistence. If a specialized reconnaissance unit is destroyed or becomes combat ineffective, commanders reassign the mission to appropriate forces.

Aggressiveness

8-11. Aggressiveness is the vigorous search for information, including the willingness to fight for it if necessary. Reconnaissance troops must collect information creatively and make maximum use of all assets and methods to ensure success on the battlefield. The OPFOR vigorously employs all available collection resources and adheres carefully to the reconnaissance plan. However, it will alter the plan when its own initiatives or enemy actions dictate.

8-12. Although reconnaissance is the primary mission, all reconnaissance units train to defend themselves. Reconnaissance troops penetrate enemy defenses, avoiding contact if possible. When required, they can ambush and raid enemy forces. They do what is necessary to fulfill the commander's information needs.

8-13. The information requirement determines the techniques used. Reconnaissance patrols by mechanized forces are not always the best means. Ambushes and raids are fruitful sources of information from captured prisoners, documents, and equipment. Such information-gathering actions are generally more important than any associated damage, but there are exceptions. Reconnaissance elements are often called upon to destroy high-value targets they find.

Timeliness

8-14. Timely information is critical on the modern battlefield. Because of the high mobility of modern forces, there are frequent and sharp changes in the battlefield situation. As a result, information quickly becomes outdated. The best intelligence is useless if it is not received in time. Timely reporting enables the commander to exploit temporary enemy vulnerabilities and windows of opportunity. He can adjust plans to fit a dynamic battlefield. The OPFOR achieves timeliness through—

- Increased automation for command and control (C2) and processing of information.
- Real-time or near-real-time aerial downlinks.
- Satellite downlinks.

This timeliness is especially critical for the success of integrated fires commands (IFCs).

Camouflage, Concealment, Cover, and Deception

8-15. The OPFOR is aware that the enemy may learn a great deal about its intentions by discovering its reconnaissance plan. Commanders understand it is often not possible to completely hide the fact that reconnaissance is being conducted. However, they make every effort to conceal the scale, missions, targets, and nature of reconnaissance missions. Specific measures can include—

- Conducting reconnaissance across a broad range of targets.
- Concealing the actions of reconnaissance elements.
- Covering and concealing assembly areas of reconnaissance elements and assets.

8-16. The OPFOR can also use C3D to "paint a picture" that confirms the enemy's stereotyped views of how the OPFOR fights. By showing the enemy what he expects to see, the reconnaissance effort can help to establish the conditions for success during ensuing combat. This is a critical part of INFOWAR (see chapter 7).

Accuracy and Reliability

8-17. The OPFOR uses all available reconnaissance means to verify the accuracy and reliability of reported information. A commander must base his decisions on reconnaissance information. So, the more accurate and complete the information, the better the decision. To maximize results, the commander's battle plan requires accurate information on the enemy's size, location, equipment, and combat readiness. The accuracy and reliability of reconnaissance information are critical to the destruction of high-value

Chapter 8

targets such as enemy weapons of mass destruction (WMD), precision weapons, attack aviation, logistics centers, C2, and communications. The OPFOR achieves accuracy through the creation of overlapping coverage and the use of improved technology.

8-18. Reconnaissance must reliably clarify the true enemy situation in spite of enemy C3D and counterreconnaissance activities. The first step is to tailor reconnaissance efforts to the tactical situation. Commanders must select and allocate reconnaissance elements in accordance with their capabilities in terms of missions and targets.

8-19. The next step is to compare, cross-check, recheck, and integrate reconnaissance reports from multiple means of acquisition. The study and integration of reconnaissance information collected by multiple sources can help in identifying and assessing false targets and other false indicators of enemy actions or intentions.

ASSETS

8-20. Tactical reconnaissance supports divisions or division tactical groups (DTGs) and below. It provides reconnaissance needed to plan and carry out tactical actions within each commander's area of responsibility (AOR). Divisions, DTGs, and below perform tactical reconnaissance using specially trained reconnaissance resources and combat troops from maneuver units. Figure 8-1 summarizes the range capabilities of the reconnaissance assets that can support tactical commanders.

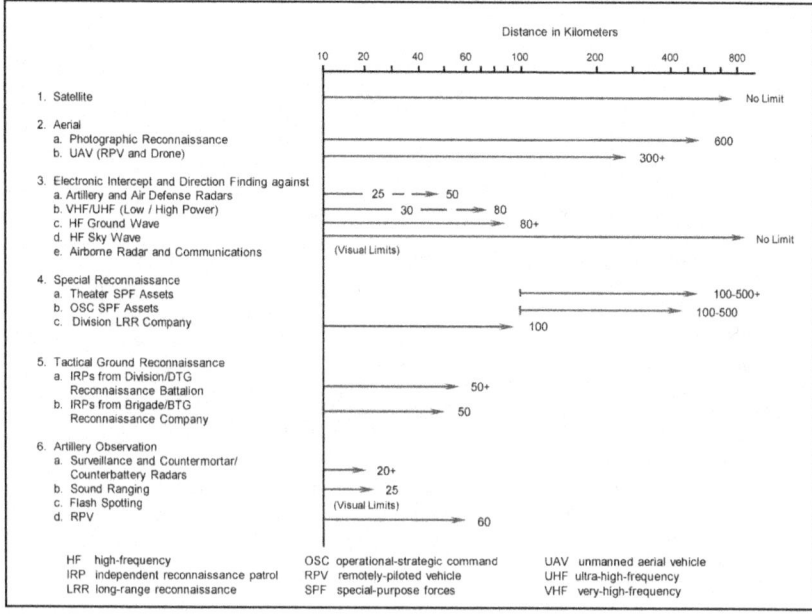

Figure 8-1. Effective ranges of example reconnaissance assets

GROUND RECONNAISSANCE

8-21. Tactical units may send out independent reconnaissance patrols (IRPs) to perform ground reconnaissance. The size of such patrols can vary, but is usually an augmented reconnaissance or combat arms platoon.

8-22. Long-range reconnaissance (LRR) units may form additional IRPs, or their personnel and vehicles can supplement patrols formed by the other reconnaissance or combined arms units. However, LRR personnel are specially trained for insertion in small reconnaissance teams at distances up to 100 km beyond the battle line.

8-23. A DTG may receive a special-purpose forces (SPF) unit to support its IFC or to perform other special reconnaissance and direct action missions. A brigade tactical group (BTG) can also be allocated an SPF unit. The SPF operate in small teams or as several teams grouped into a detachment. They can perform some of the same types of reconnaissance tasks as the LRR teams. However, the SPF receive special training and equipment that allows them to operate farther out and for longer periods. See chapter 15 for more information on SPF.

Note. Most references to SPF in this chapter also include commando units that are part of the OPFOR's SPF Command. Commandos can perform various reconnaissance missions in the disruption zone or deep in enemy territory. Commandos usually conduct reconnaissance as small teams or squads.

8-24. Signals reconnaissance assets include radio intercept and direction-finding (DF) and radar intercept and DF systems. They can also include equipment designed to exploit signals from cellular, digital, satellite, fiber-optic, and computer network systems. See chapter 7 for more detail on signals reconnaissance.

8-25. Engineer units can also dispatch one or more engineer reconnaissance patrols. This type of patrol consists of a squad or a platoon of engineer specialists sent out to obtain engineer intelligence on the enemy and the terrain. In enemy territory, it deploys as part of another ground reconnaissance element. See chapter 12 for more detail on engineer reconnaissance.

8-26. Chemical defense units establish CBRN observation posts as well as CBRN reconnaissance patrols. Chemical defense units can also attach individual chemical and radiological specialists to reconnaissance, security, or reserve elements. Their role is to—
- Identify and mark areas of CBRN contamination.
- Determine the extent and nature of any contamination.
- Find routes around contaminated areas.
- Find the shortest route through an area with low levels of contamination and select certain areas for decontamination.
- Monitor the effects of chemical or nuclear weapons and provide warning of downwind hazards.

See chapter 13 for more detail on CBRN reconnaissance.

8-27. Artillery units often have their own reconnaissance assets. These include—
- Artillery command and reconnaissance vehicles.
- Mobile reconnaissance posts.
- Battlefield surveillance radars.
- Target acquisition radars.
- Counterfire radars.
- Sound- and flash-ranging equipment.

In addition, artillery reconnaissance assets may be made available from operational-strategic command (OSC) level. (See chapter 9 for more detail on artillery target acquisition.) Artillery units can also conduct reconnaissance by fire.

Chapter 8

8-28. Affiliated irregular forces can employ a wide range of reconnaissance techniques, often quite sophisticated. Their primary ground reconnaissance means is surveillance by teams that blend carefully and completely into the local population.

RECONNAISSANCE BY FIRE

8-29. Reconnaissance by fire is a method of reconnaissance in which fire is placed on a suspected enemy position to cause the enemy to disclose his presence by movement or return fire. It is used to provoke a reaction. The OPFOR also uses a similar tactic in which individuals may brandish weapons or purposely draw suspicion in order to learn more about the enemy's rules of engagement.

8-30. At the platoon and squad level, reconnaissance by fire may also be called cover or drake shooting. This is a technique employed to quickly reveal and kill concealed enemy riflemen. Several shots are placed directly into (and through) the suspected cover. Using two- to three-round bursts, the OPFOR riflemen deliberately aim and fire low on the ground immediately to the front of the cover, raking it with fire from the left to the right. Ricochets, fragments, earth, rocks, and wood either injure the hidden enemy soldiers and/or force them to react.

AERIAL RECONNAISSANCE

8-31. Aerial reconnaissance includes visual observation, imagery, and signals reconnaissance from airborne platforms. These platforms may be either piloted aircraft or unmanned aerial vehicles (UAVs).

Rotary- and Fixed-Wing Aircraft

8-32. Attack helicopter crews report any unexpected enemy activity observed during their missions. They can report such perishable information immediately by radio to a ground command post (CP) unless such reporting would interfere with successful completion of their assigned mission. In the latter case, they report this information during post-mission debriefing. Dedicated reconnaissance helicopters, depending on equipment, can conduct visual, thermal imaging, photographic, infrared, and signals reconnaissance. Transport helicopters or fixed-wing aircraft can insert LRR elements to distances not practicable with armored reconnaissance vehicles. See chapter 10 for more information on aerial reconnaissance.

Unmanned Aerial Vehicles

8-33. The military application of UAVs has become standard practice in armies worldwide. The OPFOR is no exception. It operates UAVs at all levels, from the strategic level down through division, brigade, maneuver battalion, and some companies, as well as in specialized units (such as SPF teams). The techniques and employment of the larger and more capable operational- and strategic-level UAV platforms used by the OPFOR are similar to those employed worldwide.

Note. There are two types of UAV: the remotely piloted vehicle (RPV) and the drone. An RPV, on the one hand, can be flown by remote control from a ground station, over a flight path of the controller's choosing. A drone, on the other hand, flies a set course programmed into its onboard flight control system prior to launch.

8-34. UAV missions are planned by the chief of reconnaissance (COR) and support combat operations anywhere on the battlefield. When equipped with the proper sensors, UAVs provide imagery day and night and in all weather conditions (depending on the size and capability of the platform). UAVs are an excellent imagery asset, providing the commander with near-real-time (NRT) reconnaissance and battlefield surveillance without the possibility of risk to a manned aircraft. They provide OPFOR commanders a dedicated and rapidly taskable asset that can look wide as well as deep. The commander selects the appropriate UAV based on what is available, current mission configuration, operating range, operating radius, and endurance (flight time). During a preplanned UAV mission, changes in mission priorities or identification of new targets may occur. The OPFOR commander can then direct a UAV to support a different mission or area.

Reconnaissance

Note. The size, ease of operation, and simple design of many smaller UAVs lend them to field-expedient modification. Converting these UAVs into a munitions delivery system (improvised attack UAV) is not difficult and offers several tactical advantages. Off-the-shelf remote controlled aircraft can also provide this capability.

8-35. UAVs can provide NRT combat information about—
- Terrain.
- Disposition of enemy units.
- Battle damage assessment.
- Target recognition and detection (after which they can provide target designation and illumination).

They can assist in route, area, and zone reconnaissance.

8-36. Information gathered via UAV may be immediately acted upon, or it may be integrated with other sources to—
- Support or shape the immediate combat mission.
- Plan future operations.
- Reallocate intelligence assets.

The data may be integrated with that from ground reconnaissance, ground surveillance radar, intelligence assets, or any other information.

8-37. Units such as air defense, antitank, artillery, or logistic units, and those with stationary facilities requiring security patrols routinely use UAVs in the reconnaissance role. This allows such units to execute their missions while reducing personnel and vehicle requirements. SPF and some irregular forces can also use UAVs. See chapter 10 for more information on other UAV missions. For additional information on the capabilities and characteristics of UAVs, see the *Worldwide Equipment Guide*.

RECONNAISSANCE PLANNING

8-38. The purpose of reconnaissance planning is to thoroughly coordinate the actions of all reconnaissance organizations and levels of command. Ultimately, the planning must ensure that missions, targets, times, forms of action, zones of reconnaissance responsibility (ZORRs), and the exchange of information are fully coordinated.

ZONES OF RECONNAISSANCE RESPONSIBILITY

8-39. Each tactical-level unit, down to battalion or detachment, has one or more ZORRs. This zone is the combination of the unit's AOR and the area outside of the AOR that can be observed by the unit's technical sensors. (See figure 8-2 on page 8-8.)

Chapter 8

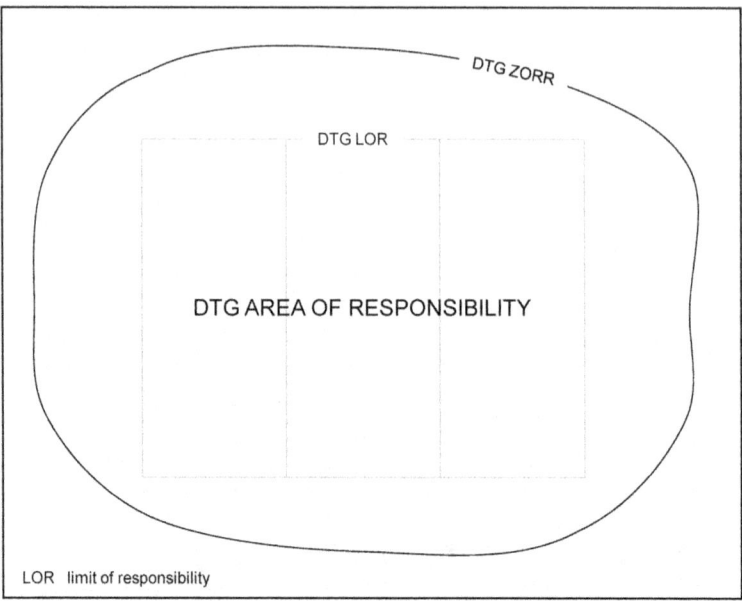

Figure 8-2. Zone of reconnaissance responsibility (example)

8-40. By definition, the ZORR extends into adjacent unit AORs. This results in overlapping coverage, which can prevent surprise and the enemy's exploitation of seams between AORs. Within this zone, the OPFOR unit must be able to monitor enemy activity sufficiently to ensure that unexpected enemy moves do not disrupt its own plans. Reconnaissance in this zone should provide early warning of potential enemy movement into the AOR from any direction.

COMMANDER

8-41. Reconnaissance planning begins with the commander. Following the receipt of a combat order from higher command, the maneuver unit commander determines what additional information is necessary to conduct his mission. To fight the battle properly, the tactical commander typically needs information on—

- Enemy positions, boundaries, and strong and weak points.
- The location of key enemy systems and installations (such as forward arming and refueling points, counterfire radars, C2 nodes, or logistics centers).
- The location and movement of enemy reserves.
- Possible axes for enemy counterattacks.
- Terrain trafficability and cover.
- The location of and approaches to obstacles.

8-42. There are several pieces of information that can be of great interest before and during the battle. General aims that guide the reconnaissance process *prior* to the initiation of combat are the timely detection or determination of—

- Enemy preparations for an attack with conventional or precision weapons, or WMD.
- Indications of the enemy's concept for upcoming action.
- Groupings of enemy forces and their preparation for combat.

- Changes in the grouping or composition of these forces.

8-43. *After* combat begins, general information requirements can include—
- The effects of precision weapons and WMD.
- Further changes in the grouping or composition of enemy forces.
- Activities and composition of enemy reserves.
- Enemy preparations to prevent the deployment of friendly troops.
- Enemy preparations to launch amphibious, airmobile, or airborne assaults.

8-44. The commander analyzes his requirements for information and determines the reconnaissance needed. At battalion and above, the commander then states broad reconnaissance instructions to his chief of staff (COS) and his intelligence officer. The intelligence officer would relay these instructions to a secondary staff officer, the COR, who actually develops the reconnaissance plan. The amount of time available for conducting reconnaissance determines the amount of detail contained in the commander's instructions. If little time is available, the commander's instructions may be very short and simple.

8-45. The commander's instructions to the intelligence officer and COR outline the overall aim or goal of reconnaissance and the priorities of the reconnaissance mission. This can include specific reconnaissance tasks assigned by the commander or by higher headquarters. The commander can also assign specific objectives, sectors, areas, or axes for concentrating the main reconnaissance effort. His instructions may specify which reconnaissance assets may or may not be used prior to combat. He defines the type of information he wants and when he needs it.

CHIEF OF STAFF

8-46. The COS interprets the commander's instructions and converts them into specific tasks. At this point, if not before, the general aims of the reconnaissance effort become specific information requirements and taskings to reconnaissance units.

8-47. The COS has overall responsibility for providing the necessary information for the commander to make decisions. At the tactical level, he has a more clearly defined role in structuring the reconnaissance effort than at higher levels.

8-48. Because reconnaissance is a combined arms task, the COS must coordinate the overall reconnaissance effort. Aside from reconnaissance troops, various other combat, combat support, and combat service support branches have reconnaissance tasks and capabilities. Thus, coordination involves not only ground reconnaissance, but also the efforts of target acquisition elements and CBRN, engineer, and signals reconnaissance, as well as any aerial reconnaissance assets allocated to support the maneuver unit's mission. The COS can ensure that the various branches report the results of all these reconnaissance efforts through the COR and the intelligence officer to the commander.

8-49. Upon receipt of the commander's reconnaissance instructions, the COS refines the requirements and passes them to the intelligence officer for the detailed development of the reconnaissance plan. The COS provides any other information available on targets and areas for concentration of the reconnaissance effort.

INTELLIGENCE OFFICER

8-50. The intelligence officer's instructions to the COR specify details of the missions identified by the commander and the method of execution. He determines the sequence for performing these tasks and the manpower and equipment necessary to complete them. He specifies the reconnaissance assets to be used for the priority reconnaissance tasks. He determines the principal means of preparing and supporting reconnaissance elements, and ensuring their interaction and coordination. He then specifies to the COR the times for preparing the reconnaissance plan and issuing combat orders to reconnaissance units.

8-51. The intelligence officer is responsible for the coordination between reconnaissance, INFOWAR, and communications requirements. He reviews the efforts of all three areas and resolves conflicts. For example, if the INFOWAR plan recommended a particular enemy C2 center for destruction, but the reconnaissance plan sought to collect vital information from it, the intelligence officer would choose the course of action that best supported the commander's intent.

CHIEF OF RECONNAISSANCE

8-52. In division, brigade, or tactical group headquarters, the intelligence officer heads the intelligence and information section of the staff, which includes a reconnaissance subsection headed by the COR. At battalion level, the platoon leader of the reconnaissance platoon serves as the COR. This staff officer is responsible for organizing reconnaissance in accordance with the commanders' general plan. The COR works for and reports to the intelligence officer. Along with or through the intelligence officer, he reports to the commander concerning the organization of reconnaissance planning.

8-53. Like other section and subsection chiefs on the division and brigade staff, the COR has a dual reporting chain. He is responsible to the commander and COS in whose headquarters he serves. However, he also receives additional instructions and guidance from his COR counterpart at the next-higher level. For example, a DTG COR coordinates with the OSC COR and with the CORs of subordinate BTGs. Through these channels, he can request reconnaissance support from higher levels or task reconnaissance elements of subordinate BTGs to perform missions for the DTG. Thus, a tactical-level COR can have access to information collected by means not directly available to him, such as aerial reconnaissance. He is also responsible for passing the results of reconnaissance both up and down the chain of command.

8-54. To the reconnaissance missions he receives from the COS, the COR adds specific instructions to complete the reconnaissance plan. His knowledge of enemy doctrine and his access to current intelligence enable him to assign precise missions to reconnaissance assets at his level of command. He establishes time constraints, reporting schedules, and reporting methods. He also establishes measures for interaction and coordination of reconnaissance actions to ensure accomplishment of all missions and objectives. He organizes and continuously monitors communications with all maneuver units and with the headquarters of subordinate reconnaissance units. He can also provide guidance to immediately subordinate maneuver units regarding their contribution to the higher unit's reconnaissance effort.

RECONNAISSANCE PLAN

8-55. The COR at division, brigade, or tactical group level develops a reconnaissance plan within the framework of the higher headquarters' mission and the higher commander's decision for combat. He combines this information with—

- The higher headquarters' instructions on conduct of the reconnaissance mission.
- Information currently available on the enemy.
- The status of reconnaissance assets.

8-56. Depending on the situation, the reconnaissance plan may include—

- The AORs of friendly units.
- The commander's concept and mission.
- All available information regarding known and suspected enemy groupings and intentions.
- A list of tasks (including obtaining new information, confirming previously available information, battle damage assessment, and calling for fire on targets of opportunity).
- A list of priority targets for reconnaissance.
- The deployment of reconnaissance assets in terms of these tasks and targets.
- The time and sequence for executing the tasks.
- Restrictions on reconnaissance actions during specific times or in certain areas.
- The method and time for reporting.

8-57. The content of reconnaissance missions depends on the commander's information requirements. These, in turn, depend on the nature of the unit's combat missions.

8-58. In the offense, reconnaissance must establish the enemy's effective combat strength, affiliation, combat effectiveness, and whether or not he has CBRN or precision weapons. It must discover firing positions for weapons, strong points, gaps, and the nature of engineer preparation of defensive positions. It is also important to locate and track enemy reserves and possible axes for counterattacks. Reconnaissance must identify terrain that may present trafficability problems for advancing OPFOR units.

8-59. In the defense, reconnaissance must cover enemy preparation for an attack and determine the possible time of the attack. It must establish the makeup of the enemy grouping and identify the axis of his main attack and the nature of his maneuver. It is especially important to determine the locations of firing positions of artillery and other weapons, locations of C2 facilities, the combat effectiveness of enemy troops, and their affiliation. The plan should include reconnaissance tasks for the entire course of defensive actions, as well as tasks that support an eventual transition to the offense.

INFORMATION FLOW AND COMMUNICATIONS

8-60. The commander's instructions, the reconnaissance plan, and combat orders to reconnaissance elements identify information requirements and specify how and when to report this information. To minimize radio traffic on command nets, the flow of information both up and down the chain of command normally is through reconnaissance channels. Commanders determine how frequently they wish to receive various types of situational data.

8-61. A reconnaissance element typically reports to the commander of its parent reconnaissance unit or to the COR (or COS) of the maneuver unit that dispatched it. In exceptional cases, however, the capability for skip-echelon communications allows the leader of a reconnaissance unit to report to a higher level if so directed in specific instructions.

Reporting

8-62. Standard procedures for reconnaissance reporting seek to ensure that the supported commander receives critical information he requires to make a decision. To reduce the likelihood of information overload, there are two different reporting categories:
- Periodic (reports submitted at a set time).
- Aperiodic (reports submitted on the staff's own initiative resulting from significant changes in the situation).

8-63. Under the direction of the COR, the reconnaissance subsection on a division, brigade, or tactical group staff evaluates and summarizes incoming information for the commander. It disseminates this information to those command and staff elements that require it, including higher headquarters and adjacent units. It is important to study information from all sources before reaching conclusions. This includes even information believed to be false because it contradicts information from other sources and does not correspond to the developing situation. The study of this false information can reveal the methods the enemy is using for deception.

Reconnaissance Report

8-64. Commanders and staffs receive reports from reconnaissance elements and/or CORs. Depending on the situation, these reports may be in the form of briefings, radio communications, or written reports.

8-65. The term *reconnaissance report* also applies to a specific document prepared by the headquarters of a brigade, division, or tactical group for reporting information about the enemy to a higher headquarters. It may be a periodic reconnaissance report forwarded every few hours at set times specified in instructions. It may also be an aperiodic report prepared at the initiative of the subordinate commander or by special request from the higher commander. In either case, it includes, at a minimum, the following:
- The general nature of enemy activities throughout the reporting unit's entire ZORR.
- The disposition and grouping of enemy forces in each area or axis within the ZORR.
- Significant changes that have occurred since the previous report.
- The reporting unit's conclusions about possible enemy actions based on these indications.
- The source of the data and the time received.

Reconnaissance Summary

8-66. The *reconnaissance summary* is a report, prepared by the headquarters of a division, brigade, or tactical group, that contains information about the enemy covering a given period of time. The reporting unit sends this summary to the higher headquarters at times established in instructions. It is normally provided once a day as a brief narrative of the highlights of the past 24 hours. It is also sent to adjacent and subordinate headquarters for information purposes. It typically includes the following:

- The general nature of enemy activities in the ZORR.
- Data about the enemy's precision weapons and WMD and their employment.
- The positions of enemy forces at the time of preparation of the summary.
- Information about the enemy's air (and naval) forces, air defense, CPs, radar equipment, logistics installations, obstacles, and field fortifications.
- The reporting unit's general assessment of the disposition, activities, and condition of enemy forces and the nature of forthcoming enemy activities.
- Information gaps to be addressed during further reconnaissance.

The summary may include the significant results of prisoner interrogation or exploitation of captured documents or equipment.

RECONNAISSANCE ELEMENTS

8-67. The general term *reconnaissance element* applies to any unit given a specific reconnaissance mission. (This is in keeping with the OPFOR practice of designating functional elements and functional forces. On those relatively rare occasions when the unit performing the reconnaissance function is as large as a battalion, it could be called a *reconnaissance force*.) Some reconnaissance elements are formed on the basis of a reconnaissance unit, but others come from maneuver units or other sources.

8-68. At the tactical level, the ground forces employ a variety of reconnaissance elements, tasked and tailored to fit the specific needs of the tactical commander in a particular situation. These elements vary in size and composition from a few scouts to a battalion.

8-69. Reconnaissance units at the tactical level may either operate independently or be task-organized with personnel from maneuver units into special types of reconnaissance elements. They may or may not have augmentation such as mechanized infantry troops, tanks, artillery, engineers, CBRN reconnaissance personnel, and other specialists.

COMMANDER'S RECONNAISSANCE GROUP

8-70. Tactical commanders conduct a personal commander's reconnaissance, where possible, as part of the planning process. The commander goes to a field site in the vicinity of planned combat actions to conduct a visual study of the enemy and terrain. He takes with him his subordinate maneuver commanders, the commanders of dedicated and supporting units, and staff officers. The purpose of this reconnaissance is to refine and verify, on the terrain, and add details to the general plan already made on a map and missions already assigned to the troops. However, it can also occur prior to making battle plans. During the reconnaissance, the commander issues an oral combat order and organizes coordination. The OPFOR takes elaborate measures to disguise the conduct of this reconnaissance and the ranks of the participants in the commander's reconnaissance group.

8-71. Prior to departure for the field site, time permitting, a commander's reconnaissance plan is drawn up. It specifies—

- The purpose and objective of the commander's personal reconnaissance.
- Principal tasks.
- The composition of the reconnaissance group.
- Routes and means of transportation.
- Halt points for reconnaissance activity.
- The principal items to be covered at each halt point.

Reconnaissance

OBSERVER

8-72. Within a squad, platoon, or company, an individual can be assigned as an *observer*. This observer can—
- Reconnoiter the ground and airspace, enemy and terrain.
- Observe the actions and position of his own unit, its subordinate units, and adjacent units.

OBSERVATION POST

8-73. An *observation post* (OP) is position within which a team is assigned the mission of conducting surveillance of enemy in a given zone or location. An OP can have literally any organization and can be drawn from any type force. Typically, OPs are kept small. An OP usually owns and/or receives sensor and communications capability that permits stealthy and rapid movement and provides the ability to locate, track, and report on its reconnaissance targets.

8-74. OPs typically operate in or near enemy-controlled areas. The reconnaissance plan includes the method by which the OP penetrates enemy security forces, eludes detection, and observes and reports on the enemy. OPs are often called upon to perform the infiltration tactical task. The INFOWAR plan often includes C3D measures that assist in preventing enemy detection of OPs. This C3D effort may include employing cover from the local population or affiliated forces.

PATROL SQUAD

8-75. A *patrol squad* is a single squad sent out with a reconnaissance mission. It can be a single vehicle (patrol vehicle or tank) or a reconnaissance or infantry squad on foot. Patrol squads may be the only reconnaissance element deployed when the risk of meeting the enemy is low. However, they can also be deployed from a larger reconnaissance element, such as any platoon-size patrol. Any maneuver company or battalion operating in isolation from the main force can send out a patrol squad, even when not performing reconnaissance missions. This occurs chiefly when the maneuver unit is on the move or when occupying an assembly area.

8-76. As a rule, the patrol squad operates off-road, moving from one suitable observation point to another. It typically reconnoiters places where an enemy unit could be concealed, such as hills, woods, or built-up areas. If it sights the enemy, the patrol squad immediately reports this to the commander or platoon leader who dispatched it, and then continues to carry out observation. In the event of a sudden meeting with the enemy, the patrol squad can open fire on him.

RECONNAISSANCE TEAM

8-77. A *reconnaissance team* is an element, usually at squad strength, formed from specially trained personnel (for example, from an LRR company). It conducts independent actions in enemy-held territory to discover precision weapons, WMD, C2 facilities, reserves, airfields, and other priority targets. A reconnaissance team may be inserted on foot or in an armored reconnaissance vehicle. If the team leaves its vehicle behind, insertion could also be by helicopter or by parachute landing from fixed-wing aircraft.

RECONNAISSANCE PATROL

8-78. A *reconnaissance patrol* (RP) is generally a platoon-size tactical reconnaissance element with the mission of acquiring information about the enemy and the terrain. While the RP is generally platoon-sized, it can be smaller or larger depending on the commander's requirements, forces available, and the operational environment. The OPFOR distinguishes among various types of patrols that fit under the general descriptive term *reconnaissance patrol*. These specific types of reconnaissance include the independent reconnaissance patrol (IRP), officer reconnaissance patrol, and fighting patrol (FP). The generic term also includes engineer reconnaissance patrols and CBRN reconnaissance patrols (see chapters 12 and 13, respectively).

Chapter 8

8-79. Other than as a generic descriptor for these specific types, the OPFOR also uses the term *reconnaissance patrol* to describe a tactical reconnaissance element dispatched from a reconnaissance detachment in the process of accomplishing its mission. This type of RP is not "independent," because it is a subordinate of a larger reconnaissance element.

8-80. It is difficult to distinguish among the various types of RP by their strength, composition, or position on the battlefield. The size of each patrol is up to a platoon, augmented when necessary. A patrol in this configuration could be an RP, FP, or IRP. They all accomplish their missions through observation, ambushes, raids, and—when necessary—combat.

8-81. In the event of unexpected contact with the enemy, all types of RP try to break contact and then reach a position from which to identify and report the strength, composition, and location of the enemy force. If the patrol discovers the enemy in an unexpected position, the patrol leader immediately executes the actions on contact battle drill (see chapter 5).

8-82. If a patrol observes enemy reconnaissance or security elements, its task is to avoid contact and continue on to locate the main force as rapidly as possible. In the event of a surprise encounter with a small enemy force, when evasion is impossible, the patrol acts decisively to destroy the enemy, capture prisoners, if possible, and continue its mission.

INDEPENDENT RECONNAISSANCE PATROL

8-83. A tactical-level command, battalion or larger, may send out *independent reconnaissance patrols* (IRPs) with a specific mission to conduct reconnaissance of the enemy and terrain. Each IRP is usually a reconnaissance or combat arms platoon, often augmented with engineers and CBRN specialists. The size of each patrol depends on several factors including the terrain, forces available, enemy strength, the operational environment, and the importance of the axis or objective. IRPs often move on multiple axes, although the main axis receives the primary reconnaissance effort.

8-84. In the offense, an IRP is assigned either an axis or an objective. In defensive situations, the IRPs are used to scout enemy reserves moving toward the battle zone or attacking on an open flank. An IRP can also support antilanding defense during an airborne or amphibious landing by the enemy.

8-85. As with other types of RP, the IRP accomplishes its missions through observation, ambushes, or raids. It may conduct reconnaissance by fire, if necessary, but becomes engaged in battle only if one of the following conditions exists:

- It cannot carry out its mission by any other method.
- It suddenly encounters the enemy.
- It detects enemy precision weapons or other high-priority targets.

Both the RP and the IRP can dispatch patrol squads to examine terrain features, detect enemy forces, or provide security.

8-86. An IRP operates at a greater distance from the parent organization than the RP and may stay out longer. The distance from the parent unit depends on—

- The nature of the mission.
- The terrain.
- The composition of the patrol.
- The ability to maintain communications with the unit that dispatched it.

OFFICER RECONNAISSANCE PATROL

8-87. A maneuver unit can send out an *officer reconnaissance patrol* when there has been an abrupt, unexpected situation change. The purpose of this patrol can be to—

- Update information on the enemy and terrain in the AOR.
- Determine the position of friendly troops.
- Check contradictory situation data.

8-88. Depending on assigned missions, this patrol can consist of one to three officers with communications equipment, and possibly two to five soldiers assigned for security. This patrol can move by helicopter, tank, IFV, APC, or other vehicle. The officer reconnaissance patrol allows the commander to oversee and maintain tight control over the maneuver of his subordinate forces or elements. These patrols usually do not go outside the area under the immediate control of that commander's unit.

FIGHTING PATROL

8-89. A *fighting patrol* (FP) is a platoon-size element, normally composed of combat troops, dispatched from maneuver battalions (and sometimes companies) or detachments. When necessary, engineer and CBRN reconnaissance troops and other specialists can be allocated to the patrol. An FP normally moves in such a way that its parent unit can provide it indirect fire support. Units dispatch one or more FPs depending on the tactical situation. This may be—
- When conducting tactical movement.
- During battle in the absence of direct contact with the enemy.
- In other cases where it is difficult for the unit to directly observe the enemy's actions.

8-90. An FP is generally deployed to reconnoiter and provide security. The main missions of the FP are—
- Timely detection of an advancing enemy.
- Locating enemy direct-fire weapons (especially antitank weapons).
- Locating minefields.

8-91. An FP employs the same techniques as other reconnaissance patrols. Because of its security function, however, it is harder for the FP to avoid becoming engaged in combat with the enemy. It may engage a weaker enemy force using an ambush, or it may avoid contact altogether, taking up a concealed observation point or maneuvering around superior enemy forces. If it encounters what it considers to be enemy scouts or security elements, it attempts to penetrate them to locate the enemy's main force. Often FPs are also called upon to fix enemy forces they encounter, to permit other security elements to maneuver to destroy them.

RECONNAISSANCE DETACHMENT

8-92. The largest element the OPFOR employs at the tactical level to supplement specialized reconnaissance is the *reconnaissance detachment* (RD). It is typically a task-organized combat arms company or battalion. The detachment often receives such assets as tanks (if it is not a tank unit), air defense, artillery, engineers, or CBRN specialists. The RD dispatches platoon-size RPs to reconnoiter specific objectives along the detachment's axis.

8-93. Although an RD typically consists of combat troops, its primary mission is reconnaissance. If it does encounter a weak enemy force, it may engage that force and take prisoners. When the detachment encounters the enemy's main forces, it—
- Assumes an observation mission.
- Attempts to determine the composition and disposition of those forces.
- Reports to the commander who sent it out.
- Then continues its mission.

8-94. The RD is employed primarily in the offense. Its mission is to acquire information on the terrain and the enemy's location or gaps in his defenses. It can also reconnoiter key objectives. It conducts reconnaissance by observation, terrain inspection, ambushes, raids, and—only when necessary—by combat.

8-95. In the defense, in the absence of close contact with the enemy, a division, brigade, or tactical group may send out an RD into the disruption zone to determine the enemy's composition and main avenue of attack. The role of the RD is to establish contact with an advancing enemy force and monitor its progress. An RD can also reconnoiter enemy airborne or amphibious landing forces in support of an antilanding reserve.

RECONNAISSANCE METHODS

8-96. Reconnaissance elements collect information by various methods. For example, RPs can gather information using a number of standard methods, including—
- Observation.
- Raids.
- Ambushes.
- Reconnaissance attack (see chapter 3).

8-97. Other tactical reconnaissance elements may use some of the same techniques, as well as—
- Listening (eavesdropping).
- Imaging.
- Interception of transmissions and DF of electronic resources.
- Questioning of local inhabitants.
- Interrogation of prisoners of war and defectors.
- Study of documents and equipment captured from the enemy.

8-98. Information is also acquired during combat by maneuver units. Tactical units may also receive information on the enemy from higher headquarters and adjacent units.

OBSERVATION

8-99. Observation is the coordinated inspection of the enemy, terrain, weather, obstacles, and adjacent friendly forces during all types of combat activity. This type of reconnaissance, performed by troops conducting direct observation of the objective, is the most common method of gathering reconnaissance information. It is also one of the most reliable and accurate methods. In many cases, it is the only source of information.

8-100. The OPFOR has great confidence in the utility of observation, but it also recognizes the limitations. It is often difficult to determine enemy intentions through observation alone. To supplement observation, the OPFOR conducts raids and ambushes to capture information that can give a clearer picture of enemy strengths and intentions.

RAIDS

8-101. The raid is more aggressive than most methods of reconnaissance because it involves the active search for and engagement of selected enemy targets. A raid can occur in any terrain, in any season, at any time of day or night, and under various weather conditions. However, it is generally conducted at night or under conditions of limited visibility. Reconnaissance tactics involve two methods of conducting raids. The difference is in the purpose of the raid, the depth of the target, and the type of reconnaissance element performing it.

Reconnaissance Raid

8-102. The primary goal of a *reconnaissance raid* is to obtain information. Any damage or destruction of enemy installations is incidental. The raiding element is usually a reconnaissance or maneuver unit up to platoon size, with some augmentation. The reconnaissance raid consists of—
- The covert approach of the raiding element to a preplanned and previously studied target (objective).
- A surprise attack to capture prisoners, documents, and equipment.
- A swift withdrawal to friendly positions.

8-103. The reconnaissance raid normally takes place in enemy-held terrain, typically during preparation for an attack. The depth of the raid is limited to the enemy's forward edge or his immediate tactical depth. Typical targets are individual soldiers or small groups of soldiers. These might be—

- Isolated firing positions, OPs, and observers.
- Isolated sentries and guard posts.
- Couriers.
- Small, isolated work details.
- Staff elements.
- Communications centers.

Other Raids

8-104. Most raids are conducted for the purpose of capturing, destroying, or disabling a high-value target or possibly just confusing or deceiving the enemy. In contrast to a reconnaissance raid, which is conducted silently where possible, the basis for this type of raid is a skillful combination of surprise, firepower, and violence. However, a secondary or incidental result may be the securing of reconnaissance information by taking prisoners, documents, and combat equipment. The raiding element or force could be as large as a combat arms detachment that has penetrated the enemy's forward edge. However, it could also be an SPF team or a dedicated reconnaissance element inserted deep into enemy territory. See chapter 3 for information on the execution of raids in general.

AMBUSH

8-105. Reconnaissance by ambush (*reconnaissance ambush*) is a method of reconnaissance accomplished by surprise attack, from cover, for the purpose of seizing prisoners, documents, and samples of weapons or equipment. The ambush is similar to the raid, but is more of a passive tactic. The ambushing unit selects a concealed position along a probable route of enemy travel and attacks enemy units when the situation is favorable. The ambushing unit can consist of a specialized reconnaissance patrol or infantry unit.

8-106. Typical targets for ambush are solitary enemy soldiers or small groups moving on foot or in vehicles. The most favorable conditions for finding such isolated targets are when the enemy is preparing for an attack or when he is regrouping or relieving his forces. In preparing for an attack, the enemy sends out reconnaissance elements and small groups of engineers looking for passages in obstacles. There is also increased movement within the enemy position. During regrouping or relief, newly assigned enemy personnel who are unfamiliar with the terrain and situation may become isolated.

8-107. Information collection is the most common purpose of an ambush conducted by reconnaissance patrols. However, patrols also may execute an ambush to delay reserves or to inflict damage on a target of opportunity. Reconnaissance ambushes can occur in all kinds of battle, on any terrain, at any time of year or day, and under various weather conditions. For information on the execution of ambushes, see chapter 3.

RECONNAISSANCE ATTACK

8-108. A *reconnaissance attack* is a tactical offensive action that locates moving, dispersed, or concealed enemy elements and either fixes or destroys them. The purpose of the reconnaissance attack can be to find the enemy and to attack him (sometimes referred to as "search and attack"). However, the purpose can be to find the enemy but not attack him. Instead of attacking, the OPFOR may use this opportunity to gain information that answers important questions about the enemy's location, dispositions, military capabilities, and quite possibly his intentions.

8-109. The use of the term *reconnaissance* in the title of this method reflects the emphasis on the use of reconnaissance assets inherent in the organization's mission. Even if the reconnaissance attack is executed by a battalion, company or detachment, that organization may dispatch platoon-size patrols or squad-size reconnaissance teams to capture prisoners, documents, and equipment. However, a platoon- or squad-size element is not well suited for attacking the enemy.

8-110. The reconnaissance attack is the most ambitious—and least preferred—method to gain information. When other means of gaining information have failed, a reconnaissance detachment (or any detachment) can undertake a reconnaissance attack. The reconnaissance objectives may be force-, terrain-, or facility-oriented, but the overall objective of a reconnaissance attack is force-oriented.

8-111. The OPFOR recognizes that an enemy will take significant measures to prevent the OPFOR from gaining critical intelligence. Therefore, quite often the OPFOR will have to fight for information, using an offensive operation to penetrate or circumvent the enemy's security forces to determine who and/or what is located where or doing what. (See chapter 3 for additional information on the reconnaissance attack.)

Chapter 9

Indirect Fire Support

Modern battle is, above all, a firefight—in which indirect fire plays a decisive role in the effective engagement of the enemy. Uninterrupted and very close cooperation with the maneuver of supported combined arms units is the basis of the actions of indirect fire support units.

FIRE SUPPORT CONCEPTS

9-1. Fire support is the collective and coordinated use of target acquisition, indirect fire weapons, and aircraft, integrated with other lethal and nonlethal means, to engage enemy forces in support of a battle plan. The goal is to synchronize all available fire support systems to achieve the most effective results, thereby maximizing combat power. Effective fire support enables OPFOR ground forces to attack successfully and quickly to exploit weaknesses. Commanders try to accomplish their missions using a combination of maneuver, fires, and information warfare (INFOWAR).

9-2. The OPFOR stresses that fire support should combine air assets, surface-to-surface missiles (SSMs), and artillery into an integrated attack of enemy targets throughout the area of responsibility (AOR). The combined arms commander always seeks to increase the effectiveness of air and missile strikes and artillery fire to destroy enemy formations, weapon systems, or key components of an enemy's combat system. (See Systems Warfare later in this chapter and in chapter 1.) This ensures continuous fire support for maneuver units throughout the AOR.

9-3. The OPFOR believes that fire support must be integrated with INFOWAR. INFOWAR provides a nonlethal alternative or supplement to attack by fire and maneuver. It is integrated into the overall concept of the battle, to confuse, deceive, delay, disrupt, disable, and disorganize the enemy at all levels. Fire support can play a role in the physical destruction element of INFOWAR.

9-4. The integration of air, artillery, SSM, and nonlethal assets into a unified fire support plan is a major task for the combined arms commander. Integration is a decisive element, fundamental to the success of any tactical action on the modern battlefield. The OPFOR does not consider itself to be an artillery-centric force. Rather, it views itself as using various forms of fire support to achieve success during offensive and defensive combat. In the offense, fire support is important to the success of any attack. It can destroy key enemy systems; disrupt, immobilize, or destroy enemy groupings; and repel counterattacks. Fire support is also the cornerstone of any defense, blunting attacks at the crucial point in the battle. It disrupts enemy preparations for the attack and repels forces.

FIRE SUPPORT PRINCIPLES

9-5. The principles of fire support are the framework for a thought process that ensures the most effective use of fire support assets. These principles apply at all levels of command, regardless of the specific fire support assets available:

- Plan early and continuously.
- Exploit all available reconnaissance, intelligence, surveillance, and target acquisition (RISTA) assets.
- Consider airspace management and the use of all fire support (lethal and nonlethal) means.
- Use the lowest level of command capable of furnishing effective support.
- Avoid unnecessary duplication of effort.

Chapter 9

- Use the most effective means to accomplish the mission.
- Provide rapid and effective coordination.
- Provide for flexibility of employment.
- Provide for safeguarding and survivability of OPFOR fire support assets.
- Attempt to achieve surprise when possible.
- Deliver highly accurate and effective fire.
- Integrate fire support with maneuver and INFOWAR at all levels.

SYSTEMS WARFARE

9-6. The foundation of OPFOR planning is the systems warfare approach to combat. Thus, the OPFOR analyzes its own combat system and how it can use the combined effects of this "system of systems" to degrade or destroy the enemy's combat system. In systems warfare, the subsystems or components of a combat system are targeted and destroyed individually. Once a favorable combat situation has developed, the targeted enemy subsystem is quickly destroyed in high-intensity battle, thus making the enemy's overall combat system vulnerable to destruction or at least degrading its effectiveness. (See Systems Warfare in chapter 1 for further information.)

9-7. Within the systems warfare approach, the OPFOR employs a fire support concept centered on a phased cycle consisting of—

- Finding a critical component of the enemy's combat system and determining its location with RISTA assets.
- Engaging it with precision fires, maneuver, or other means.
- Recovering to support the fight against another part of the enemy force.

The primary reason for attacking an enemy with fires is to destroy one or more key components of the enemy's combat system and/or to create favorable conditions for destroying other parts of his combat system.

TECHNIQUES TO EXPLOIT ENEMY VULNERABILITIES

9-8. The OPFOR seeks to avoid its enemy's strengths and exploits his vulnerabilities. When the OPFOR is operating from a position of relative strategic weakness, it seeks to tactically outmaneuver, overwhelm, and outpace the enemy. It also seeks to deny the enemy any sanctuary on the battlefield, as well as in the local theater or in his strategic depth.

9-9. The OPFOR will use all fire support means (primarily aviation, SSMs, and long-range rocket strikes) to attack targets in the homeland of an opponent in the region. In a strategic campaign, the OPFOR may use various fire support assets in access-limitation operations and attack of the enemy's lines of communications and rear. It will attack the most vulnerable parts of the enemy's combat system. This includes strikes on the infrastructure and even civilian targets. Such OPFOR attacks will be coordinated with perception management efforts to convey the view that these terror tactics are no worse than enemy bombing campaigns.

9-10. The OPFOR will also leverage the effects of its available fire support means by integrating them into an integrated fires command (IFC) in organizations down to division or division tactical group (DTG) level. The IFC (described in detail later in this chapter) synchronizes and focuses the efforts of RISTA and fire to destroy key enemy formations or systems—or key components of an enemy's combat system. Destroying such targets can not only shift the balance of combat power in the OPFOR's favor, but also undermine enemy morale and resolve.

TARGET DAMAGE CRITERIA

9-11. Target damage is the effect of fires on a given military target. It results in total, partial, or temporary loss of the target's combat effectiveness. The OPFOR categories of target damage are annihilation,

Indirect Fire Support

demolition, neutralization, and harassment. Of these categories, the first three fall under the general term *destruction*.

Annihilation

9-12. Annihilation fires render targets completely combat-ineffective and incapable of reconstruction or token resistance. For a point target such as an antitank guided missile launcher, the OPFOR must expend enough munitions to ensure a 70 to 90 percent probability of kill. For area targets such as platoon strong points or artillery firing positions, the OPFOR must fire enough rounds to destroy from 50 to 60 percent of the targets within the group. These fires result in the group ceasing to exist as a viable fighting force.

Demolition

9-13. The OPFOR uses the term *demolition* in reference to the destruction of buildings and engineer works (such as bridges, fortifications, or roads). Demolition requires enough munitions to make such material objects unfit for further use.

Neutralization

9-14. Fire for neutralization inflicts enough losses on a target to—
- Cause it to temporarily lose its combat effectiveness, or
- Restrict or prohibit its maneuver, or
- Disrupt its command and control (C2) capability.

To achieve neutralization, the OPFOR must deliver enough munitions to destroy 30 percent of a group of unobserved targets. The expectation is that the target is severely damaged but could again become capable of coordinated resistance after the fire is lifted. The term *neutralization* applies only in an artillery context.

Harassment

9-15. The OPFOR uses a limited number of fire support systems and munitions within a prescribed time to deliver harassment fires. The goal of these fires is to put psychological pressure on enemy personnel in locations such as defensive positions, command posts (CPs), and logistics installations. Successful harassment fire inhibits maneuver, lowers morale, interrupts rest, and weakens enemy combat readiness.

Note. The OPFOR carefully calculates fire support requirements in terms of weapons and munitions needed to produce a required effect on enemy targets. If insufficient fire support or ammunition is available to achieve the necessary result, the OPFOR does not fire less and hope for the best. Rather, if necessary, it engages fewer targets, adjusting the tactical, or even operational, fire support plan.

INDIRECT FIRE SUPPORT WEAPONS

9-16. In addition to aviation means (discussed in chapter 10), OPFOR indirect fire support weapons consist of mortars, cannon systems, multiple rocket launchers (MRLs), and SSMs. These systems can be either towed or self-propelled.
- **Mortars.** All OPFOR infantry, motorized infantry, and mechanized infantry battalions contain constituent 120-mm mortars. Smaller mortars are also available. Guerrilla and other organizations may have them as well.
- **Cannon systems.** Cannon artillery includes field guns, howitzers, and hybrid systems.
- **Multiple rocket launchers.** The OPFOR categorizes MRLs as medium-caliber (100- up to 220-mm) and as large-caliber (220-mm and larger).
- **Surface-to-surface missiles.** SSMs include tactical- through strategic-level ballistic missiles and land-attack cruise missiles using warheads ranging from conventional to nuclear.

For additional information on indirect fire support weapons and available ammunition types, see the *Worldwide Equipment Guide*. For information on these weapons in OPFOR organizations, see FM 7-100.4.

9-17. The majority of OPFOR artillery (152-mm and above) and large-caliber MRLs are capable of firing nuclear munitions. The majority of artillery and MRL units can fire chemical munitions. However, only select units will be issued the nuclear or chemical munitions. Nevertheless, continued improvements in conventional munitions, especially precision munitions, increase the likelihood that the OPFOR can achieve operational- or tactical-level fire superiority at the desired location and time without resorting to CBRN weapons.

COMMAND AND CONTROL

9-18. OPFOR tactical fire support is designed to be controlled at the lowest possible level. This ensures flexibility, survivability and the proper level of support to the tactical commander. OPFOR commanders allocate fire support assets and means to subordinates in direct correlation to their need based on the scheme of maneuver. The OPFOR does not retain assets at a higher level simply to preserve flexibility at that level. If a subordinate needs an asset to accomplish a mission, every effort is made to ensure he has it.

COMMAND AND SUPPORT RELATIONSHIPS

9-19. Units that provide indirect fire support for maneuver units may have one of three command and support relationships: constituent, dedicated, or supporting:
- Commanders of indirect fire support units in a subordinate (constituent or dedicated) status report directly to the commander of the maneuver unit or IFC to which they are subordinate. Units in a dedicated status continue to receive logistics support from their parent indirect fire support headquarters.
- Commanders of indirect fire support units in a supporting status are commanded by their parent organization but receive missions from their supported headquarters for the duration of the relationship.

See chapter 2 for detailed discussion of command and support relationships.

INTEGRATED FIRES COMMAND

9-20. The IFC is a combination of a standing C2 structure and a task organization of constituent and dedicated fire support units. All division-level and above OPFOR organizations possess an IFC C2 structure—staff, CP, communications and intelligence architecture, and automated fire control system (AFCS). The IFC exercises C2 of all subordinate (constituent and dedicated) indirect fire support assets retained by its level of command. This includes army aviation, artillery, and SSM units. It also exercises C2 over all RISTA assets allocated to it.

> *Note.* Based on mission requirements, the division or DTG commander may also allocate maneuver forces to the IFC commander. This is most often done when he chooses to use the IFC CP to provide C2 for a strike (see chapters 1 and 3). However, it can also be done for the execution of other missions. One possibility would be for the IFC CP to command the disruption force, the exploitation force, or any other functional force whose actions must be closely coordinated with fires delivered by the IFC.

9-21. In combat, the IFC forms the framework for the C2 of indirect fires in the division or DTG. A division or DTG always has an IFC, even if it receives no additional fire support units during task-organizing. There is one IFC per division or DTG, to support the tactical battle plan. However, each IFC is capable of engaging designated operational and strategic targets, if necessary.

Indirect Fire Support

9-22. The division or DTG deputy commander (DC) is the IFC commander. Through his IFC commander, the division or DTG commander exercises C2 over fire support and associated RISTA assets retained in the IFC. The following procedures apply to this process:
- The division or DTG commander specifies the organization of forces for combat and the tasks for indirect fire support assets.
- The IFC commander conducts and coordinates fire support planning. He also coordinates with the division or DTG chief of reconnaissance and the reconnaissance subsection for targeting data.

9-23. The IFC commander can also control (but not command) fire support and RISTA assets allocated to the division or DTG in a supporting relationship. He can give them mission priorities, but they are still commanded by their parent organization.

9-24. Fire support assets that are allocated to a DTG and not used in the IFC are allocated, in a constituent or dedicated relationship, to subordinate brigade tactical groups (BTGs). Fire support units remaining under IFC command may provide fires for maneuver brigades or BTGs in a supporting relationship. The supporting relationship allows the IFC commander the flexibility to task fire support assets to engage key enemy targets throughout the AOR.

9-25. The number and type of fire support and RISTA units allocated to an IFC is mission-dependent. The IFC is not organized according to a table of organization and equipment, but is task-organized to accomplish the missions assigned. Figure 9-1 shows the possible components of an IFC at DTG level.

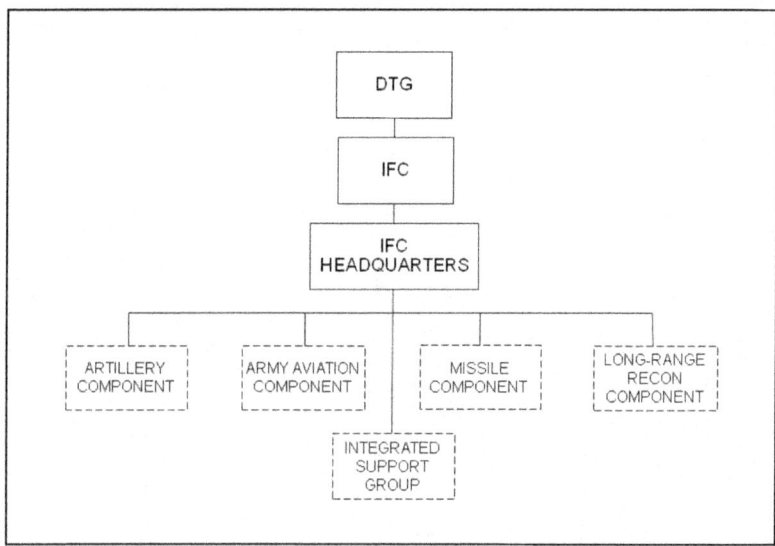

Figure 9-1. Possible IFC components in a DTG

Chapter 9

9-26. The mission of the IFC is to execute all fire support tasks required to accomplish the division or DTG mission. It is designed to—
- Exploit the combat power inherent in carefully integrated ground and air fire support actions.
- Reduce to the absolute minimum the amount of time from target acquisition to engagement.
- Permit fire support assets to mass their effects without having to operate in concentrated formations.
- Ensure the optimal fire support asset(s) are assigned any given mission.
- Ensure adherence to the commander's priorities for fire support.
- Integrate the effects of fires from units placed in support of the division or DTG.
- Act, if necessary, as the division's or DTG's alternate command structure.

IFC Headquarters

9-27. The IFC headquarters component is composed of the IFC commander and his command section, a RISTA and INFOWAR section, an operations section, and an integrated support section. (See figure 9-2.) The DC of the division or DTG serves as IFC commander. The RISTA and INFOWAR section provides the complete spectrum of intelligence support to the IFC as well as INFOWAR support for the headquarters component. The operations section provides the control, coordination, and communications for the headquarters component. Located within the operations section is the fire support coordination center (FSCC). The integrated support section provides control and coordination of various support functions. The IFC headquarters relies principally on direct liaison among subordinate units to ensure the necessary coordination of fire support. A division's standing IFC headquarters should also be capable of exercising C2 over additional assets allocated to augment the IFC in a DTG.

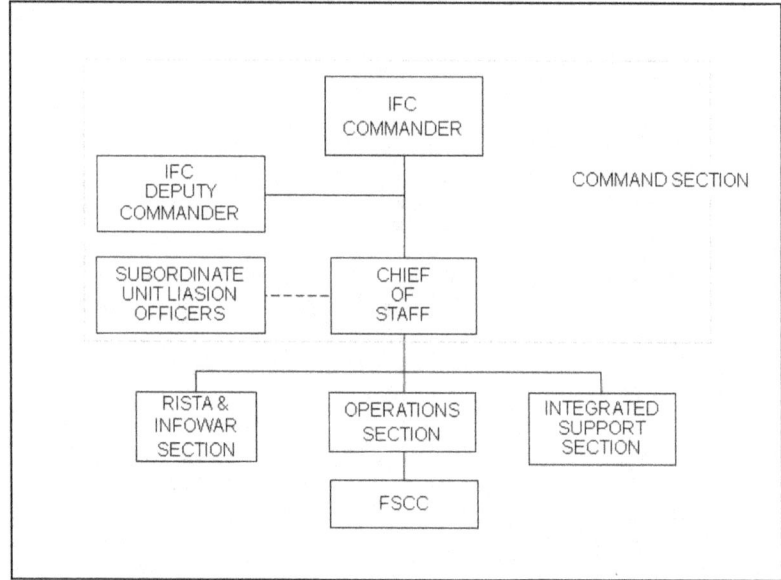

Figure 9-2. IFC headquarters

Indirect Fire Support

Note. A division or DTG will still retain a core IFC headquarters, even if it loses its originally subordinate fire support assets to another command during task-organizing. This facilitates C2 of any fire support assets that might be reallocated to that division or DTG during some subsequent phase of combat.

Artillery Component

9-28. The artillery component is a task organization tailored for the conduct of artillery support during combat. In a division IFC, it consists of all assets of the division's artillery brigade that are not allocated in a constituent or dedicated relationship to subordinate maneuver brigades. In a DTG, it is organized around—

- Assets of the artillery brigade of the division that served as the basis for the DTG (except for assets suballocated for the task organization of subordinate maneuver brigades).
- Any additional artillery units that are allocated to the DTG and remain subordinate to the DTG in a constituent or dedicated relationship rather than being further suballocated to subordinates.

Such additional assets usually come to the DTG through the operational-strategic command (OSC), when the OSC decided not to retain them at its own level of command. The artillery component includes appropriate target acquisition, C2, and logistics support assets.

9-29. The number of artillery battalions assigned to an IFC varies according such factors as mission of friendly units, the enemy situation, and terrain. However, the number of artillery units also can vary based on the capabilities of the supporting AFCS. For example, an MRL brigade AFCS can have enough command and staff vehicles for the brigade commander and his chief of staff, as well as the subordinate commanders of battalions and up to 18 batteries (6 battalions). An AFCS supporting a cannon, MRL, or mortar battalion may consist of enough command and staff vehicles to support 3 to 4 batteries (each consisting of 4 to 8 systems).

Army Aviation Component

9-30. The army aviation component is a task organization tailored for the conduct of tactical-level aviation missions. It is a flexible and balanced air combat organization capable of providing air support to the DTG commander. It includes attack aviation capability as well as requisite ground and air service support assets. The IFC commander exercises control through facilities provided by the airspace operations subsection of the DTG staff and/or the army aviation unit(s).

Missile Component

9-31. The missile component is a task organization consisting of long-range missiles or rockets capable of delivering conventional or CBRN munitions. It is organized around an SSM or rocket battalion or brigade and includes the appropriate logistics support assets. An OSC may or may not allocate SSMs or rockets to support a DTG IFC in a constituent or dedicated relationship.

Long-Range Reconnaissance Component

9-32. The long-range reconnaissance (LRR) component normally consists of assets from the division's constituent LRR company. Personnel of this company are specially trained for insertion in small reconnaissance teams at distances up to 100 km from the rest of the division. An OSC may allocate additional LRR assets to the DTG, if the OSC has received such units from the administrative force structure (AFS).

9-33. The LRR component provides the OPFOR the ability to reconnoiter and attack enemy forces throughout their tactical depth. The LRR assets conduct actions to achieve tactical objectives. Such actions may have either long-range or immediate impact on the enemy. The OPFOR concept of LRR missions includes reconnaissance, direct action, and diversionary measures.

9-34. If an OSC has received special-purpose forces (SPF) units, it may further allocate some of these units to supplement the LRR assets a DTG has in its IFC. However, the scarce SPF assets normally would remain at OSC level. Even when allocated to a DTG, probably in a supporting status, the SPF would pursue tactical goals in support of operational objectives. SPF are trained for missions similar to those of the LRR company, but may be inserted to even greater distances.

Integrated Support Group

9-35. The integrated support group (ISG) is a compilation of units performing logistics tasks that support the IFC. Other combat support and combat service support units may be grouped in this component for organizational efficiency although they may support only one of the major units of the IFC. For a DGT IFC, the ISG typically is a task-organized composite of division-level units and other units suballocated from the OSC. The ISG is discussed in detail in chapter 14.

CHIEF OF INTEGRATED FIRES

9-36. Within the operations section of the division or DTG staff, there is a *chief of integrated fires*. This officer is responsible for coordinating and advising the commander on the effective integration of C2 and RISTA means with fire support means (including precision fires) to support the overall battle plan. He controls, but does not command, the fire support units subordinate to or supporting the division or DTG. He advises the division or DTG commander on how best to use available fire support assets.

CHIEF OF FIRE SUPPORT COORDINATION

9-37. On the staff of a maneuver brigade or BTG, there is an officer responsible for planning and coordinating indirect fire support. At this level, his title is *chief of fire support coordination* (CFSC). The CFSC controls, but does not command, the indirect fire support units subordinate to or supporting his maneuver unit. He advises the maneuver commander on how best to use available fire support assets. The CFSC heads the fire support coordination element of the functional staff.

FIRE SUPPORT COORDINATION CENTER

9-38. An FSCC is established at each organizational level from maneuver brigade to IFC. The FSCC is the staff element responsible for the planning and coordination of fires to support the respective maneuver unit. It performs the following battle coordination functions:

- Acquire and identify high-payoff targets (HPTs).
- Recommend targets.
- Use target value analysis to identify target priorities.
- Determine fire support needs.
- Expedite fire support.
- Assess fire support effects.
- Change fire support plans.
- Coordinate the timing of fire support attacks (sometimes in support of INFOWAR objectives).
- Recommend the use of aviation.

The FSCC also disseminates information on fire support to commanders and staffs of maneuver forces and/or aviation units operating in the same AOR, in order to reduce potential conflicts. This information includes firing positions, targeted areas, and fire support plans.

BRIGADE-LEVEL FIRE SUPPORT

9-39. The brigade or BTG command group consists of the commander, DC, and chief of staff, and functional staff. The primary difference is that the DC does not serve as IFC commander, since there is no IFC at this level of command. At this level, there is also no chief of integrated fires in the operations section of the staff; instead, there is a CFSC.

9-40. A brigade that does not receive augmentation has whatever fire support unit was constituent to the brigade in the AFS. The constituent fire support is directly under the C2 of the brigade commander, who is advised by his CFSC and the FSCC in the operations section of his staff.

9-41. A brigade that becomes a BTG may receive additional fire support units (artillery, SSMs, or army aviation) in a constituent or dedicated relationship. Each of these additional units, along with whatever fire support unit was originally constituent to the base brigade in the AFS, is under the direct C2 of the BTG commander, advised by his CFSC and FSCC. However, it is possible that the BTG's higher command may have reallocated some or all of the base brigade's original fire support assets to some other subordinate during the task-organizing of its fighting force structure.

9-42. Any brigade or BTG may also receive one or more additional fire support units allocated to it in a supporting relationship. In this case, the supporting fire support unit(s) remain under the command of their parent headquarters (be that a fire support headquarters in some higher command's IFC or some higher headquarters remaining in its original status from the AFS). The supporting unit may or may not be located in the supported brigade's or BTG's AOR. However, the supported brigade or BTG commander can give mission priorities to these supporting fire support units and (if the supporting units are in his AOR) position these assets to carry out such missions.

FIRE SUPPORT BELOW BRIGADE LEVEL

9-43. A brigade or BTG can allocate some of its constituent or dedicated indirect fire support assets to a maneuver battalion in a constituent or dedicated relationship—in which case the battalion would become a detachment. More commonly, however, the brigade or BTG could employ some of its constituent or dedicated fire support units (or parts of units) to provide fire support for a particular battalion or detachment in a supporting relationship.

9-44. A maneuver battalion staff does not include an FSCC or a functional staff with a fire support coordination element. Nevertheless, there is still a chief of fire support coordination (CFSC) responsible for that function. The maneuver battalion's assistant operations officer functions as the CFSC in those units (such as a tank battalion) that do not have a mortar battery but still require fire support. Otherwise, the commander of the mortar battery also serves as the CFSC. The mortar battery headquarters contains a fire control section to coordinate battalion fires. (See FM 7-100.4.)

FIRE REQUESTS

9-45. Requests for supporting fires may originate at any organizational level. They are initiated when one or more of the following conditions exist:

- Constituent or dedicated fire support means at that level are fully engaged.
- The target is beyond the range of constituent or dedicated fire support means.
- The constituent or dedicated fire support means have suffered combat loss.

9-46. There are two methods of requesting supporting fires. (The following explanation illustrates how these methods work when a brigade or BTG requests fires from division or DTG level.) The preferred method is for the request to be forwarded from the brigade or BTG commander to the integrated fires subsection in the division or DTG headquarters. An alternate method is for the brigade or BTG commander to request supporting fires from the division or DTG commander. The division or DTG commander either approves or denies the request. If the request is approved, the division or DTG commander tasks the IFC to provide the requested support.

NAVAL FIRE SUPPORT

9-47. Naval fire support, when available, is not allocated to a DTG as part of its IFC, since a DTG is not a joint command. Rather, naval assets may be allocated to an IFC at OSC or theater level. Naval fire support (which includes shipborne gunfire and sea-launched cruise missiles) can give the OSC commander another means of long-range indirect fires. A division or DTG (or a separate brigade or BTG) can request naval fire support through OSC channels.

9-48. A theater or OSC that receives naval fire support assets in a constituent or dedicated relationship may further allocate such naval assets to a division or DTG in a *supporting* relationship. However, such naval assets remain under the command of the theater- or OSC-level IFC.

9-49. Another option is for naval fire support assets to remain under the command of the Navy but to provide support for ground operations. During the course of such a supporting relationship, if enemy actions threaten naval operations, the target attack priorities of the ship may cause it to suspend or cancel land fire missions. Once the threats have subsided, the fire support assets resume their support of the ground maneuver force.

9-50. A naval fire support liaison team augments the operations section of the division- or DTG-level IFC staff when naval fire support is required to support a ground maneuver force, even in a supporting relationship. The liaison team provides special staff representation and advice on naval fire support to the IFC commander. Additionally, it coordinates requests for naval fire support and operates the naval fire support nets in the IFC FSCC.

9-51. Members of the naval fire support liaison team are specially trained in the conduct of naval gunfire. However, the observer procedures are simplified and standardized so that any observer from the supported ground force unit can effectively adjust the fires of a supporting naval vessel with a minimum of additional training.

CONTROL OF FIRE SUPPORT OBSERVERS

9-52. The FSCC has three control options available to it when monitoring observers' requests for fire. (See figure 9-3 on page 9-14 for various methods of reporting targets for attack, starting from the point of detection by a human observer or other sensor.) After considering the tactical mission, the degree of training of the observers, and the availability of fire support assets, the FSCC determines which option is best suited for the mission.

Decentralized Option

9-53. The observer may call for fire from any fire support assets available to support the mission. This is the most responsive request, but allows the FSCC the least amount of control. Since the observer is allowed to determine which asset should engage each target, this option generally requires a highly trained observer.

Predesignated Option

9-54. The observer is assigned a particular fire asset from which he may request support, and he operates on that fire unit's radio net. If the observer thinks that the target requires a different fire support asset, he must request permission from the FSCC to change assets. Permission is granted on a case-by-case basis. Under this option, fire support is highly responsive, if the FSCC determines that the asset is suitable to the type of target.

Centralized Option

9-55. The observer must contact the FSCC for each call for fire. Then the FSCC refers the observer, or relays his request, to an appropriate fire support asset. This option is the least responsive for the observer but offers the highest degree of control to the FSCC. This option is generally used when a maneuver commander acts as an observer.

Tailoring

9-56. Since the level of training and the tactical situation vary for each observer, the FSCC may assign the appropriate option to each supported unit. For example—
- An SPF or reconnaissance unit may be predesignated.
- A maneuver unit may be centralized.
- An observer from an indirect fire support unit may be decentralized.

Indirect Fire Support

FIRE SUPPORT PLANNING

9-57. Fire support planning is the determination of the content, manner, and sequence of delivery of fire on the enemy in a battle or operation. The fire support planning process includes—
- Target acquisition.
- Requirements for allocation of weapons and units (task-organizing of forces for combat).
- Assignment of tactical fire support missions to IFCs, units, and weapons.
- The manner and procedure of delivery of fire during the performance of missions.
- Determination of ammunition requirements by missions.
- Organization of coordination and C2.
- Preparation of appropriately detailed fire support plans at various levels.

9-58. In the OPFOR's "top-down" approach to the planning and allocation of indirect fire support, fire support planning occurs at the highest level possible. The IFC commander at the OSC and division or DTG levels or the CFSC at brigade or BTG level plans and coordinates indirect fire support, always under the direction of the maneuver commander. The highest level of participating units coordinates and approves the fire support plan, with input from subordinate units. OSC and division or DTG headquarters perform general fire support planning. Detailed planning occurs in maneuver brigades or BTGs, IFCs, and indirect fire support units. The fires of all indirect fire support units within a brigade or BTG are incorporated into the brigade or BTG fire support plan. In turn, brigade or BTG fire support plans become part of division or DTG fire support plans. Division or DTG fire support plans become part of OSC fire support plans.

9-59. In its simplest form, fire support planning is the process of determining the best way to engage all of the enemy's units with fires. It must ensure that the required level of damage is inflicted in a manner consistent with the commander's concept of the battle. Above all else, this means that the fire support plan must match his concept for the sequence in which the battle will develop. The focus of fire support planning is on establishing and maintaining fire superiority over the enemy. Therefore, timing is critical.

ESTIMATE OF SITUATION

9-60. The planning process begins with an estimate of the situation. This estimate includes consideration of the following:
- The scheme of maneuver of supported forces.
- The enemy force to receive fire.
- The locations and types of individual targets within the designated enemy force.
- The required or desired level of target damage.
- Fire support assets available, both delivery systems and ordnance.

9-61. The commander of an indirect fire support unit at any level coordinates the fires under his control. He determines new requirements and missions and, with the IFC commander or brigade CFSC, makes suggestions to the maneuver commander about adjustments in tactical organization as the situation develops.

IFC PLANNING

9-62. The division or DTG commander, his IFC commander, and other staff members establish the basis for fire support planning during the commander's reconnaissance of the area of anticipated action. During this reconnaissance, the commander refines the organization of forces for combat and the means of coordination. The division or DTG commander gives the IFC commander the information base to determine the following:
- Targets for indirect fire weapons to engage and fire upon.
- Priority of each target.
- Sequence in which to attack targets.
- Time to attack each target.

9-63. An IFC commander and members of his staff conduct their planning in coordination with the rest of the division or DTG staff, concurrently with developing the battle plan. Planning considerations include target type, dimensions, degree of protection, mobility, and range to the target.

FIRE SUPPORT COORDINATION MEASURES

9-64. Fires pose a potential hazard to friendly maneuver forces and aircraft activities. (See chapter 10 for more information on air and artillery coordination measures.) To reduce potential conflicts between indirect fires and maneuver forces or aircraft, information pertaining to firing positions, targeted areas, and fire support plans is distributed to commanders and their staffs. The fire support plan includes a map with graphics outlining the following control lines:

- **Coordinated fire line.** A line beyond which indirect fire systems can fire at any time within the AOR of the establishing headquarters without additional coordination.
- **Final coordination line.** A line established by the appropriate maneuver commander to ensure coordination of fire of converging friendly forces. It can be used to prohibit fires or the effects of fires across the line without coordination with the affected force. For example, this line may be used during linkup operations between an airborne or heliborne insertion and converging ground forces.
- **Joint fire line.** A line established by the appropriate OSC-level and above commander to ensure coordination of fire not under his control but which may affect his operations. The joint fire line is used to coordinate fires of air, ground, or sea weapons systems using various types of ammunition against surface targets.
- **Safety line.** A line that denotes the fragmentation footprint of indirect fire munitions or of bombs or rockets released from aircraft. This indicates the minimum distance between the impact area and the nearest friendly troops.

ASSIGNING FIRE MISSIONS

9-65. When assigning missions, indirect fire support commanders and planners consider several factors, depending on the situation. These factors include—

- Type of target (for example, equipment or personnel, deliberate or hasty defensive positions, hard- or soft-skinned vehicles, point or area targets).
- Deployment of target (dug-in or in the open).
- Whether the target is stationary or moving.
- Whether the target is under direct observation during an artillery attack.
- Range to the target.
- Type, caliber, and number of weapons engaging the target.
- Types of ammunition available.
- Time available to prepare for firing.

TARGETING

9-66. Targeting (selecting and prioritizing targets) requires constant interaction between maneuver, reconnaissance, fire support, and INFOWAR at all levels. Target value analysis is an analytical tool that is used in the targeting process by which the supported maneuver commander—

- Provides focus for his target acquisition effort.
- Identifies priorities for the engagement of enemy targets that will facilitate the success of his mission.
- Identifies the target damage criteria.
- Permits planning for identified contingencies based on enemy options available when the enemy operation fails.

Indirect Fire Support

TARGET ANALYSIS

9-67. *Target analysis* is the examination of potential targets to determine military importance, priority for attack, and weapons required to obtain a desired level of damage or casualties. Targets are not attacked indiscriminately but are part of an overall scheme or plan to destroy an enemy complex. A target complex is a series of interrelated or dependent target elements that together serve a common function. The target could also be part of the infrastructure or a particular part of the enemy's combat system.

9-68. The OPFOR considers the following five factors for selection of targets in a particular target complex:
- Criticality.
- Vulnerability.
- Accessibility.
- Recoverability.
- Effect on the local population.

9-69. The FSCC uses diagrams, maps, photographs, and other intelligence to analyze a target complex and select targets for attack that offer maximum timeliness and effect. The analysis enables the IFC or maneuver commander to select the appropriate system or mechanism to conduct an attack. Some of the simplest operations can either cause or create favorable conditions for great damage to the enemy.

HIGH-VALUE AND HIGH-PAYOFF TARGETS

9-70. High-value targets (HVTs) are assets the enemy commander requires for the successful completion of his mission. High-payoff targets (HPTs) are whose loss to the enemy will significantly contribute to the success of the OPFOR mission. While target value is usually the greatest factor contributing to the target payoff, other considerations include the following:
- Sequence or order of occurrence.
- Ability to locate and identify the target.
- Degree of accuracy and identification available from the acquisition system.
- Ability to engage and defeat the target in accordance with the established target damage criteria.
- Resource requirements necessary to accomplish all of the above.

The loss of HVTs or HPTs would be expected to seriously degrade the effectiveness of the enemy's combat system.

9-71. Based on a battlefield analysis, the maneuver commander, with advice from his IFC commander or CFSC, selects HVTs and HPTs and establishes a prioritized list of them. The list identifies the targets for a specific point in the battle in the order of their priority for acquisition and attack.

TIME-SENSITIVE TARGETS

9-72. Time-sensitive targets are those targets requiring an immediate response. The reason for the urgency is that they either pose (or will soon pose) a clear and present danger to the OPFOR or are highly lucrative, fleeting targets of opportunity.

TARGET ATTACK METHODOLOGY

9-73. The vast array of targets anticipated on the battlefield can generate competing demands for fire support. These demands could exceed the capability of fire support assets to adequately respond to all requirements. Critical to the success of OPFOR combat actions is the ability to plan, detect, deliver, and assess fire (in accordance with the commander's target damage criteria) against targets throughout the AOR. Therefore, the OPFOR uses the target attack methodology of plan, detect, deliver, and assess.

Chapter 9

Plan

9-74. The *plan* phase provides the focus and priorities for the reconnaissance collection management and fire planning process. It employs an estimate of enemy intent, capabilities, and vulnerabilities in conjunction with an understanding of the OPFOR mission and concept of battle. During the plan phase, the maneuver commander, with advice from his IFC commander or CFSC, makes a determination of—

- *What* HVTs and HPTs to look for.
- *When* and *where* they are likely to appear on the battlefield.
- *Who* (reconnaissance or target acquisition assets) can locate them.
- *How* the targets should be attacked.

Detect

9-75. During the *detect* phase, the reconnaissance plan is executed. As specified targets are located, the appropriate command and observation post (COP), fire control post, or delivery system is notified to initiate the attack of the target.

9-76. Figure 9-3 illustrates the varying methods of reporting targets for attack from the point of detection by a sensor through delivery. The figure displays the methods along a range from the least to the most responsive.

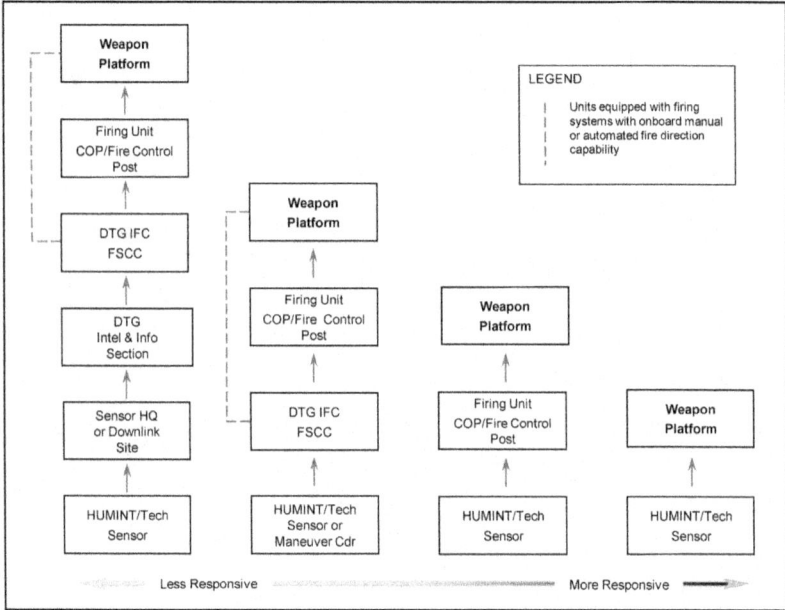

Figure 9-3. Target report flow

Indirect Fire Support

Deliver

9-77. The *deliver* phase is rapidly executed by having designated attack systems respond to the maneuver commander's guidance when HVTs or HPTs are observed. Timely, accurate delivery is the culmination of synchronized fire support.

Assess

9-78. Following the attack of the target, RISTA assets are cued to determine if the target has been defeated in accordance with the established target damage criteria. If it is determined that the target damage criteria are not achieved, delivery assets re-engage the target until the desired target damage has been achieved.

TARGET ACQUISITION AND RECONNAISSANCE

9-79. Indirect fire target acquisition is the process of detecting, identifying, and locating elements of the enemy to be engaged. This includes acquiring enemy mortar, cannon, and rocket units with sufficient accuracy, reliability, and responsiveness for counterfire and counterbattery fire to be directed against the enemy unit. Advances in RISTA and fire control systems provide the OPFOR a capability to rapidly disseminate information on suspected enemy targets within one minute or less. This includes the time from acquisition to firing data computation and the initial transmission of data to a firing battery.

RISTA ASSETS

9-80. The following are some examples of RISTA assets that can provide target information for indirect fires:

- **Weapon-locating radars.** Detect targets following a ballistic path.
- **Sound ranging.** Determine the precise location of hostile artillery by using data from the sound of its guns, mortars, or rockets firing. A series of microphones capture the sound. A computer factors the intersection of the bearings and provides the location of the firing unit.
- **Battlefield surveillance radars.** Detect enemy activity or observe point targets to detect movements. They can detect and recognize moving targets including personnel, vehicles, watercraft, and low-flying aircraft and determine accurate locations (azimuth and range) of such targets. These radars can confirm targets sensed by other types of sensors or be used to cue other sensors and weapons. They can also determine the effectiveness of the attack on a target.
- **Unmanned aerial vehicles (UAVs).** Provide increased range and offer increased accuracy and responsiveness depending on the sensor suite chosen. The OPFOR has UAVs from strategic to company and specialized team level.
- **Visual observation.** Human intelligence (HUMINT) may consist of observation posts (OPs), artillery reconnaissance patrols, SPF, or tip-offs from maneuver elements or others. HUMINT may also include information from sympathetic or affiliated civilians or guerrillas.

For information on the systems see the *Worldwide Equipment Guide*. Figure 9-3 on page 9-14 describes various methods of reporting targets for attack, starting from the point of detection by a human observer or other sensor.

OBSERVATION POSTS

9-81. The OPFOR uses an extensive system of OPs to provide fire support to the maneuver forces. These OPs are mobile in order to accompany rapidly moving forces. They may be in wheeled or tracked vehicles, or in the air. The configuration depends upon the level of command and the type of units.

9-82. After establishing a functional OP, scout observers can construct a deception OP to confuse the enemy about the actual position of the OP. Figure 9-4 on page 9-16 shows an example of the deployment of some of the most common types of OPs discussed below. Other vehicles serve as fire control posts and mobile reconnaissance posts (MRPs). Artillery commanders can also send out artillery reconnaissance patrols.

Chapter 9

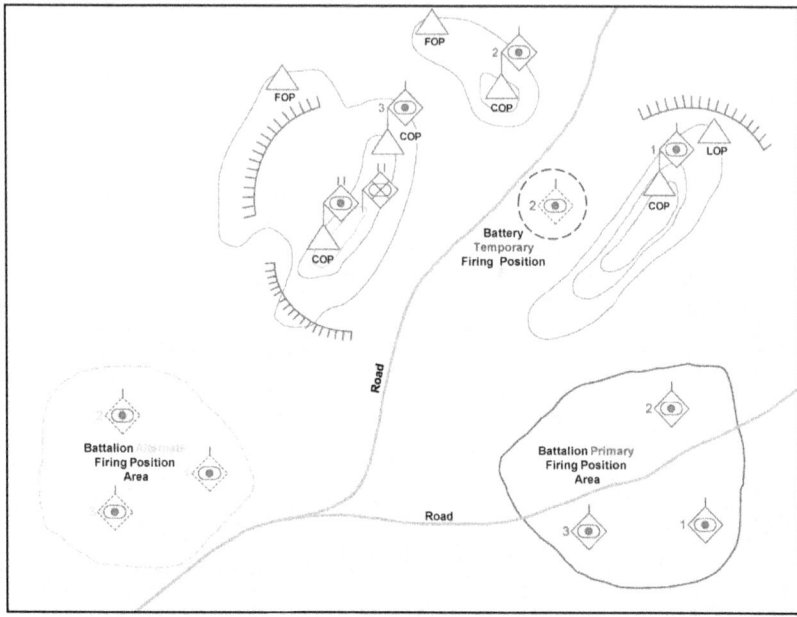

Figure 9-4. Observation posts in the battle formation of an artillery battalion (example)

Command and Observation Post

9-83. In indirect fire support battalions and batteries, the COP serves as both an OP and CP. From it, the fire support commander—
- Controls the fire and maneuver of his subordinates.
- Conducts reconnaissance of the enemy and terrain in his zone or sector of fire.
- Observes the actions of friendly combined arms units and maintains coordination with them.

Since the artillery battalion is the basic firing unit, its COP is the place where decisions are made and from which orders stem.

9-84. The battalion COP normally contains the battalion commander, chief of reconnaissance, and chief of communications. The battery COP normally includes the battery commander and the control platoon leader (who is responsible for reconnaissance and signal functions). In addition to these officers, COPs at both battalion and battery level include fire direction, communications, and reconnaissance personnel. In either case, these personnel can operate the COP either on the ground or mounted in an artillery command and reconnaissance vehicle.

Note. Depending on the type of combat action, the chief of communications may be part of the battalion fire control post, rather than the COP.

9-85. In most cases, the commander of a constituent or dedicated artillery unit colocates his COP with the main or forward CP of the maneuver unit commander to which the artillery unit is constituent or dedicated, or at least locates it near that CP. From that position, both the maneuver and artillery commanders should be able to observe the zone of responsibility or sector of fire. When the artillery unit is neither constituent nor dedicated, but merely supporting and remains directly subordinate to the senior maneuver or IFC commander, its COP is positioned near that commander's CP.

9-86. Other OPs and artillery reconnaissance patrols send reconnaissance data to the COP. The artillery commander determines which targets are to be engaged, and the COP relays target data to the firing position.

Forward Observation Post

9-87. Artillery commanders can establish one or more forward observation posts (FOPs) to supplement the COP. The purpose of the FOP is to conduct reconnaissance of the enemy and observe the terrain directly in front of the forward maneuver units. It can locate and adjust fires against targets that the COP is not in a position to observe. It also assures continuous close fire support for the maneuver forces when the COP is displacing. An FOP may be with the supported unit commander or with one of the advance maneuver elements. This enables it to maintain closer communication and coordination with supported maneuver forces.

9-88. At the battalion and battery levels, the FOPs often contain the battalion chief of reconnaissance (or the battery's control platoon leader), a scout, and a radio operator. The FOP can deploy on foot or mounted in an MRP vehicle (see below).

9-89. In either offense or defense, an artillery reconnaissance patrol can also set up an FOP behind enemy lines to adjust artillery fire and to report on enemy organization and deployment. Its primary mission is to locate enemy artillery units.

Lateral Observation Post

9-90. An artillery commander may establish a lateral observation post (LOP) to cover areas not observable from the COP or FOPs. The LOP is usually on the flank of the supported unit and should have a good view of the artillery unit's zone of responsibility. An LOP can work with a COP or FOP to conduct bilateral observation of a target area for improved accuracy. At battalion level and higher artillery echelons, the LOP accurately locates targets, reference and registration points, and can adjust fire. The parent artillery unit or the division's IFC may send reconnaissance and communications personnel to form the LOP. The LOP can deploy on foot or in a vehicle, such as an MRP. An LOP is generally smaller that a COP, manned by two to three reconnaissance specialists who communicate back to the COP.

Mobile Reconnaissance Post

9-91. An MRP is an armored, tracked vehicle with a battlefield surveillance radar and other observation and rangefinding equipment. This vehicle is designed to operate near—or even across—the battle line. It has a data transmission system for passing target information and fire missions directly to associated COPs or fire control posts. There is typically one of these vehicles per artillery battalion, one in the artillery brigade's headquarters, and one in its target acquisition battery. However, the brigade typically uses an MRP to support its own COP. At battalion level, an MRP may function as an FOP or LOP. However, it can also remain near the COP in a forward position or within the artillery battalion firing position area. The artillery brigade or battalion commander designates the position and the sector of observation for the MRP.

9-92. In the offense, the MRP may advance closely behind or within lead mechanized infantry or tank units. It can conduct reconnaissance and fire missions on the move or during short halts. During movement, the MRP can move as part of an artillery reconnaissance patrol. This single vehicle can perform reconnaissance and adjust artillery fire on targets while located with these units. In the defense, MRPs may form part of the combat security outposts in the disruption zone.

Aerial Observation Post

9-93. The artillery commander may use an aerial observation post (AOP) to supplement FOPs and LOPs. The AOP is generally established to cover rapidly moving forces in areas larger than can be covered by a ground OP. The AOP is especially effective during heliborne assaults.

METHODS OF FIRE

9-94. The success of OPFOR combat actions often depends on the ability to deliver timely and effective fire against key parts of the enemy's combat system. Targets could be the enemy's combat and combat support forces (units and weapon systems), as well as the C2, RISTA, and logistics components his combat system. The focus is a systems warfare approach to combat, where the objective of the combat action is to deny the enemy's combat system its synergistic capabilities. Thus, the OPFOR is able to compel enemy forces into multiple and rapid tactical transitions and to create opportunity by keeping them off balance, breaking their momentum, and slowing movement. The OPFOR uses various types of fire against the enemy. The methods of fire may have different purposes in the offense and defense.

RECONNAISSANCE FIRE

9-95. *Reconnaissance fire* is the integration of RISTA, fire control, and weapon systems into a closed-loop, automated fire support system that detects, identifies, and destroys critical targets in minutes. This integration capability normally exists only in an IFC. One reason for this requirement for accelerated engagement is that HVTs and HPTs may expose themselves for only fleeting periods. Reconnaissance fire is primarily designed to attack and destroy key enemy capabilities and/or set the conditions for a strike (see chapters 1 and 3).

9-96. Reconnaissance fire enables the OPFOR to deliver rotary-wing air, SSM, cruise missile, and artillery fires (including precision munitions) on enemy targets within a very short time after acquisition. The OPFOR can use reconnaissance fire in offensive and defensive phases of combat. Assets designated for reconnaissance fire are under control of the IFC commander, and control remains centralized for planning, analysis, and evaluation of reconnaissance data, and for execution of the reconnaissance fire mission. This type of arrangement allows the assets to execute other missions or taskings until the desired HVTs or HPTs are detected. The IFC commander may establish a window of time for assets tasked to support reconnaissance fire (based on an intelligence assessment of when the enemy targets should be in designated kill zones).

9-97. The division or DTG commander selects and establishes the target priority and target damage criteria of the combat system component or components to be attacked in order to force a favorable condition. The IFC staff and fire support component commanders develop the fire support plan designed to conduct reconnaissance fire necessary to create the favorable condition. The IFC commander then briefs the fire support plan to the division or DTG commander to ensure compliance with the overall battle plan. The IFC executes reconnaissance fire in accordance with the approved fire support plan.

CLOSE SUPPORT FIRE

9-98. *Close support fire* is fire used to support maneuver forces and attack targets of immediate concern to units such as battalions and brigades. The requirement is to provide a quick response time and accurate fires capable of either neutralizing or defeating all types of targets.

INTERDICTION FIRE

9-99. *Interdiction fire* is fire placed on an area or point to prevent the enemy from using the area or point. It is designed to attack targets in depth (such as logistics sites or assembly areas) and to prevent enemy follow-on or reserve forces from reinforcing or influencing a battle or situation. Technological improvements such as course-corrected rockets, projectiles, and fuzes facilitate long-range precision targeting.

9-100. The OPFOR employs long-range strike assets (operating from dispersed areas) to continuously engage targeted forces and systems. Operational and tactical RISTA systems direct them.

COUNTERFIRE

9-101. *Counterfire* is fire intended to destroy or neutralize enemy weapons. Includes counterbattery and countermortar fire. *Counterbattery fire* is a type of counterfire that accomplishes the annihilation or neutralization of enemy artillery batteries. It enables ground forces the ability to maneuver on the battlefield with little to no suppression by enemy artillery. However, combat with enemy artillery requires more than counterbattery fire. It requires the destruction of the enemy C2 centers as well as his artillery support structure.

FINAL PROTECTIVE FIRE

9-102. *Final protective fire* is an immediately available preplanned barrier of direct fire designed to impede enemy movement across defensive lines or areas. When the enemy initiates his final assault into a defensive position, the defending unit initiates final protective fire with both direct and indirect fire weapons.

RECONNAISSANCE BY FIRE

9-103. *Reconnaissance by fire* is a type of reconnaissance in which fire is placed on a suspected enemy position to cause the enemy to disclose a presence by movement or return fire. (See chapter 8 for more detail. This is not to be confused with reconnaissance fire.)

FIRE SUPPORT OF MANEUVER OPERATIONS

9-104. The fire support of maneuver operations is characterized by the use of all available fire support to carry out the commander's plan. The OPFOR believes that fire support must be flexible to meet all contingencies during combat. The OPFOR masses fires against an enemy objective with available fire support assets, with the goal of achieving the commander's specified target damage criteria in the shortest time possible.

INDIRECT FIRE SUPPORT TO A STRIKE

9-105. Indirect fire support to a strike involves the employment of a wide variety of ammunition types (such as standard, course-corrected, advanced, and precision munitions) to destroy an enemy formation after typically setting the conditions for its destruction through reconnaissance fire. IFC indirect fire support units are assigned interdiction fire missions to support the maneuver component throughout the strike. Constituent and dedicated indirect fire support units (allocated to the maneuver component) provide close support fire throughout the battle. Thus, indirect fire support to a strike incorporates other methods of fire. The autonomous weapon attack lends itself to supporting a strike. (See chapters 1 and 3 and FM 7-100.1 for more information about a strike.)

OFFENSE

9-106. Fire support considerations for the offense apply to all types of offensive action discussed in chapter 3. The OPFOR plans and executes fires to support the attack and complete the destruction of the enemy. The use of selected lines or zones controls—
- The shifting of fires.
- Displacement of fire support units.
- Changes in command and support relationships between fire support units and maneuver units.

9-107. In the offense, fires are planned to—
- Suppress enemy troop activity and weapon systems.
- Deny the enemy information about friendly forces.
- Prevent the enemy from restoring fire support, C2, and RISTA systems neutralized during previous fire support missions.
- Deny the enemy the ability to use reserve forces to conduct a counterattack.
- If necessary, create favorable conditions for the conduct of a strike.
- Support the exploitation force.

Defense

9-108. Fire support considerations for the defense apply to all types of defensive action discussed in chapter 4. Key is the application of fire support as early as possible throughout the AOR in support of the defensive battle plan. Emphasis is placed on having RISTA assets locate enemy formations and attack positions, with the goal of determining the direction and composition of the enemy main attack. Carefully analyzing the terrain over which the enemy will advance and canalizing his movement into kill zones can create conditions for fires in the defense.

9-109. In the defense, fires are planned to—
- Deny the enemy information about friendly forces.
- Develop the situation early by forcing the enemy to deploy early and thus reveal the location of his main effort.
- Maximize the effect of obstacles as combat multipliers.
- Create favorable conditions for the conduct of a strike and/or counterattacks.

9-110. Close support fire is directed against advancing enemy maneuver units. Close support fire includes fires within friendly defensive positions that are initiated after the enemy has successfully penetrated them. Indirect fires are used against enemy forces that have become wedged against defensive positions. The indirect fires may be massed or concentrated (point). The intent is to annihilate enemy forces in kill zones, thus preventing continuation of enemy offensive operations. Counterbattery fires also will be used to neutralize enemy artillery supporting the attack.

9-111. Final protective fire is planned along the most likely avenue of approach into the defensive position(s). Because the likely direction of attack can change as the enemy situation develops, the final protective fire section of the battle plan is reviewed and updated as required.

TACTICAL DEPLOYMENT

9-112. Two factors govern the deployment of indirect fire support units: continuity and dispersion. The need for continuity of fire support leads to indirect fire support units being deployed in positions to support the maneuver force throughout the battlefield. Unplanned movements to alternate firing positions deny the maneuver force the amount of fire support it requires. Therefore, the OPFOR adheres to the principle of flexibility of employment in order to ensure the delivery of highly accurate and effective fires. OPFOR units disperse batteries and battalions so that the enemy cannot destroy them with a single fire strike. Counterfire continues to be the greatest threat facing indirect fire units.

Battalion Firing Position Areas

9-113. According to their purpose, firing position areas may be primary, alternate, or temporary areas. These have applications in both offense and defense. Figure 9-5 shows an example of an artillery battalion disposition with all these types of firing position areas and the relationships of the batteries within the battalion disposition.

Primary

9-114. The primary firing position area is designated for carrying out the primary fire missions in all types of battle. Its distance from the battle line of friendly units depends on—
- The battalion's place in the supported unit's formation.
- The range of artillery systems.
- The nature of the terrain.
- Other conditions.

Within the battalion firing position area, each battery has a primary firing position and possibly one or two alternate positions.

Alternate

9-115. An alternate firing position area is usually designated in a defensive situation for battalion or battery maneuver and to carry out fire missions during an intentional or forced abandonment of the primary firing position area. A battalion usually has one or two alternate firing position areas to the flanks of the primary area or in the depth of the defense. An alternate area can be several kilometers from the original location.

Temporary

9-116. A temporary firing position area can be designated for carrying out individual fire missions. It could be forward of the battle zone, for support of maneuver units defending in the disruption zone or for firing on distant targets. It could also be for carrying out missions as roving units. Other missions could include supporting the commitment of an exploitation force or commitment of a reserve to a spoiling attack or counterattack.

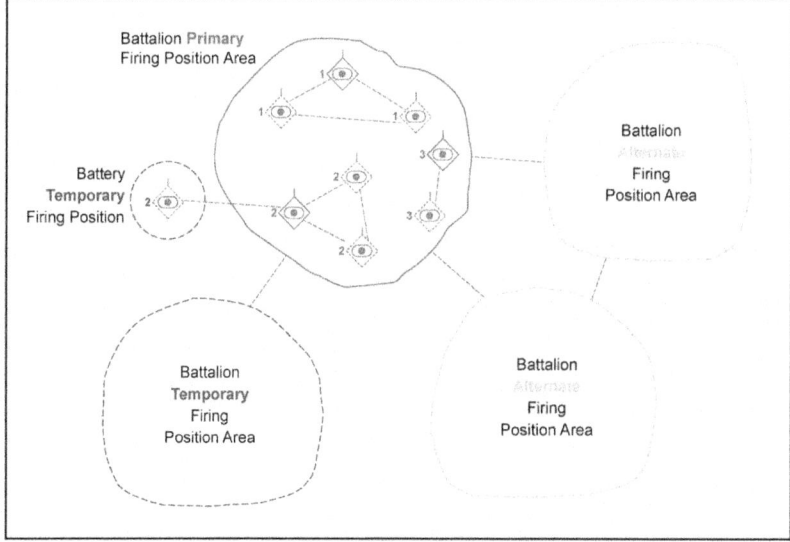

Figure 9-5. Artillery battalion and battery disposition (example)

BATTERY FIRING POSITIONS

9-117. As with battalion firing position areas, battery firing positions may be primary, alternate, or temporary. In the offense, an artillery battery can use any or all of those, and possibly create deception firing positions. The defense can require primary, alternate, temporary, and deception positions. The functions of primary and alternate firing positions are much the same as for battalion firing position areas.

9-118. Cannon, mortar, and MRL firing positions are similar. While SSMs may be fired from both fixed and mobile launchers, the OPFOR prefers to fire SSMs from mobile launchers.

Temporary

9-119. A temporary firing position can allow a battery to accomplish special, short-term, or emergency missions. In the defense, a battery can use a temporary firing position near the battle line or forward of the battle zone to support maneuver units defending in the disruption zone or to fire on a distant target. A temporary position can also be for use by a roving battery or platoon. Although temporary, these firing positions can be prepared and camouflaged.

Deception

9-120. A battery (or battalion) may prepare *deception* firing positions and COP sites on its own or as part of the senior commander's deception plan. Their purpose is to mislead the enemy as to the actual deployment of artillery units. Their preparation and camouflage must not differ sharply from that of actual positions and sites. A roving unit may periodically deliver fire from the deception firing position.

BATTERY DEPLOYMENT TACTICS, TECHNIQUES, AND PROCEDURES

9-121. The OPFOR employs indirect fire support tactical concepts that include a variety of battery tactics, techniques, and procedures (TTP) for effectiveness and survivability. The plans for the employment of the battery are thorough and cover—

- Mission.
- Location of firing positions.
- Method of fire.
- Number of rounds to be fired from each position.
- Movement schedule of the battery.
- Duration of the battery mission.

9-122. The TTP are applicable to all indirect fire units. These techniques also provide the battery commander with more flexibility to conduct multiple fire missions simultaneously, since the battery can organize into more than one distinct firing unit. The techniques include—

- Fire from varied formations.
- Fire from dispersed locations.
- Fire from fixed locations.
- Fire and decoy.
- Shoot and move.
- Autonomous weapon attack.

Fire from Varied Formations

9-123. For increased survivability, indirect fire support units use formations that increase and/or vary the interval between weapons in a firing position and disperse the weapons in depth with the aid of computers. Weapons may be in a wave formation, a forward or reverse wedge, or a semicircle.

Indirect Fire Support

Fire from Dispersed Locations

9-124. The OPFOR employs indirect fire support weapons with a variety of dispersed battery techniques applicable to mortar, cannon, and MRL units. Two effective techniques are the split-battery (two-platoon) and dispersed-platoon formations.

Split Battery

9-125. *Split battery* is a tactic designed to increase the survivability of OPFOR artillery against enemy counterfire and counterbattery fire. The battery is split into several fire units (usually two platoons), which may deploy dispersed over an extended area. As a countermeasure to precision munitions and submunitions, the increased dispersion can keep one platoon out of the seeker footprint of munitions employed against the other platoon. This can force the enemy to either employ more munitions over a larger target area or increase the number of targets to be serviced. In most cases, the battery COP can control the fires of both platoons. When necessary, however, the battery fire control post could control the fires of one of the platoons.

Dispersed Platoon

9-126. *Dispersed platoon* is another tactic designed to increase the survivability of OPFOR indirect fire support weapons against enemy counterfire and counterbattery fire. This also facilitates the employment of single firing systems or pairs in multiple small areas that would not accommodate larger groups of firing systems.

9-127. The dispersed platoon is clearly the most survivable technique against enemy counterfire. The tactic requires highly trained personnel capable of executing a very complex, decentralized type of operation. The tactic also works best with firing systems incorporating onboard position location (such as GPS), fire direction, and survey systems.

Fire from Fixed Locations

9-128. The *fire-from-fixed-locations* technique is generally employed where there is limited movement in areas such as mountains, jungles, or urban areas. The firing battery occupies dispersed pre-surveyed positions and may use hide sites as measures of both survivability and force preservation for the conduct of future battles and operations.

Fire and Decoy

9-129. The OPFOR employs *fire-and-decoy* techniques to increase survivability as well as to deceive the enemy of the actual firing unit location. The techniques include roving gun, roving units, deception battery, and false battery.

Roving Gun

9-130. *Roving gun* is a technique designed as a countermeasure against an enemy that has a sophisticated target acquisition capability. The goal is for the enemy to detect and engage this target, thinking that it is an entire unit, expending munitions that would otherwise have been used on an actual target. Enemy units that are particularly susceptible to this technique are those capable of accurately detecting units as soon as they begin firing and then attacking the target within a matter of minutes.

Roving Unit

9-131. *Roving unit* is another technique designed as a countermeasure against an enemy possessing a sophisticated target acquisition capability. It is similar to the roving-gun technique. The difference is that roving unit involves the displacement of the firing unit versus the individual indirect fire weapon system. Taking advantage of the mobility of self-propelled artillery, the OPFOR can move artillery batteries or platoons to alternate or temporary firing positions within an assigned firing position area to escape enemy counterbattery fire.

Chapter 9

Deception Battery

9-132. The *deception battery* is a technique where the OPFOR creates an additional battery in an attempt to deceive the enemy of the actual battalion location. This technique is also referred to as the "fourth battery" technique.

9-133. The OPFOR may use two methods to create a deception battery. The preferred method is for a battery to split into two platoons. Additional weapon systems are allocated to the two platoons from the remaining two batteries to provide each platoon a signature of a battery. A second method is for the battalion commander to issue instructions for each firing battery to provide one to two weapons systems to create the deception battery.

False Battery

9-134. The *false battery* (or decoy battery) is a technique that involves the use of active and decoy weapon firing positions to give the appearance of a battery firing position. Depending on conditions such as the terrain, enemy situation, and mission, the battery commander may employ up to two indirect fire support weapons in each platoon position with the camouflaged decoys or nonoperational equipment in the primary firing position to create the impression of use. The remaining indirect fire support weapons move to a hide site a distance away from the decoy position.

Shoot And Move

9-135. *Shoot and move* is a technique that involves the rapid displacement of a firing unit from a firing position immediately after completion of a fire mission. It is an effective countermeasure in protecting indirect fire support assets from enemy counterfire and counterbattery fire.

Autonomous Weapon Attack

9-136. *Autonomous weapon attack* is a technique designed for individual indirect fire systems to operate independently in dispersed locations, from which they can attack single or multiple targets. They deliver devastating fires at precisely the right time and place on the battlefield with minimal risk to themselves. This technique exploits the capability of indirect fire systems incorporating onboard position location (such as GPS), fire direction, and survey systems. While integrated and digital communications facilitate this TTP, they are not required. It is simply a matter of coordinated timing and targeting. Completely dispersed indirect fire systems deliver ordnance to targets.

9-137. Although tactical in execution, autonomous weapon attacks can contribute to operational and/or strategic objectives and have both immediate and long-range effect upon the enemy. While this technique readily lends itself to supporting a strike, it may also be used at lower levels when the weapons or unit have self-locating equipment and/or a ballistic computer, and are provided with appropriate targeting information.

9-138. Autonomous weapons attacks can be executed by one or more weapons of the following types (with above capability):
- Cannon artillery (field gun, howitzer, or hybrid system).
- Mortar.
- Rocket (single or multiple launched).
- Missile.
- Any combination of the above.

9-139. The OPFOR may establish hidden ammunition storage locations or caches along access routes near the weapon firing sites. SPF, local sympathizers, or irregular forces may establish and service caches. If a cache is discovered by enemy troops, the FSCC or CFSC will direct the firing unit to another cache.

Indirect Fire Support

TACTICAL MOVEMENT

9-140. Movement is particularly important during offensive actions, when the indirect fire support unit must keep pace with the advance of supported maneuver units. Fire support planners strive to maintain continuous support from the initiation of preparatory fire until the accomplishment of the offensive mission, including the commitment of an exploitation force. As indirect fires shift successively deeper into the enemy defenses, displacement of indirect fire support units becomes necessary. Thus, after the initial fires in support of the attack, indirect fire support units supporting or subordinate to fixing and assault forces begin to displace. This displacement is preplanned to accommodate the advance of the attacking maneuver forces.

Movement by Battalion

9-141. The movement of an indirect fire support battalion can follow several different patterns depending on such factors as enemy situation, mission, terrain, weather, and visibility. Once the battalion has reached the assembly area and completed its organization for combat, it may move by battalion or by battery.

9-142. Movement by battalion is possible only when the battalion has not been committed to battle or when there are other units available to perform any required fire missions while the battalion is moving. All elements of the battalion displace at the same time (based on a movement schedule) and are typically expected to be in their new positions at the same time.

Movement by Battery

9-143. In the offense or defense, the most common movement technique is for an indirect fire support battalion to move by battery. The battalion moves its batteries individually by bounds. Depending on the route and the pace of combat, there may be a temporary halt to rearm and refuel during the movement. Once a battery is in its new position and ready to fire, the next battery starts to displace. Typically, the battalion fire control post displaces with the center battery.

Movement by Bounds

9-144. An indirect fire support unit normally displaces by bounds, attempting to retain two-thirds of its weapons in positions within range to provide continuous support for the attacking or withdrawing force. In planning deployment of their units, indirect fire support commanders follow the "rule of a third." For example, when only a third of the maximum range of their indirect fire support weapons remains in front of the attacking OPFOR troops, they move a third of their guns forward. Once redeployment starts, no more than a third of the available guns is moving at any one time. This leaves two-thirds of the weapons in position to support tactical maneuver actions.

LOGISTICS

9-145. The OPFOR applies the "*push forward*" concept of logistics. Units do not request ammunition; rather they are allocated ammunition in the fire support plan to support the maneuver battle. Ammunition has the highest priority within the OPFOR supply system. The determination of required expenditures is the responsibility of the IFC commander or CFSC, while the chief of logistics is responsible for delivery.

AMMUNITION RESUPPLY

9-146. The OPFOR uses standard cargo trucks as resupply vehicles for cannon and mortar systems. For towed systems, the trucks also serve as prime movers. The only dedicated ammunition resupply vehicles are for some MRL systems. These vehicles have the same chassis as the rocket launcher and are fitted with racks to hold the rockets during transport. Resupply vehicles for large-caliber MRLs have cranes for reloading the launcher. To the maximum extent possible, the ammunition remains loaded on resupply vehicles to maintain mobility.

9-147. In most cases, the ammunition packaging is designed so that two men can easily move any single item. This lessens the requirement for materiel-handling equipment at ammunition transfer points and in the firing position.

9-148. To facilitate the movement of ammunition, general practice is to establish ammunition transfer points for each IFC or maneuver brigade. Under normal circumstances, an indirect fire support battalion sends its resupply vehicles to this point to pick up ammunition and deliver it to the firing unit. Transport units may skip an echelon, if necessary, to keep units resupplied.

BATTERY RESUPPLY

9-149. Depending on the threat, time, terrain, and other conditions, the battery commander may accept the risk of conducting a resupply in the firing position. Whenever feasible, the transport unit offloads in the firing position any ammunition the firing unit will consume prior to repositioning. However, this is the least preferable method of resupply and, if at all possible, should not be attempted.

9-150. When necessary, during movement, the battery will make one halt to reload the indirect fire support weapons, refuel (if required), and conduct necessary maintenance. Normally, this halt is short in duration, and the unit will proceed to its next firing position once resupply and maintenance actions are completed. The resupply point is normally a location that is covered and concealed along the route to the next firing position. If the size of the resupply point is large enough, all of the battery's indirect fire support weapons are resupplied simultaneously. If not, the maximum number of weapons that can be occupy the site at one time are resupplied followed by the remaining weapons. The resupply action is normally done in the order of movement. All of the vehicles remain in the area until the resupply action is completed. If there is sufficient time available, an advance party proceeds forward to prepare the next firing location.

RECONSTITUTION

9-151. Restoring combat effectiveness of subordinates is one of the most important duties of indirect fire battalion and battery commanders. It includes—
- Determining the degree of combat effectiveness of subordinates.
- Detailing missions to subordinates that are still combat-effective.
- Withdrawing units from areas of destruction or contamination.
- Providing units with replacement personnel, weapons, ammunition, fuel, and other supplies.
- Restoring disrupted C2.

9-152. The OPFOR makes an effort to keep some units at full strength rather than all units at an equally reduced level. Usually, the unit with the fewest losses is the first to receive replacement personnel and equipment. However, once the casualties or equipment losses are sufficient to threaten the total loss of combat effectiveness, the commander may apply the concept of composite unit replacement. The composite unit concept involves a unit formed from other units reduced by combat action.

Chapter 10
Aviation

The ability of the OPFOR to employ its aviation assets will depend on the level of airspace dominance the OPFOR possesses. When fighting a weaker opponent, the OPFOR expects to establish and maintain air superiority and thus to employ its aviation with relative ease. When faced with a superior enemy, however, the OPFOR will alter aviation missions to ensure the most effective use of its air power without the unnecessary loss of assets. In either situation, the OPFOR makes maximum use of unmanned aerial vehicles (UAV) at all levels. This chapter addresses the OPFOR aviation tactics of fixed- and rotary-wing aircraft and UAVs. For information on the impact of strategic concepts on aviation operations and airspace dominance, see FM 7-100.1.

COMMAND AND CONTROL

10-1. Aviation forces are allocated to specified levels of command to meet mission requirements. Organizational structures are designed to maintain the appropriate level of centralized control to ensure the limited number of assets are available at the right place and time. For more information on the organization of aviation units at the operational-level or above, see FM 7-100.1. However, even aviation units that are part of organizations above the tactical level can perform missions that have tactical-level impact and must, therefore, be addressed in a tactical context.

DECENTRALIZED VERSUS CENTRALIZED CONTROL

10-2. The OPFOR will task-organize aviation assets to tailor the force for the specific mission. Thus, it is possible to task-organize a fixed-wing aviation unit from the Air Force to an operational-strategic command (OSC), which is the lowest level of joint command. However, army aviation rotary-wing assets can be found allocated not only to an OSC but also to a division tactical group (DTG) or a brigade tactical group (BTG). The OPFOR is more likely to task-organize its aviation to the lowest levels against a weaker opponent, when it has established air superiority. This decentralized control allows greater flexibility and responsiveness from OPFOR aviation assets in support of ground commanders.

> *Note.* A *tactical group* is a task-organized unit organized around the baseline, administrative structure of a division or brigade. Throughout this chapter, the terms *DTG* or *BTG* will be used to identify that level of command, since a maneuver division or brigade does not include constituent or dedicated aviation assets unless it has been task-organized as a tactical group. The terms *division* or *brigade* will be used only to highlight differences (when they occur) from a tactical group.

10-3. Against a superior force, however, the OPFOR is apt to maintain control of its helicopters and airplanes at OSC and theater level, respectively. This centralized control allows the OPFOR to better protect its assets, more thoroughly plan missions, and improve reaction time during the limited windows of opportunity.

COMMAND AND SUPPORT RELATIONSHIPS

10-4. The OPFOR employs its aviation assets using its standard command and support relationships (see chapter 2). Since army aviation assets are not found below the operational level in the administrative force structure, it is the OPFOR practice to augment tactical maneuver units by allocating aviation assets in one of three command and support relationships: constituent, dedicated, or supporting.

Constituent

10-5. A *constituent* command relationship is the assignment of a unit to a headquarters. The headquarters has the authority for its employment and the responsibility for all of its logistics support. An example of this type of command relationship would be a medium-lift helicopter battalion assigned to a DTG to provide transportation capabilities for its ground forces.

Dedicated

10-6. A *dedicated* command relationship is similar to constituent with the exception of logistics support. The subordinate unit still receives logistics support from its parent aviation unit. An example of a dedicated relationship is an attack helicopter battalion dedicated to a DTG. The battalion continues to receive logistics support from its parent combat helicopter brigade, while the DTG has sole employment authority of the battalion.

Supporting

10-7. A *supporting* aviation unit remains under the command of its parent organization. It also receives all of its support from its parent unit. It executes missions according to the supported unit's priorities. The supported unit plans and employs the asset for the time allotted by the higher headquarters. The principle advantage of this is to the parent commander, who retains maximum control of his most flexible assets. An example of a supporting relationship is the employment of theater or OSC attack aircraft in the direct air support (DAS) role at division or brigade level.

FIXED-WING AVIATION

10-8. Fixed-wing assets of the Air Force are not task-organized in a constituent or dedicated relationship below the OSC level, since that is the lowest level of joint command. However, Air Force units retained at higher levels of command might have a supporting relationship with a division, DTG, brigade, or BTG. Subject to the approval of the theater or OSC commander to whom they are subordinate, they can also respond to mission requests from tactical-level units (see the section on Request Process later in this chapter).

ROTARY-WING AVIATION

10-9. Helicopters are employed across the battlefield to support the ground commander in the combined arms fight. Because of their flexibility, maneuverability, speed, and firepower, they have the capability to execute missions down to the BTG level. Helicopters can be called upon to execute any mission to support both the offense and defense. Based on mission, command and support relationship, and availability of aircraft, the OPFOR organizes its helicopters using three methods:

- Attack helicopters and possibly some combat support (CS) and combat service support (CSS) helicopters with missions related to fire support can be part of a DTG's integrated fires command (IFC).
- Other CS and CSS helicopters can be directly subordinate to the DTG commander.
- Still other attack, CS, and/or CSS helicopters can be subordinate to a BTG commander.

In a DTG IFC

10-10. The IFC is a command and control (C2) structure with a task organization that allows rapid employment of aviation systems with other ground systems. The assignment to the IFC may be in either a constituent or a dedicated relationship, but it is always tailored for the specific mission of the organization it supports. Figure 10-1 shows an example of an IFC at DTG level.

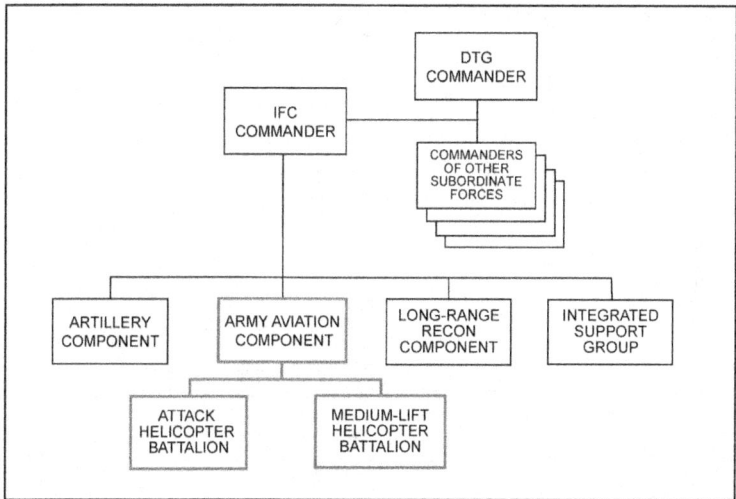

Figure 10-1. Example of aviation in a DTG IFC

10-11. An IFC may be organized to include an army aviation component. The missions assigned can include attack, DAS, and reconnaissance for an attack helicopter unit. The IFC may also employ CSS helicopters for troop movement, resupply, and C2 platforms. The command and support relationship to the IFC is based on the type of mission, available assets, and duration of the mission. If the IFC requires continuous lift capabilities to rapidly employ forces, a lift helicopter battalion may be constituent or dedicated to the IFC. On the other hand, if the movement of troops is a one-time requirement, the helicopter battalion may not become part of the IFC, but instead may have a supporting relationship for the duration of the mission while remaining under the control of the parent aviation unit. The same applies to the attack helicopter units. See chapters 2 and 9 for more details on the IFC.

In a DTG Other Than in IFC

10-12. Because the IFC is tailored for fire support missions, not all aviation assets are organized under the IFC headquarters. Army aviation units that are constituent or dedicated to a DTG, but not associated with fire support, are directly subordinate to the DTG commander or perhaps to a BTG within the DTG. Figure 10-2 on page 10-4 shows an example of how an aviation unit might be outside the IFC in a DTG organization.

Chapter 10

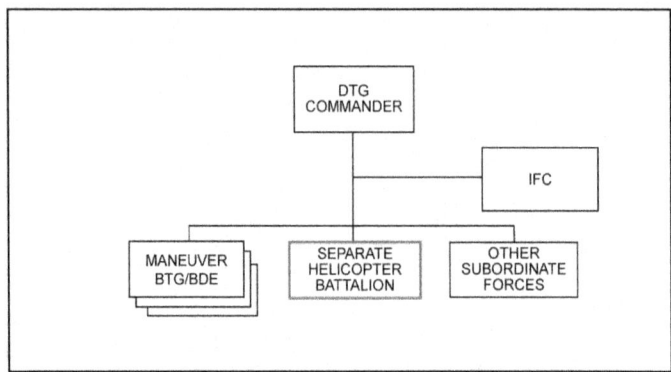

Figure 10-2. Example of DTG-level aviation other than in the IFC

10-13. A combat helicopter brigade or one or more of its battalions may become constituent or dedicated to a DTG. In this case, the attack helicopters are most likely to be employed in the DAS, attack, reconnaissance, and security roles outlined later in this chapter. In the first two of those missions, they would most likely be part of the DTG's IFC (unless allocated to a subordinate BTG). In reconnaissance and security roles, however, they could be employed outside of the IFC unless those roles are specifically related to fire support.

10-14. As an exception to the rule, a highly-trained unit equipped with modern attack helicopters may be employed as a maneuver element in the ground commander's scheme of maneuver. In this role, the attack helicopter unit can be used as a disruption, fixing, assault, or exploitation force in the offense, or serve as a disruption or counterattack force in the defense. In either offense or defense, it could serve as a reserve or deception force. Such missions would require thorough planning and rehearsals to be successful.

10-15. For CS and CSS helicopters units, the various missions are assigned primarily with a supporting relationship. However, some units that rely on routine support may be allocated a helicopter battalion or company with a constituent or dedicated command relationship.

10-16. If allocated to a DTG in a constituent or dedicated relationship, a combat helicopter brigade's lift helicopter and reconnaissance helicopter battalions (or companies from them) would normally be in the DTG's IFC only if they perform missions associated with fire support. Otherwise they would be outside the IFC, under either the DTG commander or one of his BTG commanders. This would also be true of CS and CSS helicopters from separate helicopter battalions that were not part of a combat helicopter brigade.

10-17. If a DTG is allocated an entire combat helicopter brigade (or major parts of one), that brigade's headquarters would typically come under the IFC headquarters, especially if that is where most of its battalions are employed. If most of its subordinate battalions are employed outside the IFC, the brigade headquarters could be directly under the DTG headquarters.

In a BTG

10-18. Attack, CS, and/or CSS helicopters can be directly subordinate to a BTG commander. An example of this would be a motorized infantry BTG conducting heliborne assaults. Such a BTG may include a medium-lift helicopter battalion to insert infantry units and an attack helicopter battalion to provide security and armed escort for the troop-carrying helicopters or to prepare the landing zone (LZ) by fire (see figure 10-3).

Aviation

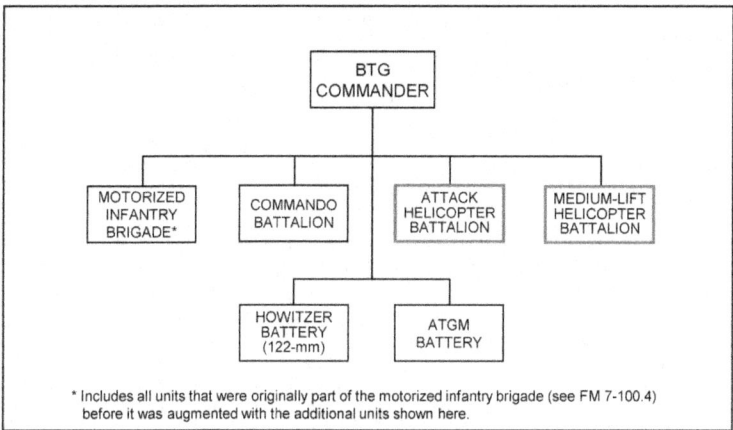

Figure 10-3. Example of aviation in a BTG

UNMANNED AERIAL VEHICLES

10-19. The military application of UAVs has become standard practice in armies worldwide. The OPFOR is no exception. It operates UAVs at all levels, from the strategic level down through division, brigade, maneuver battalion, and some companies, to special-purpose forces teams.

10-20. UAV units may be employed at the echelon where they are assigned in a constituent or dedicated status. However, smaller units, down to individual UAV teams, may be task-organized in a supporting status to support lower-level units as dictated by the mission. Both are standard practices. Which method the commander selects depends on the military situation.

AIRSPACE MANAGEMENT

10-21. The OSC is the lowest level of joint command with control of both Army and Air Force units. On the staff of an OSC, under the operations officer, the chief of airspace operations (CAO) is responsible for airspace management issues and procedures. The CAO maintains the airspace control net for controlling the command's airspace. OSC headquarters typically receive liaison teams from all constituent, dedicated, and supporting Air Force, army aviation, and air defense units associated with the command. All these units and their liaison teams are on the airspace control net. For additional information on airspace management, see FM 7-100.1.

10-22. The OPFOR assigns its organizations an area of responsibility (AOR) that typically includes not only the surface area of a defined geographic space but also the associated airspace. The coordinated use of battlefield airspace and aerial delivery of ordnance close to friendly troops are two problems any combined arms force faces. OPFOR doctrine stresses the need to provide maximum aviation support to ground force commanders. Therefore, aviation control and communications are closely aligned with those of the ground force to ensure effective and continuous communications.

10-23. To reduce air-to-ground coordination problems during the execution of missions, the OPFOR employs proactive staff elements and control measures. Planners can use attack helicopters, fixed-wing ground-attack aircraft, UAVs, and artillery simultaneously in the same part of the AOR only if coordination measures exist and controlling elements are working in conjunction with each other to ensure deconfliction. This deconfliction includes those assets allocated or employed by higher headquarters, including attack helicopters, fixed-wing ground-attack aircraft, UAVs, artillery, and surface-to-surface missiles (SSMs).

CHIEF OF AIRSPACE OPERATIONS

10-24. Air and ground force commanders and staffs work out coordination procedures between aviation elements, air defense elements, and ground forces before the launch of combat air missions. These procedures are the responsibility of the CAO at all levels of command down to brigade or BTG, even when no aviation units are subordinate to that headquarters.

10-25. It is imperative that air defense units be notified when friendly aircraft (or UAV) are flying within the air defense umbrella. Failure to coordinate with these elements will result in unnecessary fratricide. The senior air defender in the command will notify air defense units. For additional information on the coordination of air defense units, see chapter 11.

10-26. At every level of command, the CAO is responsible for airspace deconfliction. To assist in that function, he has a staff at his disposal for coordination and deconfliction of air missions. He and his staff make up the airspace operations subsection (AOS) under the operations officer. This staff subsection includes liaison officers from all subordinate units requiring airspace deconfliction. This ensures that the aviation, fire support, and air defense units continually coordinate all operations with each other. Since aviation assets are not constituent or dedicated to the pure division or brigade, the primary functions of the CAO and his staff there are to request and monitor employment of higher-level aviation assets allocated to the division or brigade in a supporting role.

AIRSPACE OPERATIONS SUBSECTION

10-27. The overall mission of the AOS is to advise commanders and staffs on the use of all air assets and to deconflict airspace use. There is an AOS at each level of command down to and including maneuver brigades. These AOSs all perform the same mission, but vary in size and complexity.

10-28. The AOSs form a vertical and horizontal channel through which airspace coordination requirements, plans, orders, and information are coordinated, disseminated, and synchronized with the battle plan. They—
- Transmit air support requests to higher-level AOSs and aviation organizations.
- Coordinate all air support.
- Maintain communication with and provide deconfliction for all aircraft in the AOR.

10-29. An AOS may divide into two or more cells. The primary cell is located in the main command post (CP), while smaller AOS cells may be in the forward CP and/or IFC CP.

Theater Level

10-30. For issues related to air support and interface, the theater-level CAO coordinates with aviation assets within the theater, including the theater air army CP (Air Force), the army aviation CP, and elements of the subordinate air defense, artillery, and SSM units. The AOS at theater level consists of several dozen individuals with aviation, artillery, SSM, or air defense coordination experience filling permanent staff positions and interfacing with their respective subordinate units.

10-31. The AOS is the theater commander's primary means of turning his guidance into a comprehensive plan for air operations. It allocates resources and tasks forces through the publishing of the aviation support plan (ASP). For more information, see the Aviation Support Plan later in this chapter.

10-32. The AOS establishes vectoring and target designation posts (VTDPs) as necessary to exercise control of aircraft in a designated AOR. These posts are air traffic control facilities that support the movement of aviation assets within an AOR and can also direct aircraft to ground targets. The VTDPs are primarily ground-based and serve as an intermediate air traffic control facility between the aircraft's parent unit and the forward air controller (FAC). (See below under BTG Level.) They accomplish direct coordination among helicopters, ground-attack and fighter aircraft, ground-based air defense units, and

FACs, primarily through VHF voice transmission. These posts are equipped with radar, communications, and automated equipment used for identification and tracking of both friendly and enemy aircraft.

10-33. Occasionally the OPFOR can employ airborne C2 aircraft to perform the same intercept function as a VTDP. These aircraft are referred to as airborne control stations (ABNCSs), and may be used to augment or replace VTDPs within the OPFOR AOR.

10-34. In mountainous terrain with VTDP radar dead space, visual observers (VOs) are used. These observers are connected into the VTDP network via VHF communications. Each observer section is equipped with radios, binoculars, and sound detection devices.

10-35. If the OPFOR uses ABNCSs or VOs, their employment is no different than that of a VTDP. They control the flow of friendly aircraft, and provide enemy intercept data to OPFOR counterair aircraft and air defense units.

OSC Level

10-36. The AOS at OSC level is manned and equipped similar to the theater-level AOS. When the theater only has one OSC, the theater AOS functions are performed by the OSC AOS.

DTG Level

10-37. At the DTG level, the AOS has some personnel filling permanent staff positions and some liaisons from subordinate units. Air support coordination is controlled by the interaction between staffs within the fire support coordination center, army aviation CP, and subordinate air defense unit CP. These staffs provide deconfliction for all aircraft operating within their AOR by monitoring radar and radio communications.

10-38. Since Air Force aviation units are not constituent or dedicated to a ground forces division or DTG, the supporting aviation regiments or squadrons normally colocate a CP with the division or DTG main CP. This facilitates the close coordination required by the AOS.

BTG Level

10-39. The BTG-level AOS is located with the BTG main CP to assist the commander and staff in all tasks associated with planning and employing air support assets. The AOS is responsible for coordinating air support by serving as the primary—

- Liaison between the BTG staff and the DTG's AOS.
- Liaison between ground forces and supporting fixed- or rotary-wing aircraft.
- Director for attacking aircraft by passing messages directly to the flight leader about targets.

10-40. The CAO is responsible for the operation of the AOS. He coordinates with the BTG commander to ensure proper integration of air missions into the overall scheme of maneuver. If a BTG employs a forward CP, a subelement or representative of the AOS may locate forward with the commander, if required. These representatives are also qualified to perform the duties of a FAC if necessary, but this is not preferred.

10-41. The BTG AOS is responsible for the coordination of all airspace and air routes within the BTG's AOR. It coordinates with the air defense units, aviation units, and the chief of fire support coordination. The AOS serves as the central point of contact for all actions between the ground force and aviation units. It continually monitors the status of ongoing and planned missions and the availability of air support.

10-42. FACs may colocate with the maneuver battalions when air strikes or support missions are planned, or when the brigade or BTG commander expects the battalions to require immediate or on-call air support. The FAC is a senior helicopter pilot experienced in combat helicopter brigade support procedures. The FAC's goal is to employ fixed- and/or rotary-wing aircraft simultaneously in the same area, and coordinate aircraft employment with artillery fires. If successful, impacts coincide in time, with different target sectors allocated.

10-43. The FAC arrives at the maneuver battalion's CP prior to a mission with his own radio set for communications with helicopters and/or fixed-wing aircraft. The type of radio is based on the type of aviation he supports, since fixed-wing and rotary-wing missions use different frequencies for communication. The radio is either VHF or UHF. Provision is made in the brigade or BTG headquarters for a FAC vehicle, and it has unique mounts for these radio sets.

10-44. A FAC serves as the ground commander's direct liaison with aviation support. He—
- Plans air missions to support the ground commander's scheme of maneuver (based on the sortie allocations from higher headquarters).
- Establishes control procedures.
- Orchestrates mission execution.

Battalion Level

10-45. A maneuver battalion seldom has a staff member dedicated to serve as an air representative and rarely receives a dedicated FAC. The brigade or BTG may allocate a FAC to a battalion when air support is planned specifically in its AOR. In such cases, a FAC works in conjunction with the commander, artillery observer, and battalion chief of fire support to coordinate the actions of attack aircraft with the artillery fires and the ground scheme of maneuver. When the maneuver battalion is not allocated a dedicated FAC, the battalion's chief of fire support is the primary coordinator to facilitate DAS at the battalion level. The platoon leader of the battalion's man-portable air defense system (MANPADS) platoon (or the senior air defender) also coordinates with the appropriate staff member, or FAC if present, to deconflict any possible fratricide issues.

10-46. Air support providing reconnaissance is coordinated by the platoon leader of the reconnaissance platoon, who serves as the battalion chief of reconnaissance. He in turn keeps the battalion intelligence officer abreast of reconnaissance activities and findings. The intelligence officer then coordinates with appropriate staff personnel.

AIRSPACE CONTROL MEASURES

10-47. The purpose of airspace control measures is to maximize the effectiveness of combat missions. Airspace control measures are established so that ground and aviation units may apply timely, efficient, and mutually supporting combat power while minimizing the risk of fratricide. This is accomplished through two methods: positive control and procedural control. In the airspace coordination order (ACO) portion of the ASP, the CAO delineates all positive and procedural airspace control measures.

Positive Control

10-48. *Positive control* is a method of airspace control that relies on electronic means such as positive identification, tracking, and aircraft vectoring, done by radar control or electronic monitoring. Positive control is established by air traffic control services around airbases and in the support zone. As aircraft depart these areas, they are handed off to subordinate airspace coordination facilities (such as a VTDP, ABNCS, or VO) and then finally to the FAC as they approach his AOR.

Procedural Control

10-49. *Procedural control* relies on previously coordinated and disseminated orders or procedures to control the operation and flow of air traffic. These procedures, coupled with the OPFOR emphasis that combined arms forces must be generally familiar with each other's tactics and equipment, help alleviate problems that arise in coordination during combat.

10-50. The OPFOR employs coordination procedures that separate airspace horizontally, vertically, or both. This buffer zone minimizes the possibility of fratricide while maximizing ordnance effects. Figure 10-4 depicts the different airspace coordination procedures available to the OPFOR.

Aviation

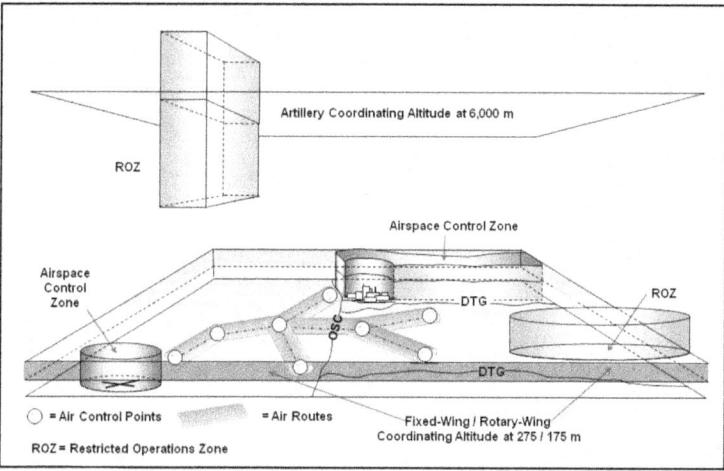

Figure 10-4. Airspace procedural control measures (example)

Coordinating Altitudes

10-51. A coordinating altitude outlines an arbitrary altitude below which fixed-wing aircraft do not fly, and above which rotary-wing aircraft do not fly. Artillery coordinating altitudes exist to deconflict artillery shell trajectories and fixed-wing traffic at high altitudes. A buffer zone may exist between coordinating altitudes to allow small altitude deviations. For example, coordinating altitudes are generally no higher than 175 m above ground level (AGL) for helicopters and no lower than 275 m AGL for fixed-wing aircraft. Deviating from these altitudes requires further coordination. Artillery coordinating altitudes are generally established at 6,000 m AGL or higher. Fixed- or rotary-wing aircraft planning extended penetration of the coordinating altitudes must notify the appropriate AOS, but prior coordinating altitude deviation approval is not required.

Airspace Control Zones

10-52. These zones define airspace that is characterized by a high density of aircraft or a high concentration of usage. An airspace control zone has defined dimensions that coincide with geographic or manmade features and extend vertically to a given altitude. The requesting authority, such as a brigade or division commander, dictates air defense weapon control status within the airspace control zone.

Restricted Operations Zones

10-53. A restricted operations zone (ROZ) is established to define a volume of airspace for a specific mission or purpose, such as a drop zone(DZ), landing zone (LZ), UAV flight pattern, or electronic warfare (EW) aircraft flight route. An ROZ is used to restrict some or all airspace users until termination of the mission. It may restrict airspace horizontally and/or vertically and by time of usage. An ROZ, for example, may be set up to restrict airspace from 1,500 to 3,000 m AGL, 5 km in all directions from a given point, from 0200 to 0600 hours, for the purpose of UAV overflights. The requesting authority, such as a brigade or division commander, controls air defense weapon control status within the ROZ.

Air Routes

10-54. Air routes are made up of air corridors and air control points (ACPs). These control measures are implemented to control the travel of aircraft through friendly airspace and to prevent friendly forces from firing on friendly aircraft. ACPs are predetermined points over the ground at a given altitude where the air route changes direction or links with another air route. An air corridor is the path of linked air control points starting at the initial point (IP) and ending at the release point.

10-55. Some air routes may include the use of mandatory reporting points. These points serve to control and monitor the flow of air traffic by requiring radio calls to the controlling authority stating the aircraft's position. An air route, for example, may dictate: returning aircraft fly above 1,500 m AGL, outbound aircraft fly below 1,500 m AGL, all helicopters below 30 m AGL. All aircraft should see and avoid other aircraft and remain within 500 m of the corridor centerline for safe transit.

10-56. Every level of command down to BTG has a unique airspace structure supporting the movement of aircraft within its AOR. Air routes run from the supporting airfields, through the theater and/or OSC airspace controlled by the VTDPs, to a "crossing checkpoint" at the DTG boundary. The aircraft then follows the airspace structure unique to that particular DTG until it reaches a BTG boundary. The BTG will provide routes that support the mission, taking the aircraft to the DZ, LZ, pick-up zone (PZ), or to an initial contact point where control is assumed by the FAC. Using the ASP, the routes are published and distributed to each level of command. Each air defense element is responsible to disseminate the information to the troops within those boundaries to prevent the fratricide of friendly aircraft. The ASP incorporates all forward arming and refueling points (FARPs) and all planned LZ and PZs for helicopters. The ASP may change the route structure on a daily basis.

Air Defense Control Measures

10-57. To coordinate the use of aviation assets with ground forces, the OPFOR utilizes different types of air defense weapons control status and procedural controls. Primarily, it employs a system of identification, friend or foe (IFF) between aircraft and air defense systems. To protect friendly aircraft from fratricide from non-IFF-capable systems, strict procedural controls are enacted. These control measures (described above) are disseminated daily using the ASP through AOS channels and aviation unit headquarters elements.

10-58. The air defense coverage may be "switched off" to allow friendly aircraft to pass, on a mission planned in advance, and then "switched on" as they exit the area. For other missions, air defense coverage may allow aircraft to transit only on "safe corridors" based on air routes or other procedural methods. If aircraft deviate from these coordinated areas, they risk being shot down by friendly ground force units. The OPFOR views the possible loss of aircraft through fratricide as a lesser risk than allowing gaps in its radar and air defense coverage that the enemy might exploit. See also chapter 11 for more information on air defense asset employment.

Fire Support Coordination Measures

10-59. Fires from mortars, cannon and rocket artillery, and SSMs pose a potential hazard to friendly aircraft activities. The highest probability of conflict between aircraft and surface-to-surface indirect weapons fire occurs at relatively low altitudes in the immediate vicinity of firing positions and targeted areas. To reduce these potential conflicts between indirect fires and aircraft, information pertaining to firing positions, targeted areas, and fire support plans is provided to the AOS at each level of command. See chapter 9 for more information on artillery employment and coordination measures.

MISSIONS

10-60. The OPFOR considers the ability of its aviation assets to provide responsive and continuous fire support to ground forces a tremendous influence on the battlefield. It emphasizes that aviation must be employed early to achieve the following goals:
- Early attainment of air superiority.
- Effective reconnaissance and targeting.
- A coordinated attack on enemy targets at all tactical and operational depths.
- Employment in mass during all phases of combat.

Aviation

- Survivability and responsiveness using effective planning and preparation.

10-61. Aviation assets perform numerous other missions to support ground forces in combat and logistics roles. Many of these missions are performed by elements located at the operational or strategic level. However, tactical ground force commanders may feel their impact.

AIR FORCE

10-62. As enemy air and ground forces are introduced into an AOR, the Air Force must concentrate missions to gain the desired degree of airspace dominance. However, the operational situation dictates the amount of aircraft dedicated to the attainment of air dominance versus support of ground forces.

10-63. Initially, most theater air assets conduct strategic- and operational-level missions. Examples of these higher-level missions are strategic bombing, counterair, air interdiction, theater air reconnaissance, EW, and CBRN delivery.

10-64. Early operational and tactical aviation missions—such as air interdiction and attacks (air strikes) on ground targets—may allow the OPFOR to attain air superiority from the outset. The degree of airspace dominance dictates aircraft employment throughout the theater at the strategic, operational, and tactical levels.

Note. Air Force or army aviation helicopter units and mixed aviation units can also perform some of the missions.

Degree of Airspace Dominance

10-65. The degree of airspace dominance has the following affects on the missions of the Air Force and how it supports the ground force:
- Aircraft sortie rates change.
- Aircraft missions may be restricted.
- Depth and distance of mission execution may be limited.
- Aircraft may assume other roles than those for which they are specifically designed.
- Aircraft ordnance changes.

10-66. The OPFOR uses standardized terms to define the degree of airspace dominance: air supremacy, air superiority, local air superiority, or air parity. This allows planners to best employ assets in the theater to satisfy the requirements to support ground forces.

Air Supremacy

10-67. *Air supremacy* is defined as the condition when the enemy air force is incapable of effective interference. Through the complete destruction of the enemy air forces, this condition is the ultimate goal of air operations. Yet, this condition may be difficult or even impossible to achieve. It may occur, however, through the establishment of a diplomatic "no-fly zone." Under the condition of air supremacy, the OPFOR commander employs all of his aircraft at will.

Air Superiority

10-68. *Air superiority* is defined as the condition when the conduct of operations is possible at a given time and place without prohibitive interference by the enemy. The most efficient method of attaining air superiority is to attack enemy early warning, C2, and ground-based air defense sites, and enemy aviation assets close to their source of maintenance and launch facilities.

10-69. The OPFOR expects to be capable of achieving air superiority against a weaker opponent. However, if faced with a superior enemy, the theater commander may be forced to hold more aircraft in reserve and to redirect aircraft from ground support to air defense operations. This will increase the burden on rotary-wing assets to fill the ground support role.

Local Air Superiority

10-70. Regardless of the scope and time of air superiority, if correctly exploited by the OPFOR, this window of opportunity can produce a devastating impact against the enemy. Even though the OPFOR hopes to attain (overall) air superiority, it recognizes the potential for only *local air superiority* to exist. Purely geographic in nature, this condition is characterized by well-timed aviation missions to coincide with enemy aircraft downtime, returning sorties, aircraft rearming, or gaps in air defense coverage. This condition may also occur in areas across the theater where the OPFOR or the enemy may not have adequate assets available to ensure air superiority. In certain situations or against certain enemies, local air superiority for a specified period of time may be a more realistic goal but just as lethal.

Air Parity

10-71. *Air parity* is defined as the functional equivalency between enemy and friendly air forces in strength and capability to attack and destroy targets. Under the condition of air parity, where neither side has gained superiority, some enemy capabilities affect friendly ground forces at times and places on the battlefield. Air parity manifests itself to the commander primarily in the amount of fixed-wing aircraft used for DAS of ground forces. More aircraft are dedicated to interdiction and attack missions to gain air superiority.

Counterair

10-72. Counterair missions integrate offensive and defensive actions to establish and maintain the desired degree of air dominance. For the mission of countering enemy air forces, the OPFOR is heavily reliant on VTDPs as well as friendly air defense assets. OPFOR aircraft survivability and success in counterair missions depend on the ability of the VTDP network to identify enemy targets and redirect fighters in flight to the proper location at the most opportune time for a successful engagement. This mission primarily falls on the assets at the operational or theater level. OPFOR ground force commanders may feel the effects of this, because assets needed to support counterair missions may detract from the ability of the theater or OSC to support the tactical maneuver.

Reconnaissance and Targeting

10-73. The theater or OSC commander's staff prepares an overall reconnaissance plan detailing tasks for all aviation reconnaissance assets. Operational-level air reconnaissance is a principal method to gather deep target intelligence. Yet, the information the aircrews obtain from those missions is analyzed and disseminated to tactical commanders.

10-74. Specifically equipped aviation assets (such as a reconnaissance aviation regiment) have the primary responsibility for air reconnaissance. They provide reconnaissance support for tactical combat actions by transmitting target information to ground CPs via radio from specially equipped reconnaissance aircraft. The division or DTG conducts its own tactical reconnaissance primarily through ground reconnaissance and UAV assets. Aircrews at all levels of command returning from missions are instructed to report sighted enemy locations and activities. The classification and location of targets obtained through intelligence gathering is the basis for planning air interdiction and attack missions.

Interdiction

10-75. The theater air forces conduct air interdiction missions to annihilate, or neutralize the enemy's military potential before it can be used to inflict damage on friendly forces. These missions are flown to the extent of the enemy's operational width and depth, and they require little integration between friendly air and ground assets.

10-76. Interdiction missions are flown to attack targets beyond the range of friendly surface weapons. These missions are usually planned and conducted at an operational level by the OSC to achieve theater and/or OSC objectives. Therefore, the tactical ground force commanders provide very little input to target selection and little or no assistance during the mission execution. Maneuver commanders may notice the impact of these missions and factor the results into their planning process.

Aviation

10-77. Air interdiction missions are planned at the highest level to synchronize, complement, and reinforce the ground force scheme of maneuver. Typical targets include bridges, roads, railroads, airfields, and large troop support facilities such as supply depots or logistics bases.

Attack

10-78. The OPFOR considers air strikes within the enemy's tactical depth to be attack missions. These are deliberate missions to attack priority enemy targets such as assembly areas, supply routes, artillery or antitank positions, multiple rocket launcher (MRL) positions, forward air bases, and reserves.

10-79. With attack missions, the ground force commander nominates targets to facilitate his scheme of maneuver. (For more information, see Planning and Preparation later in this chapter.) Targets are classified as single, multiple, line, or area. Table 10-1 shows the OPFOR classification of targets and attack techniques.

Table 10-1. Classification of attack targets

Classification	Example Target	Attack Technique
Single (or Point)	An MRL, tank, or armored vehicle; parked aircraft or helicopter; radar, observation post, or bunker	Single pair of aircraft using lower-level or dive delivery of ordnance
Multiple	Group of 10-20 single targets, occupying an area of 1-1.5 km^2	Attack by a small group of (2-8) aircraft
Line	Tactical march column (usually 1 km or longer), a train, or a runway	Attack by a single aircraft or small group of aircraft along the long axis of the target, or flanks
Area	Assembly areas of battalion or larger unit, supply depot, large C2 center, or airfield	Massive and concentrated air attacks delivered from various altitudes and directions

10-80. The AOS plans attack missions to ensure coordination between the aviation force and the ground force and to ensure survivability. Aircraft sorties and ordnance types are requested to achieve the desired results, based on the target classification. The missions are well planned with triggers to signal aircraft launch. Procedures for airspace deconfliction are enacted prior to launch.

10-81. The ground force commander uses attack missions to shape the battlefield. By attacking priority targets, these missions should prepare the conditions for his success over the ensuing 24 hours or reinforce successful attacks by his ground forces. Attack missions can help create penetrations, cover withdrawals, and guard flanks, and can be most effective when employed at decisive points in a battle. The ground force commander plans an energetic scheme of maneuver to complement attack missions and trap or destroy major elements of the enemy force.

Direct Air Support

10-82. The objective of DAS is to disrupt and destroy enemy forces in proximity to friendly forces. Although DAS is the least efficient application of air forces in terms of damaging enemy capabilities, it is the most critical to ensuring the success and survival of ground forces. These missions have the greatest potential to make an especially important contribution to the ground force commander's plan. He must be ready to exploit the effects of DAS through rapid maneuver, either by closing with and destroying the enemy or by bypassing enemy forces.

10-83. Due to the proximity of these missions to friendly ground forces, extensive care is taken to minimize fratricide. Effective DAS requires reliable air-to-ground communications and flexible, responsive C2. (See Airspace Control Measures earlier in this chapter.) It requires aviation components to appreciate the capabilities, limitations, and risks to ground forces. It also requires the ground component to understand the capabilities and limitations of DAS.

10-84. The OPFOR normally conducts DAS with fixed-wing ground-attack aircraft and rotary-wing attack aircraft. These missions typically extend only to the range of friendly ground-based systems. That is, OPFOR aircraft are covered by the fire of friendly weapon systems, and under the air defense coverage of friendly systems. They target objects of immediate concern to the ground force commander when the fires of his constituent or dedicated assets are not capable of engaging the enemy or when a mass concentration of fire is required.

10-85. DAS missions are entirely controlled by the FAC. Once the AOS or the VTDP notifies the FAC that aircraft are inbound to his location, he establishes communication with the aircraft and provides the necessary data for the aircraft to complete their mission. A FAC controls all aspects of their mission. FAC control procedures include—

- Establishing an IP.
- Establishing attack positions (APs), normally at maximum effective weapons range.
- Issuing control graphics.
- Identifying and marking friendly troop locations.

10-86. As the aircraft travel inbound from the IP, the FAC provides the pilots with—

- Target location (either in grid coordinates or in relation to a predetermined reference point).
- The exact time to execute the attack.
- Information on the ground situation.

He may also give the flight leader a signal to direct the flight to climb, acquire the target, and attack.

10-87. The primary responsibility of pinpointing the target is left up to the flight leader. He orders the flight into different formations, divides the target, and assigns individual sectors to the aircraft in his flight. The FAC assesses damage and adjusts the flight for successive target runs if necessary. So, the FAC must maintain visual contact with the target while the aircraft are on station.

Transport

10-88. Transport missions for airlift, airborne insertion, airdrop, and aerial resupply are all Air Force fixed-wing transport aircraft missions that are performed by operational-level assets. (Air Force or army aviation helicopter units and mixed aviation units can also perform some of these missions.) They may, however, have impacts on the tactical ground force commander, by limiting his maneuver. For example, if a forward airbase or an airdrop site is set up by the operational-level commander to resupply adjacent tactical units, a large area is dedicated to the Air Force for the mission. This area may present an obstacle or a restriction to the ground scheme of maneuver for the tactical commander.

ARMY AVIATION

10-89. Army aviation is a component of the ground forces and is intended for actions directly in the interests of combined arms organizations. Based on the type of missions performed, army aviation is divided by predominate aircraft capabilities into attack, CS, and CSS helicopters.

10-90. Attack helicopters are the primary assets used to provide firepower to ground forces. These assets can perform armed reconnaissance or fire support in all types and phases of ground combat. They can also provide fire support for heliborne landings. Other helicopters can also conduct heliborne landings, lay minefields, or perform a variety of logistics, reconnaissance, liaison, and communications functions in accordance with the plans of the supported combined arms organizations. Some helicopters are capable of performing in multiple roles.

Attack Helicopters

10-91. Attack helicopters (also referred to as fire support helicopters), rather than fixed-wing aircraft, provide the preponderance of the support to the ground force and provide an excellent fire and maneuver

Aviation

capability to the ground commander. The primary categories of tactical missions for attack helicopters are attack, DAS, and reconnaissance and security. Some attack helicopters may be modified to perform air-to-air combat roles. The majority of OPFOR attack helicopters are equipped for all-weather and night operations.

10-92. Attack helicopters generally have integral cannons, miniguns, and/or automatic grenade launchers. They also have the provisions to mount antitank guided missiles (ATGMs), rockets, bombs, or other ordnance on fuselage or under-wing hardpoints. Most employ target acquisition and sighting systems (such as laser, thermal, or infrared).

10-93. The OPFOR may employ multirole helicopters in the same capacity as a pure attack helicopters, but generally with less firepower. These aircraft have the provisions to carry a limited number of passengers and may have mounts for a cannon, rocket pods, or a few ATGMs. They are small, relatively quiet, and easy to conceal from radar and visual detection when silhouetted against background clutter.

Helicopter Attack

10-94. Helicopter attack missions are conducted within the enemy's tactical depth. Similar to the fixed-wing attack mission, the purpose of helicopter attack missions is to destroy priority enemy targets such as artillery or antitank positions, MRL positions, and reserves. The OPFOR may employ army aviation helicopters to attack counterattacking enemy armor columns or enemy columns moving forward to reinforce engaged units.

10-95. Deep autonomous attack helicopter missions in the disruption zone are the exception rather than the rule. However, they may occur against an extremely high-priority target. The commander understands the risks involved in missions such as these and realizes the high probability of loss of aircraft and crews.

10-96. For these disruption zone attacks, the OPFOR will launch the minimal number of aircraft (two to four) to accomplish the mission. Suppression of air enemy defenses (SEAD) is normally executed in support of the mission. The focus of the SEAD is to destroy, degrade, or neutralize enemy air defense systems in a specific area through either attack or electronic jamming. The depth of these helicopter attacks will be limited primarily by the range or endurance of the aircraft. Consideration is given for planning additional contingency time for the aircrews to react to unexpected actions in enemy territory. The distance may also be limited by the range of the artillery. Normally, if attack helicopters are operating deep, they are operating as part of an IFC at OSC or DTG level.

10-97. The primary deep mission in the disruption zone for attack helicopters is in support of heliborne landings. These helicopters can provide security and armed escort for troop-carrying helicopters. They may prepare the LZ by fire and remain after the insertion to provide DAS to the ground force. The number of aircraft employed depends on the size of the heliborne force, the degree of protection desired, and expected enemy resistance. For more information on the heliborne landings, see Combat Support Helicopters.

10-98. Like fixed-wing attack missions, helicopter attack missions are planned by the AOS to ensure coordination between the aviation force and the ground force, and minimize the risk of the mission. Based on the target, number of available helicopters, and required ammunition, missions are planned to achieve the desired results. The missions are planned in detail with triggers to signal aircraft launch. Rehearsals are performed to identify any problems and increase the probability of mission success.

Direct Air Support

10-99. The disruption and battle zones provide opportunities to the commander to effectively employ attack helicopters when the enemy presents numerous targets in the open. Armed with ATGMs and rockets, helicopters provide DAS for the advance of the ground forces by flying behind OPFOR ground forces and firing over them. This places the helicopters out of friendly direct fire ranges and behind or under friendly artillery trajectories.

10-100. Since army aviation serves as the ground force commander's primary asset for air support, DAS is the most common type of mission. In the DAS role, helicopters can augment fixed-wing DAS, ground-based artillery, and direct fires from ground forces. This fire support is conducted throughout the disruption and battle zones. Attack helicopters destroy tanks, antitank weapons, and other armored targets located in proximity to friendly units.

10-101. DAS missions use two to eight aircraft per mission. They are flown using the wingman concept with a minimum of two aircraft. The wingman has the responsibility to provide local security while the lead is focused on the target. Helicopters firing ATGMs may be exposed and vulnerable during missile flight, depending on the type of missile. To minimize exposure time, the helicopters can also employ rockets or the main gun in lieu of ATGMs, but with less effectiveness.

10-102. While in proximity to friendly forces, attack helicopters are afforded the protection of air defense assets and the covering fire of ground systems. Using the integrated fires of tank or mechanized forces, artillery, and attack helicopters, the commander creates corridors through the enemy's forward ground forces. These corridors, coupled with SEAD, allow further employment of all other types of air assets.

10-103. In the defense, helicopters can be used to counterattack tank or mechanized forces while serving as the commander's antitank reserve. The commander may employ them to independently execute a counterattack into the flanks of an enemy formation. Armed with ATGMs and rockets, the helicopter force seeks routes allowing undetected approach to the flanks of the enemy force. If terrain variations do not provide adequate concealment for the force, the helicopters may use smoke to conceal their approach. The helicopter formation then engages enemy targets from APs along preplanned attack routes.

Reconnaissance and Security

10-104. Attack helicopters are used for armed reconnaissance when visibility is limited, target information is incomplete, or enemy flanks are unprotected. In these circumstances attack helicopters, by flights of two, conduct high-speed, low-altitude penetration of enemy lines. Targets of opportunity such as radars, communication nodes, missile launchers, and antitank weapons are engaged at the discretion of the flight leader. Because these missions are considered hazardous, they are normally reserved for very experienced pilots, and therefore are quite risky to the ground force commander.

10-105. Commanders also use attack helicopters to provide assistance with the ground force counterreconnaissance battle. These helicopters are launched in small numbers to positions in the disruption zone to engage enemy ground reconnaissance assets as they approach friendly positions. The commander's intent is to deny the enemy the reconnaissance information that may expose weaknesses in his scheme of maneuver.

10-106. Attack helicopters may be employed to protect the flanks of a tank or mechanized column in the attack or counterattack or in a tactical movement by screening the column from the enemy. The aircraft protect the column by flying along the route or maneuvering by bounds using the cover and concealment of the terrain along the route. Similarly, they may serve as convoy escort.

Counterair

10-107. Helicopter air-to-air combat modifications are commonly available on the open market, and some newer helicopters may be designed with the capability. While attack helicopters are the likely candidates for this role, other types of helicopters could be configured to mount air-to-air weapons. Helicopters can employ from external weapon racks some of the same missiles used as surface-to-air missiles in the ground forces. Several ATGMs are able to engage other aircraft from aerial platforms, and mounted automatic weapons may also be employed. Helicopters equipped with such weapons (if available) are the only form of air-to-air engagement available to support the tactical ground force commander.

Combat Support Helicopters

10-108. CS helicopters serve in numerous roles. They are designed with troop- or cargo-carrying capabilities and can be armed with miniguns or machineguns fired by crewmembers other than the pilots. They have provisions to carry external loads such as fuel tanks, ATGMs, rockets, or EW equipment on external hardpoints or underslung on cargo hooks. Their primary function is to act as transport aircraft in a heliborne landing and to serve in other supporting roles. Thus, the resulting cargo weights limit the type and amount of armament used. If the OPFOR lacks a dedicated attack airframe, these helicopters may perform both roles. However, they would be less effective than designed attack helicopters, because they lack an integral fire control, sensor, and optic systems.

10-109. The OPFOR launches a heliborne landing for the purpose of inserting a ground force or reconnaissance assets, usually in the disruption zone. This normally occurs under the cover of darkness and up to 2 to 6 hours prior to a planned ground attack. LZs are selected beyond the range of enemy direct fire weapon systems. Prior to insertion, LZs are targeted with artillery (if within range) or escorting attack aircraft. After troop insertion, the CS helicopters depart, and the attack helicopters may remain. Forces remaining in position longer than 24 hours are resupplied by helicopter.

10-110. In addition, CS helicopters are called upon to transport antitank squads or perform electronic jamming. CS aircraft can also supplement obstacle detachments by laying mines along threatened flanks and gaps, and assist in the preparation of complex battle positions by providing logistics support. They may also fill a variety of other support or logistics functions.

Combat Service Support Helicopters

10-111. Helicopters providing CSS are large and lightly armed (if at all). They have large cargo areas with provisions to load freight and fuel internally or carry them underslung on cargo hooks. Their movement is usually limited to conducting resupply missions in the support zone, yet they may be employed in the battle and disruption zones in some circumstances.

10-112. These aircraft may be employed to transport an airborne or heliborne force. Attack aircraft may escort them to the DZ or LZ. The forces they carry are used to augment the prior insertion of a heliborne force by CS helicopters once the objective is secured.

10-113. Some CSS helicopters can be fitted with extra fuel tanks and pressurized refueling hoses and may be employed to establish a FARP prior to a heliborne insertion or attack mission. They do not perform this mission in enemy territory.

10-114. These helicopters may also be used in search and rescue, and downed-aircraft recovery roles. Missions such as these are escorted by two to four attack helicopters.

Forward Arming and Refueling Points

10-115. The flight services elements of the army aviation units have the personnel and equipment to establish FARPs. The OPFOR does not place as much emphasis on FARP employment as do military forces of some other nations. This is due to the lack of deep autonomous attack helicopter missions that would require FARPs.

10-116. FARPs would normally be established within friendly territory to support helicopter missions. FARPs are placed near open areas to allow for landing sites, but with nearby terrain that affords cover and concealment from the enemy. FARP operations will move to an alternate site if compromised. The flight services element may set up temporary or deception FARPs based on supporting the ground force scheme of maneuver. A combat helicopter brigade has the ability to place one FARP per attack helicopter battalion. The FARP includes four to six refueling points and an area for rearming. Under reasonable conditions, a flight of four aircraft can expect to be replenished with fuel and ammunition in 45 minutes. In maximum employment conditions, this time increases due to logistics constraints and a finite number of refueling points. Also, in adverse weather and at night, these times increase. Some aircraft may perform area security while others in the flight are refueled and rearmed. Upon completion of the air support mission, the FARPs are moved or removed, while aircraft recover to their holding areas or airfields.

UNMANNED AERIAL VEHICLES

10-117. The Air Force and the Army both use UAVs. This chapter primarily discusses UAVs in the tactical role. The techniques and employment of the larger and more capable operational- and strategic-level UAV platforms used by the OPFOR are similar to those employed worldwide.

Capabilities

10-118. UAVs can support combat operations anywhere on the battlefield. When equipped with the proper sensors, they provide imagery day and night and in all-weather conditions (depending on the size and capability of the platform). UAVs are an excellent imagery asset, providing the commander with a near real time (NRT) reconnaissance and battlefield surveillance without the possibility of risk to a manned aircraft. They provide OPFOR commanders a dedicated and rapidly taskable asset that can look wide as well as deep. During a preplanned UAV mission, changes in mission priorities or identification of new targets may occur. The commander selects the appropriate UAV based on what is available, current mission configuration, operating range, operating radius, and endurance (flight time) of the UAV. The OPFOR commander can then direct a UAV to support a different mission or area.

Note. The size, ease of operation, and simple design of many smaller UAVs lend them to field expedient modification. Converting these UAVs into a munitions delivery system (improvised attack UAV) is not difficult and offers several tactical advantages. Off-the-shelf remote controlled aircraft can also provide this capability.

10-119. UAV teams can launch UAVs from either improved or unimproved airstrips. Small UAVs can be hand-, canister- (vehicle), or tube-launched. Many UAVs are used in various roles to support destruction of enemy systems and suppression of enemy missions. Those roles vary from target acquisition to direct attack with an impact kill by the UAV.

10-120. Some air defense, antitank, artillery, SSM, littoral, logistics, and other units with stationary facilities requiring security patrols can use UAVs to execute the mission while reducing personnel and vehicle requirements. SPF, commandos, and some paramilitary forces (such as insurgents and guerrillas) can use UAVs.

10-121. UAVs can provide NRT combat information about terrain, disposition of enemy units, and battle damage assessment. They can assist in recognition, detection, designation, and illumination of targets. They can also assist in route, area, and zone reconnaissance.

Note. A GPS jammer the size of cigarette pack transmitting 4 Watts, can effectively deny use of GPS in an area ranging as far as 150-200 km. It is extremely simple to install one of these lightweight GPS jammers into a small UAV. Off-the-shelf remote controlled aircraft can also be modified to provide this capability.

Missions

10-122. UAVs are capable of locating, recognizing, and possibly engaging enemy forces, moving vehicles, weapons systems, fixed structures, and other targets. Some example OPFOR missions using UAVs include—

- NRT reconnaissance and surveillance (see chapter 8.)
- Target acquisition.
- Direct attack (used as a mini-cruise missile or other weapons delivery system).
- Laser designator. Some UAVs can be fitted with laser designators to mark targets, and others may be armed.
- EW (such as deception, GPS jamming, spoofing, meaconing [rebroadcast real GPS signals], or intercept).
- Communications relay.
- Security.
- Vectoring.
- Cargo transport.

Aviation

10-123. Information gathered via UAVs may be immediately acted upon, or it may be integrated with other sources to support or shape the immediate combat mission, to plan future operations, or to re-allocate reconnaissance assets. The data may be integrated with that from combat reconnaissance, ground surveillance radar, intelligence assets, or any other information.

PLANNING AND PREPARATION

10-124. Ground commanders can employ air support, integrated with other forms of fire support, throughout the AOR to attack the greatest threats to successful ground combat. Mission planners are responsible for incorporating the most current information on enemy and friendly positions, current weather, terrain, fire support plans, and EW targets to plan air support missions that complement the ground maneuver plan.

10-125. Planned missions afford ground maneuver commanders greater freedom of movement and flexibility by allowing them to mass firepower at decisive points to annihilate or neutralize enemy forces. At every level of command from battalion to OSC, ground commanders nominate targets for air support assets to attack. Assets are requested, forces are allocated, an ASP is produced, and pre-mission planning is performed to maximize effects and minimize risk.

TARGET SELECTION

10-126. At theater and OSC levels, the targets are selected based on strategic or operational-level goals. At tactical levels, targets are selected to shape the battlefield for the success of the ground forces. The targeting process is mostly preplanned, based on integrating the fires of ground assets (such as artillery, MRLs, and SSMs) and aviation assets. It is a continuous, ongoing process designed to exploit current intelligence and attack high-priority targets in all phases of the battle to best achieve the commander's scheme of maneuver. As the tactical battle continues, targets are selected from the existing targeting database, or new ones emerge as windows of opportunities develop.

10-127. Target lists are categorized and prioritized based on depth into the enemy forces. The OPFOR attempts to plan targets for its attack aircraft which shape the battlefield versus reacting to ground maneuver forces that require immediate support. However, the following priorities are established:

- Enemy forward positions, maneuver units, artillery, and C2 nodes.
- Deeper artillery, C2 nodes, reserves, assembly areas, supply routes, artillery or antitank positions, MRL positions, and forward air bases.
- Deeper reserves, lines of communication, airbases, and troop support and logistics facilities.

10-128. From these target lists, requests for artillery fire and air support are generated at every level down to battalion. The targeting responsibilities of the ground force do not end with target nomination. Commanders and their AOSs planning fires must continue to refine and update target information until the desired results are achieved. The forum for this is the targeting meeting held within the AOS at each level of command. These AOSs correlate the ground force commander's targeting priorities with actual targets, plan attack positions, incorporate FAC input from prior missions, and discuss mobile targets. The latter is particularly important, since mobile targets represent the most difficult problem facing ground force commanders. When considering mobile targets, commanders may employ one of three methods to control the timing of the air attack: on-call, immediate, or preplanned. For more information on this subject, see Aviation Support Plan below.

10-129. Commanders plan for targeting contingencies during the course of a battle. When a target of opportunity presents itself, the commander—through his AOS and FACs—has the ability to redirect his air support to attack the new target. Additionally, pilots have the capability of acquiring targets in the performance of their mission. This presents an ability to exploit targets of opportunity that present themselves to the pilots, provided the targets are included on the commander's targeting list.

REQUEST PROCESS

10-130. Formally, the lowest command level capable of requesting aviation support is the brigade or BTG. Battalion commanders input requests to the brigade or BTG. However, as every commander plans and conducts combat actions, he identifies situations where aviation attacks or DAS can be employed to enhance mission accomplishment. The brigade or BTG AOS also assists in nominating targets and integrating aviation into the overall scheme of maneuver. This same procedure occurs at each level of organization by the supporting AOS.

10-131. Air support requests from ground maneuver forces are screened at every level of command to determine whether or not—

- Ground support missions can be supported while meeting strategic- or operational-level air requirements.
- The level of air support to ground forces meets operational and tactical requirements for achieving the goals of ground battle plan.
- Alternate systems (such as artillery, MRLs, or SSMs) would be more effective to accomplish the mission.
- Air requests are supportable based on current available aircraft.
- Planned airspace usage, artillery fires, and intelligence requirements can be met.

10-132. All requests for aviation support are compiled and submitted through AOS channels for approval by the theater and OSC commander. The DTG commander will approve missions for rotary-wing aircraft constituent or dedicated to his level of command. The AOS must divide the requests between those supportable by rotary-wing assets and those supportable by fixed-wing assets. Helicopter missions are ranked by assigned priority and precedence, and given to the executing army aviation headquarters for planning. Some air support requests continue on to be filled by theater or OSC fixed-wing assets. If approved, these requests are also assigned a priority and precedence. Requests for air support are submitted as early as 72 hours prior to the requested aircraft on-station time and no later than 24 hours prior to the start of the ASP.

10-133. *Preplanned*, *immediate*, and *on-call* refer to the requests themselves. Preplanned requests are those submitted in time to be included in the published ASP. Immediate requests fill operational or tactical requirements that are too late to be published in the ASP. On-call mission requests do not state a specific aircraft time-on-target. They involve aircraft placed on an appropriate alert status and employed when requested by the supported unit. Aircraft used to fill immediate requests may come from on-call missions established for this purpose.

10-134. Starting at the brigade or BTG level, the CAO submits the air support requests. He submits *preplanned* requests through ground command and staff channels, or *immediate* requests through AOS channels. *On-call* requests are transmitted by the FACs or the AOS to the division or DTG AOS using VHF communications.

10-135. The AOS at every level is of key importance in the processing of immediate air support requests. This type of request is primarily passed via the FACs and AOSs to the level of command that controls the required aircraft. If an OSC aviation unit can support the requesting ground force, it fills the requirement. If not, the request will be passed up to the theater AOS. Once an immediate request is approved at the theater or OSC level (depending on the type of supporting aircraft), the AOS tasks on-call missions or diverts scheduled missions to satisfy that request.

10-136. Once a request is approved at either theater or OSC level, it is forwarded to the aviation unit to determine if it is supportable based on the projected sortie generation rate and operational tempo. If disapproved at any level, the requests are returned to the originator through AOS channels with an explanation.

10-137. For all requests, the higher aviation command or IFC provides the required information (including target, location, required on-station time, and radio frequency) to the tasked unit. Each aviation unit then conducts its own mission planning and coordinates directly with the ground maneuver unit. The approved

Aviation

missions and enacted airspace procedures are disseminated to all levels of ground and aviation commands through the ASP.

SORTIE GENERATION

10-138. The aviation units are able to manipulate and predict to some extent their ability to launch and sustain aircraft. This information is compiled and forwarded up to the AOS at theater and OSC level where it is reconciled with the commander's scheme of maneuver and the requests for air support. There, the commander determines the number of air assets to best fit into his plans for the operation. The decision is made how to employ all of the available air assets to accomplish the theater or operational goals, including support to ground forces. The resulting product is known as the maximum sorties available in a single 24-hour period.

10-139. The OPFOR defines an aircraft sortie as a flight by one aircraft in an air action. Across the theater, the maximum aircraft sortie rates are determined daily incorporating many factors, at every level of organization. Table 10-2 contains a generic formula that holds true for both fixed- and rotary-wing aircraft. It can incorporate many of the factors involved, which are listed in the paragraphs following the figure.

Table 10-2. Calculation of aircraft sorties (example)

Maximum Theoretical Sorties/Day =
(Total Aircraft Available − Attrition) x Allocation x OR Rates x Aircraft Sorties/Day
An example calculation of sortie generation rate follows.
Conditions:
110 total aircraft.
10 lost yesterday through combat action.
The commander wants 40 % dedicated to DAS, 30 % to attack, and 30 % to interdiction.
OR rate today = 80 % based on logistics sustainability and maintenance posture.
Aircraft sorties/day are: 3.5 DAS, 2.0 attack, and 2.5 interdiction.
(110 − 10) x 40 % x 80 % x 3.5 = 112 DAS sorties today
(110 − 10) x 30 % x 80 % x 2.0 = 48 attack sorties today
(110 − 10) x 30 % x 80 % x 2.5 = 60 interdiction sorties today

Total Aircraft Available

10-140. Total aircraft available, or "flyable," is calculated to incorporate all aircraft regardless of type or mission. This number can be calculated to account for aircraft by specific mission type.

Attrition

10-141. Attrited, or "non-flyable," aircraft is the number of aircraft losses due to combat, fratricide, or irreparable enemy damage since the last sortie generation calculation. Attrited aircraft may be returned to service for future sortie generation cycles.

Allocation

10-142. Allocation or "how flying," is the ground force commander's intent on how sorties should be allocated to individual missions, such as interdiction, attack, DAS, counterair, airlift, or transport. As hostilities develop in the region, the OPFOR balances the strategic- and operational-level goals against the tactical air support requirements to determine how to best allocate the aircraft to specific missions to attain the desired effect or change upon the enemy. At the strategic level, this allows the ground commander to account for the air objectives and ground force objectives. At the operational and tactical levels, it allows ground force commanders to allocate percentages of air support assets to best fit the ground scheme of maneuver.

10-143. Depending on how the OPFOR perceives the air situation, the allocation may differ. (See Degree of Air Dominance earlier in this chapter.) Multirole aircraft prove most valuable in considering allocation. They can quickly and easily be tailored to perform different missions based on the commander's needs. Commanders may also elect to keep a number aircraft as a reserve ready to serve if needed for unexpected contingencies.

Operational Readiness Rate

10-144. The operational readiness rate (OR rate), or "ready-to-fly rate," refers to the capability of a unit, equipment, or weapon system to perform the mission or function for which it was organized or designed. This encompasses the ability of the OPFOR to sustain its aerial forces. Factors considered include—
- On-hand major end items.
- Spare part availability.
- Scheduled aircraft maintenance.
- Logistics and resupply procedures.
- Transportation capabilities.
- Aircraft cannibalization and/or transfer procedures.

Initially, an OR rate in excess of 85 to 90 percent is considered normal. As hostilities continue, this rate can diminish considerably, based on the above-listed factors.

Aircraft Sorties per Day

10-145. The number of aircraft sorties each day, or "turns per day," varies with each type of aircraft. It is primarily a function of mission duration and the time required to refuel and rearm the aircraft for the next mission. This also can incorporate the human factors of pilot-to-aircraft ratio, aircrew availability, proficiency, endurance, and training level. It can also encompass the availability of fuel and proper munitions for the intended mission. If the aircraft is fueled, properly armed, and mission-ready, it cannot fly the planned number of sorties per day without a qualified, prepared crew to man the cockpit.

10-146. Commanders must balance their ability to regenerate their aviation assets against their willingness to allow that ability to be degraded through loss of assets. Planning rates allow aviation units to operate at a certain sortie rate for a certain period of time, normally 30 days, without resupply. Units may elect to operate in a "sustain mode" with a slower operational tempo, planned maintenance, and a normal logistics flow. This allows them to operate at a higher rate over a longer stretch of time.

10-147. Alternatively, units may elect to conduct "surge operations." This is characterized by a higher than usual operational tempo, and neglecting preventive maintenance and scheduled services for 1 to 2 weeks. This gives units the ability to fly more sorties than normal in a short period of time. Compared to sustain mode, surge operations actually force a slower operational tempo over the long term, since more extensive maintenance needs to be performed on these aircraft. Eventually logistics stocks are depleted and fatigue increases. Following an extended surge, a unit must recover by performing the maintenance that has been neglected. If the unit returns to surge rate prior to recovery, its sortie generation capability may continue to fall, and future recovery time increases.

AVIATION SUPPORT PLAN

10-148. The theater or OSC AOS publishes a daily document called the aviation support plan (ASP). This document has two parts: the air tasking order (ATO) and the airspace coordination order (ACO).

10-149. The ATO is the portion that outlines all approved fixed-wing, rotary-wing, and UAV missions to include interdiction, attack, DAS, counterair, reconnaissance, airlift, transport, or aerial refueling. The ATO development process is continual and starts with requirements for air support that are submitted as requests. These requests are changed, refined, or reviewed at each day's targeting meeting. (See Target Selection earlier in this chapter for more information.) All requests must be finalized and submitted no later than 24 hours prior to the beginning of the next ATO cycle. The ATO is published 12 hours prior to going into effect, which occurs in the early morning hours and continues for 24 hours.

Aviation

10-150. To publish the ATO, the theater or OSC AOS reconciles air support requests from all levels of command with sortie generation capabilities and command objectives for the allocation of air assets. The AOS does so with the assistance of the air army (Air Force) CP, the army aviation CP, and the theater or OSC IFC CP. Ground commanders are advised to submit preplanned requests for on-call missions to ensure availability of sufficient sorties with appropriate ordnance to respond to immediate air support requests. If more aircraft are available on a given day than required for combat operations, the excess are either assigned missions to augment the air missions already planned or held in reserve. Conversely, if there are more air support requests than available aircraft, missions are filled based on the priority assigned to each request.

10-151. Once the ATO is published, it is an execution order. All published missions occur for that 24-hour period. Ground force commanders may not know which unit or what type of aircraft will support them, but they are assured the support.

10-152. The second portion of the published ASP is the ACO. This is also an execution order that delineates all positive and procedural airspace controls enacted to best accomplish the ATO. The ACO controls the combined efforts of all aviation assets, and missile and artillery forces. The ASP is disseminated to all air and ground force unit's AOSs.

10-153. The OPFOR uses three types of air support missions to meet the needs of ground force commanders. The names are based on the types of request and on the timing of the air support. They are *preplanned, immediate,* and *on-call*. (See also Request Procedures, above.)

Preplanned

10-154. DAS missions are primarily preplanned. The ground force commander identifies the targets, times, and desired damage for the missions. The IFC commander determines the force, size, ordnance, and technique that can accomplish the mission. The IFC staff plans these missions in great detail and integrates them with other forms of fire support. The target selection process identifies possible kill zones for the application of aviation assets. The sortie generation process, coupled with the ASP cycle, assigns aviation assets to the highest-priority missions to attack targets, allowing the ground force commander to achieve his scheme of maneuver. Aircraft are allocated, prepared with the proper ordnance and countermeasures, and launched to attack a target at a specific time and place as a part of an integrated ground and air scheme of maneuver.

10-155. After ASP confirmation of preplanned requests for air support, a ground force commander consults his IFC staff to finalize detailed plans for the coordinated air and ground scheme of maneuver in his AOR. If they plan to use attack helicopters, the planners coordinate directly with the army aviation unit to ensure target deconfliction and to limit fratricide. The planned attack allows the ground force to update targets, current enemy and friendly situation, and disposition of enemy air defenses just prior to the mission.

10-156. Additional detailed pre-mission planning and coordination done prior to a preplanned mission by the ground force and the aviation force specifies—

- Target description and desired results.
- Type and number of assets required to accomplish the mission.
- Time.
- Location.
- Attack technique.
- Ordnance required.
- Communication frequencies and codes.
- Approach and departure routes.
- EW support.

10-157. Once airborne, the aircraft proceed to a designated checkpoint behind friendly lines and confirm their target assignment with VTDPs controlling their transit through the AOR to their APs. En route to the IP, the flight receives target updates from the VTDP or the FAC.

10-158. Preplanned missions are similar for CS and CSS helicopters. Most of the missions flown by these types of helicopters are preplanned in nature. The lack of time-critical constraints allows the aviation unit and the maneuver unit to conduct the greatest amount of coordination before the mission even commences. This coordination can cover issues such as LZ/PZ preparation, equipment preparation, pick-up and drop-off times, airspace management, and communications. Preplanned missions also allow the aviation units to take all possible steps to minimize risk throughout the course of the mission.

Immediate

10-159. Most air support missions are preplanned, but immediate missions also are used extensively. Ground force commanders can request them through AOS channels for inclusion in the ASP. By doing this, ground force commanders identify general times and places where they believe air support is required, but without finalizing the intricate details as in a preplanned mission. An immediate mission allows the ground force commander to have air support assets readily available to employ at a given time against targets. This type of request is used primarily for attack aircraft. If a CS aircraft is needed for this type of mission, it locates in the vicinity of the requesting CP and assumes more of an on-call role.

10-160. The ASP allocates air support assets for immediate missions. Some pre-mission planning and coordination occurs between the supported and supporting forces to ensure aircraft survivability. Aircraft designated for immediate missions can be airborne or on the ground at airfields. Before takeoff, pilots are briefed on a checkpoint to proceed to, and possible target type and location. Aircraft are prepared with the ordnance and countermeasures for the most probable target they may encounter. As the ground force commander decides he needs the air support to engage, he notifies the AOS or the FAC to pass the request. The request is passed to the attack helicopter battalions or to the fixed-wing units. See figure 10-5 for details on the immediate DAS request process.

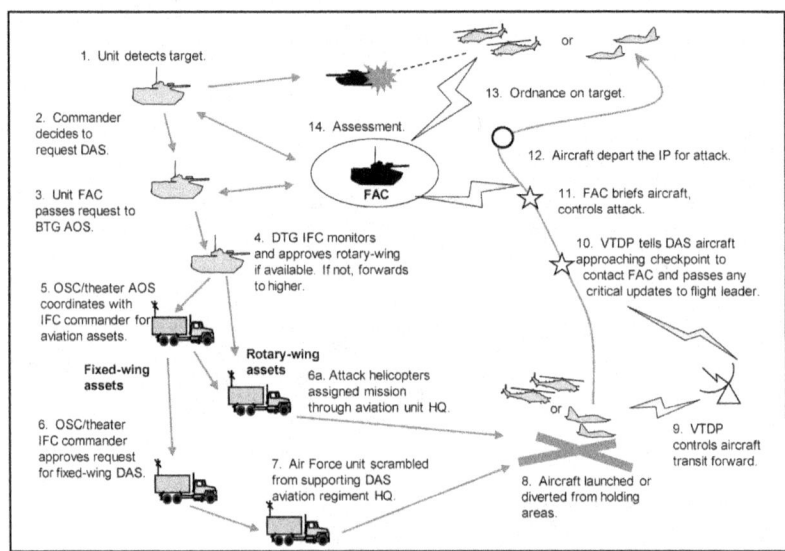

Figure 10-5. Immediate DAS request process

10-161. The aircraft are launched from airbases, or released from holding areas. The flight establishes communications with the VTDP as it moves forward. En route, the aircraft are briefed on the frequency

Aviation

and contact point to establish communications with the FAC, and receive critical mission updates. With communication established, the FAC gives them final target designation and confirmation, and the aircraft depart the IP for the attack.

> **Note.** A *holding area* is a site (on the ground or in the air) located between airbases or FARPs and IPs that may be occupied for short periods of time by aircraft while coordination is being made for movement to the IPs.

10-162. In some cases, there may be an excess of aircraft available as compared to requested missions outlined in the ASP. Then the theater or OSC IFC commander can establish a pool of aircraft to be available for immediate missions if unforeseen contingencies develop. He may also designate that multirole aircraft be configured for DAS. These extra aircraft may be armed and prepared for "generic" targets and launched to holding areas. Once in the holding area, they serve as an available asset to subordinate ground force commanders that require immediate air support. The ground commander's FAC requests these aircraft and briefs them on the target en route. Sometimes the OPFOR refers to this practice as "push" DAS.

On-Call

10-163. Planners may allocate aviation assets for on-call missions where the ground force commander can time the attack at his discretion. He bases the trigger for the attack on the enemy's reaction to the OPFOR scheme of maneuver at planned decision points. This allows him to mass his fires at decisive places and times on the battlefield and to refrain from employing the attack assets if the target no longer threatens. Additionally, he may elect to shift the assets to attack another target. If not supporting a preplanned mission, CS and CSS helicopters assume an on-call role. By using these methods of employment, the commander can conserve his air assets for use when needed, or avoid the need for the ground forces to halt their attack to wait for an unnecessary preplanned air support mission.

10-164. On-call mission requests do not state a specific aircraft time-on-target. They involve aircraft placed on an appropriate alert status and then employed when requested by the supported unit. Prior to the employment of air assets, the ground force will have established communications or liaison with the aviation unit. On cue from the ground force, the aviation unit performs the necessary preparations to launch when the ground force requires the air support.

READINESS CONDITIONS

10-165. The OPFOR recognizes three levels of combat readiness for fixed- and rotary-wing aircraft and aircrews. Aviation unit CPs use these three categories to describe varying levels of alert status or readiness conditions and thus to guide the units to total mission preparedness for the most probable launch execution order. Aircraft in categories one and two can be expected to respond to on-call missions timed by the ground force commander. Table 10-3 on page 10-26 lists the categories and shows their duration and time before assets can be in the air. Based on his decision points, the ground force commander tells the designated aviation unit when to increase the level of readiness. Under these conditions, both parties prepare to execute the air support mission when needed.

Table 10-3. Levels of combat readiness

Category	Crew and Aircraft	Duration of Readiness	Time Before Takeoff
One	Aircraft are fully serviced and armed. Combat crews are briefed on their mission and are in the aircraft ready to start engines. Ground personnel are assisting the combat crews.	1-2 hours	3-5 minutes
Two	Aircraft are fully serviced and armed. Combat crews are briefed and are on standby in the vicinity of the aircraft, ready to take off within a specified short period of time after receiving a mission order.	2-4 hours	15 minutes
Three	Aircraft are refueled and serviced. Cannons are loaded. External systems (such as bombs, rockets, missiles, and fuel tanks) are not loaded. Combat crews are designated, but not on standby; they have not been briefed on the air and ground situation, but will be before takeoff.	2-4 days	1-2 hours

10-166. Army aviation tends to operate from FARPs or holding areas. A flight of helicopters held at the highest state of readiness should reach its target in 15 to 20 minutes; a full battalion or squadron requires up to 25 minutes. Preparation of a follow-on attack could take as little as 15 to 20 minutes depending on the number of helicopters involved and refuel and rearm procedures. Fixed-wing aircraft, at the highest readiness state, should reach their target in 20 to 30 minutes after launch, since the aircraft are based further away, but travel at much faster speeds. Ground commanders are aware of these approximate aircraft transit times and factor them into the launch trigger.

RISK MANAGEMENT

10-167. Effective pre-mission planning and preparation are paramount to increasing the survivability of aircraft in combat. Prior to any aviation mission, aircrews must have a detailed intelligence picture of the battlefield. They use this information to plan every aspect of the mission. This helps crews to determine ingress and egress routes through gaps in enemy air defense coverage and to plan tactical maneuvers. Similarly, they try to use routes that afford cover and concealment from the enemy, while allowing the protection of friendly air defense assets.

Deception and Surprise

10-168. The OPFOR emphasizes the importance of deception and surprise to paralyze hostile air defenses and enhance aircraft survivability. Aircraft approach target areas at the lowest permissible altitude, given weather and terrain restrictions. They maintain minimum radio transmissions, emanating only the minimum required communications and sensor signals. The OPFOR exploits detected gaps in enemy radar coverage and often uses decoy flights in advance of attacking aircraft to distract enemy air defense systems. If more than one pass is necessary to destroy the target, attacking formations approach the target from different directions or from bright sunlight, minimizing air defense effectiveness, visual detection, and recognition.

Suppression of Enemy Air Defenses

10-169. Whenever possible, the combined arms commander includes a plan for SEAD. He can employ other aviation assets, artillery, MRLs, SSMs, and EW assets to prepare the ingress and egress routes for helicopters and fixed-wing aircraft air support missions. The entire purpose behind SEAD is to disarm or disable enemy early warning radars and to destroy or reduce enemy air defense assets that may come to bear on friendly aircraft. If SEAD is employed, it precedes the approaching aircraft by 20 seconds to 1 minute. SEAD may be employed along the flight route to cover areas where the aircraft are unprotected by terrain or friendly weapon systems. It may also be employed to prepare the AP prior to aircraft occupation.

10-170. The effectiveness of air support may be increased through the use of artillery to suppress enemy air defenses. Additionally, suppression of electronic systems that provide early warning, target acquisition,

Aviation

fire control, communications, and data support for air defense systems is a high priority. Specially equipped airborne and ground-based EW systems target both the radar and C2 networks used by enemy air defense. Both fixed- and rotary-wing aircraft, particularly the most advanced, employ a mix of radar warning receivers, self-protection jammers, flares, and chaff.

FLIGHT TACTICS

10-171. To obtain the full potential desired from an air assets, different tactics are employed by fixed-wing and rotary-wing aircraft. The OPFOR prefers to employ fixed-wing aircraft more frequently on missions with previously reconnoitered routes, fixed or semifixed targets, and greater depths. Fixed-wing aircraft are vulnerable to ground-based air defenses when executing ground attacks. This necessitates a low-altitude, high-speed target approach and minimum time in the target area. Under such conditions the pilot's ability to visually acquire and properly identify the target may be extremely limited.

10-172. The OPFOR prefers to use helicopters for time-sensitive attacks close to friendly forces. Helicopters have reduced logistics requirements compared to fixed-wing aircraft, allowing their deployment close to the battle zone. This proximity enhances their ability to respond to requests for air support. Tactically, helicopters have two advantages over fixed-wing aircraft:
- Their ability to maneuver relatively undetected.
- Systems that allow the pilots to rapidly evaluate and react to battlefield conditions.

FIXED-WING TACTICS

10-173. Fixed-wing assets can be employed at the strategic, operational, and tactical levels simultaneously. The tactics employed by fixed-wing assets to support tactical ground force battles are designed to ensure aircraft survivability in a high-threat air defense environment and provide supporting fires to the scheme of maneuver. Fixed-wing aircraft are employed much less often than attack helicopters to support the ground force commander. Yet, when effectively employed, these aircraft have the ability to give friendly forces great advantages in firepower, mobility, and shock effects.

ROTARY-WING TACTICS

10-174. Many of the tactics and techniques used by the OPFOR are similar to tactics employed worldwide. Helicopter design dictates the capabilities and limitations of each aircraft and, to a large extent, their employment. Some minor variations among models can cause similarly designed helicopters to differ in hovering capabilities, cargo and load capacities, and employment characteristics.

Flight Modes

10-175. All OPFOR helicopters can employ any of three differing flight modes: nap-of-the earth (NOE), contour, or low-level.

Nap-of-the-Earth

10-176. NOE is flown at varying airspeeds and altitudes as close to the earth's surface as possible while following the contours of the terrain. It is a weaving flight path that orients along the axis of movement and takes advantage of terrain masking.

Contour

10-177. The contour mode is flown at relatively low altitudes, conforming generally to the contours of the terrain. The flight is characterized by varying altitudes and varying airspeed. This mode of flight is most often employed with helicopters offering limited maneuverability. Because altitudes are higher than NOE, aircrews are able to fly at higher airspeeds to reduce exposure times. The aircraft may begin in support zones at contour altitudes and then reduce altitude and airspeed to NOE flight as the probability of enemy contact increases.

Low-Level

10-178. The low-level mode is flown at low altitude, with constant heading, airspeed, and altitude to facilitate speed and ease of movement while minimizing detection. It is used only in areas where enemy contact is not likely.

Attack Helicopter Employment

10-179. Employment of attack helicopters varies according to scheme of maneuver and the desired results in an attack. Ground force commanders may elect to subdivide helicopter units and employ them at varying strengths as needed. This allows for longer asset employment and accounts for variations in strength of asset coverage. For example, an attack helicopter battalion has 20 attack aircraft arranged in five companies of four aircraft each (assuming that the unit has not been attrited and has no aircraft that are in scheduled maintenance). Any number of companies may be employed on a mission.

Employment Methods

10-180. The battalion commander typically employs his aircraft as companies, unless conditions dictate employment as a battalion. Subordinate company commanders decide employment within the companies. They may choose employment as a company or in pairs. The OPFOR categorizes these employment methods as maximum, alternating, and continual. The descriptions below apply to the employment of a battalion; however, with minor adjustment they also apply to employment of companies.

10-181. **Maximum.** In the maximum method, the entire helicopter force is launched simultaneously to different APs to produce a great force multiplier and shock effect for the OPFOR to rapidly defeat the enemy. The aircraft remain in the APs as long as their fuel and ammunition last, and as long as they are afforded the security of friendly air defense coverage. This employment method allows the OPFOR a great advantage in the battle, yet it removes supporting aviation assets from the battle for several hours after they attack, since they must return to rear areas for refueling and rearming. This high number of aircraft requiring service at the FARP at the same time places a heavy demand on the logistics system.

10-182. **Alternating.** The alternating method allows for some helicopter companies to attack the target, while the others wait in a holding area or at the FARP. For example, as the two companies engaging the enemy begin to break contact, the third and fourth fly in to continue the attack. As the latter leave station, they are replaced by the fifth company to continue the attack, but at a diminished rate. The alternating method allows the OPFOR to achieve a moderate amount of shock effects and force multiplier initially, but it allows for a longer engagement than the maximum method. This method also does not strain the logistics system as much. Therefore, the aircraft serviced at the FARP have the potential to return to the battle quicker and replace the company or companies in the AP.

10-183. **Continual.** The continual method employs only one helicopter company at any given time. While one company is in the AP, another is waiting in a holding area to replace it, a third is waiting in the FARP, and the other two flights are in transit between any two of the points. As the first breaks contact to return to the FARP, the others rotate forward. One moves in to continue the attack, while another assumes its position in the deployment area. This method allows the OPFOR the opportunity to keep constant pressure on the enemy with supporting aviation assets. It places little strain on the operation of the FARP, so individual companies can expect a quick turnaround time. This method allows the engagement to continue indefinitely, based only on the logistics capabilities to resupply the FARP, and the fatigue of the aircrews.

Attack Helicopter Formations

10-184. Attack helicopters utilize several formations in the attack, and the OPFOR stresses their use in flights consisting of companies or pairs. The most common formations used are—

- Line abreast.
- Echelon (left or right).
- Trail.

Aviation

In any of the three formations, separation between aircraft can be up to 90 m horizontally. Altitude may vary between helicopters in a pair. Separation depends on terrain, visibility, aircrew proficiency, and the enemy air defense threat.

10-185. Once the FAC provides the flight with target location, the flight leader executes attack control. Inbound to the AP from the IP, he orders the appropriate formation, divides the target, assigns individual target sectors, and determines the movement technique and engagement profile.

Movement Techniques

10-186. Approaching the AP from the IP, the flight commander may employ movement techniques based on the likelihood of enemy contact. The techniques are—

- Traveling.
- Traveling overwatch.
- Bounding overwatch.

The traveling overwatch and bounding overwatch techniques can be used by all armed helicopters, not just attack helicopters.

Engagement Techniques

10-187. OPFOR engagement techniques, like much of the helicopter doctrine and employment techniques used worldwide, are based on the capabilities and performance of the aircraft and the ordnance they carry. The flight leader selects the appropriate engagement profile for his flight as determined by the situation. Either a hovering fire or a running fire is employed. The design of some helicopters makes them more conducive to the employment of the hovering technique, versus other helicopters that may require a running technique. Even two of the exact same model helicopters flying similar profiles under the same atmospheric conditions may perform differently based on gross weight. One may be able to hover because of a lighter fuel load and no cargo. However, the other may have a full fuel tank and be loaded to its maximum gross weight. This would cause the latter helicopter to require a running attack technique.

10-188. **Hover Fire.** The OPFOR employs hover fire in the attack if the capabilities of the aircraft allow. APs are chosen so that surrounding terrain provides cover and concealment for attack helicopters. They should also afford good, unrestricted fields of fire as the OPFOR attempts to engage the enemy at the maximum effective range of its weapons. These APs are near the ground forces, within the range of friendly air defense assets, and within range of friendly artillery. In the defense, the OPFOR prefers this technique, rather than running fire, and chooses APs to provide flanking fire on advancing enemy formations.

10-189. In the APs, the aircraft form into the attack pair, and mask themselves behind covering terrain. They unmask vertically or horizontally, fire their ordnance, and then remask. Based on a quick battle damage assessment, the FAC then directs the aircraft to engage the enemy again or to disengage. Because the AP was compromised, standard practice is to move to another location within the AP or to an alternate AP before firing again.

10-190. In hover fire, the helicopter may either be stationary or moving slightly. In either case, the pilot must keep the aircraft stable, for most accurate delivery of ordnance. In moving hover fire, the helicopter deliberately makes horizontal movement, which may be in any direction. However, movement is always below effective translational lift airspeed. The pilot can allow the helicopter to drift with the wind, if the threat situation and terrain permit.

10-191. If their weapons afford them a greater standoff range than enemy air defense systems can range, attack helicopters may not be concerned about masking. This allows them to employ their ordnance at the maximum range capable, and with a clearer shot at the target. These conditions also facilitate use of stationary hover fire.

10-192. **Running Fire.** If not employing a static AP with hover fire, attack helicopters can perform an attack using running fire. They can attack as a company or break down into pairs. Both simultaneous and successive attacks can be conducted from either one or two directions depending on the situation and target area.

10-193. The running fire profile is flown with an altitude that is terrain-dependant and is characterized by an increase in altitude prior to weapons release in order to acquire line of sight to the target. Airspeed varies between 90 and 280 km/h. The forward airspeed during running fire adds stability to the helicopter and thus increases the accuracy of ordnance delivery, especially for rockets. The dive to engage the target also results in fewer rotor down-wash effects on munitions, further increasing accuracy.

10-194. Engagements using running fire begin with a high-speed, low-altitude run using one of the movement techniques from the IP to the AP. The flight leader selects an IP about 8 to 10 km from the target. The IP is typically an easily identifiable terrain feature along the desired route to the target. Beginning at the IP, the helicopters move toward the target, usually at contour altitudes, using terrain to mask the approach. Once the AP is reached, the flight leader directs the helicopters to climb and acquire the target. When the target is identified, the helicopters execute a shallow dive toward the target and engage it. Depending on the range of the weapon system to be used for the engagement, they may level off for a short distance, between acquiring the target and beginning a dive toward the target.

10-195. The distance at which the helicopters begin the dive and begin firing depends on the type of ordnance to be used. The helicopters begin firing during the dive, as they reach the most effective range for their munitions. The wingman maintains his position during the firing run and releases his ordnance simultaneously with the lead. The running fire profile can be used for delivery of either guided or unguided munitions.

10-196. At the end of the firing run (regardless of the types of munitions delivered), all aircraft break off and dive down and away from the target area, leaving at minimum altitude and using terrain masking. If more than one pass is needed, helicopters may approach from another direction, or from the sun, to hinder visual identification. Target identification and engagement distances are shorter when using this technique at night.

CS and CSS Helicopter Employment

10-197. CS and CSS helicopters primarily perform preplanned missions to support the ground force commander. In this role, these helicopters are employed individually or in pairs across the width and depth of the battlefield, but primarily in the support zones, to perform their individual missions as required. If they are flying in areas where enemy contact is likely, they operate with attack helicopter escort.

10-198. The CS and CSS helicopters fly to an LZ/PZ established by the ground force unit. It is carefully planned to ensure a landing area clear of debris and with minimal slope. The takeoff and landing direction is into the wind, and landing spots within the LZ/PZ are carefully marked to allow proper spacing and safe operating distances between aircraft. The marked sites should allow for both larger and smaller helicopters to maneuver in the LZ/PZ without their rotor downwash interfering with the operation of each other.

10-199. An airfield or LZ/PZ without an air traffic controller would have an assigned air-to-air radio frequency. Inbound flights make a call on that frequency approximately 10 km away. In this radio call, the flight leader states his intentions, requests information on the wind direction and the established landing direction, and passes all pertinent information: number of aircraft, formation used, and loads carried. If other aircraft are already operating in the LZ/PZ, they return the call stating their intentions, size of helicopter, and their location and number within the LZ/PZ. They also notify the incoming flight of the landing direction in use, and the inbound flight adopts the same procedures.

10-200. Similarly, if the helicopter or flight is arriving at a specific unit's LZ/PZ, the flight leader makes an initial radio call to the ground force point of contact on a predetermined frequency upon entering the AOR. Another radio call is made 3 to 5 km away from the LZ/PZ. If there are no other aircraft operating in the vicinity, the flight leader states his intended landing direction. Once communication is established with the ground force and landing is assured, the aircraft requests a frequency change from the VTDP, and continues with no positive air traffic control into the LZ/PZ.

Aviation

10-201. On takeoff, similar procedures are followed. The flight leader announces his intentions to taxi and takeoff. If operating alone in the LZ/PZ, he can set his own procedures. If operating in conjunction with other aircraft, he uses the procedures already in effect. Another radio call is made to notify the LZ/PZ traffic the flight is clear of the LZ/PZ. The helicopter or flight lead then reports when leaving the ground forces unit's AOR.

WEATHER AND NIGHT CAPABILITIES

10-202. Night systems, infrared, radar, or avionics upgrades are readily available for procurement on the open market. Regardless of modern systems capabilities, the OPFOR still expects pilots to navigate by land, search for targets visually, and determine distances to targets. The effectiveness of air support depends on the ability of aircrews to positively identify targets in prevailing weather and light conditions.

10-203. The OPFOR realizes that system upgrades and improvements are financially more attainable and easier to procure in smaller numbers. It also understands that every aircraft in the inventory does not require the same modification. Similarly equipped units or higher-capability aircraft working in conjunction with unimproved aircraft can still present a definitive edge to OPFOR aviation and the ground force commander. Even a limited number of upgraded aircraft may have a significant impact on the battle.

10-204. Currently, flights in poor weather or at night are primarily conducted by helicopters, since they are routinely employed in marginal weather conditions, well below those acceptable to fixed-wing aircraft. They navigate through the use of instruments. This forces air assets to fly at a higher altitude and at slower airspeeds. Although this allows the OPFOR to accomplish missions in less than ideal conditions, it exposes the aircraft to greater danger. They are no longer afforded the cover and concealment of terrain, and may be unsupported by direct fire coverage from friendly ground force units.

10-205. Older or unmodified OPFOR aircraft are not likely to have any night-fighting capabilities without the aid of artificial illumination. Artificial illumination is still not adequate to fire ATGMs using day-only visual sighting, although guns and rockets can be effective under these conditions. If employed, illumination (flares or illuminating rounds) is fired from artillery or aerial platforms to assist friendly forces in engaging the enemy during periods of darkness or limited visibility.

10-206. The use of precision munitions offers a higher probability of the ordnance hitting the target than conventional projectiles or rockets that have ballistic trajectories. Precision munitions may be used for surgical air attacks in minimal weather conditions against targets such as bridges, small targets (weapon emplacements or armored vehicles), and specific buildings. External stores racks may allow OPFOR aircraft to carry precision munitions, yet most aircraft do not have the systems to aim or deliver these weapons to hit their intended targets. The munitions must be guided by other ground-based sources. Newer or recently modified aircraft may be able to deliver precision munitions in bad weather or at night.

This page intentionally left blank.

Chapter 11

Air Defense

Air defense is an integral component of combined arms combat. The OPFOR system of air defense includes the strategic, operational, and tactical levels. This chapter concentrates on tactical-level air defense. It discusses operational-level air defense only when it contributes to an understanding of tactical air defense and the relationship between the two. For more detailed information on air defense at the strategic and operational levels, see FM 7-100.1.

AIR DEFENSE SYSTEM

11-1. OPFOR air defense supports combined arms combat by the comprehensive integration of a large number and variety of weapons and associated equipment into an effective, redundant air defense system. Employment of this system pursues the basic objectives of air defense by employing certain concepts and principles. This is best accomplished by establishing an integrated air defense system (IADS). Overall, the OPFOR employs a three-phase approach to air defense, in which tactical air defense is primarily part of the third phase.

OBJECTIVES

11-2. The objective of OPFOR tactical air defense efforts is to reduce the effectiveness of enemy air attacks and prevent enemy air action from interfering with maneuver force operations. This objective can be accomplished by any of the following means:
- Destroying enemy aircraft.
- Forcing the aircraft to expend their munitions before reaching the optimum or effective range.
- Diverting the aircraft before reaching their targets.
- Mitigating the effectiveness of the attack.
- Forcing the enemy to break off and/or discontinue the air attack.

CONCEPTS

11-3. OPFOR air defense doctrine emphasizes three key and interrelated concepts. The first is that every unit is immediately responsible for defending itself from aerial observation and air attack by whatever means are available. All units conduct air surveillance whenever aerial threat is imminent.

11-4. The second concept is that air defense is an integral part of combined arms combat. A maneuver unit commander who disregards the enemy air threat or fails to properly plan for defending against it risks mission failure. All units are required to report the presence of enemy aerial systems on detection.

11-5. The third concept is that air defense weapons, radars, and associated equipment cannot be regarded as single pieces of equipment or even units engaged in combat actions but as parts of an IADS. Proper integration of these assets as both a system and integral part of mission planning and execution for mission accomplishment is the ideal way the commander can effectively deal with the enemy air threat.

Chapter 11

PRINCIPLES

11-6. The OPFOR follows several basic principles when conducting air defense: surprise, firepower, mobility, continuity, initiative, coordination, and security. Of these, the element of surprise is the most critical.

Surprise

11-7. Achieving surprise is fundamental to any successful air defense battle. At the tactical level, surprise can be achieved through a variety of means, including—
- The positioning of air defense systems in unexpected locations.
- The use of camouflage, concealment, cover and deception (C3D).
- The use of non-air defense systems in conjunction with air defense systems.

Firepower

11-8. The OPFOR force structure includes a wide variety of air defense weapons (missiles and guns). This mix of capabilities gives ground force commanders outstanding firepower for air defense. It is important that air defense planning consider and employ all assets available, across all arms, to achieve maximum firepower. Almost all tactical vehicles and many support vehicles have guns for self-defense against aerial systems. Other weapons, even improvised weapons, can be used against some aircraft.

Mobility

11-9. Air defense assets must have mobility comparable to the ground forces for which they provide cover. When planning air defense, the commander must always consider the mobility of air defense weapons and the time required for their deployment. The ground forces, for which air defenses provide cover, are quite mobile and frequently change formation as they deploy. The air enemy is mobile and can attack from many directions or altitudes. Therefore, the commander must use to the maximum the mobility and firepower of his assets, creating optimum groupings and fire plans. Improvements in mobility and fire control now allow more air defense weapons and sensors to operate and engage air targets while moving on difficult terrain. Mobility contributes directly to continuity.

Continuity

11-10. Air defense forces must provide continuous protection of critical organizations and assets. This includes keeping up with dynamic maneuver elements to ensure comprehensive coverage. That requires constantly moving air defense units with adequate logistics support (or self-contained logistics). They must provide air defense day or night in all weather conditions. Shorter emplacement, displacement, and response times, and radars that can operate while moving can support the requirement. Most air defense systems have integrated fire control, with local sensors needed for autonomous operation.

Initiative

11-11. The modern battlefield is a fluid and volatile environment where air defense unit commanders must respond to constant changes in the situation. This demands aggressive action, initiative, and originality. If the supported unit receives a modified mission, the commander must reevaluate his own unit's deployment in light of the new requirements. He also must be aware of changes in the tactics enemy air forces employ.

Coordination

11-12. The OPFOR stresses coordination between supported maneuver and supporting air defense units, between air defense units, and with other arms. Commanders must operate efficiently even when communications with other air defense units fail. All tactical-level air defense weapons must coordinate precisely with supported and flanking units, with senior airspace management, and supporting aviation assets (if available).

Air Defense

Security

11-13. The OPFOR recognizes that enemy air assets can attack from any quarter. Therefore, it must provide security for units at any depth and from any direction. Air defense units are positioned to assure radar security and overlapping coverage of sectors. Because of the threat from enemy ground elements, air defense units must coordinate with supported maneuver units to ensure sufficient ground security.

Integrated Air Defenses and the Tactical Fight

11-14. OPFOR air defense weapons and surveillance systems at all levels of command are part of an integrated air defense system (IADS). This ability provides a continuous, unbroken (usually overlapping) umbrella of air defense coverage and presents a significant threat to any potential enemy air activity.

Organization

11-15. Each level of command with air defense assets has its own IADS. This system is capable of passing early warning, acquisition, tracking, and firing data—
- Upward to higher-echelon IADS.
- Horizontally to adjacent IADS.
- Down to the lowest levels of air defense radars and maneuver units.

11-16. Use of IADS enables the OPFOR to mass the effects of air defense assets from dispersed sites to protect the most critical targets. It also facilitates the use of passive air defense techniques, including dispersal, deception, and camouflage. The ability of the OPFOR air defense and maneuver units to receive early warning, target acquisition, tracking, and firing data remotely from the dispersed radars significantly reduces the physical and electronic signature of air defense systems.

Sectors

11-17. The OPFOR recognizes that it is unlikely to be able to defend its entire airspace adequately. Therefore, it must establish priorities to ensure denser coverage of key assets or areas. These priorities may change during the course of combat, as the tactical, operational, or strategic situation changes. The OPFOR is prepared to adapt its air defense operations and tactics to use IADS at sector levels. Within sectors, the OPFOR may be able to challenge even the most modern air forces, at least initially, and perhaps temporarily prevent them from attaining air supremacy.

11-18. The OPFOR offsets limitations of sector defense by overlapping sector coverage, and by employing an IADS at each echelon above battalion level. A division or division tactical group (DTG) IADS divides its area into overlapping sectors for subordinate brigade or brigade tactical group (BTG) IADS. In most cases, a tactical air defense sector will overlap other tactical IADS sectors, and will be within the larger sector of an operational-strategic command (OSC).

11-19. In choosing to fight within sectors, the OPFOR accepts risks, since air defense sectors present seams in the defenses and may be unable to provide mutual support. On the other hand, sector air defense can help reduce the physical and electronic signature of defensive systems. It can also enable the OPFOR to mass the effects from dispersed sites to protect critical targets within sectors. Within air defense sectors, the OPFOR develops air defense ambushes along the most likely air avenues of approach.

Phases

11-20. Essential to integration and successful employment at the strategic and operational level is the use of three phases. The phases are defined by where the enemy aircraft are and what they are doing:
- **Phase 1.** Actions against enemy aircraft and control systems on the ground before they are employed. This phase is conducted using *primarily strategic- and operational-level assets* of the Army and the Air Force.

- Phase II. Actions against enemy aircraft while in flight but before they enter the airspace over OPFOR ground maneuver forces. Again, this mission is performed primarily at the *strategic and operational levels*.
- Phase III. Actions against enemy aircraft that have penetrated into the airspace over OPFOR ground maneuver forces. Thus, the "target area" consists of the area where enemy aircraft have penetrated over the OPFOR disruption, battle, and support zones. OSC-level tactical fighters and the short- to medium-range surface-to-air missiles (SAMs), antiaircraft (AA) guns, and other weapons of the ground maneuver units execute this phase. It is in this phase that *ground-based tactical air defense* plays its primary role.

For additional information on the phases of the air defense, see FM 7-100.1.

COMMAND AND CONTROL

11-21. The intent of IADS is for air defense forces at all levels of command to create a continuous, unbroken umbrella of air defense coverage. An integrated communications system provides target information and early warning to air defense and ground maneuver units. Integration is both vertical and horizontal. Vertical integration is between the strategic, operational, and tactical levels, while horizontal integration is within each of these levels. Enemy capabilities may present a situation where a totally integrated system at the strategic and or operational level is neither possible nor even desirable (see FM 7-100.1).

11-22. At the tactical level, the commander normally strives to achieve horizontal integration. His ability to integrate or be integrated vertically will depend on the air defense course of action taken at the next-higher level.

CENTRALIZED VERSUS DECENTRALIZED AIR DEFENSE

11-23. Air defense command and control (C2) relationships are subject to conflicting pressures for centralization and decentralization. Factors favoring the former include greater efficiency and effectiveness of centralized target detection systems and the increased ranges of modern SAM systems. Factors favoring the latter include the need for flexibility to support fast-paced operations by maneuver units and the many contingencies that can arise in local situations.

11-24. Centralized control is necessary, especially during defense, to ensure that the coverage of air defense units is mutually supporting and comprehensive. Without centralization at some level, the air defense umbrella does not exist, and target tip-off will not be received. At the same time, decentralized control is required, since OPFOR air defense commanders are expected to demonstrate aggressive action and originality, responding to changes in the tactical situation and operating effectively when cut off from communications with other air defense units.

11-25. Even with a decentralized control, the ability to receive information concerning inbound enemy aircraft is essential. This may be accomplished by air defense data link or other automated communications such as battlefield management systems, or simply radio or telephone communications passing essential information on enemy aircraft. With or without centralized control, information from human intelligence (HUMINT) sources can be quite valuable.

DUTIES AND RESPONSIBILITIES

11-26. At all levels above the maneuver battalion, air defense is directed by the chief of airspace operations (CAO). However, the unit commander is ultimately responsible for the success or failure of these operations. For example, during the planning phase, a division or brigade commander (assisted by the CAO and force protection staff officers) personally directs the deployment of his air defense weapons to support his mission and establishes priorities and procedures for logistics support. At the maneuver battalion level, the battalion commander has overall responsibility for the organization and conduct of air defense.

Air Defense

11-27. On the primary staff at division, DTG, brigade, and BTG level, the operations section is responsible for air defense. Within that section is the airspace operations subsection (AOS), headed by the CAO. This subsection does the planning and insures that those plans are executed within the commander's intent. The AOS is assisted in this effort by the force protection staff element. It is the force protection staff element that receives liaison teams from constituent, dedicated, and supporting air defense units associated with the division, DTG, brigade, or BTG.

11-28. At division, DTG, brigade, and BTG, the commander of the air defense units at that level has the following duties and responsibilities:

- Organize, plan, and conduct the air defense of the organization.
- Prepare recommendations on the employment of air defense assets.
- Contribute to the maneuver commander's decisionmaking process.
- Know the situation, status, and capabilities of air defense units at any stage of the battle.
- Issue orders to air defense units and staffs of subordinate units.
- Direct the regrouping of forces during the battle.
- Coordinate logistics support of air defense units.
- Establish coordination between air defense units.
- Organize communications.
- Provide liaison to the division, DGT, brigade, or BTG staff.
- Monitor the execution of orders.
- Assist subordinate units and staffs.

11-29. In most situations, an OSC commander can direct the employment of the air defense assets of at least his immediate tactical-level subordinates—divisions and DTGs or separate brigades or BTGs. The OSC may also allocate air defense assets down to DTG, BTG, or even battalion-size detachment if the conditions warrant. Brigades and BTGs that are part of a division or DTG provide coverage for their own units and vertically integrate with division or DTG coverage. When not part of a division or DTG, they vertically integrate with OSC-level coverage, which would be part of their next higher level of command. There may be skip-echelon situations when the OSC will specify how divisional maneuver brigades or BTGs employ their air defense battalions. Normally, however, the division or DTG will dictate that.

11-30. The division or DTG can dictate how maneuver brigades or BTGs employ their air defense battalions and/or may allocate air defense assets down to cover gaps. Finally, the maneuver brigade's or BTG's air defense commander has overall responsibility for the coordinated air defense coverage and administrative control of the man-portable air defense system (MANPADS) platoons constituent to the maneuver battalions.

COMMAND POSTS AND COMMUNICATIONS

11-31. Every air defense unit above platoon level (and sometimes at that level) has an air defense command post (CP). That CP serves as the mechanism for linking surveillance, fire control, weapons, and support activities. The division air defense CP normally colocates with the division staff at the division main CP. Many CPs also have colocated air surveillance radar, sometimes mounted on a command vehicle, to perform air defense battle management on site. A communications vehicle and staff vehicles are usually colocated with the CP for support. Most air defense batteries and some air defense platoons have CPs (often armored command or command and reconnaissance vehicles).

11-32. The division, DTG, brigade, or BTG CAO and his staff in the AOS normally colocate with the main CP at that level of command. The AOS is responsible for airspace management. (See chapter 10 for more information on airspace management, CAO, and AOS.) An AOS staff member is located at the forward CP to represent the CAO and advise the maneuver commander.

11-33. The OPFOR IADS includes an integrated communications system that provides early warning and targeting information to all air defense and ground maneuver units. If dedicated air defense communications with other air defense units fail, commanders switch to other communications means and use their own initiative and flexibility, in order to adapt to frequent changes in the ground or air situation.

11-34. Enemy jamming of dedicated data links does not necessarily stop the IADS from passing necessary information to the air defense and ground maneuver units. Information obtained by components of the IADS is generally directly transmitted using the IADS network. In a backup situation, however, it can be also be transmitted by numerous other methods such as—
- Automated battlefield information systems.
- Radio (voice and/or Morse code).
- Satellite communications.
- Data transmission.
- Cellular phone.
- Telephone.
- Fiber optic cable or hard wire if located in proximity (or relayed via switchboard).
- Retransmission.
- Any other real time (or even near-real time) methods.

11-35. Sufficient early warning or tracking data used in a backup situation only needs to be basic information about the enemy aircraft. The air defense unit only requires actionable information such as time, bearing, range, speed, altitude/height, and aircraft type (if possible). This simple data is sufficient for the receiving radar to determine exactly where and when to look and the appropriate time to turn on his radar in order to track and fire. Air defense assets may also be positioned close enough together to be hardwired.

11-36. The OPFOR has the ability for its lowest air defense and maneuver units to receive air defense information remotely relayed from the most powerful high-level surveillance systems. This enables OPFOR air defense units to operate with the radar turned off and still receive sufficient information to track and fire on approaching enemy aircraft within their respective sectors. It provides several levels of redundancy, which prevents the enemy from breaking the systems integration by merely knocking out one (or several) radar and/or communications means.

11-37. The basic rule for the establishment of communications between supported and supporting unit is that the higher command allocates landline, radio relay, and mobile communication means, while radio equipment is allocated by both higher and subordinate levels. This ensures proper coordination of communications. If communication is lost, the commanders and staffs of all units involved are responsible for the immediate restoration of communication. Redundant communications systems with multiple operating frequencies are often available to assure communications integrity even under electronic warfare (EW) conditions. A multi-aspect attack warning system sends immediate alarm of incoming enemy aircraft to maneuver units, the staffs, and logistics units. The warning is communicated through signal equipment that is specially allocated for this purpose. Within air defense organizations, alerts are sent via acoustic signal and graphic computer display, and on portable azimuth displays (plotting boards).

AIRSPACE MANAGEMENT

11-38. The OSC is the lowest level of joint command with control of both Army and Air Force units. Under the operations officer on every staff from OSC down to brigade and BTG level, the CAO is responsible for airspace management issues and procedures. The CAO maintains the airspace control net for controlling the command's airspace and all related matters. These headquarters typically receive liaison teams from all constituent, dedicated, and supporting Air Force, army aviation, and air defense units associated with the command. An OSC headquarters allocated Air Force assets would also receive an Air Force liaison team. All these units and their liaison teams are on the airspace control net. For additional information on airspace management, see FM 7-100.1.

11-39. To reduce air-to-ground coordination problems during the execution of tactical missions, the OPFOR employs proactive staff elements and control measures. Primarily, the OPFOR employs a system of identification, friend or foe (IFF) between aircraft and air defense systems. To protect friendly aircraft from fratricide from non-IFF-capable systems, strict procedural controls are enacted that separate airspace horizontally, vertically, or both. This buffer zone minimizes the possibility of fratricide while maximizing ordnance effects. For specifics concerning airspace coordination procedures and zones, see chapter 10.

Air Defense

11-40. Unless otherwise notified, air defense weapons consider the airspace a "free fire" zone and will fire on *all* aircraft. "If you fly, you die" is the OPFOR default. The air defense coverage may be temporarily "switched off" to allow friendly aircraft to pass on a mission planned in advance and then "switched on" as they exit the area. For other missions, air defense coverage may allow aircraft to transit only on "safe corridors" based on air routes or other procedural methods. If aircraft deviate from these coordinated areas, they will be shot down by friendly ground force units. The OPFOR views the possible loss of aircraft through fratricide as a lesser risk than allowing gaps in its radar and air defense coverage that the enemy might exploit. See also chapter 10 for more information on airspace coordination and management.

11-41. Airspace coordination is critical to those ground maneuver units and others using unmanned aerial vehicles (UAVs). Otherwise, they stand a very good chance of losing the UAVs. Due to the proliferation of UAVs in recent years, coordination responsibility has reached very-low-level tactical units. If an OPFOR air defense unit acquires an unidentified UAV, it will shoot it down. That is because the enemy is also likely to be using UAVs. The unit launching the UAV must coordinate with the appropriate AOS or CAO prior to launching. Maneuver battalions may have a forward air controller (FAC) assigned to coordinate with supporting air elements and air defense units. In the event the maneuver battalion does not have a designated FAC, the battalion operations officer will facilitate the coordination. Maneuver companies will notify their parent battalion when the company plans to launch UAVs.

11-42. It is imperative that air defense units be notified when friendly aircraft (or UAVs) are flying within the air defense umbrella. Failure to coordinate with these elements will result in fratricide. The senior air defender in the command is responsible for airspace deconfliction and the notification of air defense units.

AIR SURVEILLANCE

11-43. Air surveillance is the key factor in guaranteeing the earliest possible warning of impending enemy air attack. It is conducted by electronic and electro-optical means and by visual observation. Radar is used for technical surveillance, providing all-weather detection capability. EW systems, acoustic systems, unattended ground and aerial sensors, and other assets are used to provide early warning of aircraft activities and alert air defense systems to engage air targets. Although this chapter primarily addresses ground-based air defense at the tactical level, the addition of UAVs in tactical organizations provides organic aerial surveillance for various roles, which can include air defense.

11-44. Ground-based and airborne air surveillance assets at the operational level play a major role in gathering, integrating, and disseminating information to tactical units. The objective is to establish a system that not only provides the earliest possible warning of approaching enemy aircraft but also develops target information sufficient to plan and conduct effective air defense.

11-45. OPFOR tactical air defense units receive preliminary early warning target data passed from higher-level or adjacent radar units to air defense commanders and their firing batteries via automatic data links or other communications. This practice reduces the vulnerability of battery radars and radar-equipped gun carriages and missile launchers to jamming or destruction. Ideally, only those aircraft that have been positively identified as hostile will be engaged.

SENSORS

11-46. Sensors are a critical component of air defense systems. They perform surveillance and tracking functions against fleeting air targets. The primary target detection and acquisitions means for air defense units are radars. Radars can more easily detect and track aircraft with less operator input than other sensors. However, many detection and acquisition packages are sensor suites using multiple sensors, including acoustics, optics, and electro-optics.

Radars

11-47. Air defense units employ a mix of radar systems operating at different frequencies, in varied intervals and with overlapping coverage. Radars fall into the general categories of surveillance and fire control. Surveillance radars include early warning, target acquisition, and height-finding radars, while some perform all of these roles. Air defense unit target acquisition radars can acquire and track targets and assign them to the fire control system for engagement. Some fire control radars also have a limited target acquisition capability. Dual-mode radars perform both functions simultaneously. (For additional information on the technical capabilities of air defense radars, see the *Worldwide Equipment Guide*.)

11-48. The OPFOR is fielding more modern mobile radar systems with the ability to quickly employ radars or operate radars while moving. Early warning radars with long-range capability detect approaching aircraft and cue the IADS. The IADS identifies air targets and assigns its own target acquisition radars to acquire and track aerial targets. Then it assigns an air defense unit to engage those targets. Some air defense surveillance radars can perform both early warning and target acquisition roles.

11-49. Units containing older radars (requiring some operational down time for maintenance) generally use at least two radars that are set up at critical terrain points to insure continuous overlapping coverage. Usually only one radar will move at a time. To reduce the likelihood of detection by enemy electronic intelligence (ELINT), the radars on tactical air defense systems may not be operated unless the requirement for their use outweighs the risk of detection. Radars are emplaced to provide integrated overlapping cover to prevent air attack against any single radar. Overlapping coverage ensures any aircraft attacking a radar will be covered by at least one radar and possibly several radars. The data transmitted by the covering radars allows several air defense systems to fire on the attackers.

11-50. An example of the integration of multiple sensors is the air defense brigade subordinate to a division. It contains an early warning/target acquisition battery that includes—

- Early warning, target acquisition, and possibly height-finding radars.
- IFF interrogators.
- Communications vehicles.
- CPs.

The surveillance section of the battery provides redundancy in that it can take over the CP function in an emergency. If required (especially while covering a moving unit), the commander may receive additional radars from higher level. One is usually placed at the forward portion of the moving unit and the other with the main body.

Other Electronic and Electro-Optical Sensors

11-51. Some OPFOR air defense weapons are integrated with passive sensors, such as—

- Optics.
- Electro-optics.
- TV cameras.
- Night-vision sights.
- Auto-trackers.
- Laser rangefinders.
- Acoustic sensors.

Multiple units are simultaneously alerted on aircraft approach for overlapping sector coverage. Azimuth warning systems, such as azimuth displays and plotting boards of dismounted guns or MANPADS teams are also alerted.

11-52. Acoustic sensors include acoustic arrays, both stationary and vehicle-mounted. Passive sensor systems can also include acoustic-triggered unattended ground sensors.

11-53. Air approach alarms are available for tactical ground force units, and may be linked to MANPADS teams. Most of these alarms provide bearing and range. Night sights are also now common on MANPADS, AA guns, and other air defense equipment. The OPFOR uses all types of infrared devices to detect "hot spots" and subsequently tip off other acquisition means to acquire and begin tracking the aircraft. For additional information on equipment, see the *Worldwide Equipment Guide*.

Air Defense

VISUAL OBSERVATION

11-54. An effective system of visual surveillance often provides the first warning of an enemy air attack. This is especially true of attacks conducted by low-flying aircraft or armed helicopters using nap-of-the-earth (NOE) techniques. When operating close to enemy forces or in areas where enemy air attack is considered likely, all units post air observers to continually observe the sky. Observers may also use hand-held or vehicle-mounted optics, electro-optics, and laser rangefinders, and unattended or remote sensors. Despite the presence of a technologically advanced early warning system, the OPFOR continues to stress the importance of visual surveillance. This is especially true at the small unit levels.

11-55. In the defense, air observation posts (air OPs) are set up at suitable locations, usually on terrain offering good visibility, near CPs, and/or close to air defense units in firing positions. During tactical movement and during both the defense and offense, observers are posted on each vehicle. Observers are changed frequently to reduce fatigue and maintain their effectiveness.

11-56. Visual air surveillance is conducted on a 360-degree basis, and observers are assigned sectors of airspace to monitor. OPFOR air defense units realize that an aircraft can be visually detected at ranges of 2 to 5 km when the observer is assigned a 60- to 90-degree sector of observation, and at ranges of 6 to 7 km when assigned a 30-degree sector. Naturally, these distances are affected by terrain and lighting conditions. The use of binoculars can increase detection ranges to approximately 12 km. Aircraft can be observed much further (30+ km) when using modern electro-optical equipment. Aircraft flying at high altitudes may be detected at ranges up to 50 km when more sophisticated optical rangefinding equipment is used.

11-57. To visually observe activity at enemy airfields in or near the area of responsibility (AOR), whenever possible, the OPFOR makes extensive use of—
- Special-purpose forces (SPF) teams.
- Human intelligence (HUMINT) agents.
- Sympathetic civilians.
- Affiliated irregular forces.
- Any combination of these.

These observers report by radio or telephone the number and types of aircraft taking off or seen, and their direction of travel. Other observers stationed along probable approach routes can monitor and report the progress of the enemy aircraft en route to their targets. The OPFOR also prefers to establish complete early warning "sky watch/air observation" networks using local civilian personnel. This information, combined with electronic tip-off from radar units, enables OPFOR air defense units to leave their radars turned off and still be able to detect, track, and (with some systems) fire on incoming enemy aircraft.

11-58. Every tactical air defense battalion has a subordinate air observer platoon (AOP). This is a specialized high-mobility unit designed to fill or close gaps in tactical air defense coverage. This platoon becomes especially critical during a dispersed fight. While the AOP typically engages with enemy units only in self-defense, it is equipped with a laser designator to lase high-value targets such as forward arming and refueling points (FARPs). Another common tactic is for air defense ambush teams (MANPADS or AA guns) to accompany the AOP while they are en route to conduct an ambush. Prior to deployment, the AOP conducts a map terrain analysis of friendly radar coverage. The resultant analytical overlay provides the AOP critical locations to surveil based on radar terrain masking. Based on this, the air defense ambush teams are dropped off to provide air coverage while the AOP continues farther away to provide early warning. This platoon and/or its subordinate squads are routinely suballocated to maneuver units.

11-59. Every maneuver brigade has a subordinate reconnaissance company. A primary mission of this reconnaissance unit is to provide early warning of any enemy air activity. Within this reconnaissance company are several specialized long-range and high-mobility reconnaissance platoons. Each of these platoons is designed to range across the disruption and battle zones and report any enemy air activity. For specific details on the organization structure, see FM 7-100.4.

Chapter 11

TACTICAL ASSETS

11-60. Air defense assets available to the tactical commander are a blend of air defense units and combined arms units using weapons well suited in the air defense role. Commanders must properly integrate these assets into a system (IADS) and make them an integral part of mission planning and execution.

ORGANIZATIONS

11-61. The OPFOR ground force structure includes air defense units. These units are equipped with a variety of systems having the firepower, mobility, and range to fully support fast-moving tank and mechanized forces in dynamic offensive operations. For information on organizational assets and equipment above the tactical level (OSC and above) and IADS, see FM 7-100.1. For specifics on tactical organizations, see FM 7-100.4.

Divisions

11-62. Most maneuver divisions contain at least an air defense brigade. The brigade is fully capable of providing air defense coverage for the entire division. Divisions can be assigned to create task-organized division tactical groups (DTGs). When this occurs, consideration must be given to allocating additional air defense assets to ensure protection of the augmented force. In some situations, the DTG may be allocated assets normally associated with operational-level organizations. The division or DTG commander also has the option of further allocating or task-organizing some, or all, of these assets to subordinate units.

Divisional and Separate Brigades

11-63. Like the division, maneuver brigades may be task-organized as brigade tactical groups (BTGs) and may require additional air defense assets to protect newly allocated units. Divisional maneuver brigades contain organic air defense assets, usually an air defense battalion. Separate brigades may contain a more robust air defense battalion. Depending on a number of circumstances, BTGs formed from either type of brigade may also have additional assets either at the brigade level or allocated down to their assigned battalions. However, separate brigades typically contain a more robust capability than divisional brigades, even without augmentation. Thus, air defense augmentation of a separate brigade (task-organized as a BTG) may be the equivalent of an air defense brigade.

Battalions

11-64. Maneuver battalions can be task-organized as detachments to perform a specific mission. When assigning air defense assets to battalions forming detachments, brigade planners need to pay special attention to the command and support relationships they assign (see chapter 2).

11-65. Maneuver battalions typically have a MANPADS platoon for self-protection. Other air defense assets may also be allocated to the battalion. The platoon leader of the MANPADS platoon (or the senior air defender) also coordinates with the appropriate staff member, or FAC if present, to deconflict any possible fratricide issues.

Companies

11-66. The MANPADS platoon at battalion may be retained at battalion level, or the battalion may allocate its MANPADS squads down to maneuver company level. Most OPFOR tactical vehicles are equipped at least with a 7.62-mm general-purpose machinegun that can engage enemy aircraft in addition to ground targets. All OPFOR units receive training in the employment of massed small arms weapons fire to engage low-flying enemy aircraft. This technique is routinely practiced by troop units and is usually employed under the supervision of the company commander when he has been notified that an enemy aircraft is approaching. For additional information, see All-Arms Air Defense below.

Air Defense

WEAPONS

11-67. The OPFOR force structure includes a wide variety of weapons providing ground force commanders outstanding firepower for air defense. Air defense planning must consider and employ all assets available, across all arms to achieve maximum firepower.

Air Defense Weapons

11-68. The OPFOR inventory of tactical ground-based air defense weapons includes a variety of missiles, guns, and support equipment. Tactical-level air defense includes short- and medium-range SAMs, short-range AA guns, and MANPADS. Tactical air defense assets are increasingly using combination AA gun and missile systems, offering added flexibility and quick and lethal engagement of all aerial targets, especially low level targets. The OPFOR's tactical air defenses support the need to protect ground forces and the desire to seize any opportunity to shoot down high-visibility (flagship) enemy airframes.

11-69. The assets contained in the division's air defense units are capable of providing the commander with area defense, point defense where required, and ground to medium-altitude coverage. Their mission is to protect the division's maneuver elements and other units within its AOR. Assets redeploy as necessary to maintain coverage of advancing forces. Many systems are capable of providing air surveillance on the move and launching from a short halt to respond to detected enemy aircraft. They can also displace by pairs or as batteries and halt in intervals to provide coverage of the force as it moves. Some air defense weapons, such as MANPADS, can be fired while on the move.

11-70. Nearly all self-propelled AA guns can fire on the move with passive electro-optical fire control systems. Some still only have optical sights. Others have onboard radars that can operate on the move, or in seconds with a short halt. Most have an alert system with sufficient warning to ambush approaching aircraft. Aside from short-range air defense, AA guns also can be employed against all but the heaviest of enemy ground force systems, as well as against personnel, with devastating effects. Some typical infantry weapons such as the automatic grenade launcher, machinegun, and recoilless rifle are also equipped with ballistic fire computers and radars. All are extremely lethal in the air defense role.

11-71. A variety of relatively new systems, which significantly enhance air defense capabilities, have entered the OPFOR inventory. These include but are not limited to remote helicopter infrared sensing devices and passive acoustic acquisition systems. Antihelicopter mines are widely available and increase the OPFOR ability to deny low-altitude approaches, firing positions, and landing sites to enemy rotary-wing aircraft.

All-Arms Air Defense

11-72. The OPFOR recognizes that air defense is an all-arms effort. Thus, all OPFOR units possess some type of an organic air defense capability to differing degrees, depending on the type and size of the unit. The OPFOR continuously looks for new and adaptive ways of employing not only air defense systems but also systems not traditionally associated with air defense. Many OPFOR weapons not designed as air defense weapons will also damage and/or destroy tactical aircraft when within range.

11-73. Throughout maneuver units, there are a number of systems designed for air defense and other systems that can be used in an air defense role. The heavy AA machineguns on tanks are specifically designed for air defense. Machineguns on APCs and automatic cannon on IFVs can engage both ground and air targets. Most antitank guided missiles (ATGMs) are extremely effective against low-flying helicopters. Several ATGM manufacturers offer antihelicopter missiles and compatible fire control, which are especially effective against low-flying rotary-wing aircraft. Field artillery and small arms can also be integral parts of the air defense scheme. All these weapons can be extremely lethal when used in this role.

11-74. Many maneuver units have modified selected infantry vehicles into fire support vehicles, specially equipped for multiple-role use with cannons, ATGMs, and MANPADS. These vehicles are employed in air defense or antitank platoons, which carry dismount teams with missiles to engage aerial and other targets. Some MANPADS and their vehicle-mounted versions are capable of antiarmor roles as well as air

Chapter 11

defense. The OPFOR attempts to adapt these systems and develop new tactics that may help to fill the void when a specific capability is denied by a more sophisticated enemy.

Note. Some air defense systems also lend themselves to multiple roles. An example of this capability is the Starstreak High Velocity Missile, which is available in vehicle-mounted and man-portable configurations. Although designed as a hypervelocity MANPADS with a range of over 7 km in an air defense role, the Starstreak can also penetrate vehicles with over 4 inches of armor at the same range. It is a high-precision missile and is countermeasure resistant, with laser beam-rider guidance. Starstreak has a very high probability of hit against less maneuverable aircraft, especially helicopters conducting terrain flying such as NOE. Hypervelocity speed permits destruction of an aircraft at 7-km range in about 5 seconds, and denies the aircraft time to engage targets or evade or counter the missile. This combination gives multipurpose weapons like the Starstreak a permanent place in the ground role as well as air defense weapons.

11-75. The OPFOR considers every soldier with a MANPADS to be an air defense firing unit. These weapons are readily available at a relatively low cost and are widely proliferated. Therefore the OPFOR is acquiring as many MANPADS as possible and issues them in large numbers to a wide variety of units. It can also disseminate them to selected affiliated forces. The small size and easy portability of these systems provides the opportunity for ambush of enemy airframes operating in any area near OPFOR units. The OPFOR also employs them to set ambushes for enemy helicopters, especially those on routine logistics missions.

11-76. To counter the helicopter threat, a wide variety of tactical and combat support vehicles have MANPADs or machineguns with AA sights to engage aircraft. Two of the greatest advantages of helicopters are weapons stand-off and ability to use terrain cover on approach. Many ground force and air defense weapons can match the stand-off and inflict damage to force aircraft to disengage. When flying in an NOE mode (20 to 25 ft above ground level), a helicopter rotor is approximately 40 ft off the ground. A helicopter flying NOE cannot easily engage targets or evade missiles, and it is easily targeted by ground weapons. Nearly all SAMs, small arms, direct-fire crew weapons, ATGMs, antitank grenade launchers, automatic grenade launchers, and machineguns can engage it.

11-77. Anti-helicopter mines can be placed on likely enemy helicopter firing positions. This area can then be left unattended. The technique can be used at sights to economize assets. For additional information on anti-helicopter mines see, Air Defense Ambushes and Roving Units below.

11-78. OPFOR maneuver squads and above are routinely trained to use their weapons to engage tactical aircraft and have incorporated the engagement techniques into their tactics, techniques, and procedures. Below are typical examples of these weapons and their air defense capabilities:

- ATGMs (out to 5,500 m).
- Antitank grenade launchers (800 m +).
- 35- and 40-mm automatic grenade launchers (2000 m +) (ballistic computers; some may be radar-guided).
- Machineguns (7.62-mm 1,200 m; .50-cal 2,000 m +) (ballistic computers; some may be radar-guided).
- Antimateriel rifles, .50-cal or 14.5-mm (2000 m).
- Sniper or marksman rifles, .30 cal or 7.62-mm (600 m).
- Recoilless rifles, 73- to 106-mm (1,100 m +) (ballistic computers; some may be radar-guided).
- Air-to-surface rockets (improvised from air-to-surface pods), example 57-mm (3,000 m).
- Antitank disposable launchers.
- Infantry rocket flame weapons (thermobaric).
- Mini-UAVs and micro-aerial vehicles, with or without warheads (can attack or harass rotary-wing aircraft).
- Volley fire by squad, platoon, or company with assault rifles when aircraft are within range and passing overhead.

Air Defense

- Tank main guns (laser-guided missiles offer precision beyond 5 km).

For additional information on weapons capable of damaging and/or destroying tactical aircraft, see the *Worldwide Equipment Guide*.

EMPLOYMENT

11-79. The details of the employment of air defense assets are not templated, carbon-copy solutions. Employment options depend on several factors, some examples of which are—
- Missions assigned.
- Scale of the missions.
- Availability and capability of systems.
- Enemy air order of battle.
- Priority of the protected target.
- Conditions under which combat is waged and the type of combat.
- Specific terrain and meteorological conditions.

11-80. Whatever the nature of combat being conducted by maneuver forces, the actions of supporting air defense units are, as the term implies, inherently defensive. Assigning specific missions to air defense units requires an understanding of the types of missions, the planning considerations involved, and the engagement procedures used.

MISSIONS

11-81. The primary mission of OPFOR ground-based air defense systems is to protect maneuver units and installations from attack by fixed- and rotary-wing aircraft. This reduces the availability of enemy air assets to influence the development of the ground battle. As part of the overall air defense effort, these forces perform a variety of missions, including the following:
- Timely detection of incoming aircraft, continuous tracking of airborne targets, and warning troops of attacking aircraft.
- Protection of the support zone with a primary emphasis on protecting targets that play key roles in supplying troops.
- Prevention of observation and reconnaissance by aircraft or UAVs.
- Destruction of airborne or air assaults during overflight, airdrop, or landing.
- Prevention of deeper penetration by enemy aircraft, in cooperation with adjacent air defense units.
- Prevention of reinforcement or resupply of encircled enemy forces.
- Protection of units or forces from attack by unmanned combat aerial vehicles and attack UAVs.
- Protection of units or forces from attack by missiles and artillery rockets.
- Countering of enemy aerial system activities, such as jamming, communications transmission, and infiltration or exfiltration.

PLANNING CONSIDERATIONS

11-82. A number of factors determine the appropriate employment of an air defense unit. Prior to the employment of the unit, consideration is given to—
- **Effective range and altitude.** If the effective range of the air defense unit's specific weapon system(s) does not exceed 10 km, the air defense unit is not assigned a mission covering 30 km. The same applies to altitude. A unit of small-caliber AA guns is not assigned a mission of independently protecting an object that can be successfully hit by an air strike from altitudes over 2,000 m.

- **Probability of kill.** Effectiveness of fire of an air defense unit's weapon usually is described by its probability of destroying an air target. If the probability is small, the object may be covered by several air defense weapons.
- **All-weather capability.** If enemy aircraft are capable of striking the object during any weather condition, units without fire control radar are not capable of providing adequate air defense.
- **Mobility.** An air defense system chosen to defend a given object must have at least the same mobility as the object or unit defended.
- **Supporting systems.** This includes the availability and capability of early warning and target acquisition radars, weapons, electronic jamming and electronic protection measures, and the operating requirements of these systems.

11-83. The essence of an air defense unit's combat mission can be expressed in two words: "to cover." Combat orders indicate—
- The object(s) or combat unit(s) to be covered.
- The starting time and duration of the air defense mission.
- The degree of readiness and procedures for conducting fire.
- Procedures for organizing early warning, target acquisition, and communications.
- Routes of movement to the fire or launch position area.
- Coordination between ground troops and friendly aviation, control, communications, and logistics elements.
- Other applicable instructions.

11-84. Air defense planning is not strictly limited to considerations for the employment of air defense systems. It also includes coordination with other arms. Air defense planners should view air defense as pulling together all aspects and potential contributions of other arms to supplement and complement the air defense plan. Airspace management is one obvious requirement. However, there is also the need to identify likely air avenues of approach and hovering sites for enemy rotary-wing aircraft. The hovering sites would be submitted through artillery channels as preplanned targets available on-call. Similar coordination is also required with EW elements and engineers.

Objects and Units Covered

11-85. The characteristics of the objects to be defended are the determining factors in the tactical employment of air defense units. Primary among these are the combat function and location of the defended unit or object. This is determined primarily by the role and location of ground combat units, logistics units, and the current tactical situation. Other factors have a considerable influence on a decision as to the type and quantity of air defense units assigned, such as—
- **Sensitivity of the target.** A fuel dump is more sensitive to air strikes than a fuel depot. A mountain road is considerably more critical than a road on a plain, since damage to the road surface would force troops to halt in the first case, but not necessarily in the second.
- **Geometric dimensions of a target.** The larger the target, the greater the probability of its being hit. If a target's dimensions are large, it can be attacked from horizontal flight at medium-to-high altitudes. In most cases, targets of smaller dimensions would be attacked from very low altitudes or by diving aircraft.
- **Mobility of a target.** Targets maneuvering on the battlefield are harder to locate and attack than are fixed targets. Therefore, the enemy will most likely attempt to destroy them immediately upon detection. Stationary or immobile targets are not necessarily subject to air strikes immediately upon their detection, but as the tactical situation warrants.
- **Weather and visibility.** Various kinds of weather and nighttime conditions can affect the possibility of attacking aircraft locating a target. Reference points on the approaches to the target, and at the target's location, can be used for navigational fixes to more accurately acquire the target.

Air Defense

11-86. The tactical importance of units and facilities is not constant but changes during the course of combat as assigned missions are accomplished. For example, the role of a battalion advancing along the main axis and that of a battalion removed to the reserve are not of equal importance with respect to successful accomplishment of the combat mission. Also, a water-crossing site loses value after the main body of troops has crossed it. Over time, there is a systematic and continuous reappraisal of the role and function of combat units and support facilities. Their role and significance in accomplishing the overall mission can change, and therefore their priority for protection can change also. In certain cases, there may be insufficient air defense assets to cover all targets. In those cases, air defense units are relieved from covering targets that have become of secondary importance. Instead, they are assigned to cover new, more important targets to ensure the combined arms forces can complete their missions without interference from enemy air action.

Zones

11-87. Air defense units of an OSC conduct an overlapping sector-based area defense, engaging enemy aircraft at some distance from the supported maneuver divisions and themselves. In the best-case situation, a division or DTG will have sufficient assets to provide coverage over its entire AOR, primarily conducting area coverage. However, there is a significant element of point defense in support of the division's or DTG's maneuver brigades or BTGs.

11-88. At maneuver brigade or BTG level, there is a significant element of point protection in support of subordinate units and brigade-level assets. This is due to the nature of the units defended and the relatively short range of air defense weapons at this level. Batteries of the brigade- or BTG-level air defense battalion and the MANPADS platoons of the maneuver battalions generally conduct a point defense protecting high-value targets. These targets include radars, EW systems, main CPs and communication nodes, key material support, engineer equipment, and artillery. Other instances of point defense include use at key air approaches shielded by terrain from other air defense units.

11-89. Ideally, brigades or BTGs should also be able to provide coverage for their own units and vertically integrate with divisional coverage. The degree to which these assumptions apply depends on mission, assets available, and enemy capabilities. Employment of air defense varies among the three basic zones that make up a supported organization's AOR.

Disruption Zone

11-90. Air defense in the disruption zone—in either offense or defense—should provide area coverage to defend forces in the zone and provide point defense for key assets involved in conducting fires. It is essential that air defense assets assigned have mobility and survivability equal to those they are defending. Even the systems providing point defense must be highly mobile and capable of moving with units as they displace to hides or new firing positions or conduct survivability moves.

11-91. Paramount to the success of air defense in the disruption zone is participation in the counterreconnaissance effort. This effort must be both creative and aggressive. Early warning, tracking, and remote cueing are key. When necessary, the OPFOR will move air defense assets normally located in the battle zone well into the disruption zone to accomplish this and to assist in area coverage. This, in conjunction with the well-planned use of other arms to achieve air defense missions, allows the OPFOR to attack air platforms in the disruption zone and beyond.

11-92. Coverage for maneuver forces in the disruption zone is a priority. However, air defense units may be assigned missions that are offensive in nature and not directly tied to the defense of a specific organization or site. The extensive use of air defense ambushes located along likely routes of ingress and egress is essential. Missions can include an integrated effort to destroy FARPs or aircraft using FARPs, and actions to destroy UAVs before they reach the battle zone.

11-93. The commander may create a disruption zone that extends well into enemy-held territory. In this case, disruption forces operating in that area may or may not have air defense coverage. SPF and affiliated irregular forces may rely strictly on C3D for protection from enemy air attack. Regular forces should have sufficient man-portable assets to provide protection for the force. Some of the stay-behind forces may be air defense teams, equipped with man-portable assets and assigned pre-planned targets to ambush.

Battle Zone

11-94. Air defense in the battle zone requires assets that provide coverage and have the mobility to move with supported forces. In the battle zone, air defense emphasis is on protecting the fighting forces. This is accomplished through a combination of area and point defense. Elements of the maneuver brigade's or BGT's air defense unit deploy to cover its maneuver battalions. Brigades and BTGs that are part of a division or DTG vertically integrate with division or DTG coverage. (When not part of a division or DTG, they vertically integrate with OSC-level coverage, which would be their next-higher level of command.) Where necessary, divisional assets are assigned supporting missions to brigades or BTGs. Brigade or BTG assets are pushed down to battalions when required. MANPADS from brigade or BTG level can augment the maneuver battalions, to close gaps in the coverage or establish ambushes. The use of other arms to attack enemy helicopters at their firing positions is also part of the air defense effort. Artillery using proximity fuzes is especially effective against these helicopters while they are hovering or slow moving, such as preparing to fire.

11-95. In the offense, most air defense assets would normally be within the battle zone. Their main role is to allow friendly ground forces the freedom to maneuver as the situation develops. Air defense can create the window of opportunity for offensive action.

11-96. In support of defense in the battle zone, priority for air defense assets is the protection of those forces where an enemy penetration is expected or those assigned to kill zones. Protection of long-range fire systems and reserves are the next priorities.

11-97. In fluid battle conditions, portions of the battle zone can become part of the disruption zone. In such cases, some air defense assets may be designated to stay behind and move to hide positions until activated to conduct air defense ambushes from within the enemy's depth.

Support Zone

11-98. In the offense or defense, some air defense units may be deployed in the support zone. Their role is to help keep this zone free of significant air action and thus permit the effective logistics and administrative support of forces. Compared to the battle and disruption zones, the commander can afford to assign less mobile air defense assets here. The use of point protection is increased relative to the other zones within the AOR. Throughout the support zone, the OPFOR makes extensive use of passive air defense measures, including maneuver, dispersal, and C3D (particularly the use of deception positions).

Deployment and Redeployment

11-99. The location of the air defense unit is critical, whether it is to protect a high-value asset or to accompany and provide air defense of a tactical unit. There is no fixed pattern of deployment. The decision for deployment depends primarily on the supported unit's mission, the terrain, and the ground and air tactical situations. As the supported unit performs its assigned missions, its location and combat formation can change. The air defense unit commander must respond to these changes and redeploy his weapons in a timely manner to provide continuous and effective coverage to the supported unit. Deployment and redeployment take into account the requirements for—

- Maintaining mutual support among air defense units.
- Covering the main threat.
- Providing comprehensive coverage to all elements within the AOR.

In addition to maneuver units, coverage must include headquarters, artillery units, and logistics units. C2, terrain mobility, and dispersion to reduce vulnerability are also considered in both deploying and redeploying.

11-100. Both the commander of the supported maneuver unit and the commander of the supporting air defense unit usually conduct terrain reconnaissance. A preliminary map reconnaissance can tentatively

Air Defense

identify positions for deployment of air defense weapons in defensive positions, along movement routes, or in areas to be seized by advancing forces. Significant emphasis is placed on identifying all potential attack routes for low-flying enemy aircraft of all types. Routes of approach suitable for armed helicopters and positions from which these helicopters might fire ATGMs are of special concern.

OFFENSE

11-101. The employment of air defense units in the offense depends on—
- The situation.
- Missions of the supported units.
- The effective range of air defense systems.
- Their maneuver capabilities.

The two basic types of OPFOR tactical offensive action are the *attack* and the *limited-objective attack*.

Attack

11-102. There are two forms of attack: the *integrated attack* and the *dispersed attack*. For either form, an attack from the air represents the greatest danger when ground forces are moving and deploying for the attack. Therefore, mobile air defense units should be able to accompany maneuver forces at a distance close enough to maintain effective coverage. It is especially important to prevent the enemy from conducting air reconnaissance and locating the movement of forces and assets. During the attack, the positions of supported maneuver units continue to change, and the supporting air defense units must also redeploy. Air defense must also provide point or area coverage for other units in the AOR, including headquarters, artillery, and logistics units.

Integrated Attack

11-103. An integrated attack may be conducted from positions in direct contact with the enemy, or require a tactical movement forward from behind forces in contact. In the former case, the forces in contact constitute the bulk of the fixing force. While protection of the fixing force is important, the situation may permit the allocation of fewer air defense assets there in order to provide greater protection for other enabling or action forces.

Dispersed Attack

11-104. In the dispersed attack, there may be times when dispersion is so great that it is not possible to provide coverage for all units in the supported command's AOR. In these cases, the commander must allocate and position his air defense assets in those areas where the air threat is perceived to be the greatest. Priority is also given to providing protection for those maneuver units most critical to the success of the attack.

11-105. According to these priorities, maneuver brigades or BTGs normally have their organic air defense assets augmented by divisional air defense assets. If these batteries are equipped with medium-range SAMs, they need not operate in the maneuver brigade's or BTG's formation. The range capability of their radars and missiles allows them to provide support from positions farther away. This provides an additional advantage to the commander in enabling him to more quickly shift priorities of air defense coverage in the event the enemy increases his air attacks in other areas.

11-106. In trying to protect dispersed forces, commanders may have to accept some risk in certain areas. In those areas, commanders should plan for increased C3D and the increased use of other arms to assist in air defense.

Limited-Objective Attack

11-107. Air defense systems or organizations have a key role in conducting a limited-objective attack. Since such attacks are generally conducted against a stronger enemy, they may be extremely vulnerable to enemy air attack. Air defense may also be the principal means of destroying certain airborne flagship systems. Proper air defense planning and task-organizing to support the maneuver and fire support units conducting these attacks are critical to mission success. There are two types of tactical limited-objective attack: *spoiling attack* and *counterattack*.

11-108. Those units conducting either type of limited-objective attack may not have enough constituent or dedicated air defense assets to protect themselves. So, coverage may have to come from assets located at higher levels of command. The higher command may have to reposition some air defense units so that they can provide coverage for the limited-objective attack, while hopefully still being in position to contribute to area coverage of the higher command. Air defense units also must be prepared to shift their priorities on short notice, to support the limited-objective attack if it is within range from their position or with a minimum of repositioning.

Spoiling Attack

11-109. The spoiling attack allows little time for the commander to allocate additional assets to the force conducting the attack. This problem can be largely mitigated when initially allocating assets to the forces that typically conduct this mission. The commander must also be prepared to shift the priorities of other available air defense units on short notice. Forces that deploy by means of helicopters would be equipped with man-portable systems, and the requirement for a quick reassignment of priorities of other air defense units within range takes on extra importance.

Counterattack

11-110. The counterattack also allows little time for the commander to allocate additional assets for air defense. Compared to a spoiling attack, however, larger forces are generally involved due to difference in purpose. Commanders must be prepared to make rapid shifts in priority assignments. They should also begin anticipating requirements to support a rapid transition of the remainder of the force to offensive actions. Fixing forces involved in the counterattack should not require extensive changes in air defense mission assignment. The assault force, if one is used, needs sufficient assets to allow it to effectively engage helicopters that will likely be part of any mobile forces committed against it. The action force requires sufficient assets to defend against air attack once discovered and to allow it to consolidate its gains.

DEFENSE

11-111. As in the offense, the division and its subordinate brigades ideally have sufficient assets to provide air defense coverage for all of their units during defensive actions. In cases where a DTG or BTG has been formed, additional air defense assets should be allocated to satisfy increased requirements. In a situational defense, there is limited time available for allocating or reallocating air defense assets. *Maneuver defense* and *area defense* are the two basic types of defensive actions.

Maneuver Defense

11-112. The key to air defense support of the maneuver defense lies in mobility. Air defense units must be positioned to cover defending forces but capable of displacing with rapidly moving ground maneuver forces. Air defense covers the contact and shielding forces as they maneuver from one defensive array to another. Ideally, most air defense assets can be positioned with the shielding force and provide adequate coverage for the contact force. Distances between the two forces are key in determining if systems can provide coverage at the ranges required. In any case, it is essential that sufficient mobile assets be allocated to the contact force to cover its movement away from the enemy and to cover the flanks.

11-113. In the maneuver defense, air defense units must displace more frequently than in the area defense. This displacement requires units or parts of units to move by alternating bounds. One element continues cover while the other moves. This means reduced coverage for at least part of the time. Taking this into account, additional assets could be allocated to make up the difference. In any case, moves should be planned in detail and every effort made to reduce the number required. In the maneuver defense, the need for frequent displacement often mitigates the requirement for survivability moves.

Air Defense

11-114. As the contact force initiates its movement to begin the hand-off to the shielding force, there are two options for air defense assets positioned with the original shielding force to cover the contact force. One is to remain with the shielding force as it becomes the new contact force. Another is to begin movement by bounds to the defensive array where the former contact force will take up its position as the new shielding force. In many situations, it may be possible to position longer-range systems where they can cover the initial positions of both the contact and shielding forces—and possibly of the new shielding force. This increases the time and continuity of coverage. Again, units could displace to positions where they can cover subsequent contact or shielding force positions. It is essential that planners allocate and position systems that are capable of responding to a highly fluid situation. Commanders should take advantage of range, mobility, and creative means of positioning to allow sufficient standoff to prevent their systems from being destroyed by enemy direct fire systems.

11-115. Air defense planners need to take into account the requirement to support rapidly executed ground counterattacks. Assets supporting and moving with counterattacking forces could be MANPADS. Longer-range systems are positioned with defending forces where they can cover the counterattacking force.

Area Defense

11-116. In an area defense, air defense considerations in the disruption zone are similar to those found in the maneuver defense. The disruption force must be capable of rapidly attacking the enemy or shifting to a maneuver defense or a combination of the two. Frequent displacement is the rule. In many cases dispersed ambush forces and precision weapons systems will require point protection. Area coverage is desirable in trying to attack enemy aerial reconnaissance assets and preventing effective employment of ground-attack platforms. The disruption zone will require a relatively high density of MANPADS and a well thought-out and executed air defense ambush plan.

11-117. In the battle zone, the main defense force occupies battle positions set in complex terrain. This presents a whole set of problems that the air defender must solve. Complex terrain limits the capabilities of line-of-sight systems. This includes acquisition, tracking, and firing systems. A detailed terrain analysis, which takes into account the masking features of the terrain, is essential. Although not desirable, some acquisition systems will have to be positioned on high ground to be effective. The use of C3D is key to mitigating the vulnerabilities of systems so sited. No matter what techniques are used, there will be gaps in coverage. Some of these gaps may be areas in which the commander chooses to take risks. Others can be covered by shorter-range but more suitable systems. The use of ambushes from hides could be particularly effective as the enemy attempts to exploit these gaps. Often aerial observers will be employed, with links to air defense units for rapid response in the area.

TECHNIQUES

11-118. Whatever the nature of combat actions conducted by maneuver forces, the actions of supporting air defense units must prevent enemy aircraft from successfully attacking maneuver forces. The OPFOR uses a variety of techniques to ensure the survivability of air defense units and their ability to protect maneuver forces and other key assets during either offensive or defensive actions.

Movement of Air Defense Units

11-119. Ideally, air defense units conduct major movements at night or in adverse weather. The OPFOR seeks to maintain effective air defense coverage by ensuring that not all elements relocate at the same time. This typically involves leaving one or more firing batteries in their positions to provide coverage while others move.

11-120. If the enemy air threat is serious, air defense unit(s) supporting a maneuver unit will usually move as part of that unit, integrated into its march column. If the air threat is not imminent, the air defense unit may move separately to its new position. In this case, the air defense battery commander usually conducts an initial map reconnaissance and designates the movement route and tentative firing positions in the new area. He sends out a reconnaissance patrol that normally consists of one of the firing platoon leaders, several soldiers, and a vehicle. This patrol conducts route reconnaissance, identifies temporary firing

Chapter 11

positions along the movement route, and conducts limited CBRN reconnaissance. The reconnaissance patrol then confirms the suitability and location of the new positions. The reconnaissance patrol can operate as part of a supported maneuver unit's reconnaissance patrol, as part of the reconnaissance patrol of the next-higher air defense unit, or it can carry out its mission independently.

11-121. Where time permits, the OPFOR reconnoiters and prepares alternate positions in advance. Ideally, every air defense unit should have two to three alternate positions. Movement to them can be carried out at night or under conditions of limited visibility whenever possible. Air defense units would most likely move to alternate positions under one of the following conditions:

- Immediately after enemy reconnaissance aircraft have overflown their current position.
- After an air strike has been repulsed.
- After units have been at a single position for an extended period.

For divisional air defense assets, this extended period of time would consist of approximately 4 to 6 hours, after which they would move to alternate positions. This time could obviously be reduced when there is a high threat of air or precision attack. In some cases, given the systems' capability to do so, moves could take place as often as every 10 minutes. For more information, see Maneuver and Dispersal, below.

11-122. The total time for movement of air defense units includes the time for leaving the position, moving to the area of the new position, and occupying this position. It is the mission of commanders and staffs to reduce this time to the minimum, since during this period the unit is removed from battle. However, a necessary condition for air defense effectiveness is the destruction of enemy aircraft on the approaches to the supported units. This must be taken into account, along with survivability considerations, when determining the frequency of changing positions. The procedure and time periods for movement, and occupying and preparing positions, are determined during planning for combat.

Local Security and Self-Defense

11-123. Air defense units at division and below usually deploy close to enemy ground forces, where vulnerability to both ground and air attack is significantly greater. When in proximity to supported ground units, air defense units often rely on them for their local ground security. However, this is not always the case.

11-124. Self-defense against air attack is accomplished through the use of the unit's primary weapons and small-caliber AA guns and MANPADS. Units equipped with AA guns can defend themselves against ground attack, to some extent, through the employment of their systems in a direct fire role. Air defense unit personnel armed with light antitank weapons can augment local ground defense capabilities. If SAM batteries are threatened by ground attack, they can often move to more secure positions without seriously degrading their capability to continue their primary mission.

Air Defense of Tactical Movement

11-125. The OPFOR anticipates that units conducting tactical movement may be subjected to intense attacks by both fixed-wing ground-attack aircraft and armed helicopters. These attacks can occur anywhere on the battlefield. Accordingly, units engaged in movement are protected by their organic assets and, in many cases, are allocated additional air defense assets from their parent unit.

11-126. In general, air defense units are integrated into moving tactical units and are ready to fire. Many SAM and AA gun systems can be fired on the move. However, stationary engagements are preferred. SAM units that require setup time may move along separate routes by bounds, alternating moves by platoon or battery. Tactical units may also receive air defense coverage from air defense units of higher echelons and possibly adjacent units. In the interests of secrecy, air defense radars and associated communications systems of the moving unit are placed in a standby and receive-only mode, respectively, unless absolutely required to engage enemy aircraft.

11-127. Enemy air attacks are considered particularly likely at—

- Obstacles (such as river crossings).

Air Defense

- Choke points (bridges, mountain passes, defiles, or places where off-road movement is restricted, such as in swampy areas).

To ensure air defense coverage for units moving through such areas, a portion of a unit's air defense weapons may be dispatched ahead of the unit to deploy in and around the obstacle or choke point to provide effective coverage as the unit passes. The remainder is spread throughout the supported unit.

11-128. If adequate coverage of the unit can be maintained by higher-echelon and adjacent unit assets, then the entire air defense unit may be sent forward of the parent unit formation. Alternatively, individual batteries or sections may be sent ahead. If the restricted terrain area is of such size as to exceed weapon and or sensor coverage of the air defense weapons, then air defense elements may move by bounds ahead of each other to provide continuous coverage by the parent unit.

11-129. Reconnaissance and air surveillance are vital to protecting moving units from air attack. Air observers are posted on all vehicles. Air defense elements, including MANPADS gunners, remain ready to engage targets at all times. If the tactical situation requires and terrain conditions permit, surveillance and target acquisition radars may be set up at suitable locations adjacent to the movement routes to provide continuous radar coverage. As in other tactical situations, MANPADS gunners are assigned specific sectors of observation and fire to preclude several gunners engaging one target. Vehicle-mounted weapons are also employed. For example, AA machineguns on tanks are specifically designed for this purpose. Missile-firing tanks have a capability against helicopters.

11-130. Brigade air defense weapons play a major role in the defense of units on the move. These systems are normally employed in mutually supporting pairs. System range determines the distance that can be maintained between them. The systems must also ensure that they maintain sufficient distance from other vehicles to ensure an unobstructed field of fire in engaging low-flying aircraft. Whenever a column stops, even for brief periods, brigade systems pull off to the side of the road with the rest of the column and remain ready for action. Battery personnel who are not operating with the firing platoons may be directed to engage enemy aircraft with small-arms fire.

11-131. In some circumstances, a supported unit may continue to move while the air defense elements halt to engage enemy aircraft. This is not the recommended course of action, because it leaves the supported unit with reduced or no coverage while the air defense elements are engaged.

Engagement Procedures

11-132. Aircraft posing the greatest threat are engaged on a priority basis. Aircraft are engaged with as many weapons as possible and in the shortest time possible in order to achieve the greatest destructive and deterrent effect. The preferred engagement technique is to continue firing at an already engaged target rather than to switch from target to target, unless a later-acquired target seriously threatens the air defense unit itself or a high-priority target. The OPFOR would rather engage an enemy aircraft prematurely and waste some ammunition than wait too long and allow the aircraft to gain a favorable attack position. Suspected enemy aircraft are fired on as long as they remain within range. Air observers and weapons crews outside the attacked sector maintain continuous observation and readiness to fire in order to prevent the enemy from conducting a successful attack from several directions simultaneously.

11-133. The OPFOR emphasizes that air defense units do not have to destroy aircraft to accomplish their mission, although such destruction is obviously desirable. The mission is accomplished if air defense units prevent enemy aircraft from conducting successful air activities. For example, air defense units can force enemy aircraft to break off their attacks or to expend their ordnance inaccurately without having to destroy the aircraft. In fact, the mere presence of active and effective air defense weapons systems can reduce the effectiveness of enemy air activities by forcing aircraft to avoid the systems or to operate using less than optimum procedures.

11-134. The OPFOR prefers to either leave target acquisition and tracking radars off, or turn them off, to preclude exposing their presence and location. This is especially effective when the unit or system with the air defense radar is receiving real-time information via either direct data transmission from radar to radar or other communications. If required, the radar will turn on at the last few seconds to acquire the target before launching and/or firing.

Chapter 11

11-135. OPFOR AORs are free-fire zones, unless OPFOR aviation missions are scheduled. Free-fire zones do not require IFF checks. The air defense weapons usually launch or fire on first detection.

Air Defense Ambushes and Roving Units

11-136. The OPFOR recognizes the disproportionate effects that sudden, unexpected destruction of an aircraft or small group of aircraft can have on enemy tactics and morale. For example, the surprise destruction of one or two lead aircraft on what the enemy thought was a clear avenue of approach could cause an enemy air assault to be called off or seriously disrupted. Air defense ambushes and roving air defense units can cause the enemy to believe that significant air defense units are located in areas where actually there are only a few weapons. This can reduce the effectiveness of enemy reconnaissance and the likelihood of enemy air attack in the area concerned. Ambushes and roving units can also employ antihelicopter mines.

Air Defense Ambushes

11-137. Air defense ambushes may set up at temporary firing positions to surprise and destroy enemy aircraft and disorganize enemy fixed-wing aircraft and rotary-wing operations. Typical missions include defending—
- Maneuver units.
- Possible air avenues of approach.
- CPs.
- Reserves.
- Artillery and missile units.
- Other air defense units in firing positions.
- River-crossing sites.

11-138. Air defense ambushes are often employed when there is a perceived inadequacy of air defense assets. Tactical air defense ambushes usually comprise a single AA gun or MANPADS team, section, platoon, or battery with the mission of engaging enemy aircraft from a hidden or unexpected position. The senior air defender (or perhaps an AOP) conducts a map terrain analysis of friendly radar coverage prior to deployment. The resultant analytical overlay provides the air ambush unit(s) critical locations to surveil based on radar terrain masking. Based on this, the air defense ambush teams are dropped off to provide air coverage.

11-139. **Deployment.** Air defense ambushes are placed—
- In the disruption zone.
- On secondary and tertiary air avenues of approach.
- Along flanks, forward, behind and in gaps between units.
- In terrain that offers poor fields of observation, to fire "window shots."
- In valleys or defiles likely to be used as ingress or egress routes by infiltrating aircraft, or on adjacent heights to shoot down onto them.
- Just behind a crest to catch aircraft from behind as they clear a ridge.

Single-launcher MANPADS ambushes may be set up on wooden platforms built in treetops to catch aircraft flying low over a forest.

11-140. A typical OPFOR air defense ambush might be a MANPADS team dropped off by a reconnaissance patrol enroute to conduct a reconnaissance mission. The MANPADS team may or may not be picked up by the returning patrol. If not, they may exfiltrate on their own and possibly rendezvous at a different location. These MANPADS ambushes may be single teams, but more likely the teams will be numerous and dispersed widely throughout the battle zone or disruption zone. Air defense systems that lend themselves to multiple roles are especially appropriate for air ambushes.

11-141. The unit or weapon assigned to an air defense ambush usually occupies the site under cover of darkness or poor visibility conditions. The unit or weapon is carefully camouflaged and keeps all its

Air Defense

emitters off or in "dummy load" until ordered to engage a target. It may assume a hide position and establish local ground security and air observers. Depending on the unit or weapon involved and the situation, it may be able to receive automated surveillance and target tracking data from its parent unit. More than one air defense ambush, involving more than one weapon type may be established along an air avenue of approach. These may work independently or in concert, depending on the situation.

11-142. **Preparation.** Air defense ambushes may be planned and executed on short notice with little preparation. However, they may also involve elaborate preparation and camouflage. Preparation can include tracking enemy aircraft over several days to discern operational patterns and possibly weaknesses, and optimum weather patterns for a specific ambush site. These ambushes are the most effective when they are coordinated with local combined arms units. The fires from the combined arms units may either assist in deterring the attacking aircraft or vector them into the air defense ambush kill zone.

11-143. Detailed preparations can involve removal of tracer ammunition from AA gun ammunition belts so that near misses do not alert the target aircraft. They can also involve construction of "tree-stands" in remote locations for air observers or MANPADS gunners, with provision made for the alert of the ambush unit through wire, visual, or radio signals. Decoys or derelict weapons may be placed to draw the attention of enemy aircrews, causing them to enter the ambush zone or fire at the wrong target. Visual decoys can be made more believable through the use of imitative communications or even decoy emitters. The key to a successful pre-planned ambush is creating a credible target or set of targets that the enemy will attack.

11-144. **Execution.** Target engagement decisions may be left up to the ambushing unit commander. Special engagement techniques may be used, such as delaying radar illumination until the last possible moment, coupled with a favorable cloud base and remote tracking information from other sensors. Occasionally AA guns may choose not to employ their radars, using strictly electro-optical sights. This tactic takes into account the capability of modern aircraft, including attack helicopters, to detect radar and infrared systems.

11-145. When a target is detected, the ambushing weapon or unit prepares to engage. This may involve removal of some camouflage or a short movement from its hide position to its firing position. The ambushing unit fires on the target until it is destroyed or until the target moves out of its firing zone. After ensuring that it is safe to move, the ambushing unit or weapon immediately displaces via a concealed route to a new ambush site or returns to its parent unit.

Roving Air Defense Units

11-146. Employment of roving units is similar to that of air defense ambushes. The primary difference is that, while an ambushing unit lies in wait for approaching enemy aircraft, a roving unit moves to the most likely areas of enemy air attack and occupies a series of predesignated positions in the supported unit's AOR. The commander of the roving unit identifies these positions during his terrain reconnaissance and coordinates them with the air defense and maneuver unit commanders. The roving unit occupies these positions according to a prearranged schedule or on order of the air defense unit commander. Roving units terminate their missions and return to previously designated primary firing positions upon direction of their commander.

Antihelicopter Mines

11-147. The OPFOR routinely employs antihelicopter mines to support air defense ambushes and roving units. Areas protected by antihelicopter mines can also be left unattended and therefore preserve needed combat power. The intent of these mines is not so much to destroy helicopters as to accomplish one or more of the following:

- Force low-flying helicopters to rise or change course, exposing them to more lethal means.
- Alert air defenders to trigger the ambush.
- Distract pilots while engaging them with ground weapons.
- Cause the attacking aircraft to break off and/or discontinue the attack.

11-148. The OPFOR emplaces antihelicopter mines in locations it believes the enemy will use as firing (battle) positions for attack helicopters or in possible landing zones for lift helicopters. Antihelicopter mines use a combination of sensors and fuzes to acquire the helicopter and initiate the mine once the helicopter enters the lethal zone. More advanced mines use a fairly sophisticated data processing system to track the helicopter, aim the ground launch platform, and fire the kill mechanism toward the target. As the helicopter nears the mines, the acoustic sensor activates or cues an infrared or millimeter-wave sensor. This second sensor initiates the mine when the helicopter enters the lethal zone of the mine. These mines may have multiple-fragment warheads that are more than capable of destroying tactical aircraft. Alternate warhead designs include high-explosive warheads and single or multiple explosively formed penetrators. For addition information on antihelicopter mines, see the *Worldwide Equipment Guide*.

Defensive Measures Against Unmanned Aerial Vehicles

11-149. The OPFOR recognizes the increasing importance of UAVs on the battlefield, to both its own forces and those of the enemy. They can perform high- and low-altitude missions, collect the full spectrum of intelligence, and immediately downlink the data to a ground station. They have the capability to loiter or to fly deep and can collect against a predetermined target or look for targets of opportunity. Their construction can make them difficult or easy to detect.

11-150. Typically, the enemy conducts reconnaissance missions using UAVs operating in the window between low-flying helicopters and higher-altitude fixed-wing aircraft. This altitude window is between 300 and 4,000 m. The most common technique is to approach the target area at high altitude. Then, once at the target area, the UAV drops down to a lower altitude that optimizes the capabilities of the sensor package on board. Once the mission is complete in the target area, the UAV climbs to higher altitude and departs the mission area. Countering tactical UAVs requires not only an IADS but also an integrated all-arms approach.

11-151. Most UAV systems consist of three basic subsystems:
- The air vehicle.
- The ground station.
- The launcher.

In some cases the latter two may be one vehicle, or they may all be man-portable. There are also a variety of communication data links between the ground station and the air vehicle. Some systems also include satellite links.

11-152. Air defense commanders and planners should view the three UAV subsystems as three separate targets that can be countered through a variety of means. This view reinforces the concept of an integrated system that includes coordination with others arms to ensure all targets are addressed. These means are both passive and active.

Passive

11-153. Since the mission the UAV is executing may not be apparent, actions should be taken to counter all possibilities. The integrated use of the passive measures described below under Nonlethal Air Defense Measures can reduce the effectiveness of UAVs. The use of a variety of decoys provides a false picture of the mission area to the enemy and to a large extent can deny information or distort the information collected by the UAV.

11-154. Other passive measures may include monitoring UAV data links. Often UAV satellite up- and downlinks use commercial telecommunications. Many of these communications links, including satellites, are not protected and can be easily intercepted through readily available commercial software. Monitoring these various links can be very effective because they provide valuable information such as flight data, system capabilities (location as well as collection), and attack priorities and/or intentions.

11-155. Signals reconnaissance units may also be able to determine the general location or direction of UAV flight by intercepting the downlink (UAV transmission) and using direction finding equipment. Based on this intercepted data, they may also be able to determine the number of UAVs in flight. In this event, the signals reconnaissance unit may be able to tip off air defense or other units in the projected path of the UAV(s) for further action.

Air Defense

Active

11-156. A wide variety and large number of active measures are available to the OPFOR to counter UAVs. The effectiveness of air defense radars can vary depending on the radar cross section (RCS) and altitude of the vehicle. Of course, this does not preclude the use of radar, since these factors are considerations in detecting any aircraft. The relative small size of many UAVs obviously reduces their RCS.

11-157. Sound-ranging systems are available, which can provide early warning and azimuth of an approaching UAV. This in turn provides air defense weapons and maneuver unit weapons an opportunity to prepare for the vehicle's approach and to put up a large volume of fire, provided the UAV can subsequently be visually detected. The early warning provided by sound ranging increases the probability that visual observers will be able to spot the vehicle.

11-158. The location of UAV ground stations and launchers is a high priority for reconnaissance. UAVs can support targeting and fires from enemy artillery, multiple rocket launchers, or aircraft that can quickly engage targets once the information is received. Therefore, their priority for destruction increases. SPF operating in the enemy rear can also be a valuable asset in locating launchers and ground stations. They can either take direct action to destroy the targets or relay location information to allow other means to be employed against them.

11-159. Jamming can counter UAV system data links. The effectiveness of this procedure varies depending on the UAV system being attacked. High-power spot or barrage noise jammers can effectively mask ground targets from side-looking airborne radars. Supporting satellites and infrastructure can also be jammed to some degree, depending on the type of UAV employed. Airborne jammers may serve as the ideal platform for jamming UAVs. GPS jammers, preferably mobile, are also effective in interfering with UAV guidance.

11-160. Additional techniques for countering UAVs could also include targeted air, SSM, or artillery strikes. Ground targets could be as simple as bombing a UAV airfield runway or destroying a petroleum storage facility. Effective air targets could include destroying an airborne C2 platform, which would disable the UAVs. An example of this is to destroy relay and retransmission UAVs that are electronically tethered to other airborne UAVs. Destruction of retransmission stations or vehicles can either limit the range of a UAV or completely disable it, causing it to crash.

11-161. Some of these techniques are limited to UAVs requiring improved runways and petroleum fuel. Others are limited to UAVs that fly at medium to high altitudes. Also, some of these techniques and targets will not be effective against micro, mini, or other tactical UAVs.

Air Defense in Urban Areas

11-162. In urban areas, AA guns or blinding lasers could be set up within the top or middle floors of buildings to fire laterally or even down on low-flying aircraft while remaining unseen from almost every angle. Weather conditions may also facilitate the use of an air defense ambush. For example, low cloud bases may force enemy aircraft down into the envelope of a particular weapon. Ambushing units may work in concert with smoke or aerosol dispensing units, or ground-based jammers that jam a low-flying aircraft's terrain-following radar, forcing it up into the ambush weapon's optimum engagement envelope.

11-163. Some air defense systems may prove useful in close combat in urban areas. Air defense weapons usually have a very high angle of fire, allowing them to target the upper stories of buildings. The high-explosive rounds allow the weapons to shoot through the bottom floor of the top story, successfully engaging enemy troops and/or equipment located on rooftops. The accuracy and lethality of air defense weapons also facilitates their role as a devastating ground weapon when used against personnel, equipment, buildings, and lightly armored vehicles.

Air Defense at Night

11-164. Air defense is conducted at night and during other periods of reduced visibility almost as effectively as during periods of normal visibility. This is because of the numerous surveillance and fire control radars in air defense units and the inherent limitations of enemy aircraft maneuvers and coordination at night. Commanders prefer to move or realign units at night or during other periods of reduced visibility to reduce the likelihood of detection by the enemy, and such moves must be covered by air defenses.

11-165. Air defense units frequently deploy closer to supported units at night. Like other units, they are more likely to conduct maintenance and resupply at night. The OPFOR recognizes, however, that the increasing night capabilities of aircraft and the proliferation of night strike packages requires that air defenses be ready 24 hours a day. In order to meet this evolving threat, the OPFOR continually upgrades its night air defense capability.

11-166. The OPFOR employs various "clip-on" image-intensifier night sights, which are fitted to some MANPADS launchers. Most radar-guided SAMs have electro-optical tracking backup that can be used in daylight hours in the event the radar is jammed, and a night low-light-level television channel for engagements in clear nighttime conditions. On some systems, the electro-optical mode is considered the primary mode for target engagement, with radar-only engagement mode being the second (or nighttime) choice. Many newer OPFOR air defense systems have thermal sights, or some older systems have been upgraded with them. For additional information, see the *Worldwide Equipment Guide*.

Air Defense of Assembly Areas

11-167. While assembly areas are most commonly used in the offense, the air defense employment methods used there also apply to any situation in which a tactical unit requires assembling for any reason. Brigade or BTG assembly areas are essentially composed of a series of battalion assembly areas. The brigade or BTG commander assigns air defense elements, normally a platoon of AA gun or gun-missile systems, the mission of supporting a particular battalion for a specified period of time. This period can begin before the battalion moves into its assembly area, in which case the platoon provides protection to the battalion during movement to the assembly area.

11-168. The air defense unit may also join the maneuver battalion in the assembly area. However, it is preferable to have both arrive at the same time, in order to ensure the battalion is not exposed to possible enemy air attack. In either case, the air defense platoon leader reports to the maneuver battalion commander, and direct communications are established between the two units. The platoon continues to maintain communications with its parent air defense battery and/or battalion. It also receives information from the divisional air defense target identification and warning network. This ensures timely receipt of information on the tactical situation.

11-169. While in the assembly area, the maneuver battalion commander uses all available C3D techniques to reduce the likelihood of detection. Additionally, a 360-degree surveillance of the surrounding airspace provides early warning. The maneuver battalion commander and the supporting air defense platoon leader work closely to integrate their weapons into an effective air defense plan. The battalion commander provides guidance for the placement of all air defense systems, while the air defense platoon leader supervises the details of the placement of his weapons. Proper placement of air defense weapons increases the engagement envelope to the maximum extent possible. As is the case in most tactical situations, the platoon leader must ensure that his AA gun-missile or gun systems are kept within mutually supporting distance. As a rule, one crew in each pair of systems remains alert, except when warning of an air attack is received. Any attached or organic MANPADS supplement the defense, and the attached air defense platoon leader may be given some degree of control over the MANPADS gunners. Radio silence and light discipline are observed. If the supported unit is to remain in the assembly area overnight, the air defense systems are dug in.

11-170. Air OPs and firing positions are often colocated. This is especially true in the case of MANPADS. The OPs and firing positions should be positioned to provide comprehensive observation and interlocking fires on the most likely approach routes for low-flying fixed- or rotary-wing aircraft. All other weapons, to include AA machineguns on tanks, general-purpose machineguns, ATGMs, and anything else that may cause attacking aircraft to discontinue an attack are further integrated. Even planning for the use and integration of massed small-arms fire, antitank grenade launchers, and other weapons is essential to an effective air defense at the tactical level.

11-171. Proper planning and execution at maneuver battalion level and further integration into the overall air defense scheme of the brigade, division, and higher levels of command should result in enemy aviation having to pass through overlapping coverage to attack the assembly area. Attacking enemy aircraft must first penetrate the OSC and division engagement envelopes formed by their respective missile units. The aircraft then come within range of the maneuver brigade and battalion systems. Missile, gun, and gun-missile systems and any other weapons the commander deems appropriate engage enemy aircraft as soon as they come within range. MANPADS gunners engage aircraft that maneuver to avoid these systems or pass over the MANPADS positions. Finally, small arms (in volley) and vehicle-mounted weapons engage enemy aircraft that pass over the maneuver unit's positions.

Innovative Techniques

11-172. The OPFOR will use any means and methods to mitigate enemy air capabilities. This often involves the use of innovative and adaptive methods and techniques. The OPFOR views these as part of the overall air defense effort. The extent to which creative techniques can be applied is limited only by—
- The commander's and staff's knowledge of the enemy air threat.
- The capabilities of their own systems.
- Their ability to apply that knowledge to come up with innovative solutions.

11-173. An adaptive method proven to be effective is for engineers to string cables across convergent air avenues of ingress or egress and potential landing zones. These cables are especially lethal against helicopters flying NOE. Their use is generally tied to or supported by an air defense ambush. The use of artillery in conjunction with antihelicopter mines against attack helicopter firing positions is another example of an adaptive approach.

11-174. SPF teams can infiltrate MANPADS close to airfields or along identified and potential flight routes. These teams are best employed in pairs. Early warning can be achieved to some extent through the use of a team watching an airfield and transmitting enemy aircraft departure time and direction to SAM teams.

11-175. SPF teams equipped with man-portable ATGMs can target aircraft on the ground, maintenance vehicles and facilities, and even air traffic control and communications vans. Infiltrated or stay-behind SPF and infantry can conduct on-order raids against airfields and ground support facilities away from the airfields. These raids can be timed in conjunction with other methods so that they assist in keeping the target under constant pressure. SPF can also conduct small team ambushes along lines of communications (LOCs) with the specific purpose of destroying certain types of vehicle or equipment related to air operations.

11-176. Affiliated irregular forces, possibly using terror tactics, can intimidate (enemy) host country civilian contractors to force them to sabotage enemy operations they are supporting. Examples include contaminating fuel and lubricant supplies, placing bombs on generators or ground support equipment. Irregular forces or local civilians can be contacted and/or supplied and trained to perform missions that support the objectives of air defense operations.

NONLETHAL AIR DEFENSE MEASURES

11-177. In order to meet the progressive increase in enemy air capability, the OPFOR realizes it must maximize both the lethality and the effectiveness of its air defense. The overall effectiveness can be enhanced by the use of nonlethal air defense means, especially when employed in conjunction with lethal SAM and AA gun systems.

ACTIVE

11-178. The OPFOR deploys active jamming assets, in conjunction with lethal systems, to defend what the OPFOR has identified as high-value assets. Examples of these include air bases, logistics centers, critical LOCs and choke points, and higher-level CPs.

Air Defense Jammers

11-179. Air defense units at division or DTG and below only have air defense jammers if they have been task-organized down from operational level. When available, the OPFOR uses such jammers to limit enemy air advantage in the tactical fight.

11-180. Air defense jammers target the onboard emitters enemy aircraft use for terrain-following, navigation, and radar-aided bombing. They can also target airborne radar reconnaissance systems. The goal of jamming these systems is twofold. The primary goal is to force the attacking enemy aircraft to alter their flight profile, bringing them into the targeting umbrella of SAMs or AA guns. Jamming the terrain-following radars or radar altimeters employed by attacking aircraft does this by forcing low-flying aircraft to gain altitude. The secondary goal is to cause the aircraft to miss their target or abort the mission through the disruption of radar-aided bombing and target acquisition systems.

GPS Jammers

11-181. The OPFOR also can employ low-cost GPS jammers to disrupt aircraft navigation and precision munitions targeting. GPS jammers are also effective against UAVs and cruise missiles.

PASSIVE

11-182. In addition to active air defense, the OPFOR practices a variety of passive air defense measures. When conducting actions against a superior foe, the OPFOR must seek to operate on the margins of enemy technology and maneuver during periods of reduced exposure. Other passive measures include C3D and the use of maneuver and dispersal to degrade the effects of enemy systems.

Camouflage, Concealment, Cover, and Deception

11-183. The OPFOR emphasizes the use of natural terrain and vegetation, camouflage netting and other artificial materials, smokescreens, and decoy equipment to provide C3D. These measures include, but are not limited to, vehicle-mounted camouflage nets that reflect radar and reduce vehicle thermal signature. Similar materials can be used as screens for personnel and equipment, to reduce detection and identification by aircraft.

11-184. Deception includes dummy positions, mockups, decoys, and false electronic, acoustic, and thermal signatures. The OPFOR can use quick-setup, high-fidelity decoys; derelict vehicles; radar emitter decoys; quick-hardening foams; and many other types of manufactured and field-expedient means. It also employs simple heat sources to confuse infrared sensors and weapons seekers.

11-185. The dispersion measures discussed below should be employed with consideration of the protective and screening properties of natural and artificial screens, and would be combined with thermal camouflage and engineer preparation of positions. Natural screens consist of vegetation, terrain folds, populated areas, and local features or objects. Artificial screens include camouflage nets that would enhance natural screens, and radar-opaque screens using local features, radar nets, metallic nets, and corner reflectors. Concealment would be combined with the use of dummy positions, using decoy equipment and activities. Like real positions, dummy positions would be changed periodically. Dummy emitters and jammers would be used to attract enemy reconnaissance and targeting.

Maneuver and Dispersal

11-186. Maneuver and dispersal of air defense assets, both emitters and other types of equipment, is important for their survival in both combat and march formations. Sudden maneuver and periodic changes of position are simple and effective means to counter enemy reconnaissance and precision weapons. These measures are planned and implemented at the tactical level.

11-187. All, or only a portion of, the elements of an air defense unit may maneuver to alternate positions. When and how they do this depends on such factors as the degree of air threat, time of day, and meteorological conditions. The first elements to shift positions are those that have performed combat alert duty for an extended period, or that have been deployed in the position they currently occupy since before

Air Defense

the onset of combat. The optimum configuration for shifting to alternate positions involves no more than one-third of the assets of a given unit shifting at one time, in order to maintain adequate air defense coverage.

11-188. Distances related to dispersion and distances of air defense units from supported units vary with the situation and the threat. Of special concern is the enemy ATGM and precision weapon capability. If it is high, the OPFOR increases the spacing between SAM launchers, AA guns, or gun-missile systems and the distances of air defense systems from the battle line. Ideally, the degree of dispersal for units would be the same whether the enemy is employing conventional or precision weapons or even tactical nuclear weapons. A general rule for the degree of dispersion is that the enemy strike should not destroy two adjacent units simultaneously. A maximum of one-third of a unit should be vulnerable to a single precision weapon or nuclear strike.

SECURITY MEASURES

11-189. The OPFOR employs a number of air defense security measures and tactics to counter enemy air attack and suppression of enemy air defense (SEAD) operations. Measures taken to improve air defense system security include the following:

- **Signals security.** SAM, AA gun, and gun-missile system radars, which move forward to cover an initial assault, remain silent until after the assault begins.
- **Frequency spread.** Each of the air defense systems operates within separate radar frequency bands. (No one jamming system could operate simultaneously against all bands.)
- **Frequency diversity.** Tracking and guidance radars change frequencies to overcome jamming.
- **Multiple and interchangeable missile guidance systems.** Some OPFOR systems work on pulsed radar; others work on continuous waves. Some radar tracking systems also possess optical tracking for continued operations in a high jamming environment. Other systems use infrared homing.
- **Mobility.** All OPFOR tactical air defense systems and most operational-level systems are mobile. They quickly change positions after firing or after enemy reconnaissance units detect them.

Table 11-1 on page 11-30 lists additional examples of measures that can provide security for air defense units from enemy air attacks and SEAD.

Table 11-1. Example security measures

Considerations	Examples
Protection and Countermeasures	Use concealment, mixing with civilian sites and traffic. Use cover (dug-in positions, hardened facilities, and urban structures). Disperse assets and use autonomous capabilities. Relocate frequently. Use protection envelope of friendly forces. Use deception operations for convoys and river crossings.
Tactics	Conduct movement using bounding overwatch for air defense. Use passive (electro-optical) mode (radars turn on just at launch). Direct attacks against airborne warning and control system (AWACS) aircraft, SEAD aircraft, airfields, and FARPs. Engage SEAD and EW aircraft from an aspect outside of the jamming arc. Conduct beyond-border operations against air capabilities. Engage enemy air infrastructure when possible--including UAV infrastructures.
Reconnaissance, Intelligence, Surveillance, and Target Acquisition (RISTA)	Identify likely aircraft ingress and egress routes. Use passive radar and electro-optical modes. Use IADS links for target acquisition data. Use emissions control measures. Use civilians and irregular forces links. Use numerous OPs and air OPs linked to air defense units, including civilians, reconnaissance units, and forward-based SPF. Employ non-air defense sensors and units available to feed reports to IADS.
C2	Use mobile, redundant, concealed systems. Use communications and operational security measures.
Weapons	Engage aircraft, air-to-surface missiles, and antiradiation missiles beyond their firing range. Prepare all weapons to respond to aircraft. Have all units conduct air watches with weapons at ready at all times.

Chapter 12

Engineer Support

The OPFOR realizes that engineer support is vital for the successful execution of combat. Due to the fluid nature of modern combat, effective engineer support is essential for ground forces to employ or preserve combat power, as the conditions dictate. Engineer support can give combat forces the ability to maneuver quickly to exploit windows of opportunity. It can help change the nature of the conflict to something for which the enemy is not prepared.

ADAPTIVE ENGINEER SUPPORT

12-1. OPFOR engineers must be flexible enough to support two basic types of combat. The first is the fight against a less powerful neighbor, in which the OPFOR expects to dominate what is generally a traditional, conventional fight. The second is the fight against a more powerful force, a fight in which the OPFOR expects to be overmatched in at least some conventional capabilities. When the enemy is the dominant force in the region, this will likely compel the OPFOR to fight a defensive fight. In order to defeat this more powerful enemy, the OPFOR employs innovative, adaptive tactics to mitigate the enemy's advantages. An example of this innovativeness is the manner in which the OPFOR attempts to change the nature of the conflict. To accomplish this, the OPFOR attempts to place the enemy on the defense rather than offense, wherever possible. One means of accomplishing this is the constant and ubiquitous use of mine warfare. There is no sanctuary for the enemy—mines are everywhere. Two examples of this are—
- Emplacement of "toe-popper" mines on enemy foot traffic routes to produce wounds, not kills. This stresses the medical evacuation system and creates tentativeness among enemy soldiers. This could be tied in with attacks on the enemy's medical evacuation system.
- Maximum use of antihelicopter mines against possible attack helicopter firing positions or landing zones.

Other examples of adaptive methods engineers are likely to employ against a more powerful enemy (in addition to methods used in a more conventional battle) are interspersed throughout this chapter.

MISSIONS AND TASKS

12-2. The primary engineer missions performed in combat are reconnaissance, mobility, countermobility, and survivability. (See sections below on each of these missions.) Some examples of specific engineer tasks required to support those missions are to—
- Conduct engineer reconnaissance of the enemy and the terrain.
- Prepare and maintain routes of movement and supply.
- Clear passages through obstacles and areas of destruction.
- Perform demolition work.
- Establish and maintain water obstacle crossings.
- Establish and improve engineer obstacles.
- Prepare fortifications.
- Protect personnel and equipment from the effects of conventional direct and indirect fires, precision munitions, and chemical, biological, radiological, and nuclear (CBRN) strikes.
- Carry out engineer measures to eliminate the aftereffects of CBRN weapons.

Chapter 12

- Support information warfare (INFOWAR) and carry out engineer camouflage, concealment, cover, and deception (C3D) measures.
- Extract and purify water and establish water supply points.

See tables 12-1, 12-2, and 12-3 on pages 12-3 and 12-7 for how these tasks support the preparation and conduct of offense, defense, and tactical movement, respectively.

12-3. The OPFOR plans the complete integration of civilian and military engineer resources. For example, maneuver commanders may use civilian earthmoving, road-building, and construction equipment and personnel in support zones. This allows constituent combat engineer equipment and personnel to accompany maneuver forces in battle. Civilian workers or maneuver units can perform many basic combat engineer tasks, with engineers providing guidance and technical expertise.

12-4. Engineer tasks are a shared responsibility throughout the OPFOR. For instance, combat troops, as well as engineers, perform mine warfare tasks such as minelaying, minefield recording, and mine removal or breaching. Engineer and combat arms personnel also perform survivability tasks such as constructing fortifications, clearing fields of fire, and camouflage. The same is true for water obstacle crossings, where some units and equipment can ford, swim, or snorkel across with little or no engineer support. Although the highest level of engineer training and the greatest technical capabilities exist in the engineer troops, all military personnel and units train in fundamental engineer tasks.

12-5. The OPFOR's intent is to make the entire force as flexible and capable as possible while minimizing dependence on limited engineer support. This allows maneuver forces to autonomously execute rudimentary or basic engineer tasks. It also frees the engineer troops to—

- Perform engineer-specific or critical tasks supporting the maneuver commander's intent.
- Exploit and expand successful engineer effort begun by the combat troops.
- Support units that have little or no engineer capability.

OFFENSE

12-6. During preparation for the offense, the engineers focus on four major activities:

- Preparing routes for the advance and employment of combat forces.
- Providing survivability support to units in assembly areas.
- Establishing passages in obstacles and minefields.
- Establishing and maintaining crossings over water obstacles.

Table 12-1 identifies these, along with other specific engineer tasks required to support the actual conduct of offensive actions.

12-7. During the offense, the engineers' primary mission is to support the attack and assist in maintaining a high tempo of combat. Once the attack has started, engineer troops continue to perform tasks contributing to high rates of advance. Occasionally, they create obstacles to protect flanks, disrupt counterattacks, and block enemy reinforcements. Ongoing engineer reconnaissance is performed independently or in conjunction with other reconnaissance elements.

12-8. The OPFOR views commitment of exploitation forces or reserves as one of the most critical and vulnerable periods of combat. Engineer troops play a vital part in ensuring its success. They ensure the force's timely arrival on the line of commitment and provide support for its deployment and protection against flank attacks.

Table 12-1. Engineer support for preparation and conduct of the offense

Tactical Missions Requiring Engineer Support	Engineer Technical Tasks
Movement forward, deployment, and transition to the offense. Preparation of assembly areas. Crossing water obstacles. Supporting disruption and battle zones. Repelling counterattacks. Penetration of enemy defenses. Conduct of the battle. Commitment of exploitation force or reserve. Reinforcing captured positions.	Conduct engineer reconnaissance of enemy and terrain. Prepare fortifications in assembly areas. Clear passages in obstacles and perform demolition work. Establish and maintain water obstacle-crossing sites. Establish obstacles. Extract and purify water and establish water supply points. Carry out engineer C3D measures. Prepare and maintain movement routes. Eliminate aftereffects of CBRN strikes.

DEFENSE

12-9. When the enemy is the dominant force in the region, the OPFOR generally fights a defensive fight. In order to defeat a more powerful force, the OPFOR employs innovative engineer methods to mitigate the enemy's advantages, in addition to those employed in the more conventional battle.

12-10. Engineer support for the defense focuses on reconnaissance, countermobility support, and survivability support. It places emphasis on fortifying battle positions and assembly areas, performing engineer C3D measures, and adapting the terrain for defense. The defense is also conducive to the extensive use of various obstacles to interfere with the enemy's advance.

12-11. The general aims of engineer support the defense include—

- Controlling access and tempo by delaying, disaggregating, and canalizing enemy forces.
- Establishing conditions necessary for organizing the defense.
- Ensuring the integration of engineer support to INFOWAR and preparing deception positions.

Table 12-2 identifies specific engineer tasks required to support defensive actions.

Table 12-2. Engineer support for preparation and conduct of the defense

Tactical Missions Requiring Engineer Support	Engineer Technical Tasks
Repelling enemy attacks in front of the battle line. Preparation of defensive areas. Troop movement. Battle to hold positions. Repelling enemy penetrations of defense. Overcoming covering force zone. Counterattack by exploitation forces or reserve. Reinforcing lines taken in counterattack. Transition to the offense.	Conduct engineer reconnaissance of enemy and terrain. Prepare fortifications in battle positions and assembly areas. Clear passages in obstacles and perform demolition work. Establish and maintain water obstacle-crossing sites. Establish and improve obstacles. Extract and purify water and establish water supply points. Carry out engineer C3D measures. Prepare and maintain movement routes (for maneuver and supplies). Protect personnel and equipment from the effects of conventional direct and indirect fires, precision munitions, and CBRN strikes. Eliminate aftereffects of CBRN strikes.

12-12. The type and scale of engineer support depends on the tactical situation, enemy forces, and the conditions under which the OPFOR assumes the defense. If the OPFOR does so during the course of the offense, support may have to begin with the protection of threatened axes by obstacle detachments (ODs) and antitank reserves (ATRs) and the improving of routes needed for regrouping. If the OPFOR assumes a defense when not in contact with the enemy, support can begin with the creation of defensive works and the improvement of routes necessary for the OPFOR units to deploy. In both cases, engineer work supports development of the battle position by enhancing the effectiveness of OPFOR weapons and protecting personnel and equipment from the effects of conventional fire and weapons of mass destruction.

12-13. In the disruption and battle zones, the goals of engineer support are to hold up the enemy advance. In battle zones, engineer support facilitates organized withdrawal, maneuver, or counterattack by friendly forces. Defensive planning measures ensure extensive use of obstacles, integrated with preplanned direct and indirect fires, to affect the enemy's advance and facilitate his destruction.

COMMAND AND CONTROL

12-14. Engineer units allocated to a tactical group in constituent or dedicated relationships may be retained directly under the command of the tactical group commander. However, rather than keeping all organic and allocated engineer assets under his direct command and control (C2), the tactical group commander may suballocate some of his constituent or dedicated engineer units to his subordinate units. Additionally, tactical group commanders control—but do not command—other engineer assets that are allocated to them in a supporting relationship.

12-15. In the case of a division tactical group (DTG), the commander can allocate engineer units to his integrated fires command (IFC) and/or integrated support command (ISC). Some engineer units may be grouped under the integrated support groups (ISGs) that perform combat support tasks for the IFC or the DTG.

STAFF RESPONSIBILITY

12-16. In maneuver divisions, brigades, and tactical groups, engineer officers are permanent members of functional staff subsections under the chief of force protection and the chief of infrastructure management. Additionally, an engineer liaison team from each subordinate or supporting engineer unit supports the staff. These teams provide the operations officer with detailed expertise on engineer functions and provide a direct communications conduit to subordinate and supporting units executing such functions. Based on the advice of the liaison teams and coordination with the engineer units through the respective liaison teams, the functional staff chiefs advise the operations officer and/or the commander on engineer employment within their functional areas. The senior engineer team leader is designated as the *chief of engineer liaison teams* (CELT) and is the primary staff advisor to both the operations officer and the maneuver commander. In those units without liaison teams, the senior engineer serves as the CELT.

12-17. Other liaison teams may fall under the chief of current operations, to advise and assist in mobility and countermobility functions. The engineer liaison teams also coordinate, as necessary, with other staff elements, including the chief of INFOWAR. Liaison team leaders speak for the commanders of their respective units.

12-18. The maneuver commander specifies the tactical combat action(s) of his subordinate and supporting units, their start time and duration, and the area for these actions to take place. With this information, the engineer officers on his staff determine the required engineer missions to support the maneuver commander's plan. They prioritize engineer efforts to execute the technical tasks necessary to accomplish the overall mission. They can then determine the appropriate mix of troops, equipment, and materials necessary to perform the tasks under current conditions. They advise the commander and his staff on the best employment of available engineer assets to support the maneuver commander's mission, intent, and objectives.

12-19. The engineer liaison teams keep their respective engineer unit commanders informed of requirements for engineer support and pass on any guidance from the maneuver unit commander and staff on possible task-organizing. Then they monitor the execution of the directed missions. They provide input to the maneuver commander's combat orders and battle plans, the reconnaissance plan, the obstacle plan, and deception plans. They help organize the crossing of water obstacles and other barriers, and the preparation and maintenance of movement routes. They coordinate with the division, brigade, or tactical group chief of logistics regarding the preparation, improvement, and maintenance of supply and evacuation routes.

Engineer Support

12-20. The main steps that the liaison teams perform in support of combat actions are—
- Helping the engineer unit commander decide the appropriate organization of engineer support and reporting it to the maneuver commander.
- Participating in the reconnaissance conducted by the maneuver commander.
- Monitoring the completion of tasks by engineer units during the preparation for, and conduct of, combat.
- Reporting the status of engineer support to the maneuver commander.

TASK-ORGANIZING

12-21. There are no doctrinal constraints on task-organizing for mission success. The ability to allocate assets downward and to task-organize is restrained only by the availability of assets and the nature of the mission. If the necessary assets and/or capabilities are not available within the organization, the OPFOR commander will request the appropriate assets through his higher headquarters.

12-22. Engineer assets generally are constituent at no lower than brigade level. However, the OPFOR prefers to task-organize for mission success at even lower levels, when the assets are available. This may dictate that, instead of maintaining engineer units in their original composition, the commander may choose to break them down and combine them into smaller multirole engineer support elements. These engineer elements range in size from companies down to multirole platoons and engineer squads.

12-23. Engineer assets deploy throughout the battlefield and perform numerous distinct missions simultaneously during the course of the battle. In this way, route-clearing assets perform one function, while others perform demolitions, lay mines, construct obstacles, prepare battle positions, or set up water purification sites. Occasionally, the combined arms commander can also allocate to these groupings additional non-engineer assets, such as artillery, tank, or infantry troops. He can also augment maneuver elements with the engineer groups.

12-24. The following is a list of typical task-oriented engineer groupings:
- Obstacle detachment (OD, see Countermobility).
- Movement support detachment (MSD, see Mobility).
- Engineer reconnaissance patrol (ERP, see Engineer Reconnaissance).

SUPPORT TO INFORMATION WARFARE

12-25. The complete integration of engineer support of INFOWAR is critical at the tactical level, especially when fighting a powerful enemy. Deception is one of the basic elements of INFOWAR. Engineer support of the deception plan is vital for the deception to succeed. (See the subsection on C3D under Survivability.) Engineers' largest role in an integrated deception plan is that of constructing physical decoys (simulations in deception positions), enabling the enemy to see what he expects to see.

12-26. However, engineer support to INFOWAR is not limited to C3D measures. For example, engineers may support the INFOWAR campaign with psychological warfare activities to lower morale and instill a sense of tentativeness among enemy soldiers, and to undermine confidence of "enemy-friendly" populations. This can be achieved simply by the ubiquitous use of booby traps and mines. See chapter 7 for additional information on INFOWAR at the tactical level.

ENGINEER RECONNAISSANCE

12-27. Engineers conduct reconnaissance independently, or combined with chemical and reconnaissance elements. If the maneuver unit commander needs unique, specific engineer data for planning and preparation, he may order or request the use of engineer assets to form engineer reconnaissance patrols (ERPs), observation posts (OPs), and photographic reconnaissance posts. Engineer reconnaissance elements usually gather the following information:
- Enemy engineer preparation of battle positions and individual fighting positions.
- Location, type, and composition of enemy obstacles.

Chapter 12

- Conditions of roads, bridges, water obstacle-crossing sites, and routes.
- Presence of local building materials and other materials available for engineer tasks.
- Protective and camouflaging properties of the terrain.
- Enemy obstacles and demolitions created both during the preparation for the attack and during the attack.
- Movement routes and trafficability of off-road terrain for the attacking combat units.
- Locations where the enemy established obstacles during his withdrawal.
- Water obstacles on the main axis of advance.
- Local water supplies.

12-28. Water obstacles place additional requirements on engineer reconnaissance missions. See Reconnaissance under Water Obstacle Crossing, below, for a listing of the types of information required and who is likely to obtain it.

RECONNAISSANCE PATROLS

12-29. To provide engineer expertise, the OPFOR can allocate engineer specialists to accompany a tactical reconnaissance patrol dispatched by a division, brigade, tactical group, or even a battalion-size detachment (BDET). Additionally, reconnaissance elements of maneuver units can provide limited engineer-related information, although with less technical precision. However, under most conditions, the missions of all these reconnaissance elements preclude them from concentrating solely on engineer requirements. Therefore, the maneuver commander may order or request the engineer unit to form its own engineer reconnaissance elements.

12-30. A brigade or brigade tactical group (BTG)—or in some cases, a BDET—can include two or three engineer reconnaissance personnel in a regular reconnaissance patrol or security element. When engineer personnel augment other patrols in this manner, there is not likely to be a separate ERP.

ENGINEER RECONNAISSANCE PATROLS

12-31. When the engineer mission is expected to be a complicated one, however, it is better to form one or two ERPs. The use of two patrols allows the conduct of engineer reconnaissance by the leapfrog method. Ideally, the ERP(s) would begin their mission 1 to 2 hours before the main body of the brigade, BTG, or BDET starts to move. They assess the routes chosen by the staff, checking the validity of plans made from a map and reporting on—

- Obstacles and the effort required to overcome them.
- Conditions of crossing sites on water obstacles.
- The general nature of the terrain.

Engineer advice is an important element in the selection of routes and crossing points.

12-32. ERPs vary in strength from a squad to a platoon. A divisional brigade or BTG is more likely to form a squad-size patrol from its engineer company. An ERP can also include one or two CBRN reconnaissance specialists.

ROUTE RECONNAISSANCE

12-33. When engineers reconnoiter routes, one of their goals is to identify anything that could impede mobility. Taking into consideration any guidance from supported commanders and their staffs, the engineer unit commander can increase the size of his reconnaissance element and divide it into smaller teams in order to cover several points simultaneously. This allows him to assess a large number of features in the shortest amount of time.

12-34. When moving in areas where contact with enemy forces is unlikely, the engineer or maneuver commander can send an ERP ahead to obtain the required data. When anticipating enemy contact, engineer reconnaissance and data collection may be limited to reports from troop reconnaissance elements reporting on the engineer aspects observed along the route.

Engineer Support

12-35. When reconnoitering routes, engineers attempt to—
- Verify the condition of the route.
- Determine aspects of off-road terrain.
- Identify all obstacles and locate bypasses or recommended breach sites.
- Inspect bridges and dams.
- Identify suitable halt and assembly areas.

They report information on these topics to the commanders of the engineer and/or maneuver units that sent them out.

12-36. When the OPFOR route of advance encompasses potential water obstacles, ERPs try to find spots to set up ferry and bridge crossings, plus assembly or preparation areas. If bridges exist, engineers gather information on—
- The support structure.
- Load capacity.
- Necessary repairs.
- The presence of mines and demolitions on the approaches and on the bridge itself.

MOBILITY

12-37. When the OPFOR is the dominate force in the region, fighting a less powerful enemy, the OPFOR generally has freedom to maneuver wherever it wants whenever it wants. If the enemy hinders its movement, the OPFOR has alternatives because it dominates the region. However, when fighting a more powerful opponent, it is especially critical that the OPFOR maintain the ability to move unimpeded. This ability allows the OPFOR to control the access and tempo of enemy forces. As long as the OPFOR has complete access to the battlefield, it will allow no sanctuary to the enemy and determine the nature of the conflict. Engineer support can create opportunities for infiltration of small forces into unexpected locations, to inflict damage or to support INFOWAR.

12-38. Engineers are responsible for accomplishing tasks permitting the unimpeded movement of forces along the movement route, plus activities at assembly and halt areas. They also support the crossing of water obstacles. Table 12-3 lists the specific engineer technical tasks that provide the required support for tactical missions prior to and during tactical movement.

Table 12-3. Engineer support for preparation and conduct of tactical movement

Tactical Missions Requiring Engineer Support	Engineer Technical Tasks
Preparation of assembly and halt areas. Tactical movement. Crossing water obstacles.	Conduct engineer reconnaissance of enemy and terrain. Clear passages in obstacles and perform demolition work. Establish and maintain water obstacle-crossing sites. Extract and purify water and establish water supply points. Carry out engineer C3D measures. Prepare and maintain movement routes. Prepare fortifications at assembly and halt areas. Eliminate aftereffects of CBRN attacks.

MOVEMENT ROUTES

12-39. A movement route can follow any line and may include existing roads, cross-country roads, and off-road areas. After careful consideration of reconnaissance data and consultation with engineer officers on his staff, the commander specifies the particular movement route(s) his force will use. The engineer units and their liaison teams in the maneuver unit's staff are then responsible for planning and coordinating

Chapter 12

engineer support to prepare and maintain the specified movement routes. They provide input to the engineer support plan for the commander, who then issues orders, missions, and requirements to the constituent and dedicated engineer unit commanders for execution.

MOVEMENT SUPPORT DETACHMENT

12-40. The MSD is a task-oriented, temporary grouping of engineer assets to support route clearance and movement of the force in preparation for, and during tactical movement. The composition of an MSD is not fixed and varies depending upon the—
- Condition of the terrain.
- Character of enemy actions.
- Amount of work necessary.
- Assigned rate of movement for the columns.
- Availability of engineer troops and equipment.

12-41. Since its various technical tasks involve different types of equipment, the MSD frequently task-organizes into smaller elements to allow concurrent actions along the movement route. A typical MSD consists of a reconnaissance and obstacle-clearing element, plus one or two road and bridge construction and repair elements.

Reconnaissance and Obstacle-Clearing Element

12-42. Responsibilities of the reconnaissance and obstacle-clearing element include—
- Marking the movement route.
- Making immediate assessments of the terrain and obstacles.
- Identifying bypasses.
- Creating and marking passages through obstacles.
- Determining the character of destruction along the route.
- Locating building materials.

12-43. Augmenting assets from the division engineer battalion can use explosive charges or mechanical equipment to overcome rubble, rock barriers, and dragon's teeth (concrete pillars or iron posts). Engineers can breach wire obstacles after examining them for booby traps and electrification. Tree barriers may require the use of dozer blades or explosives.

12-44. The reconnaissance and obstacle-clearing element typically includes—
- An engineer unit base.
- Hand-held or vehicle-mounted mine-detection equipment.
- Explosives.
- Mineclearing vehicles such a tank with roller and plows.
- Route- or obstacle-clearing vehicles.

Road and Bridge Construction and Repair Element

12-45. The road and bridge construction and repair element usually has—
- One or more engineer squads.
- Tank- or truck-launched bridges.
- Route-clearing vehicles.
- Cranes and road graders.

The equipment varies depending on mission requirements and what was passed down from higher levels of command.

12-46. Responsibilities of the road and bridge construction and repair element include—
- Mineclearing and obstacle clearing along the route.

Engineer Support

- Reinforcement of bridges and repairs to roads.
- Construction of bypasses.
- Building and reinforcing bridges.
- Establishing fords and bypasses.
- Strengthening the route in swampy sections.
- Removing rubble.
- Repairing damage.

This element can also complete the route reconnaissance and marking of the route begun by the reconnaissance and obstacle-clearing element.

MSD Position During Tactical Movement

12-47. While moving, the MSD travels in advance of the main body, preparing the route so the main body can continue its advance unimpeded. Elements of the MSD are often performing tasks in proximity to elements of the security detachment. The location of the MSD in relation to the security detachment depends upon the possibility of enemy contact. When enemy contact is likely, the MSD may follow the security detachment. If enemy contact is unlikely, the MSD may be well ahead of the security detachment.

OBSTACLE BREACHING

12-48. The OPFOR is prepared to overcome obstacles during all phases of combat. In the offense, the OPFOR expects to cross obstacles on movement routes and throughout the enemy defense. Creating passages for the advance of the force in the face of enemy resistance is a combined arms task.

Explosive Obstacle Breaching

12-49. Explosive devices are the most significant obstacles the OPFOR expects to encounter. The OPFOR expects the enemy to use explosive obstacles and other obstructions for defensive purposes to impede the OPFOR's advance. In order for the OPFOR to conduct (or continue) an attack, maneuver units must breach these obstacles under direct and/or indirect fire. Units engaged in breaching these obstacles are extremely vulnerable to all enemy fires. Whenever possible, the OPFOR attempts breach a minefield from tactical movement, with minimum delay, and press the attack without first halting to consolidate on the far side of the obstacle.

12-50. The OPFOR may be required to breach enemy minefields when fighting a more powerful force. Although it may breach them in the more conventional manner described here, the OPFOR can also devise innovative methods to cross minefields. One such method might be to manually clear a path through the minefield through covert action. Several lanes could be cleared in this fashion. Then, at a time of the OPFOR's own choosing, dismounted troops could infiltrate through the minefield and rendezvous at a designated location on the other side, undetected by the enemy. See figure 12-1 on page 12-10.

Chapter 12

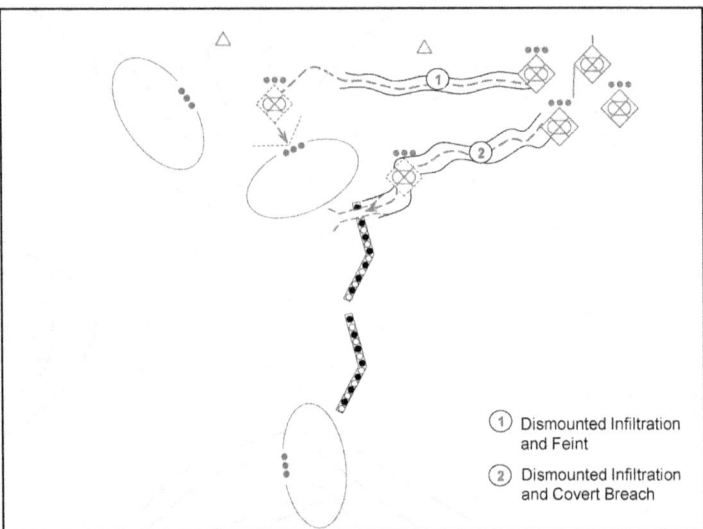

Figure 12-1. Covert breach (example)

Organizing Forces and Elements

12-51. There are three fundamental methods by which the OPFOR organizes for breach operations. First, and most preferable, is to make no special alterations to task organization for breach operations. A detachment is expected to breach as part of its situational breach battle drill (see chapter 5). Therefore, its higher commander will make every attempt to include in its task organization the means necessary to breach any anticipated obstacles without the need to deviate from the basic structure of action, support, and security elements or the need for outside assistance. The action element temporarily becomes a breaching element in order to reduce the obstacle such that it can accomplish its mission (as the action element).

12-52. Should any anticipated obstacles require significant allocation of specialist assets, the detachment commander may form a clearing element. A clearing element is a type of specialist element that penetrates obstacles permitting the action element to accomplish the detachment's mission.

12-53. Complex or extensive obstacles may require the formation of an MSD. MSDs are typically formed by tactical groups to support the movement of multiple detachments through a given zone of obstacles or to support their movement across a major water obstacle. (For more detail on MSDs, see that subsection above.)

Planning

12-54. Planning and preparation for the breaching of an explosive obstacle includes—
- Reconnaissance of the obstacle, including attempts to locate a bypass, and marking optimal breach locations.
- Infiltration of stealth breach teams, if possible.

Engineer Support

Breaching Methods

12-55. The OPFOR has three basic means to breach a minefield: explosive, mechanical, and manual:
- Explosive means such as line charges, bangalore torpedoes, and volumetric explosives all work by detonating mines through explosive pressures.
- Mechanical mineclearing plows or plow and roller combinations mounted on combat vehicles provide the main countermine capability to conventional forces. These systems detonate mines by striking them in advance of coming into contact with a vehicle or by physically moving the mines out of a defined path.
- Manual breaching requires personnel to physically displace or defuse explosive devices.

Mechanized Breaching

12-56. Mechanized and tank units make use of all three breaching methods to rapidly create lanes through obstacles with minimal delay. All OPFOR mechanized and tank units are trained, equipped, and expected to breach explosive obstacles without resorting to requests for help to higher levels of command (see figure 12-2).

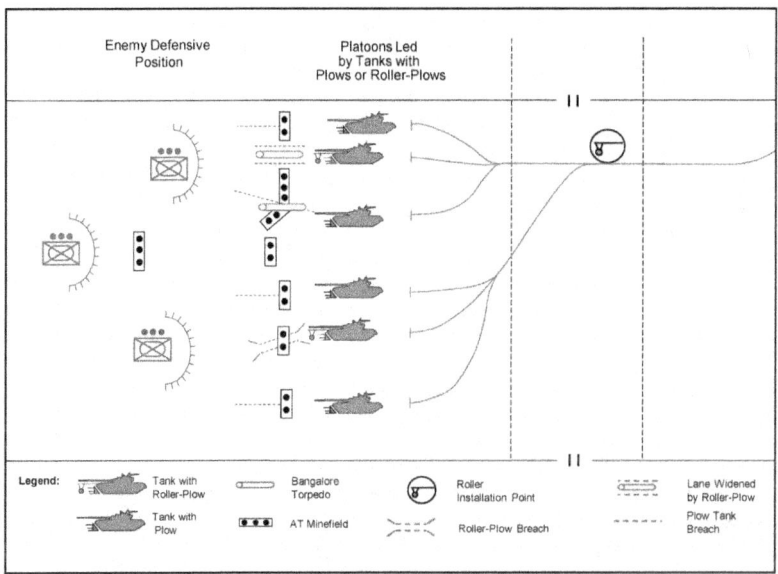

Figure 12-2. Mechanized breach (example)

12-57. Despite the advantages of mechanical means attached or integral to combat and combat engineer vehicles, it is still preferred that explosive breach means, whether mechanized or employed by infantry or engineer forces, be the primary method for executing a mechanized breach. This is because mechanical means place the combat vehicles at more risk. Mechanized explosive means are also the least vulnerable to booby traps placed in and around obstacles to make their breaching more difficult.

12-58. If at all possible, non-mechanized and/or affiliated irregular forces will breach anticipated obstacles in advance of a mechanized force. Such forces typically employ C3D to prevent detection while creating the breach.

Chapter 12

Nonexplosive Obstacle Breaching

12-59. The breaching of nonexplosive obstacles is essentially the same as breaching explosive ones with these salient differences:
- Mechanical and manual breach methods will typically take precedence.
- Significant nonexplosive obstacles (large antitank [AT] ditches, rivers, or rubble from a collapsed multi-story building) will not be rapidly breachable by manual means, if at all.

12-60. During the offense, an MSD also creates lanes through nonexplosive obstacles. In this case, the MSD may require additional engineer augmentation beyond just countermine equipment. For example, it may employ obstacle-clearing vehicles to knock down berms. It may also use truck-launched bridges to cross AT ditches.

WATER OBSTACLE CROSSING

12-61. The enemy is expected to use rivers and other water obstacles for defensive purposes. In order to conduct (or continue) an attack, OPFOR maneuver units must often cross water obstacles whose opposite banks may or may not be occupied by the enemy. *Crossing* is a generic term identifying the site of a water obstacle crossing or the act of crossing. Crossing involves using bridges, ferries, fords, or amphibious combat equipment. The OPFOR identifies two methods of overcoming water obstacles:
- *Opposed crossing* (when expecting enemy contact).
- *Unopposed crossing* (when not expecting enemy contact).

12-62. Rarely would the OPFOR attempt the classic opposed water crossing (described below under Opposed Crossing) when fighting an opponent more capable than itself. However, there may be times when the OPFOR must cross rivers in territory occupied by the enemy. Even then, it would typically only attempt the opposed crossing if convinced of success and if the enemy did not believe the OPFOR would attempt the crossing. This crossing would be integrated into the overall battle plan and the INFOWAR plan.

12-63. More likely, however, when opposing a stronger force, the OPFOR would attempt to cross the river covertly at night or during inclement weather. This would allow the OPFOR to infiltrate units—a few vehicles at a time—across the river. The units would regroup at a designated area and continue operations in enemy territory. Engineer support for this may only be engineer reconnaissance of the river and routes. Engineers could also build (undetected) an underwater bridge out of sandbags, or make rafts rigged to transport vehicles.

Note. Aside from water obstacles, crossings can involve other kinds of gaps, such as ravines. These other kinds of gap crossing can employ some of the same engineer assets and methods used to overcome water obstacles.

12-64. The OPFOR also expects to make most crossings without the advantage of an existing bridge or convenient fording site. Therefore, engineers must be prepared to provide specialized bridging and amphibious transport (tracked amphibians and ferries) to facilitate a timely crossing.

Organizing Forces

12-65. There are no doctrinal constraints on task-organizing for mission success. The ability to allocate assets downward and to task-organize is restrained only by the availability of assets and the nature of the mission. The OPFOR normally designates functional forces for a water obstacle crossing as follows.

Crossing Force

12-66. The *crossing force* is essentially the exploitation force for the obstacle crossing. It is the force whose movement the operation is designed to facilitate.

Crossing Site Force

12-67. The *crossing site force* is the enabling force of the crossing. Its mission is to enable the crossing force to move rapidly through or over the obstacle and continue its mission.

Security Force

12-68. The *security force* has the same function as that in any offensive course of action. Security forces for obstacle crossings will typically have a strong air defense capability.

Crossing Zones and Sites

12-69. A *crossing zone* is a specialized form of area of responsibility (AOR). It is the AOR of the crossing force and is commanded by the crossing commander. Its size and orientation depend on—
- The nature of the obstacle.
- The number of crossing sites.
- The size force that needs to cross.
- The ability to neutralize the enemy.

Under favorable conditions, the crossing zone of an opposed crossing may be identical with the unit's attack zone.

12-70. At tactical group level, the CELT advises the commander on selection of crossing sites within the crossing zone based on reconnaissance of the obstacle and approaches to it. The number of crossing sites within a zone depends on—
- The tactical situation.
- The nature of the water obstacle and surrounding terrain.
- The types of crossing equipment available.

There are usually separate sites for each type of crossing means: swimming, fording, snorkeling, tracked amphibian, ferry, and bridge. Especially for opposed crossings, preference is given to those sites where there are—
- Relatively weak enemy defenses.
- Concealed movement routes to the water obstacle.
- A bend toward the attackers.

12-71. Crossing site force commanders have responsibility for the conduct of the crossing and the tactical arrangement and security of the crossing zone. Crossing force units are placed in a supporting relationship to the crossing site commander while within the crossing zone.

Categories

12-72. The width of the water obstacle affects the method of crossing, the type of crossing, the need for augmentation, and the length of time to conduct the crossing. In terms of width, water obstacle categories are—
- Narrow (less than 100 m).
- Medium (100 to 250 m).
- Wide (250 to 600 m).
- Large (greater than 600 m).

12-73. In terms of depth, the categories are—
- Shallow (up to 1.5 m in depth).
- Deep (1.5 to 5 m in depth).
- Very deep (over 5 m deep).

12-74. Although canals are narrow obstacles, engineers place them in a special category because their deep water and steep banks make it difficult to use tracked amphibians, ferries, and standard bridging equipment. It is often necessary to erect piers and special constructions to negotiate them.

Reconnaissance

12-75. Depending on the situation, an ERP, the reconnaissance element of an MSD, or engineers augmenting other division and brigade reconnaissance elements can reconnoiter a water obstacle. The reconnaissance includes determining—
- The depth, width, and current velocity.
- The composition of the bottom.
- The presence of underwater obstructions or mines.
- Possible fording, ferrying, bridging, and snorkeling sites.
- The composition, height, and slope of the banks.
- Approach and exit routes.
- The camouflage potential of the area.
- The presence and nature of obstacles on the banks.
- Critical terrain features overlooking both banks.
- Information on the nature of enemy fortifications and defensive positions.

The engineers transmit this information to the CELT for planning purposes. They mark recommended crossing sites, bypasses, routes, and critical areas for the follow-on engineer elements responsible for establishing the crossing.

12-76. The division's engineer battalion has qualified divers with scuba gear; specialized vehicles and equipment to analyze soil data, stream velocities and depth; and mine-detection equipment. Commanders can also use maps, aerial photographs, engineer and fighting patrols, radars, signals reconnaissance, and human intelligence to gather data on crossing sites.

12-77. The number of ERPs depends on the width of the water obstacle and the number of required crossing sites. Patrols vary from squad to platoon size. The patrols can be equipped with one or more of the following types of vehicle:
- Amphibious scout cars, APCs, or IFVs.
- Tracked amphibians.
- Special engineer reconnaissance vehicles.

Planning and Preparation

12-78. Based on reconnaissance, the commander organizes his unit to ensure the most expedient crossing and continuation of the offense. When approaching a water obstacle, he selects his unit's formation based on the mission, enemy, and terrain. Constituent and dedicated engineer assets typically deploy well forward. Mechanized infantry units lead, while fire support and direct air support elements deploy forward to overcome expected enemy resistance on the line of the obstacle. As in an ordinary attack, this involves lateral deployment of the formation as late as possible and immediately before assaulting the water obstacle. Direct air support is more critical during water obstacle crossings than during other types of ground combat action.

12-79. Units engaged in a water obstacle crossing are extremely vulnerable to enemy aviation. Therefore, there is a need for air defense at crossing sites before a crossing is attempted. In some tactical situations, air defense assets may move across first to maximize the range of their weapons to protect subsequent units

Engineer Support

making the crossing. Placement and movement sequence of air defense assets varies as the commander assesses each crossing individually. (See Air Defense below.)

12-80. Crossing of water obstacles always requires some measure of engineer preparation, even if it is only limited to engineer reconnaissance at the crossing site. Whenever possible, the OPFOR attempts to cross water obstacles with minimum delay and press the attack into the enemy's depth without first halting to consolidate on the far bank.

Means

12-81. The OPFOR places high priority on the fielding of water obstacle-crossing systems. Any obstacle that slows tactical movement causes a concentration of forces and invites destruction. To ensure a rapid advance, many OPFOR APCs, IFVs, and reconnaissance vehicles are amphibious, as are some self-propelled artillery and tactical air defense systems. Therefore, the OPFOR employs amphibious combat vehicles and specialized water obstacle-crossing systems at the division, brigade, and tactical group level whenever possible. If APCs and IFVs are amphibious, virtually all vehicles within mechanized infantry or tank battalions would have either an amphibious or snorkeling capability. During crossings, tracked amphibians are primarily used for carrying towed artillery pieces, trucks, small vehicles, and troops. When not engaged in a crossing, they may be used as tracked cargo carriers. The OPFOR recognizes the need for tactical water obstacle-crossing assets during all types of combat and ensures sufficient assets are readily available in engineer units at higher levels.

12-82. The OPFOR crosses some narrow water obstacles by fording, by swimming with amphibious combat vehicles, or by using tank- or truck-mounted and low-water bridges. Other narrow obstacles (up to 100 m) and medium obstacles require tracked amphibians, ferries, or pontoon bridges. Wide and large water obstacles require tracked amphibians, ferries, or pontoon bridges (sometimes configured as rafts or ferries). Crossing large water obstacles may necessitate the use of heavy floating bridges or girder bridges erected by special-category engineers of the strategic-level transportation services.

12-83. The characteristics of the water obstacle mainly determine the method chosen for the crossing. (Table 12-4 lists the preferred crossing methods, depending on the type of obstacle.) However, the nature of enemy defenses, the mission, and the availability of engineer systems are also factors.

Table 12-4. Preferred water obstacle-crossing methods

Water Obstacle Characteristics	Preferred Crossing Method
Depth <1.5 m	Ford
Depth >1.5 m	Ferry or bridge
Width <20 m	Tank- or truck-launched bridge
Width 20-100 m	Pontoon bridge
Width >100 m	Ferry or tracked amphibian

Execution

12-84. Typically, a division or DTG crosses a major water obstacle with crossing forces consisting of brigades and/or BTGs operating in separate crossing zones. A typical brigade or BTG crossing zone is up to 10 km wide, with two to three detachments crossing first. A brigade's or BTG's combat elements typically can cross a significant water obstacle in approximately 2 to 3 hours. Assuming that not all subordinate brigades or BTGs can cross simultaneously, it may take approximately 5 or 6 hours for a division's or DTG's combat elements to cross a significant water obstacle.

12-85. C3D is the primary consideration in conducting a water crossing. The OPFOR is aware that even a crossing considered "unopposed" is vulnerable to air and missile attack. The OPFOR will make every effort to conceal and protect crossing units, sites, and means from detection and attack.

Opposed Crossing

12-86. Opposed water crossings are the least preferred method of overcoming an obstacle. This type of crossing requires secrecy, surprise, and high speed supported by C3D and direct and indirect fire. To preserve the secrecy of the intended crossing and its location, the OPFOR generally uses minimal preparation or construction prior to its execution. It emphasizes conducting the crossing while moving as swiftly as possible and then continuing the offense on the opposite bank.

12-87. In a mechanized crossing, the OPFOR maximizes the speed and maneuverability advantages of combat vehicles. In the initial wave of the lead elements, amphibious APCs or IFVs make a rapid amphibious crossing to seize a bridgehead on the far bank. The crossing is usually covered from the near bank by all available fires and usually takes place either at night or under a smokescreen. These fires include direct artillery and tank fires, as well as all available indirect fires. Direct air support, generally fixed-wing, is more critical during water obstacle crossings than during other types of ground combat action. Heliborne (or, less probably, airborne) forces may seize and hold a bridgehead on the far bank.

12-88. In non-mechanized opposed crossings, C3D generally takes a greater role. Feints and demonstrations may be used to confuse the enemy as to the actual crossing zone. Low-visibility conditions are also ideal for conducting both opposed and unopposed crossings.

Unopposed Crossings

12-89. After an opposed crossing, the OPFOR can move company- or platoon-size pontoon bridge units to the crossing site. If preceding units (including the security detachment of a brigade or BTG) have eliminated enemy resistance at the water obstacle, battalions or BDETs in the main body of a brigade or BTG can conduct an unopposed crossing. If the brigade(s) or BTG(s) must conduct an opposed crossing and are successful, this allows the division's or DTG's follow-on forces to conduct an unopposed crossing.

12-90. **Bridges.** Bridge crossings are a typical feature of unopposed crossings. Construction of bridges starts when the enemy is denied the ability to subject the crossing to direct or observed fire. Bridges have greater load-bearing and throughput capacities than other crossing means and are preferred in order to maintain high rates of advance.

12-91. The division or DTG commander may send out an independent mission detachment (IMD) ahead of the security detachments of his lead brigades or BTGs when there is an opportunity to seize a bridgehead over an undefended or poorly defended water obstacle or bridge. A brigade or BTG can possibly send out an IMD of its own. In either case, the IMD attempts to bypass enemy resistance forward of the water obstacle and infiltrates to the far side of the water obstacle to establish a bridgehead.

12-92. If the air situation is unfavorable, the OPFOR may only use bridges during periods of limited visibility. At other times, it would tuck the bridges into the bank and camouflage them.

12-93. The crossing commander designates traffic control points (TCPs), OPs, and work teams at the crossing site. A bridge team is assigned to—

- Inspect pontoon couplings and bank moorings.
- Evaluate and repair damage.
- Monitor entry and exit of vehicles.

12-94. Low-water bridges can free pontoon bridges, ferries, and tracked amphibians for use in other crossings. Low-water bridging is relatively permanent, using piling to provide support.

12-95. **Ferries.** Ferry crossings are used to transport nonamphibious heavy equipment across medium to wide water obstacles. This usually requires three to four ferries per site. Ferries can be joined into a pontoon bridge or can be used as individual ferries. Individual folding pontoon bridge sections can also be used as ferries. Ferries are not usually employed in the initial wave of an opposed or unopposed crossing but rather in subsequent waves of tanks and other combat vehicles.

12-96. Ferry crossings begin can 15 to 20 minutes after the start of an opposed crossing. Ferries are launched from prepared ramps, and personnel from the ferry platoon create landing ramps on the far bank. Floating pontoon bridge rafts are maneuvered and positioned by powerboats.

12-97. Engineer missions in the unopposed crossing are the same as they are in the opposed crossing, but engineers are also assigned to prepare mooring and launching sites and to assemble the ferries. Based on engineer reconnaissance data, the crossing commander selects mooring sites, determines the number and disposition of ferries or pontoon bridges used as ferries, and plans traffic control. An engineer squad or a traffic control unit directs movement to the sites.

12-98. **Tank Snorkeling.** Snorkeling is attempted when fording, bridge, or ferry crossing sites are not available or when an opportunity for surprise exists. Snorkeling crossings are used only at water obstacles that have—

- Depths of 5.5 m or less.
- Prepared entry and exit points.
- Entry slope of 47 percent (25°) or less.
- Exit slope of 27 percent (15°) or less.
- Stream velocities of 3 m/s or less.
- Hard, level bottoms with no boulders, craters, or soft spots.

12-99. **Fording.** In opposed and unopposed crossings, the OPFOR establishes fords at shallow water crossing sites. Since fordings are not as complicated as other crossings, the unit may remain in tactical movement formation. If possible, multiple units cross simultaneously on a wide frontage.

12-100. Deeper fords can be undertaken by tanks without the use of snorkels but may require partial sealing of the tanks up to the turret ring. When partially sealed, air for the engine and crew is drawn in through open turret hatches. Deep fording by tanks is limited to depths not exceeding 2.3 to 2.5 m, depending on currents.

12-101. **Assembly and Preparation Areas.** Engineers assist in the preparation of assembly areas and of boarding and preparation areas near crossing sites. As units leave assembly areas, they pass through an engineer checkpoint (ECP). The ECP is a checkpoint to ensure that vehicles do not exceed the capacity of the crossing means. At the ECP, vehicle drivers receive final instructions on site-specific procedures and information, such as vehicle speed and interval. Near the ECP is the first of a series of TCPs to direct the unit to the appropriate crossing site and avoid bunching up at the crossing site or on the approach route.

12-102. Figure 12-3 on page 12-18 shows an example of a mechanized infantry battalion crossing supported by tracked amphibians, ferries, and a pontoon bridge. In this example, two companies cross by amphibious means, while the third company and support elements are able to cross over a pontoon bridge in tactical movement formation. Normally, bridges are erected only after the far bank is secured to a depth precluding enemy direct fire on the crossing site. However, if the enemy defense has been neutralized by fire or the opposite bank has been seized by airborne or heliborne forces, bridge construction may begin along with the opposed crossing.

Chapter 12

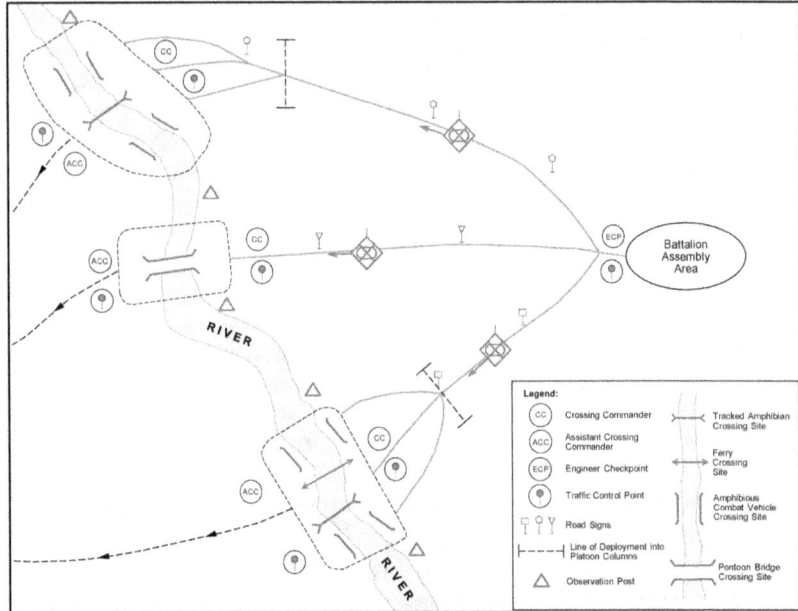

Figure 12-3. Engineer support of a mechanized infantry battalion crossing (example)

Air Defense

12-103. Forces conducting water obstacle crossings are high priority targets for enemy air strikes. Thus, the importance of air defense increases. The mission of air defense units is to protect the airspace above and around the crossing site. (For more information on air defense, see chapter 11.)

12-104. The OPFOR expects that most water obstacle crossings will be opposed by enemy air and ground defense and made without the advantage of an existing bridge, convenient fording sites, or defensive air cover. Accordingly, doctrine calls for conducting such opposed crossings rapidly, without slowing the pace of the offense. The leading maneuver battalions and BDETs (or brigades and BTGs) may have to provide their own air defense while crossing and then cover the crossing of follow-on forces.

12-105. For unopposed crossings, the maneuver unit may cross a water obstacle without deploying from its movement formation. In this case, division-, brigade-, or tactical group-level air defense units can establish firing positions on the near bank to cover the site for the time it takes the column(s) to cross. The air defense units then move to the far bank and either establish firing positions or continue to move.

Smoke

12-106. Most opposed crossings are conducted under the cover of smoke or other obscurants, which can degrade the enemy's ability to locate and target the actual crossing sites. However, smoke can also degrade or prevent visual acquisition of air threats by some OPFOR air defense systems. Therefore, wind and obscurant conditions must be accounted for in the deployment of air defense weapons for the crossing. (For more information on smoke employment, see chapter 13.)

Engineer Support

COUNTERMOBILITY

12-107. The OPFOR makes extensive use of countermobility operations to limit access and control tempo by delaying, disaggregating, and canalizing enemy forces. The obstacle plan is completely integrated with the maneuver, fire support, and INFOWAR plans.

12-108. OPFOR engineer obstacles include any actions taken to inflict losses and to delay and impede enemy movement. The creation of engineer obstacles and execution of demolition activities are critical engineer functions in all phases of the battle.

12-109. Countermobility support is extremely innovative, especially when the OPFOR fights a more powerful enemy. Minefields may be irregular-shaped and are thoroughly merged with the terrain. They also tend to be much smaller than those laid in linear operations, which may easily be over a kilometer in length. Some examples of adaptive engineer countermobility methods likely to be employed against a more powerful force (in addition to methods used in a more conventional battle) are—

- Lay mines intermittently along road or trails. This involves the enemy in prolonged, potentially dangerous and time-consuming detection and clearance operations, and requires a great deal of enemy manpower.
- Mine and re-mine enemy lines of communications (LOCs). This requires the enemy forces to constantly sweep for mines. Once a road is swept and left unsecured, the OPFOR re-mines it.
- Limit access by denying the enemy key facilities. For example, destroy airfield runways in an aerial port of debarkation (APOD) or docks in a sea port of debarkation (SPOD).
- Deny LOCs from APOD and SPODs to enemy maneuver units, staging areas, or base camps. Contain (or trap) enemy forces in specific areas such as an APOD or SPOD and built-up areas.
- Maximize the use of controlled minefields. This lets the OPFOR pass through the minefield and activate it prior to the arrival of enemy units. It can also be used to trap enemy units. This is used in conjunction with artillery as a kill zone.
- Use off-road and chemical mines whenever possible. Always use antihandling devices to slow clearing efforts.
- Target vehicle mine plows and rollers as high-priority targets.
- Use plastic mines to defeat mine detection sweeps.
- Plant underwater mines at port or ford sites.

OBSTACLE DETACHMENT

12-110. The OD is the basic building block of the OPFOR's countermobility effort. ODs are temporary, task-organized groupings composed primarily of engineer assets intended to create minefields and obstacles. Their basic equipment includes mechanical minelayers and trucks carrying mines, explosives, and other equipment. They are sometimes augmented with mechanized infantry troops for close protection and extra labor. The size and composition of the OD depend on the tactical situation and the needs of the maneuver commander.

12-111. In addition to minelayers, ODs may add other engineer resources to support critical obstacle development. The division may supplement the OD with engineers for demolition work, ditchers to create AT ditches, and other engineer systems. This augmentation does not normally occur until the earthmoving equipment completes other tasks, such as preparing fortifications.

Missions

12-112. The OD uses its ability to rapidly lay minefields and construct obstacles to—
- Protect flanks.
- Strengthen captured positions.
- Disrupt attacks, counterattacks, and other enemy activities.
- Strengthen the defense.
- Cover gaps between battle positions.

Chapter 12

- Deny the enemy access to key terrain.
- Block the axis of an enemy armored advance.
- Block enemy penetrations.
- Block enemy reinforcements, exploitation forces, or reserves.
- Channel the enemy into a kill zone and contain him there.
- Protect the flank of a counterattacking force.

12-113. In the defense, the OPFOR commander may hold the OD and other forces in reserve and can quickly employ them during an enemy attack, to mine potentially vulnerable gaps. Engineer tasks during the defense implement obstacle plans, particularly AT obstacles. Together with ATRs, ODs provide a quick-reaction AT force to block enemy penetrations.

12-114. Engineers can lay mines and construct obstacles in the disruption zone and on likely enemy armored avenues of approach. They can also lay obstacles in the depth of friendly units in the battle zone, and at subsequent defensive positions throughout the AOR. However, simultaneous obstacle construction throughout the AOR can only occur when sufficient time, equipment, and personnel are available. In any part of the AOR, minefields and other obstacles require barriers, security, and marked maneuver passages.

12-115. Engineers create obstacles on possible enemy approaches to OPFOR battle positions or artillery and air defense firing positions, in the gaps between battle positions, and on flanks. They normally construct barrier systems in coordination with the overall fire support plan.

12-116. In preparation for movement, a division, brigade, or tactical group creates one or more ODs to maximize mechanical minelaying and explosive obstacle support for maneuver forces. The OD provides countermobility support and denies key terrain to the enemy. Its mission is to alter the tactical situation by emplacing obstacles in response to enemy actions. During the tactical movement, the OPFOR commander's greatest concern is armor attacks against the flanks. Therefore, the OD emplaces AT obstacles in front of detected armor threats or along possible routes suitable for armored vehicles.

Positioning

12-117. Although the OD can act independently, the division, brigade, or tactical group often assigns it to move and act in close coordination with the ATR. Even in the latter case, the OD still reports directly to the engineer unit commander, who assigns its priorities, areas of concern, and task organization. This arrangement provides the maneuver commander with a combination of organizations capable of rapidly emplacing AT obstacles as well as covering the obstacles with AT fires. A minelaying squad of an engineer mine warfare platoon or minelayer platoon usually serves as the core of an OD.

12-118. While conducting tactical movement, the OD may travel behind elements of the security detachment and in front of the main body. Sometimes, it may move on a threatened flank or forward within the main body, ready to deploy to either flank.

12-119. Following the maneuver commander's guidance, the CELT recommends positioning of the OD so it can quickly deploy in response to enemy actions. This may be to seal a critical area or to provide time to shift forces and fires. The maneuver commander, the engineer liaison team(s), and other staff elements monitor the progress of the tactical movement and plan for possible enemy courses of action. They then identify possible deployment lines for the ATR and obstacle-emplacement locations for the OD. If reconnaissance assets report enemy activity along a given axis that confirms a course of action, the commander dispatches an OD and an ATR to the appropriate deployment line to conduct their missions.

OBSTACLE PLANNING

12-120. The obstacle plan is tailored and integrated into the overall operation plan with mutually supporting systems of fire. This integration is exemplified by the habitual association between the OD and the ATR. Just as it develops a fire support plan, the OPFOR also develops an integrated obstacle plan tailored specifically to each unique tactical situation. In the offense, obstacles protect flanks, disrupt counterattacks, and strengthen captured positions. In the defense, engineer obstacles may strengthen the defense, disrupt enemy operations, and cover gaps.

Engineer Support

12-121. To develop the obstacle plan, the CELT conducts an evaluation of the situation from an engineer perspective. From this evaluation, he determines engineer allocations and priorities, and directs obstacle development and other engineer preparation.

12-122. The OPFOR divides engineer obstacles into three categories:
- Explosive obstacles—minefields, groups of mines, and objects prepared for demolition.
- Nonexplosive obstacles—AT ditches, escarpments, abatis, wire barriers, and water obstacles.
- Combination obstacles—a combination of explosive and nonexplosive obstacles.

12-123. Of the three categories, explosive obstacles are the most common. Engineers and others can emplace minefields more easily and quickly when compared to the construction effort for nonexplosive obstacles. Additionally, the OPFOR plans for the self-destruct or self-neutralization capabilities frequently found in scatterable mines. It can also lay mines with remote-control devices to activate or deactivate the minefield at will. This minimizes the adverse effect of friendly minefields on future actions and reduces the need for the OPFOR to breach its own obstacles.

12-124. However, this is not the case with nonexplosive obstacles, which are time- and resource-intensive to install and eliminate. For these reasons, the OPFOR usually emplaces mines and other explosive obstacles first, and eventually supplements them by constructing nonexplosive obstacles. When this occurs, it creates combination obstacles, which are the next most common after the explosive type. It is extremely rare for the OPFOR to use a nonexplosive obstacle in isolation without any mines, explosives, or booby traps.

Explosive Obstacles

12-125. The OPFOR emphasizes the use of explosive obstacles. These include mines and demolitions. The widespread use of landmines on today's battlefields results from a combination of mass production, plastic mines, improved battlefield delivery systems, and development of sophisticated fuzing. Remotely delivered mines have expanded capability for changing the tempo of battle.

Mines

12-126. Mines are the most significant obstacles the OPFOR can employ and are usually emplaced in groups or in minefields. Therefore, minefields and minelaying are afforded separate sections below.

Demolitions

12-127. The OPFOR emphasizes the importance of roads as high-speed avenues of attack for both friendly and enemy forces. Therefore, it views the use of demolitions on roads as a significant way to disrupt enemy movement. Critical points at which the OPFOR might use demolitions include—
- Overpasses.
- Bridges.
- Ravines.
- Intersections.
- Bypasses.
- Approaches to water obstacles.
- Roadways through urban or other complex terrain.

Nonexplosive Obstacles

12-128. Nonexplosive obstacles fall into three categories: AT, antipersonnel (AP), and antilanding. Nonexplosive AT obstacles include ditches, dragon's teeth, and various other manmade and natural barriers. AP obstacles include concertina and barbed wire. Antilanding obstacles include dragon's teeth, AT ditches, and wire obstacles. The OPFOR uses these obstacles at potential drop or landing zones for amphibious, airborne, or heliborne assaults. The primary responsibility for the construction of nonexplosive obstacles rests with the maneuver unit. The diversion of water from rivers and streams or releasing water from dams can also cause a significant nonexplosive obstacle.

MINEFIELDS

12-129. The OPFOR frequently uses minefields during all phases of combat. There are five basic types of OPFOR minefield:
- AT.
- AP.
- Mixed.
- Decoy.
- Antilanding.

12-130. The OPFOR stresses the importance of covering minefields with both direct and indirect fires, particularly with long-range AT weapons. Minefields inflict damage on attacking enemy forces and slow and canalize enemy forces into kill zones covered by massed fires. Whenever possible, the OPFOR contains enemy forces in a window of vulnerability for the longest time possible. This facilitates the destruction of the enemy.

12-131. Conventional OPFOR minefields generally conform to doctrinal standards. This standardization ensures that engineers and combat personnel follow consistent, uniform practices. Scatterable minefields, however, are much less predictable in pattern. Maneuver commanders use combat personnel to emplace protective minefields around individual fighting positions and perhaps battle positions. Meanwhile, engineers use minefields to shape the battlefield for the maneuver commander.

12-132. Commanders of battalions, companies, or detachments emplacing mines prepare minefield records in three copies: one for the unit, one to the brigade or BTG, and one to the division or DTG. The CELT at division, brigade, or tactical group level then uses the records to prepare combined obstacle overlays for the maneuver commander at that level. Minefields are a fundamental part of the total obstacle plan that incorporates barriers and terrain features.

Antitank

12-133. AT minefields are the primary type of OPFOR engineer obstacle and serve to destroy or disable armored vehicles. They are primarily established in belts consisting of multiple rows on avenues that are favorable for tanks in front of the battle line and on the flanks. Where difficult terrain is available, minefield belts will be tied into terrain obstacles to reduce the mine requirement. They are also placed at unit boundaries and in the depths to cover artillery firing positions, command posts (CPs), and other key assets.

12-134. The OPFOR usually emplaces AT minefields on a frontage of 200 to 300 m or more and to a depth of 60 to 120 m. The mines are laid in three or four rows with approximately 20 to 40 m separating each row. The normal spacing between AT mines in the rows is 4 to 5.5 m for pressure-activated mines, and 9 to 12 m for full-width-attack mines. The normal mine outlay for 1 km of frontage in AT minefields is usually 300 to 400 full-width-attack mines, or 550 to 750 pressure-activated mines. This mine outlay can reach 1,000 or more AT mines per km of frontage on major avenues of approach. The OPFOR refers to this density of mines as a "minefield of increased effectiveness."

12-135. Figure 12-4 illustrates the general emplacement of an AT minefield with track-attack mines. Figure 12-5 shows an AT minefield with full-width-attack mines.

Engineer Support

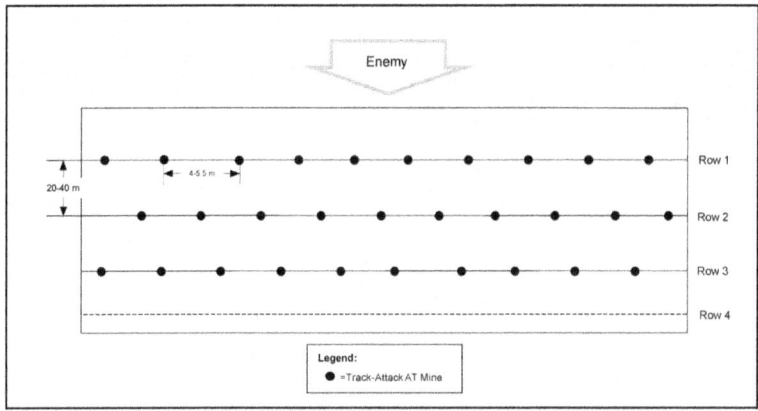

Figure 12-4. AT minefield configuration with track-attack mines (example)

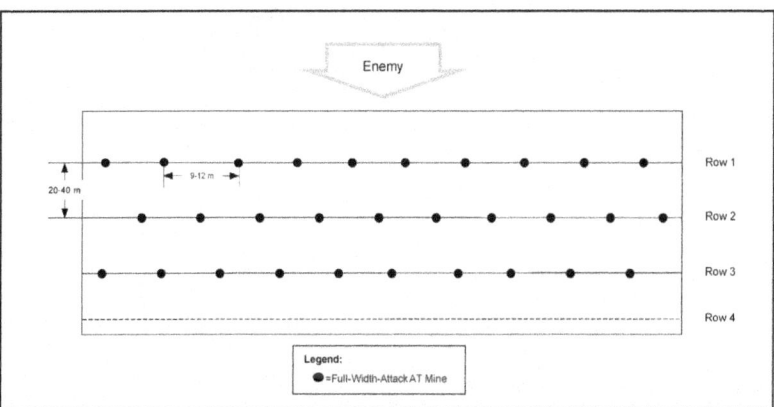

Figure 12-5. AT minefield configuration with full-width-attack mines (example)

12-136. In urban environments, the OPFOR may place groups of AT mines on narrow streets and alleys. It calculates emplacement of AT mines at the rate of one mine per 100 m of street or alley.

Antipersonnel

12-137. On the battlefield, the modern AP mine is used to—
- Inflict personnel casualties.
- Hinder soldiers in clearing AT minefields.
- Establish defensive positions.
- Deny access to terrain.

Chapter 12

12-138. The OPFOR can set up conventional AP minefields on the forward edge of friendly defensive positions, in front of AT minefields, or along dismounted avenues of approach. These minefields can consist of blast mines, fragmentation mines, or a mixture of the two. The OPFOR emplaces AP minefields on a frontage of 30 to 300 m or more with a depth of 10 to 50 m or more. It usually lays AP mines in two to four rows with a distance of 5 m or more between rows.

12-139. The OPFOR may emplace 2,000 to 3,000 blast and 100 to 300 fragmentation mines per km of frontage. An AP minefield of increased effectiveness may have as much as three times the normal outlay of AP mines. Intervals between mines in a row are at least 1 m for blast mines and up to twice their destructive radius for fragmentation mines. Figure 12-6 shows variations of the employment of AP minefields.

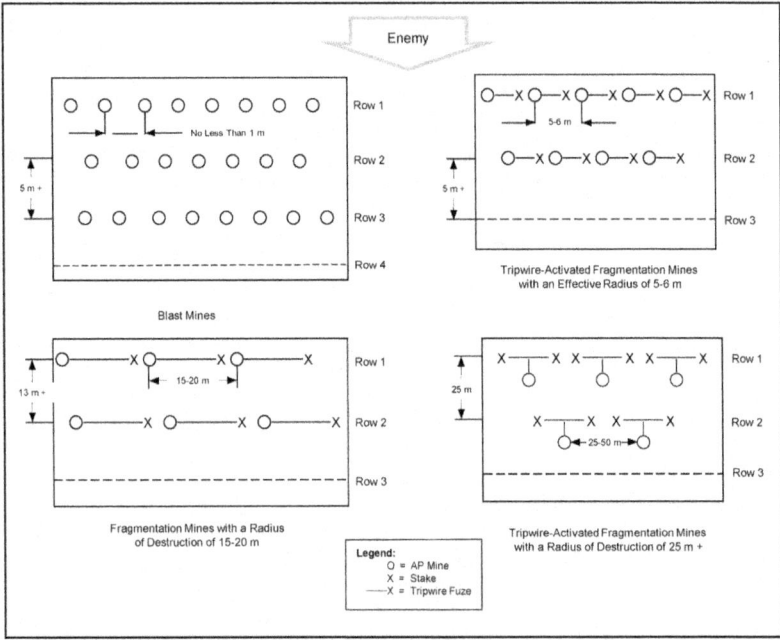

Figure 12-6. AP minefield configuration (example)

12-140. Emplacement of minefields with increased effectiveness is more likely on dismounted avenues of approach. In urban environments, the OPFOR can emplace 2 to 3 fragmentation mines for every 50 to 100 m of street. It prefers to use blast mines and fragmentation mines within buildings.

Mixed

12-141. Mixed minefields contain both AT and AP mines. A mixed minefield is generally viewed as a minefield with pure homogenous rows of either AP or AT mines. This is mainly due to the physical constraints of mechanical minelayers. They cannot lay both AT and AP mines in the same row. This does not preclude mixed minefields from having a mixture of both AT and AP mines. They can be laid manually or remotely. It is easy to remotely "seed" an area with a combination of both. However, the AT mine requirements govern the mixed minefield's parameters, outlay, and density. In areas that are not suitable

Engineer Support

for armored vehicles, AP mines constitute the majority of mixed mine obstacles. Figure 12-7 shows an example of a mixed minefield with blast AP rows between AT rows.

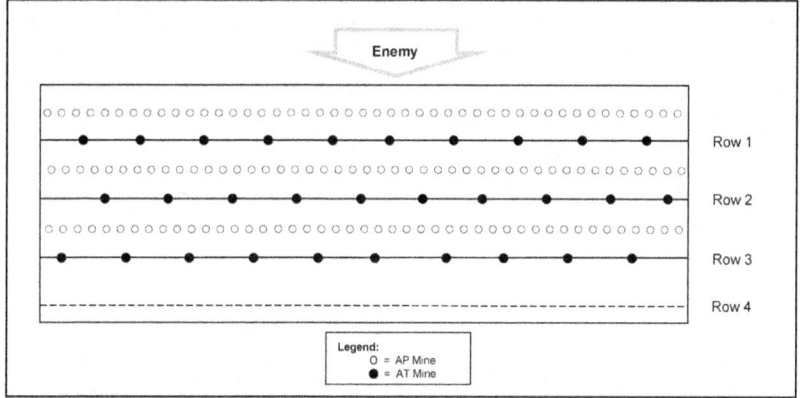

Figure 12-7. Mixed minefield with blast AP rows between AT rows (example)

12-142. Combat arms personnel set up nonexplosive and mixed minefield obstacles to cover their defensive positions. Engineers lay mixed minefields in front of the battle line and on primary avenues into the defensive depth. Mixed minefields are usually established in front of unit positions that are transitioning to the defense. Figure 12-8 illustrates an example of a mixed minefield with an AP minefield leading to full width AT minefield.

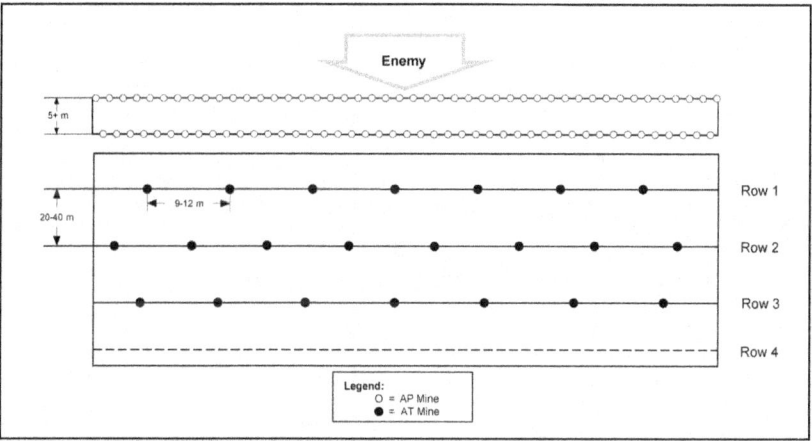

Figure 12-8. Example mixed minefield with an AP minefield leading to a full-width AT minefield

Decoy

12-143. Decoy minefields are a significant form of deception to slow movement or deceive as to true unit locations. The OPFOR uses decoy, or false minefields to mislead the enemy as to the locations of actual minefields. As part of tactical deception, units typically give the impression of minelaying activity, usually scarification of the soil, minelaying debris, minefield fences, and markers.

Antilanding

12-144. Antilanding minefields prevent landings by amphibious, airborne, or heliborne assault forces. The OPFOR uses antilanding mines at possible landing or drop zones (LZs or DZs) or when conducting combat along the seacoast or inland water features. It employs explosive, nonexplosive, and combination obstacles. Minefields established in the water consist of bottom and anchored mines and, at shallower depths, waterproof mines. The OPFOR uses all types of mines above the shoreline, emplacing them following normal minefield doctrine. At LZs and DZs, it uses fragmentation and directional AP mines. It also emplaces antihelicopter mines in locations it believes will be used as firing positions for enemy attack helicopters or in possible LZs to be used by lift helicopters.

Controlled

12-145. Many OPFOR units have the capability to lay controlled minefields. These minefields consist of landmines with electronic switches (on/off) giving the operator control over the operational status of the minefield. The operator can change the status of an area of the battlefield and either make it hazardous for the enemy or render it safe for friendly troops. This is done either by a direct hardwire link or by radio. An entire minefield can be emplaced and turned on or off, as necessary, to best support friendly operations.

12-146. On a smaller scale, select passages in a conventional minefield can contain controllable landmines, allowing for the option of clearing safe lanes for friendly use. The addition of selectable anti-removal and self-destruct features to controlled mines enhances flexibility and overall effectiveness. Controlled minefields can also be established in a maneuver defense to ensure unrestricted maneuver of units over mined areas and to cut off enemy units in pursuit.

MINELAYING

12-147. The means of emplacing minefields can be manual, mechanical, or remote. Since minelaying is a common task skill, manual emplacement is performed by anybody and is the method employed by maneuver units. However, manual minelaying is labor-intensive and requires the expenditure of more time than may be available during high-speed maneuver. Therefore, OPFOR engineers may have towed and/or tracked conventional mechanical minelaying vehicles that can quickly emplace both buried and surface-laid minefields. The engineers may also have vehicle-mounted scatterable minelaying systems. These mechanical systems to allow engineers to quickly mine an area just prior to or during the battle. Engineer resources are supplemented by remote mine delivery from artillery and aircraft. Infantry units can also have man-portable remote mine dispensers.

12-148. The methods and extent of minelaying depend on—
- The OPFOR's intentions.
- The tactical situation.
- Terrain characteristics.
- The type of mine.
- Time available.
- Available engineer support.

12-149. With the high tempo of the modern mobile battlefield, the use of remotely delivered mines is increasing. In volume, however, they do not exceed the use of conventional landmines. Conventional minefields are better suited to protecting defensive positions that the OPFOR intends to maintain for some time. In this case, it expends greater time and effort to bury and camouflage the mines and integrate the minefields into the total defensive scheme. Mine density in these types of fields is also greater. These

Engineer Support

minefields are more likely to have a mix of AT and AP mines. In setting up a fully prepared defense, troops of all units take part in preparing obstacles and laying mines.

Manual

12-150. The OPFOR manually emplaces minefields when—
- There is no contact with the enemy.
- Mechanical minelayers are unavailable.
- Use of mechanical minelayers is inadvisable because of terrain restrictions.

A mine warfare platoon can manually lay 200 to 300 AT mines in 1 to 2 hours. It can recover about 200 AT mines an hour, if the mines are not equipped with self-destruct or antihandling devices.

Mechanical

12-151. OPFOR engineers rely extensively on mechanical minelayers. These can bury or surface-lay AT mines. The layout of mechanically emplaced minefields is the same as those emplaced by hand.

12-152. The normal sequence for mechanically laying mines is to emplace the most forward minefield first and to work progressively back to friendly defensive positions. The engineers align the mechanical minelayers parallel to the battle line. The minelayers start at separate intervals. This staggers the minelayers in a 30- to 45-degree echelon formation as they travel along the battle line. This method ensures that mines in one row are not directly behind those in another when approached by the enemy. This increases the probability for a mine encounter by ensuring that, if an attacker misses the first mine, he should still encounter one in subsequent rows. Mines can also be emplaced by helicopters or vehicles with the use of chutes (slides). Mine chutes can also be used to assist manual burial emplacement or to surface-lay mines. Figure 12-9 illustrates the mechanical AT minelaying sequence.

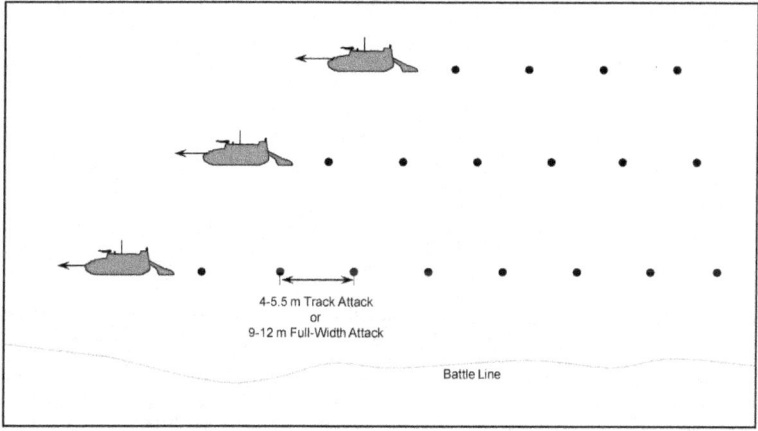

Figure 12-9. Mechanical AT minelaying sequence (example)

Remote

12-153. Remote minelaying gives the OPFOR a capability to strike targets and rapidly project mines deep in enemy territory or anywhere in the AOR. It provides increased tactical responsiveness and flexibility and reduces the manpower requirement for minelaying. It also minimizes exposure of the minelaying equipment to enemy fire. Once emplaced, these minefields can instantaneously affect the tactical situation and degrade the enemy's reaction time to the sudden appearance of the obstacle. Thus, they are ideally suited to the highly mobile, lethal battlefield of modern warfare. Since many scatterable landmines feature self-destruct and antidisturbance fuzing, they are well suited for missions that deny terrain for a specific period.

12-154. Remote delivery is useful against enemy reinforcing units, areas of troop concentration, CPs, firing positions, and other objectives. It can protect flanks or block enemy penetrations. The OPFOR prefers to trap a force inside a minefield rather than merely create an obstacle that the enemy can bypass. Therefore, remote minelaying can create two types of minefields: covering and containing. A covering minefield can block the movement route of an advancing or withdrawing enemy. A containing minefield can prevent the enemy unit from moving out of a deployment area (or firing position) or within the area.

12-155. The OPFOR uses remotely delivered or scatterable mines to reduce enemy mobility, inflict losses, and to create the conditions for the destruction of an enemy force. Scatterable mines also have a psychological effect on enemy morale. They can be laid close to friendly positions and covered by friendly fire or laid deep in enemy territory. Minefields created by scatterable mines lack precise boundaries or a definitive mine emplacement pattern, and generally remain on or near the surface of the ground.

12-156. Scatterable mines can be delivered from jet aircraft, helicopters, multiple rocket launchers (MRLs), trucks, and other ground vehicles, or by dismounted soldiers in the forward parapet of fighting positions. These mines are scattered randomly on the ground with no semblance to classical patterns. Frequently, these types of landmines also incorporate self-destruct or self-neutralization features to control and limit their active duration once emplaced. After the allotted time has expired, the terrain can again be used by friendly forces.

12-157. The OPFOR prefers to retain the element of surprise and therefore employs remotely delivered, scatterable mines immediately before combat or during the course of the battle. When dispensed, these minefields are generally laid immediately in front of attacking, reinforcing, and withdrawing enemy troops, or may be emplaced directly on enemy formations. Possible uses of scatterable mines are to—

- Isolate enemy forces.
- Disrupt the attacking forces, causing them to deploy early and expend mineclearing assets.
- Disrupt and delay enemy exploitation forces, reserves, or counterattack forces.
- Prevent enemy artillery from displacing during counterbattery fire.
- Interdict LOCs.
- Prevent the use of a logistics site.
- Obstruct a choke point.
- Protect flanks.
- Seal breaches in friendly obstacles or gaps between units.

Artillery

12-158. Some cannon artillery systems are capable of delivering both AP and AT mines. However, MRLs are the primary means of remote minelaying. The principal advantage of MRL mine delivery is its ability to quickly emplace large minefields in a single volley, while minimizing exposure to enemy targeting and weapon systems. For example, a single volley from a 220-mm MRL battery can deliver over 2,300 AT scatterable mines to a range of 10 to 35 km. With these mines, it can emplace a covering minefield approximately 3 km wide or a containing minefield about 1,200 m wide and 1,200 m deep. For additional information on artillery systems and mine delivery capabilities, see the *Worldwide Equipment Guide*.

Infantry

12-159. The OPFOR may also employ small, man-portable remote mine dispensers with its lower-level infantry units. These mine dispensers, weighing only a few pounds, can be either pulled like a handcart or carried by a soldier. Depending upon the design, infantry remote minelaying systems propel mine canisters out to approximately 100 m, while rocket-dispensed systems may go out as far as 3,000 m. The operator loads the mine-filled rockets (or a propelling charge and mine canister) into the launch tube, mounts the system on the edge of a trench or firing parapet, aims the tube, connects the wire to the tube, moves off to a safe distance, and connects the wire to the blasting mechanism. With a trained operator, it takes only a few minutes to set up the dispenser and create a minefield. For additional information, see the *Worldwide Equipment Guide*.

12-160. Infantry-fired ground dispensers are ideal for installing small, defensive, AP or AT minefields. They allow low-level units to remotely emplace minefields to protect their battle positions, flanks, and boundaries between units, or to cover firing lines and gaps in combat formations. They can quickly close breaches in existing protective minefields and increase the density of mines on armor avenues of approach.

Ground Vehicles

12-161. In recent years, the trend has been to mount scatterable-mine dispensers on ground vehicles. Both AP and AT mines can be launched from ground vehicles. This also gives the engineers the ability to re-seed or reinforce an obstacle without entering the minefield itself.

Aerial

12-162. Other than the above-mentioned ground force systems for remote minelaying, fixed-wing aircraft and helicopters can remotely deliver mines. The following paragraphs describe aerial minelaying capabilities.

12-163. Bombers or ground-attack aircraft can lay remotely delivered minefields throughout the AOR. Aircraft are only used to deliver mines beyond the battle line when they cannot be delivered (for whatever reasons) by indirect fire means. Delivering mines from low altitudes deep in enemy territory increases the risk of losing the aircraft to enemy ground fires or air defense. Therefore, fixed-wing aircraft are generally used to deliver ordnance such as scatterable mines beyond the range of OPFOR artillery systems (including MRLs). Ground-attack aircraft lay these minefields in the enemy's tactical depths.

12-164. High-performance aircraft can lay mines at a speed of 400 to 800 km/h from an elevation of 50 to 200 m. Aircraft-delivered scatterable mine canisters are dropped on parachutes. The canisters are set to burst open at a predetermined height to scatter the mines.

12-165. Helicopter minelaying systems are used to emplace large barrier minefields in the execution of offensive or defensive maneuver plans. This type of aerial minelaying is normally conducted over friendly territory—along the flanks or in the rear areas. When supporting an airborne or heliborne landing, helicopters may lay mines on enemy territory.

12-166. Aerial minelaying systems can lay both AT and AP minefields. Some attack and medium-lift helicopters and a few light helicopters have the capability to perform minelaying missions. A light helicopter does not carry armament when accomplishing these missions. Medium-lift and attack helicopters are most commonly used for aerial minelaying.

12-167. Some helicopters can dispense conventional mines by the addition of an internal minelaying chute within their cargo area. Mines are contained on custom mine racks and then fed manually or automatically onto the chute. The mines are then armed as they are dispensed. These heliborne minelaying systems can be used in formations of two, three, or four and operate relatively similar to the ground mechanical minelaying systems. The speed of minelaying can vary from 20 to 300 km/h from an altitude of 30 to 100 m. While at a near hover above the ground, the helicopters operate 20 to 40 m apart, with each laying a straight-line row. The mines in each row can be staggered and the distance between mines depends on whether the mines are pressure-initiated or full-width attack mines.

12-168. One light helicopter can deliver 60 to 80 AT mines or 100 or 120 AP mines. A medium-lift helicopter equipped with a minelaying system delivers 100 to 140 AT mines or 200 to 220 AP mines on a single sortie. To lay a minefield 15 by 30 m with AT mines takes approximately six sorties of a single light helicopter or approximately four flights of a single medium-lift helicopter.

SURVIVABILITY

12-169. The construction of battle and fighting positions is a labor-intensive process and is therefore a shared responsibility of engineers and supported units. The OPFOR makes maximum use of civilian engineer assets and personnel. Full preparation of defensive positions involves—
- Fighting positions for individual and crew-served weapons.
- Fighting trenches and communication trenches.
- Firing positions for tanks, IFVs, APCs, air defense, and other weapon systems.
- Protection for CPs and logistics sites.

12-170. Survivability activities when fighting a more powerful opponent have several unique engineer requirements. Some examples are to—
- Take full advantage of screening, protective, and C3D techniques, along with careful selection of terrain to passively deny the enemy the ability to acquire OPFOR positions for targeting.
- Make extensive use of local building materials, equipment, and work force.
- Protect CPs and logistics sites.
- Bury communications lines.
- Construct false positions, equipment, movement routes, and LOCs.
- Assimilate minefields and obstacles to the terrain.
- Prepare caves, tunnels, and tunnel complexes in which troops can live and from which they can fight.

FORTIFICATION

12-171. Although engineers have the bulk of specialized equipment for constructing sophisticated survivability positions, engineer support at brigade level and lower is very limited. Therefore, maneuver units at battalion and below exert maximum effort to develop and improve their own positions. (See Shared Responsibilities for Field Fortifications, below.) Considering the projected time of stay, the conditions of the terrain, and the upcoming combat tasks, the maneuver commander determines the amount, sequence, and time for the fortification of an area

12-172. Fortified positions increase OPFOR weapons effectiveness. They also protect personnel, weapons, and material from enemy targeting and reconnaissance assets, and from the effects of enemy attack. The OPFOR categorizes field fortifications according to purpose. It divides them into structures for—
- Firing and observation.
- Protection for personnel, equipment, and material.

Fortification Priorities

12-173. Commanders assign fortification priorities to tasks that provide the best level of protection at all times against a possible enemy attack. The normal priority is from front to rear, beginning with the primary battle and fighting positions, then the temporary positions, alternate positions, and if possible deception positions. One of the greatest factors influencing the level and sequence of fortification preparation is whether the transition to the defense occurs in contact or out of contact with the enemy.

12-174. If forward maneuver units are in contact with the enemy, they prepare their own positions as much as possible. The maneuver unit is responsible for the majority of defensive fortifications. This includes emplacing minefields and nonexplosive obstacles.

12-175. When not in direct contact with the enemy and when the situation permits, engineer excavating and bulldozer equipment may be used to—
- Dig communications and fighting trenches and tank and IFV or APC emplacements. (For the most effective use of the heavy equipment, the supported mechanized infantry or tank unit must lay out and mark the areas for ditching.)
- Fortify squad trenches and platoon battle positions.

Engineer Support

- Provide engineer reconnaissance.
- Emplace engineer obstacles.
- Prepare alternate battle and fighting positions.

Preferably, engineer preparations occur at night or under other conditions of reduced visibility.

Shared Responsibilities for Field Fortifications

12-176. Engineer tasks are a shared responsibility throughout the OPFOR. Engineer and combat personnel perform survivability tasks such as constructing fortifications, clearing fields of fire, and camouflage. Although the highest level of engineer training and the greatest technical capabilities exist in the engineer troops, all military personnel and units train somewhat in fundamental engineer tasks. The majority of defensive preparation is conducted at the maneuver unit level. The following are several examples of specific responsibilities:

- Soldiers: dig individual fighting positions and trenches.
- Combat vehicles: several hundred vehicles in a mechanized infantry division may have self-entrenching capability.
- Engineers: construct fortified positions and communications trenches; dig in critical equipment, C2 sites, and logistics facilities.

12-177. Maneuver commanders realize that fortification of battle positions or assembly area positions is a shared responsibility involving all available personnel and equipment. This process starts with infantry using shovels and armored vehicles using integral self-entrenching blades, if available. When building the positions, they must take advantage of the protective and camouflaging properties of the terrain, local building materials, engineer construction equipment, explosives, and prefabricated installations.

12-178. Meanwhile, engineers using specialized equipment dig positions for critical sites such as medical facilities and C2 centers. As scarce engineer equipment becomes available, it supports maneuver units by augmenting and improving on the work the units have already begun.

12-179. In preparation for offensive action, the primary use of field fortification is in the preparation of assembly areas. Even there, the tasks of preparation typically exceed the capability of engineers constituent to the brigade and even of those likely to be assigned as augmentation.

12-180. The goal is to prepare a separate assembly area for each battalion-size unit, using engineer equipment to construct positions for vehicles shortly after they arrive at their assigned location. Within 1 to 2 hours, engineers dig assembly area fighting positions for all personnel. The engineers may prepare prefabricated structures for battalion CPs and carefully camouflage all structures.

CAMOUFLAGE, CONCEALMENT, COVER, AND DECEPTION

12-181. The OPFOR uses various C3D measures to mislead the enemy about size and location of forces and weapon systems and about the nature of defensive engineer preparations. Defensive measures include—

- Use of screening properties of terrain, darkness, and other conditions of limited visibility during engineer preparation of defensive positions and positioning of forces.
- Camouflage painting of material.
- Use of local materials and standard-issue camouflage screens.
- Strict camouflage, noise, and light discipline.
- Construction of false battle positions, decoy positions, and decoy equipment.
- False actions to draw attention.
- Assimilation of minefields and obstacles to the terrain.

12-182. However, OPFOR applications of C3D are not limited to the defense. During the offense, C3D measures include—
- Selection of terrain for its screening effect.
- Use of obscurants (smokescreens).
- Use of artificial and natural camouflage screens.
- Simulation of characteristic defensive measures—to "mine" the terrain in view of the enemy with decoy minefields or to give the appearance of reinforced defensive positions.
- Use of concealed routes for movement of supplies and reserves.

Artificial Camouflage

12-183. The OPFOR employs artificial camouflage as a supplement when natural screens cannot provide the concealment of forces and combat material. It includes both natural and manufactured camouflage. The OPFOR uses camouflage nets and screens extensively. It improves multispectral screening by using camouflage nets, covers, and individual camouflage equipment.

Decoys

12-184. The OPFOR uses deception activities and equipment to counter enemy reconnaissance. All engineer units receive special training in constructing decoys from locally available materials. These decoys cover a wide spectrum of types and must be introduced or "discovered" in the same priority a real or existing unit would emplace them. The general priority of engineer construction is from front to rear, beginning with the primary battle and fighting positions, then the temporary and alternate positions. The time sequence in which these "appear" gives credibility to the deception. The engineers may use obsolete or derelict equipment for deceptive purposes. The emphasis is on tactical systems and measures that provide effective concealment and deception.

12-185. To aid in water obstacle crossings, engineers can construct deception crossing sites, before or at the same time they are establishing actual ones. They try to draw the enemy's attention to simulated crossing sites while real ones remain carefully camouflaged. They give authenticity to simulated crossings by using corner reflectors, by deploying vehicles on roads and other approaches to them, by moving simulated vehicles across them, and by positioning construction and bridging equipment near simulated sites.

12-186. The OPFOR plans to employ mock-ups and decoys as an integral part of battles. Simulations can obscure OPFOR intentions and cause the enemy to waste effort by destroying decoys. The engineers bear a major responsibility for constructing simulations. The OPFOR places emphasis on those engineer simulation measures that it can transport easily and construct rapidly.

12-187. The following conditions must exist in order for decoy equipment to be successful:
- Placement must be in areas where the enemy would reasonably expect to find that type of actual equipment in use.
- Dimensions of simulated equipment must approximate those of actual equipment.

12-188. The simulations that engineers construct can represent any type equipment in the OPFOR inventory. Actual equipment that is not functional due to combat damage or mechanical malfunction can be made to appear operational by repainting it to conceal damage or by constructing components to simulate destroyed parts.

12-189. Engineers can create false excavations to simulate revetments, hull-defilade vehicle trenches, or individual fighting positions. These false excavations may be only half the depth of actual excavations, although the engineers may create the appearance of greater depth by adding dark materials such as branches, grass, or soil to the bottoms. Troops can temporarily occupy these deception positions and fire from them to aid deception.

Chapter 13

CBRN and Smoke

The use of chemical, biological, radiological, and nuclear (CBRN) weapons can have an enormous impact on all combat actions. Because chemical employment is more likely than the other three types, this chapter begins by focusing on OPFOR chemical capabilities. Because the OPFOR may also have some biological, nuclear, and radiological capabilities, these also deserve discussion, despite of the lower probability of their employment. The chapter concludes with discussions of CBRN protection and employment of smoke.

WEAPONS OF MASS DISTRUCTION

13-1. CBRN weapons are a subset of weapons of mass destruction (WMD), although the latter exclude the delivery means where such means is a separable and divisible part of the weapon. WMD are weapons or devices intended for or capable of causing a high order of physical destruction or mass casualties (death or serious bodily injury to a significant number of people). The casualty-producing elements of WMD can continue inflicting casualties on the enemy and exert powerful psychological effects on the enemy's morale for some time after delivery.

13-2. Existing types of WMD include chemical, biological, radiological, and nuclear weapons. However, technological advances are making it possible to develop WMD based on qualitatively new principles, such as infrasonic (acoustic) or particle-beam weapons. In addition, conventional weapons, such as precision weapons or volumetric explosives, can also take on the properties of WMD.

PREPAREDNESS

13-3. In response to foreign developments, the OPFOR maintains a capability to conduct chemical, nuclear, and possibly biological or radiological warfare. However, it would prefer to avoid the use of CBRN weapons by either side. This is especially true of nuclear and biological weapons, which have lethal effects over much larger areas than do chemical weapons. The effects of biological weapons can be difficult to localize and to employ in combat without affecting friendly forces. Their effects on the enemy can be difficult to predict. Unlike nuclear or biological weapons, chemical agents can be used to affect limited areas of the battlefield. The consequences of chemical weapons use are more predictable and thus more readily integrated into battle plans at the tactical level. In the event that either side resorts to CBRN weapons, the OPFOR is prepared to employ CBRN protection measures.

MULTIPLE OPTIONS

13-4. Force modernization has introduced a degree of flexibility previously unavailable to combined arms commanders. It creates multiple options for the employment of forces at strategic, operational, and tactical levels with or without the use of CBRN weapons. Many of the same delivery means available for CBRN weapons can also be used to deliver precision weapons that can often achieve desired effects without the stigma associated with CBRN weapons.

13-5. The OPFOR might use CBRN weapons either to deter aggression or as a response to an enemy attack. It has surface-to-surface missiles (SSMs) capable of carrying nuclear, chemical, or biological warheads. Most OPFOR artillery is capable of delivering chemical munitions, and most systems 152-mm and larger are capable of firing nuclear rounds. Additionally, the OPFOR could use aircraft systems and

Chapter 13

cruise missiles to deliver a CBRN attack. The OPFOR has also trained special-purpose forces (SPF) as alternate means of delivering CBRN munitions packages.

13-6. The threat of using these systems to deliver CBRN weapons is also an intimidating factor. Should any opponent use its own CBRN capability against the OPFOR, the OPFOR is prepared to retaliate in kind. It is also possible that the OPFOR could use CBRN against a neighbor as a warning to any potential enemy that it is willing to use such weapons. The fact that CBRN weapons may also place noncombatants at risk is also a positive factor from the OPFOR's perspective. Thus, it may use or threaten to use CBRN weapons as a way of applying political, economic, or psychological pressure by allowing the enemy no sanctuary.

TARGETING

13-7. The OPFOR considers the following targets to be suitable for the employment of CBRN weapons:
- CBRN delivery means and their supply structure.
- Precision weapons.
- Prepared defensive positions.
- Reserves and troop concentrations.
- Command and control (C2) centers.
- Communications centers.
- Reconnaissance, intelligence, surveillance, and target acquisition centers.
- Key air defense sites.
- Logistics installations, especially port facilities.
- Airfields the OPFOR does not intend to use immediately.

Enemy CBRN delivery means (aircraft, artillery, missiles, and rockets) normally receive the highest priority. The suitability of other targets depends on the OPFOR's missions, the current military and political situation, and the CBRN weapons available for use.

STAFF RESPONSIBILITY

13-8. On the functional staff of a division- or brigade-level headquarters, the chief of WMD is responsible for planning the offensive use of WMD, including CBRN weapons. (See the subsections on Release under Chemical Warfare, Nuclear Warfare, and Biological Warfare below.) The WMD staff element advises the command group and the primary and secondary staff on issues pertaining to CBRN employment. The WMD element receives liaison teams from any subordinate or supporting units that contain WMD delivery means.

13-9. CBRN defense comes under the chief of force protection. The force protection element of the functional staff may receive liaison teams from any subordinate or supporting chemical defense units. However, those units can also send liaison teams to other parts of the staff, as necessary (including, for example, the chief of reconnaissance).

CHEMICAL WARFARE

13-10. The OPFOR is equipped, structured, and trained to conduct both offensive and defensive chemical warfare. It is continually striving to improve its chemical warfare capabilities. It believes that an army using chemical weapons must be prepared to fight in the environment it creates. Therefore, it views chemical defense as part of a viable offensive chemical warfare capability. It maintains a large inventory of individual and collective chemical protection and decontamination equipment. (See the CBRN Protection portion of this chapter.)

WEAPONS AND AGENTS

13-11. Virtually all OPFOR indirect fire weapons can deliver chemical agents. These delivery means include aircraft, multiple rocket launchers (MRLs), artillery, mines, rockets, and SSMs. Other possible delivery means could include SPF, affiliated insurgent or guerrilla organizations, or civilian sympathizers. For additional information on delivery systems, see the *Worldwide Equipment Guide*.

13-12. One way of classifying chemical agents according to the effect they have on persons. Thus, there are two major types, each with subcategories. Lethal agents, categorized by how they attack and kill personnel, include nerve, blood, blister, and choking agents. Nonlethal agents include incapacitants and irritants.

Nerve Agents

13-13. Nerve agents are fast-acting. Practically odorless and colorless, they attack the body's nervous system, causing convulsions and eventually death. Nerve agents are further classified as either G- or V-agents. At low concentrations, the sarin (GB) series incapacitates. It kills if inhaled or absorbed through the skin. The rate of action is very rapid if inhaled, but slower if absorbed through the skin. V-agents produce similar effects, but are quicker-acting and more persistent than G-agents.

Blood Agents

13-14. Blood agents block the body's oxygen transferal mechanisms, leading to death by suffocation. A common blood agent is hydrogen cyanide (AC). It kills quickly and dissipates rapidly.

Blister Agents

13-15. Blister agents, such as mustard (H) or lewisite (L) and combinations of these two compounds, can disable or kill after contact with the skin, being inhaled into the lungs, or being ingested. Contact with the skin can cause painful blisters, and eye contact can cause blindness. These agents are especially lethal when inhaled.

Choking Agents

13-16. Choking agents, such as phosgene (CG) and diphosgene (DP), block respiration by damaging the breathing mechanism, which can be fatal. As with blood agents, this type is nonpersistent, and poisoning comes through inhalation. Signs and symptoms of toxicity may be delayed up to 24 hours.

Incapacitants

13-17. Incapacitants include psychochemical agents and paralyzants. These agents can disrupt a victim's mental and physical capabilities. The victim may not lose consciousness, and the effects usually wear off without leaving permanent physical injuries.

Irritants

13-18. Irritants, also known as riot-control agents, cause a strong burning sensation in the eyes, mouth, skin, and respiratory tract. The best known of these agents is tear gas (CS). Their effects are also temporary. Victims recover completely without having any serious aftereffects.

AGENT PERSISTENCY

13-19. Chemical agents are also categorized according to their persistency. Generally, the OPFOR would use persistent agents on areas it does not plan to enter and nonpersistent agents where it does.

Persistent Agents

13-20. Persistent agents can retain their disabling or lethal characteristics from days to weeks, depending on environmental conditions. Aside from producing mass casualties initially, persistent agents can produce a steady rate of attrition and have a devastating effect on morale. They can seriously degrade the performance of personnel in protective clothing or impose delays for decontamination.

Nonpersistent Agents

13-21. Nonpersistent agents generally last a shorter period of time than persistent agents, depending on weather conditions. The use of a nonpersistent agent at a critical moment in battle can produce casualties or force enemy troops into a higher level of individual protective measures. With proper timing and distance, the OPFOR can employ nonpersistent agents and then have its maneuver units advance into or occupy an enemy position without having to decontaminate the area or don protective gear.

OTHER TOXIC CHEMICALS

13-22. In addition to traditional chemical warfare agents, the OPFOR may find creative and adaptive ways to cause chemical hazards using chemicals commonly present in industry or in everyday households. In the right combination, or in and of themselves, the large-scale release of such chemicals can present a health risk, whether caused by military operations, intentional use, or accidental release.

Toxic Industrial Chemicals

13-23. Toxic industrial chemicals (TICs) are chemical substances with acute toxicity that are produced in large quantities for industrial purposes. Exposure to some industrial chemicals can have a lethal or debilitating effect on humans. They are a potentially attractive option for use as weapons of opportunity or WMD because of—
- The near-universal availability of large quantities of highly toxic stored materials.
- Their proximity to urban areas.
- Their low cost.
- The low security associated with storage facilities.

13-24. Employing a TIC against an opponent by means of a weapon delivery system, whether conventional or unconventional, is considered a chemical warfare attack, with the TIC used as a chemical agent. The target may be the enemy's military forces or a civilian population.

13-25. In addition to the threat from intentional use as weapons, catastrophic accidental releases of stored industrial chemicals may result from—
- Collateral damage associated with military operations.
- Electrical power interruption.
- Improper facility maintenance or shutdown procedures.

These events are common in armed conflict and post-conflict urban environments.

13-26. The most important factors to consider when assessing the potential for adverse human health impacts from a chemical release are—
- Acute toxicity.
- Physical properties (volatility, reactivity, and flammability).
- The likelihood that large quantities will be accidentally released or available for exploitation.

Foremost among these factors is acute toxicity.

13-27. The following are examples of high- and moderate-risk TICs. The risk assessment is based on acute toxicity by inhalation, worldwide availability (number of producers and number of countries where the substance is available), and physical state (gas, liquid, or solid) at standard temperature and pressure:
- **High-risk.** Ammonia, chlorine, fluorine, formaldehyde, hydrogen chloride, phosgene, and sulfuric acid.
- **Moderate-risk.** Carbon monoxide, methyl bromide, nitrogen dioxide, and phosphine.

This list does not include all chemicals with high toxicity and availability. Specifically, chemicals with low volatility are not included. Low-vapor pressure chemicals include some of the most highly toxic chemicals widely available, including most pesticides.

13-28. Some of the high-risk TICs are frequently present in an operational environment. Chlorine (water treatment and cleaning materials), phosgene (insecticides and fertilizers), and hydrogen cyanide are traditional chemical warfare agents that are also considered TICs. Cyanide salts may be used to contaminate food or water supplies. Hydrogen chloride is used in the production of hydrochloric acid. Formaldehyde is a disinfectant and preservative. Fluorine is a base element that is used to produce fluorocarbons. Fluorocarbons are any of various chemically inert compounds that contain both carbon and fluorine. Fluorocarbons are present in common products (refrigerants, lubricants, and nonstick coatings) and are used in the production of resins and plastics.

Household Chemicals

13-29. The OPFOR understands that some everyday household chemicals have incompatible properties that result in undesired chemical reaction when mixed with other chemicals. This includes substances that will react to cause an imminent threat to health and safety, such as explosion, fire, and/or the formation of toxic materials. For example, chlorine bleach, when mixed with ammonia, will generate the toxic gases chloramine and hydrazine that can cause serious injury or death. Another example of such incompatibilities is the reaction of alkali metals, such as sodium or potassium, with water. Sodium is commonly used in the commercial manufacture of cyanide, azide, and peroxide, and in photoelectric cells and sodium lamps. It has a very large latent heat capacity and is used in molten form as a coolant in nuclear breeder reactors. The mixture of sodium with water produces sodium hydroxide, which can cause severe burns upon skin contact.

CHEMICAL RELEASE

13-30. Among CBRN weapons, the OPFOR is most likely to use chemical weapons against even a more powerful enemy, particularly if the enemy does not have the capability to respond in kind. Since the OPFOR does not believe that first use of chemical agents against units in the field would provoke a nuclear response, it is less rigid than forces of other nations in the control of chemical release.

13-31. Initially, the use of chemical weapons is subject to the same level of decision as nuclear and biological weapons. At all levels of command, a chemical weapons plan is part of the fire support plan. Once the National Command Authority (NCA) has released initial authorization for the use of chemical weapons, commanders can employ them freely, as the situation demands. Then each commander at the operational-strategic command (OSC) and lower levels who has systems capable of chemical delivery can implement the chemical portions of his fire support plan, as necessary.

13-32. After a decision for nuclear use, the OPFOR can employ chemical weapons to complement nuclear weapons. However, the OPFOR perceives that chemical weapons have a unique role, and their use does not depend on initiation of nuclear warfare. It is possible that the OPFOR would use chemical weapons early in an operation or strategic campaign or from its outset. It would prefer not to use chemical weapons within its own borders. However, it would contaminate its own soil if necessary in order to preserve the regime or its sovereignty.

OFFENSIVE CHEMICAL EMPLOYMENT

13-33. The basic principle of chemical warfare is to achieve surprise. It is common to mix chemical rounds with high-explosive (HE) rounds in order to achieve chemical surprise. Chemical casualties inflicted and the necessity of chemical protective gear degrade enemy defensive actions. The OPFOR also may use chemical agents to restrict the use of terrain. For example, contamination of key points along the enemy's lines of communications can seriously disrupt his resupply and reinforcement. Simultaneously, it can keep those points intact for subsequent use by the attacking OPFOR.

13-34. Nonpersistent agents are suitable for use against targets on axes the OPFOR intends to exploit. While possibly used against deep targets, their most likely role is to prepare the way for an assault by maneuver units, especially when enemy positions are not known in detail. The OPFOR may also use nonpersistent agents against civilian population centers in order to create panic and a flood of refugees.

13-35. Persistent agents are suitable against targets the OPFOR cannot destroy by conventional or precision weapons. This can be because a target is too large or located with insufficient accuracy for attack by other than an area weapon. Persistent agents can neutralize such targets without a pinpoint attack.

13-36. In the offense, likely chemical targets include—
- Troops occupying defensive positions, using nonpersistent agents delivered by MRLs to neutralize these troops just before launching a ground attack. Ideally, these agents would be dissipating just as the attacking OPFOR units enter the area where the chemical attack occurred.
- CBRN delivery systems, troop concentration areas, headquarters, and artillery positions, using all types of chemical agents delivered by tube artillery, MRLs, SSMs, and aircraft.
- Bypassed pockets of resistance (especially those that pose a threat to the attacking forces), using persistent agents.
- Possible assembly areas for enemy counterattack forces, using persistent agents.

13-37. The OPFOR could use chemical attacks against such targets simultaneously throughout the enemy defenses. These chemical attacks combine with other forms of conventional attack to neutralize enemy nuclear capability, C2 systems, and aviation. Subsequent chemical attacks may target logistics facilities. The OPFOR would use persistent agents deep within the enemy's rear and along troop flanks to protect advancing units.

DEFENSIVE CHEMICAL EMPLOYMENT

13-38. When the enemy is preparing to attack, the OPFOR can use chemical attacks to—
- Disrupt activity in his assembly areas.
- Limit his ability to maneuver into axes favorable to the attack.
- Deny routes of advance for his reserves.

Once the enemy attack begins, the use of chemical agents can impede an attacking force. It can destroy the momentum of the attack by causing casualties or causing attacking troops to adopt protective measures. Persistent chemical agents can deny the enemy certain terrain and canalize attacking forces into kill zones.

BIOLOGICAL WARFARE

13-39. The OPFOR closely controls information about the status of its biological warfare capabilities. This creates uncertainty among its neighbors and other potential opponents as to what types of biological agents the OPFOR might possess and how it might employ them.

13-40. Biological weapons can provide a great equalizer in the face of a numerically and/or technologically superior adversary that the OPFOR cannot defeat in a conventional confrontation. However, their effects on the enemy can be difficult to predict, and the OPFOR must also be concerned about the possibility that the effects could spread to friendly forces.

WEAPONS AND AGENTS

13-41. Biological weapons consist of pathogenic microbes, micro-organism toxins, and bioregulating compounds. Depending on the specific type, these weapons can incapacitate or kill people or animals and destroy plants, food supplies, or materiel. The type of target being attacked determines the choice of agent and dissemination system.

Pathogens

13-42. Pathogens cause diseases such as anthrax, cholera, plague, smallpox, tularemia, or various types of fever. These weapons would be used against targets such as food supplies, port facilities, and population centers to create panic and disrupt mobilization plans.

Toxins

13-43. Toxins are produced by pathogens and also by snakes, spiders, sea creatures, and plants. Toxins are faster acting and more stable than live pathogens. Most toxins are easily produced through genetic engineering. Toxins produce casualties rapidly and can be used against tactical and operational targets.

Bioregulators

13-44. Bioregulators are chemical compounds that are essential for the normal psychological and physiological functions. A wide variety of bioregulators are normally present in the human body in extremely minute concentrations. These low-molecular-weight compounds, usually peptides (made up of amino acids), include neurotransmitters, hormones, and enzymes. Examples of bioregulators are—
- Insulin (a pancreatic protein hormone that is essential for the metabolism of carbohydrates).
- Enkephalin (either of two pentapeptides with opiate and analgesic activity that occur naturally in the brain and have a marked affinity for opiate receptors).

13-45. These compounds can produce a wide range of harmful effects if introduced into the body at higher than normal concentrations or if they have been altered. Psychological effects could include exaggerated fear and pain. In addition, bioregulators can cause severe physiological effects such as rapid unconsciousness and, depending on such factors as dose and route of administration, can also be lethal. Unlike pathogens, which take hours or days to act, bioregulators could act in only minutes. The small peptides, having fewer than 12 amino-acid groups, are most amenable to military application.

AGENT EFFECTS

13-46. Biological weapons are extremely potent and provide wide-area coverage. Some biological agents are extremely persistent, retaining their capabilities to infect for days, weeks, or longer. Biological weapons can take some time (depending on the agent) to achieve their full effect. To allow these agents sufficient time to take effect, the OPFOR may use clandestine means, such as SPF or civilian sympathizers, to deliver biological agents in advance of a planned attack or even before the war begins.

DELIVERY MEANS

13-47. It is possible to disseminate biological agents in a number of ways. Generally, the objective is to expose enemy forces to an agent in the form of a suspended cloud of very fine biological agent particles. Dissemination through aerosols, either as droplets from liquid suspensions or by small particles from dry powders, is by far the most efficient method. For additional information on delivery systems see the *Worldwide Equipment Guide*.

13-48. There are two basic types of biological munitions:
- Point-source bomblets delivered directly on targets.
- Line-source tanks that release the agent upwind from the target.

Within each category, there can be multiple shapes and configurations.

13-49. Military systems, as well as unconventional means, can deliver biological agents. Potential delivery means include rockets, artillery shells, aircraft sprayers, saboteurs, and infected rodents. The OPFOR might use SPF, affiliated irregular forces, and/or civilian sympathizers to deliver biological agents within the region, outside the immediate region (to divert enemy attention and resources), or even in the enemy's homeland.

TARGETS

13-50. Probable targets for biological warfare pathogen attack are enemy CBRN delivery units, airfields, logistics facilities, and C2 centers. The OPFOR may target biological weapons against objectives such as food supplies, water sources, troop concentrations, convoys, and urban and rural population centers rather than against frontline forces. The use of biological agents against rear area targets can disrupt and degrade

Chapter 13

enemy mobilization plans as well as the subsequent conduct of war. This type of targeting can also reduce the likelihood that friendly forces would become infected.

BIOLOGICAL RELEASE

13-51. The decision to employ biological agents is a political decision made at the national level—by the NCA. Besides the political ramifications, the OPFOR recognizes a degree of danger inherent in the use of biological agents, due to the difficulty or controlling an epidemic caused by them.

13-52. The prolonged incubation period makes it difficult to track down the initial location and circumstances of contamination. Thus, there is the possibility of plausible deniability. Even if an opponent might be able to trace a biological attack back to the OPFOR, it may not be able to respond in kind.

RADIOLOGICAL WEAPONS

13-53. It is possible that the OPFOR may develop and employ radiological weapons whose effects are achieved by using toxic radioactive materials against desired targets. The purpose of employing radiological materials could be to achieve leverage or intimidation against regional neighbors. However, such weapons could also be used to deter intervention by extraregional forces or to disrupt such forces once deployed in the region. While they can be used as area denial, intimidation, and political weapons, radiological weapons are also considered weapons of terror.

13-54. A *radiological weapon* is any device, including weapon or equipment other than a nuclear explosive device, specifically designed to employ radioactive material by disseminating it to cause destruction, damage, or injury by means of the radiation produced by the decay of such material. Radiological weapons differ from chemical and biological weapons in that radiation cannot be "neutralized" or "sterilized" and many radiological materials have half-lives in years. Two general types of radiological weapons are radiological dispersal devices and radiological exposure devices.

RADIOLOGICAL DISPERSAL DEVICES

13-55. A *radiological dispersal device* (RDD) is an improvised assembly or process, other than a nuclear explosive device, designed to disseminate radioactive material in order to cause destruction, damage, or injury. Unlike nuclear weapons, RDDs do not produce a nuclear explosion. However, RDDs spread radioactive material contaminating personnel, equipment, facilities, and terrain. They kill or injure by exposing people to radioactive material. Victims are irradiated when they get close to or touch the material, inhale it, or ingest it. The actual dose rate depends on the type and quantity of radioactive material spread over the area, and contributing factors such as weather and terrain.

13-56. The OPFOR could disperse radioactive material using low-level radiation sources in a number of ways, such as—
- Arming the warhead of a conventional missile with active material from a nuclear reactor.
- Releasing low-level radioactive material intended for use in industry or medicine.
- Disseminating material from a research or power-generating nuclear reactor.
- Depositing a radioactive source in a water supply.

Dispersal of radioactive materials is inexpensive and requires limited resources and technical knowledge. The primary source for radioactive material used in the construction of RDDs is from nuclear power plants and radioactive materials used in hospitals.

13-57. One design of RDD, popularly called a "dirty bomb," uses conventional explosives to disperse radioactive contamination. Any conventional or improvised explosive device can be used by placing it in close proximity to radioactive material. The explosion causes the dissemination of the radioactive material. A dirty bomb typically generates its immediate casualties from the direct effects of the conventional explosion (blast injuries and trauma). However, one of the primary purposes of a dirty bomb is to frighten people by contaminating their environment with radioactive materials and threatening large numbers of people with exposure. As an area denial weapon, an RDD can generate significant public fear and economic impact. In some cases, an area may not be habitable for nonmilitary personnel, but military operations could continue, with appropriate protective measures.

RADIOLOGICAL EXPOSURE DEVICES

13-58. A *radiological exposure device* (RED) is a radioactive source placed to cause injury or death. In this case, rather than resulting from dispersed radioactive material, radioactive exposure results from discrete sources, such as a radioactive source concealed in a high traffic area. The placement of an RED may be covert in order to increase the potential dose if the source is not detected.

NUCLEAR WARFARE

13-59. The OPFOR believes a war is most likely to begin with a phase of nonnuclear combat that may include the use of chemical weapons. The OPFOR emphasizes the destruction of as much as possible of enemy nuclear capability during this nonnuclear phase. To do so, it would use air and missile attacks; airborne, heliborne, and special-purpose forces; and rapid, deep penetrations by ground forces. The OPFOR hopes these attacks can deny the enemy a credible nuclear option.

DELIVERY MEANS

13-60. If the OPFOR decides to use nuclear weapons, the nuclear delivery systems may include aircraft from both national- and theater-level aviation, and SSMs. Most artillery 152-mm or larger is capable of firing nuclear rounds, if such rounds are available. Other possible delivery means could include SPF. The OPFOR may also choose to use affiliated forces for nuclear delivery. For additional information on nuclear delivery systems, see the *Worldwide Equipment Guide*.

TRANSITION TO NUCLEAR

13-61. Even when nuclear weapons are not used at the outset of a conflict, OPFOR commanders deploy troops based on the assumption that a nuclear-capable enemy might attack with nuclear weapons at any moment. The OPFOR continuously updates its own plans for nuclear employment, although it prefers to avoid nuclear warfare. As long as it achieves its objectives, and there are no indications that the enemy is going to use nuclear weapons, the OPFOR would likely not use them either. However, it could attempt to preempt enemy nuclear use by conducting an initial nuclear attack. Otherwise, any OPFOR decision to go nuclear would have to be made early in the conflict, so that sufficient nonnuclear power would remain to follow up and to exploit the gains of nuclear employment.

13-62. If any opponent were to use nuclear weapons against the OPFOR, the OPFOR would respond in kind, as long as it is still capable. The same would be true of any nuclear-capable opponent, if the OPFOR were the first to use nuclear means. While the OPFOR recognizes the advantage of its own first use, it may risk first use only when the payoff appears to outweigh the potential costs. Therefore, it will probably avoid the use of nuclear weapons against a more powerful enemy unless survival of the regime or the nation is at stake.

13-63. The OPFOR is probably more likely to use its nuclear capability against a less powerful opponent. The likelihood increases if that opponent uses or threatens to use its own nuclear weapons against the OPFOR or does not have the means to retaliate in kind. This could account for a nuclear or nuclear-threatened environment existing at the time a more powerful force might choose to intervene.

TYPES OF NUCLEAR ATTACK

13-64. The OPFOR categorizes nuclear attacks as either massed or individual attacks. The category depends on the number of targets hit and the number of nuclear munitions used.

13-65. A *massed* nuclear attack employs multiple nuclear munitions simultaneously or over a short time interval. The goal is to destroy a single large enemy formation, or several formations, as well as other

Chapter 13

important enemy targets. A massed attack can involve a single service of the Armed Forces, as in a nuclear missile attack by the Strategic Forces, or the combined forces of different services.

13-66. An *individual* nuclear attack may hit a single target or group of targets. The attack consists of a single nuclear munition, such as a missile or bomb.

NUCLEAR RELEASE

13-67. At all stages of a conflict, the OPFOR keeps nuclear-capable forces ready to make an attack. The decision to initiate nuclear warfare is a political decision made at the national level.

13-68. After the initial nuclear release, the NCA may delegate employment authority for subsequent nuclear attacks to an OSC commander. The commander of the OSC's integrated fires command submits recommendations for the subsequent employment of nuclear and chemical weapons to the OSC commander for approval and integration into OSC fire support plans.

OFFENSIVE NUCLEAR EMPLOYMENT

13-69. Once the NCA releases nuclear weapons, two principles govern their use: mass and surprise. The OPFOR plans to conduct the initial nuclear attack suddenly and in coordination with nonnuclear fires. Initial nuclear attack objectives are to destroy the enemy's main combat formations, C2 systems, and nuclear and precision weapons, thereby isolating the battlefield.

13-70. Nuclear attacks target and destroy the enemy's defenses and set the conditions for an exploitation force. Other fire support means support the assault and fixing forces. The OPFOR may plan high-speed air and ground offensive actions to exploit the nuclear attack.

13-71. If the enemy continues to offer organized resistance, the OPFOR might employ subsequent nuclear attacks to reinitiate the offense. Nuclear attacks can eliminate the threat of a counterattack or clear resistance from the opposite bank in a water obstacle crossing. If the enemy begins to withdraw, the OPFOR plans nuclear attacks on choke points where retreating enemy forces present lucrative targets.

Planning

13-72. Although the opening stages of an offensive action are likely to be conventional, OPFOR planning focuses on the necessity of—
- Countering enemy employment of nuclear weapons.
- Maintaining the initiative and momentum.
- Maintaining fire superiority over the enemy (preempting his nuclear attack, if necessary).

13-73. When planning offensive actions, the OPFOR plans nuclear fires in detail. Forces conducting the main attack would probably receive the highest percentage of weapons. However, the OPFOR may also reserve weapons for other large, important targets. In more fluid situations, such as during exploitation, the commander may keep some nuclear weapon systems at high readiness to fire on targets of opportunity. Nuclear allocations vary with the strength of the enemy defense and the scheme of maneuver.

13-74. Since the enemy too is under nuclear threat, he also must disperse his formations, which can make him more vulnerable to penetration by an attacking force. However, the OPFOR realizes that enemy troops are also highly mobile and capable of rapidly concentrating to protect a threatened area. Therefore, it considers surprise and timing of offensive actions to be extremely critical in order to complicate enemy targeting and deny him the time to use his mobility.

Execution

13-75. Upon securing a nuclear release, the OPFOR would direct nuclear attacks against the strongest points of the enemy's formations and throughout his tactical and operational depth. This would create gaps through which maneuver units, in "nuclear-dispersed" formations, would attack as an exploitation force. As closely as safety and circumstances permit, maneuver forces follow up on attacks near the battle line. Airborne troops may exploit deep attacks.

13-76. An exploitation force would probably attack to take full advantage of the speed of advance it could expect to achieve. The aim of these maneuver units would be to seize or neutralize remaining enemy nuclear weapons, delivery systems, and C2 systems. By attacking from different directions, the maneuver units would try to split and isolate the enemy.

13-77. Commanders would ensure a rapid tempo of advance by assigning tank and mechanized infantry units to the exploitation force. Such units are quite effective in this role, because they have maneuverability, firepower, lower vulnerability to enemy nuclear attacks, and the capability to achieve penetrations of great depth.

Defensive Nuclear Employment

13-78. Primary uses of nuclear weapons in the defense are to—
- Destroy enemy nuclear and precision weapons and delivery means.
- Destroy main attacking groups.
- Conduct counterpreparations.
- Eliminate penetrations.
- Support counterattacks.
- Deny areas to the enemy.

If nuclear weapons degrade an enemy attack, the OPFOR could gain the opportunity to switch quickly to an offensive role.

CBRN PROTECTION

13-79. Due to the proliferation of CBRN weapons, the OPFOR must anticipate their use, particularly the employment of chemical weapons. OPFOR planners believe that the best solution is to locate and destroy enemy CBRN weapons, delivery systems, and their supporting infrastructure before the enemy can use them against the OPFOR. In case this fails and it is necessary to continue combat actions despite the presence of contaminants, the OPFOR has developed and fielded a wide range of CBRN detection and warning devices, individual and collective protection equipment, and decontamination equipment. The OPFOR conducts rigorous training for CBRN defense.

13-80. OPFOR planners readily admit that casualties would be considerable in any future war involving the use of CBRN weapons. However, they believe that the timely use of active and passive measures can significantly reduce a combat unit's vulnerability. These measures include but are not limited to protective equipment, correct employment of reconnaissance assets, and expeditious decontamination procedures. Other operational-tactical responses to the threat include—
- **Dispersion**: Concentrations must last for as short a time as possible.
- **Speed of advance**: If the advance generates enough momentum, this can make enemy targeting difficult and keep enemy systems on the move.
- **Camouflage, concealment, cover, and deception (C3D)**: C3D measures complicate enemy targeting.
- **Continuous contact**: The enemy cannot attack with CBRN weapons as long as there is intermingling of friendly and enemy forces.

Organization

13-81. OPFOR chemical defense units are responsible for biological, radiological, and nuclear, as well as chemical, protection and reconnaissance measures. In the administrative force structure, such units are subordinate to all maneuver units brigade and above. During task-organizing, tactical-level commands may also receive additional chemical defense units allocated from the OSC or higher-level tactical command. However, those higher headquarters typically retain some chemical defense assets at their respective levels to deal with the threat to the support zone and provide chemical defense reserves.

Chapter 13

Note. OPFOR "chemical troops" and "chemical defense" units also perform biological, radiological, and nuclear protection functions. For example, a chemical defense battalion at division or higher level contains a CBRN reconnaissance company. Likewise, a chemical defense platoon at brigade level has a CBRN reconnaissance squad.

13-82. Chemical troops are a vital element of combat support. They provide trained specialists for chemical defense units and for units of other arms. Basic tasks chemical troops can accomplish in support of combat troops include—
- Reconnoitering known or likely areas of CBRN contamination.
- Warning troops of the presence of CBRN contamination.
- Monitoring changes in the degree of contamination.
- Monitoring the CBRN contamination of personnel, weapons, and equipment.
- Performing decontamination activities.
- Providing trained troops to handle chemical munitions.

They perform specialized CBRN reconnaissance in addition to supporting regular ground reconnaissance efforts described in chapter 8.

13-83. CBRN protection functions are not limited to maneuver units. Artillery and air defense brigades have their own chemical defense units. Medical and SSM units have some decontamination equipment. Engineer troops also are important, performing functions such as decontaminating roads, building bypasses, and purifying water supplies. Of course, all arms have a responsibility for CBRN reconnaissance and at least partial decontamination without specialist support. However, they can continue combat actions for only a limited time without complete decontamination by chemical troops.

EQUIPMENT

13-84. OPFOR troops have protective clothing. Most combat vehicles and many noncombat vehicles have excellent overpressure and filtration systems. Items of equipment for individual or collective protection are adequate to protect soldiers from contamination for hours, days, or longer, depending on the nature and concentration of the contaminant. Antidotes provide protection from the effects of agents. Agent detector kits and automatic alarms are available in adequate quantities and are capable of detecting all standard agents.

13-85. Chemical troops have a wide variety of dependable equipment that, for the most part, is in good supply and allows them to accomplish a number of tasks in support of combat troops. They have specialized equipment for detecting and monitoring CBRN contamination. They have some specialized CBRN reconnaissance vehicles, and they may use helicopters for CBRN reconnaissance. Decontamination equipment is also widely available.

RECONNAISSANCE

13-86. Under the guidance of the brigade-level chief of force protection, CBRN instructors in maneuver battalions and other units subordinate to the brigade train additional combat, combat support, and combat service support troops for CBRN observation and reconnaissance missions. Medical personnel also have instruments to check casualties for CBRN contamination.

13-87. Chemical defense personnel assigned to reconnaissance elements of chemical defense units perform CBRN reconnaissance. This involves two general types of activity: CBRN observation posts (OPs) and CBRN reconnaissance patrols. Such posts and patrols may augment any maneuver unit down to company level.

CBRN Observation Posts

13-88. A maneuver battalion or higher commander normally designates a CBRN OP to locate near his forward command post (CP). At division, brigade, or tactical group level, such a CBRN OP is normally manned by four to six CBRN specialists. These would typically be specially trained specialists from one squad of the CBRN reconnaissance platoon of a division's chemical defense battalion or the CBRN reconnaissance squad of a brigade's chemical defense platoon. In some cases, the CBRN OP at division or division tactical group (DTG) level might comprise an entire CBRN reconnaissance platoon. Most likely, a brigade or brigade tactical group (BTG) commander would keep one team of the CBRN reconnaissance squad from his chemical defense platoon to man the CBRN OP near his forward CP. However, he could employ the whole squad.

13-89. Since no chemical defense units are organic below brigade level, each maneuver battalion or company normally has a small team of additional-duty CBRN specialists. These company- and battalion-level specialists perform limited CBRN reconnaissance when CBRN support is not available from brigade or BTG level. When chemical troops are not available, virtually every company- or battery-size combat, combat support, and combat service support unit can establish its own CBRN OP using its own troops trained as observers. This OP is normally near the unit's CP or command and observation post.

13-90. The functions of CBRN OPs are to—
- Detect CBRN contamination.
- Determine radiation levels and types of toxic substances.
- Monitor the drift of radioactive clouds.
- Report CBRN information and meteorological data to higher headquarters.
- Give a general alarm to threatened troops.

13-91. When stationary, a CBRN OP is normally in a camouflaged trench or a dug-in CBRN reconnaissance vehicle. During movement, it moves in its own vehicle close to the combat unit commander. The observers immediately activate CBRN detection devices after an enemy overflight, missile burst, or artillery shelling.

CBRN Reconnaissance Patrols

13-92. A CBRN reconnaissance patrol allocated to a maneuver, reconnaissance, or security element receives instructions from that unit's commander. That commander tells the CBRN reconnaissance patrol leader of any specific areas to reconnoiter, the time for doing so, and the procedures for reporting the results. A patrol may also receive reconnaissance assignments from the brigade or division chief of force protection.

13-93. When operating in CBRN reconnaissance patrols, chemical defense personnel travel in reconnaissance vehicles specially equipped with CBRN detection and warning devices. Before a patrol begins its mission, its personnel check their individual CBRN protection equipment and detection instruments. They also examine the CBRN and communications equipment located on their reconnaissance vehicle.

13-94. As a patrol performs its mission, a designated crewman observes the readings of the onboard CBRN survey meters. Upon discovering radioactive or chemical contamination, the patrol determines the size and boundaries of the contaminated area and the radiation level or type of toxic substance present. The patrol leader then plots contaminated areas on his map, reports by radio to his commander, and orders his patrol to mark the contaminated area. The patrol designates bypass routes around contaminated areas or finds routes (with the lowest levels of contamination) through the area.

13-95. The OPFOR can also use helicopters to perform CBRN reconnaissance. Helicopters equipped with chemical and radiological area survey instruments are particularly useful for performing reconnaissance of areas with extremely high contamination levels.

Augmentation to Other Reconnaissance and Security Elements

13-96. CBRN reconnaissance elements are often allocated to maneuver units, sometimes becoming part of a security element or various types of reconnaissance patrol. A maneuver battalion (particularly one acting independently) may receive chemical troops allocated from the brigade-level chemical defense company or platoon. The resulting battalion-size detachment would typically use these chemical augmentees as part of a platoon-size fighting patrol.

CBRN DETECTION AND WARNING REPORTS

13-97. The OPFOR transmits CBRN warning information over communications channels in a parallel form using both the command net and the air defense and CBRN warning communications net. Depending of what type of unit initially detected the contamination, detection reports leading to such warnings may go either through chemical defense and force protection channels or through the maneuver unit or ground reconnaissance reporting chain.

Detection Reports

13-98. When chemical defense units establish CBRN OPs or CBRN reconnaissance patrols, these reconnaissance elements report through the chemical defense and force protection chain. Upon detection of contamination, a CBRN observer or CBRN reconnaissance patrol normally transmits a CBRN *detection report* to the commander of the parent chemical defense unit that dispatched it and (if capable and directed to do so) possibly to the chief of force protection (or chief of staff) on the staff of the commander to whom the chemical defense unit is subordinate or supporting. In any case, the chemical defense unit commander transmits the report to the chief of force protection of the maneuver division, brigade, or tactical group.

13-99. When CBRN observers (whether from the chemical troops or another branch) are allocated to regular ground reconnaissance elements, security elements, or maneuver units, the CBRN element that detects contamination would initially pass the detection report through reconnaissance or maneuver unit reporting channels. Of course, they would report the detection to the commander of the unit to which they are allocated.

Note. If the CBRN augmentees have their own special CBRN reconnaissance vehicles and associated radios, they can also send CBRN detection reports back to their parent chemical defense unit or the chief of force protection.

13-100. A ground reconnaissance or security element would pass the detection report to the unit that dispatched it. For example, a reconnaissance patrol leader would transmit the detection information via reconnaissance reporting channels to his parent reconnaissance unit commander and possibly by skip-echelon communications to the chief of reconnaissance at brigade, division, or tactical group level. The reconnaissance unit commander or chief of reconnaissance would then ensure that the chief of staff and/or chief of force protection at his level receives the CBRN detection report and takes appropriate action. When a maneuver unit chief of staff or chief of reconnaissance receives a CBRN detection report through his own channels, he immediately passes it to the chief of force protection at that level (or to the next-higher level that has one).

13-101. Similarly, upon detection of contamination, a CBRN observer in a fighting patrol (FP) would transmit a CBRN detection report to the FP leader. The FP leader would use the battalion command net to transmit the report to the maneuver battalion commander who sent out the FP. Using the brigade command net, the maneuver battalion commander would inform the maneuver brigade commander (or chief of staff) of the CBRN detection report. The brigade commander (or chief of staff) would consult with his chief of force protection. Based on the finding of the chief of force protection, both he and the brigade commander (or chief of staff) could then disseminate the CBRN warning report using their respective communications nets.

Warning Reports

13-102. The chief of force protection and his staff evaluate the CBRN detection report and determine whether it warrants the issuing of a warning. If it does, they inform the maneuver commander (or his chief of staff). At this point, the CBRN detection report changes to a CBRN *warning report*. Then, for example, the maneuver brigade commander (or chief of staff) disseminates a CBRN warning report via the brigade command net to all subordinate unit commanders, and via the division command net to the division commander and commanders of other brigades and other division subordinates. Simultaneously, the brigade chief of force protection disseminates the same CBRN warning report to all the brigade's units over the air defense and CBRN warning communications net. He would also inform the division chief of force protection. The desired goal is to rapidly disseminate CBRN warning reports as soon as possible to all affected units.

13-103. The division or brigade chief of force protection (and/or the chief of staff) may issue an advance CBRN warning based on the predicted development of a CBRN situation. CBRN protective measures would change or be rescinded based on subsequent CBRN detection reports or on warning reports from higher, lower, or adjacent units. Changes in the CBRN protective measures are disseminated by the maneuver division or brigade commander (or chief of staff) and the chief force protection using their respective communications nets.

DECONTAMINATION

13-104. The OPFOR distinguishes between two types of decontamination of personnel and equipment: partial and complete. It tries to perform one or both as soon after exposure as possible. It also conducts decontamination of terrain and movement routes.

Partial Decontamination

13-105. OPFOR doctrine dictates that a combat unit should conduct a partial decontamination with its own equipment no later than 1 hour after exposure to contamination. This entails a halt while troops decontaminate themselves and their clothing, individual weapons, crew-served weapons, and vehicles. When forced to conduct partial decontamination in the contaminated area, personnel remain in CBRN protective gear. Following the completion of partial decontamination, the unit resumes its mission.

Complete Decontamination

13-106. The commander of a maneuver unit directs complete decontamination when the unit has already completed its mission but is still tactically dispersed. This type of decontamination involves the decontamination of the entire surface of a contaminated piece of equipment. This usually requires special decontamination stations established by decontamination units. Under some circumstances, however, it may be accomplished by troops using individual decontamination kits.

13-107. Chemical defense troops usually perform complete decontamination of maneuver units. Decontamination units of chemical defense platoons, companies, and battalions can operate either as a whole or in smaller elements. For example, the decontamination company of a chemical defense battalion may function independently. It may be separated from the rest of the battalion by as much as 20 km to decontaminate elements of a maneuver battalion or brigade-size force.

13-108. Decontamination units deploy to uncontaminated areas where contaminated units are located. They set up near movement routes or establish centrally located decontamination points to serve several troop units. Before deploying his equipment, the commander of a decontamination unit dispatches a reconnaissance element to select a favorable site, mark areas for setting up the equipment, and mark entry and exit routes.

13-109. Site selection depends on local features such as nearby roads, cross-country routes, and sources of uncontaminated water. Another selection criterion is the availability of camouflage, cover, and concealment. If natural concealment is insufficient, smokescreening of the site can provide camouflage. After decontamination stations are set up, the decontamination unit commander establishes security. The supported unit can also assign personnel to assist in providing security for the site.

Terrain and Route Decontamination

13-110. Decontamination of maneuver routes is necessary if OPFOR units cannot safely bypass or cross the contaminated routes. The first priorities for decontamination are routes (cross-country and roads) on the primary axes of advance. The OPFOR decontaminates terrain by either removing or covering the contaminated soil or by spraying liquid decontaminate by specially designed decontamination vehicles.

13-111. For radiological decontamination, engineer earthmoving equipment can remove the contaminated top layer of soil. An alternative is to cover the area with uncontaminated materials (soil, wood, or other surfacing materials). Similarly, one means of decontamination and disinfection of chemical and biological agents would be to remove a 3- to 4-cm layer of contaminated ground.

13-112. Another means of chemical or biological decontamination is the use of terrain decontamination vehicles. These truck-mounted systems can decontaminate or disinfect an area 5 m wide and 500 m long with a single load of decontamination solvent. After decontaminating a route or area, chemical defense troops must decontaminate or disinfect their own equipment.

13-113. For terrain sector decontamination, chemical defense units equipped with decontamination trucks assemble near the contaminated area. When decontaminating or disinfecting sectors of terrain, they divide the sector into strips. The vehicles move on parallel axes using an echelon-right or echelon-left formation with the lead vehicle downwind from the others. Individual vehicles move at a distance of 30 to 50 m behind one another, but to the left or right just far enough that the covered strips slightly overlap. See figure 13-1.

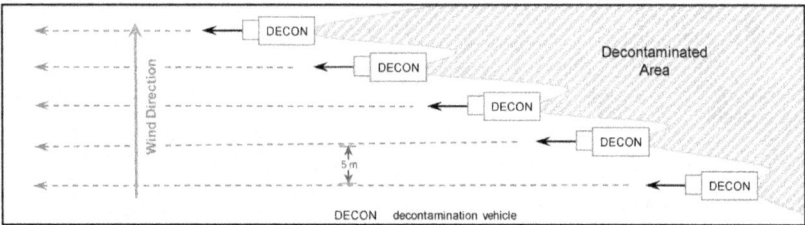

Figure 13-1. Decontamination trucks using echelon-left formation

13-114. For road decontamination, the decontamination vehicles form a column, with each vehicle assigned a sector of road. If the road is more than 5 m wide, then two or three vehicles form an echelon-right or echelon-left formation (as used in terrain sector decontamination). Jet engine-type decontamination vehicles are useful for decontaminating hard-surface roads or runways.

RECOVERY ACTIVITIES

13-115. Commanders at all levels plan for restoring units that fall victim to CBRN attacks. This plan includes—
- Restoring C2.
- Reconnoitering the target area.
- Locating and rescuing casualties.
- Decontaminating personnel and equipment.
- Evacuating casualties.
- Evacuating weapons and combat equipment.
- Repairing vehicles.
- Clearing obstructions.
- Extinguishing fires.

13-116. To perform these tasks, the maneuver unit commander forms a recovery detachment. Depending on the situation and availability of forces, recovery detachments either come from subordinate units or from the reserves of a higher headquarters. If formed from subordinate units, recovery detachments generally are formed from the maneuver unit's reserve. Regardless of origin, they include chemical reconnaissance and decontamination assets, engineers, medical and vehicle-repair personnel, and infantry troops (for labor and security).

13-117. CBRN reconnaissance patrols normally reach the area of destruction first, to establish the nature and extent of contamination. Priority for decontamination and recovery help goes to personnel and equipment easily returned to combat.

13-118. The recovery detachment commander selects locations for setting up a medical point, CBRN decontamination station, damaged vehicle collection point, and an area for reconstituting units. He also designates routes to and from the area, for reinforcement and evacuation. He then reports to his next-higher commander on the situation and the measures taken. Meanwhile, engineers assigned to the detachment clear rubble, extinguish fires, rescue personnel, and build temporary roads.

13-119. The final step consists of reconstituting units and equipping them with weapons and combat vehicles. While the recovery detachment performs its mission, unaffected elements provide security against any further enemy activity.

SMOKE

13-120. The OPFOR plans to employ smoke extensively on the battlefield whenever the situation permits. Use of smoke can make it difficult for the enemy to conduct observation, determine the true disposition of OPFOR troops, and conduct fires (including precision weapon fires) or air attacks. The presence of toxic smokes may cause the enemy to use chemical protection systems, thus lowering his effectiveness, even if the OPFOR is using only neutral smoke.

ORGANIZATION

13-121. In the administrative force structure, army groups, armies, and corps typically have smoke companies in their chemical defense battalions and/or smoke battalions. In either case, the smoke companies each consist of 12 smoke-generator trucks and 12 smoke-generator trailers. The generators carried on the trailers can be towed by combat vehicles or may be mounted on other vehicles. The smoke companies also have assorted smoke pots, drums, barrels, and grenades. The decontamination company subordinate to a chemical defense battalion has some systems that can also generate large-scale protective smokescreens as a secondary mission and may augment a smoke battalion or company when required. These smoke-generating assets are often allocated to OSCs, which can then suballocate them to tactical-level subordinates. For additional information, see FM 7-100.4.

AGENTS

13-122. The OPFOR employs a mix of smoke agents and their delivery systems, as well as improvised obscurants to generate obscuration effects. Smoke agents may be either neutral or toxic. Neutral smoke agents are liquid agents, pyrotechnic mixtures, or phosphorus agents with no toxic characteristics. Toxic smokes (commonly referred to as combination smoke) may include tear gas or other agents. They degrade electro-optical (EO) devices in the visual and near-infrared (IR) wavebands. They can also debilitate an unmasked soldier by inducing watering of the eyes, vomiting, or itching.

13-123. Some of these smokes and other obscurants contain toxic compounds and known or suspected carcinogens. A prolonged exposure to obscurants in heavy concentrations can have toxic effects. The toxic effect of exposure to fog oil particles is uncertain and, as is true of all smokes, depends heavily upon dose, time, and frequency of exposure.

13-124. The more common obscurants used by the OPFOR include—
- **Petroleum** smokes (fog oil and diesel fuel).
- **Hexachloroethane** (HC) or hexachlorobenzene (HCB) smoke.
- **Aluminum-magnesium** alloy smoke (Type III IR for bispectral effects—visual/IR bands).
- **Phosphorus:** white phosphorus (WP), red phosphorus (RP), WP/butyl mix (PWP).

- **Metallic** (including graphite or brass) smokes for millimeter wave (MMW) band and multispectral effects.
- **Improvised obscurants** including colored signal smokes, dust, and burning tires and oil wells.

13-125. The OPFOR recognizes the need to counter target acquisition and guidance systems operating in the IR and microwave regions of the electromagnetic spectrum. It has fielded obscurants, including chaff, capable of attenuating such wavelengths. Table 13-1 shows several example EO systems defeated by obscurants.

Table 13-1. Electro-optical and other systems defeated by obscurants

Spectral Region	Optical/Electro-Optical System	Type of Obscurant
Ultraviolet 1 nm-0.4 µm	Developmental Sensors and Weapon Systems	Developmental Obscurants
Visible 0.4 µm-0.8 µm	Viewers: - Naked Eye - Day Sights, Optics - Camera Lens - EO Systems, Including Charged-Coupled Device (CCD) cameras - Battlefield TV and CCD TV ATGMs with Daylight Beacons	All NOTE: Obscurants can counter or degrade nighttime use of visual band illumination—including spotlights, flares, flashlights, and vehicle lights.
Near-IR 0.8 µm-1.3 µm	Viewers: - SACLOS ATGM Trackers - Night Vision (Image Intensifiers, IR) - CCD, aka Low-Light-level (LLL) TV Sensors: - Laser Designators, ND-Yag Laser - Laser Rangefinders, ND-Yag Laser	All NOTE: Obscurants can counter or degrade nighttime use of IR band illumination—including spotlights, flares, and night vision systems.
Short-Wave IR 1.3 µm-2.5 µm	Sensors: - Laser Rangefinders, Other Lasers - Laser Designators, Other Lasers	WP, PWP, RP, Dust, Type III IR Obscurant
Mid-IR 2.50 µm-7 µm	Viewers and Sensors (3-5µm): - Thermal Imagers - Terminal Homing Missiles	WP, PWP, RP, Dust, Type III IR Obscurant
Far-IR 7 µm-15 µm	Viewers and Sensors (8-12µm): - Thermal Imagers - Terminal Homing Missiles	WP and PWP (Instantaneous Interruption Only), Dust, Type III IR Obscurant
Millimeter Wave-Lower Frequency 300 GHz-30 GHz	Radars Communication Systems	MMW Band and Multispectral Obscurants
ATGM antitank guided missile	GHz gigahertz	ND neodymium nm nanometer
SACLOS semiautomatic command-to-line-of-sight guidance		TV television µm micrometer

13-126. As shown in table 13-1, smokes may operate in more than one band of the spectrum. The OPFOR is capable of employing obscurants effective in the visible through far-IR wavebands as well as portions of the MMW band. These obscurants are commonly referred to as multispectral smoke. The OPFOR uses a number of different smoke agents together for multispectral effects. For instance, an obscurant such as fog oil blocks portions of the electromagnetic spectrum more fully when seeded with chaff. The vast quantities of WP on the battlefield also suggest that random mixtures of this agent with other obscurants (both manmade and natural) could occur, by chance or design.

CBRN and Smoke

DELIVERY SYSTEMS

13-127. The OPFOR has an ample variety of equipment for smoke dissemination. Its munitions and equipment include—
- Smoke grenades.
- Vehicle engine exhaust smoke systems (VEESS).
- Smoke barrels, drums, and pots.
- Mortar, artillery, and rocket smoke rounds.
- Spray tanks (ground and air).
- Smoke bombs.
- Large-area smoke generators (ground and air).
- Improvised means (such as setting fires in forests, urban areas, fields, or oil soaked ground).

Although not designed for this purpose, some decontamination vehicles with chemical defense units can also generate smoke.

13-128. Smoke grenades include hand grenades, munitions for various grenade launchers, and smoke grenade-dispensing systems on armored vehicles. These grenades can provide quick smoke on the battlefield or fill gaps in smokescreens established by other means. Some armored fighting vehicles have forward-firing smoke grenade dispensers that can produce a bispectral screen up to 300 m ahead of vehicles.

13-129. All armored fighting vehicles can generate smoke through their exhaust systems. With these VEESS-equipped vehicles, a platoon can produce a screen that covers a battalion frontage for 4 to 6 minutes.

13-130. Smoke-filled artillery projectiles, smoke bombs, spray tanks, and generator systems are also common. Artillery can fire WP rounds (which have a moderate degrading effect on thermal imagers and a major one on lasers). The OPFOR makes considerable use of smoke pots emplaced by chemical troops, infantrymen, or other troops. The OPFOR still uses smoke bombs or pots dropped by fixed- or rotary-wing aircraft. For additional information on smoke systems, see the *Worldwide Equipment Guide*.

TYPES OF SMOKESCREENS

13-131. The OPFOR recognizes three types of smokescreens: blinding, camouflage, and decoy. Classification of each type as frontal, oblique, or flank depends on the screen's placement. Smokescreens are either stationary or mobile depending on prevailing winds and the dispensing means used. Each basic type can serve a different purpose. However, simultaneous use of all types is possible.

Blinding

13-132. *Blinding* smokescreens can mask friendly forces from enemy gunners, OPs, and target-acquisition systems. They can restrict the enemy's ability to engage the OPFOR effectively. Delivery of WP and plasticized white phosphorus (PWP) is possible using MRLs, artillery, mortars, fixed-wing aircraft, or helicopters. The OPFOR lays blinding smoke directly in front of enemy positions, particularly those of antitank weapons and OPs. Blinding smoke can reduce casualties significantly, since it can reduce an enemy soldier's ability to acquire targets by a factor of 10.

13-133. Blinding smokescreens are part of the artillery preparation before an attack and the fires in support of the attack. Likely targets are enemy defensive positions, rear assembly areas, counterattacking forces, and fire support positions. The screening properties of a blinding smokescreen can couple with dust, HE combustion effects, and the incendiary effects of phosphorus. This can create an environment in which fear and confusion add to the measured effectiveness of the smoke.

Camouflage

13-134. The OPFOR uses *camouflage* smokescreens to support all kinds of C3D measures. Such screens can—

- Cover maneuver.
- Conceal the location of units.
- Hide the nature and direction of attacks.
- Mislead the enemy regarding any of these.

13-135. The camouflage smokescreen is useful on or ahead of friendly troops. These screens are normally effective up to the point where forces deploy for combat. The number, size, and location of camouflage smokescreens vary depending on terrain, weather, and type of combat action. Camouflage also forces enemy attack helicopters to fly above or around a screen, thus exposing themselves to attack. Camouflage smoke can also cover assembly areas, approaches of exploitation forces, or withdrawals. Smokescreens can also cover a wide surface area around fixed installations or mobile units that do not move for extended periods.

13-136. Establishing camouflage smokescreens normally requires use of a combination of smoke grenades, smoke barrels, smoke pots, vehicles (and trailers) mounting smoke generating devices, and aircraft. Some decontamination vehicles also have the capability to generate smoke.

13-137. Two smoke-generator vehicles can lay a smokescreen of sufficient size to cover a battalion advancing to the attack. For larger smokescreens, the OPFOR divides the smokescreen line into segments and assigns two vehicles to each segment. Doctrinally, camouflage smokescreens should cover an area at least five times the width of the attacking unit's frontage.

13-138. The threat of enemy helicopter-mounted antitank systems concerns the OPFOR. Consequently, its doctrine calls for advancing forces to move as close behind the smokescreen as possible. The higher the smokescreen, the higher an enemy helicopter must go to observe troop movement behind the smokescreen, and the more vulnerable it is to ground-based air defense weapons. Depending on weather and terrain, some large-area smoke generators can produce screens up to several hundred meters high. There is considerable observation-free maneuver space behind a screen of this height. Conversely, smoke pots provide a screen 5 to 10 m high. This screen masks against ground observation but leaves the force vulnerable to helicopters "hugging the deck" and popping up to shoot.

13-139. The protection produced by camouflage smoke also interacts as a *protective* smoke. Just as smokescreens can degrade enemy night-vision sights, the protective smoke can shield friendly EO devices from potentially harmful laser radiation. This protective effect is greater with a darker smoke cloud because of the better absorption capability of that cloud. Protective smokescreens are also a good means of reducing the effects of thermal radiation from nuclear explosions. A protective smokescreen is useful in front of, around, or on top of friendly positions.

Decoy

13-140. A *decoy* smokescreen can deceive an enemy about the location of friendly forces and the probable direction of attack. If the enemy fires into the decoy smoke, the OPFOR can pinpoint the enemy firing systems and adjust its fire plan for the true attack. The site and location of decoy screens depend on the type of combat action, time available, terrain, and weather conditions. One use of decoy smoke is to screen simultaneously several possible crossing sites at a water obstacle. This makes it difficult for the enemy to determine which site(s) the OPFOR is actually using.

AREA SMOKESCREENS

13-141. Area smokescreens can cover wide surface areas occupied by fixed or semifixed facilities, or by mobile facilities or units that must remain in one location for extended periods. Screens set down on a broad frontage can also cover maneuver forces. The OPFOR uses area smokescreens to counter enemy precision weapons and deep attacks.

13-142. The means of generating area smokescreens can be either subordinate or supporting chemical units or the use of smoke pots, barrels, grenades, and VEESS. As the situation dictates, the objects screened by area smokescreens can include—
- Troop concentrations and assembly areas.
- CPs.
- Radar sites.
- Bridges and water obstacle-crossing sites.

13-143. The OPFOR can also screen air avenues of approach to such locations. It tries to eliminate reference points that could aid enemy aviation in targeting a screened location. To create an effective smokescreen against air attacks, the OPFOR must establish an effective air defense and CBRN warning communications network so that a smokescreen can be generated in time to degrade reconnaissance and targeting devices on incoming aircraft. Units using smoke must maintain reliable communications and continuous coordination with air defense early warning units and air defense firing positions.

13-144. The OPFOR follows the following basic principles for generating area smokescreens:
- Screening should include not only the protected object but also surrounding terrain or manmade features so as to deny the enemy reference points.
- The protected installation should not be in the center of the screen.
- The smoke release points must not disclose the outer contours of the screened object.
- Screening must be initiated early enough to allow the area to be blanketed by the time of the enemy attack.
- If possible, decoy smokescreens should be used.
- For larger objects (such as airfields and troop concentrations), the screen should be at least twice as large as the object.
- For smaller objects (such as depots, small crossing points, and radar sites), the screen should be at least 15 times as large as the object.

Depending on terrain, smoke release points are set up within a checkerboard pattern, in a ring (circle), or in a mix of the two patterns that covers the area to be screened.

Checkerboard Area Smokescreen

13-145. A checkerboard pattern is a rectangle that is divided into 4-km² squares with smoke release points distributed evenly within each square (see figure 13-2). This pattern is useful if the terrain is contoured or covered with buildings, trees, or other obstructions that prevent the precise distribution of smoke points.

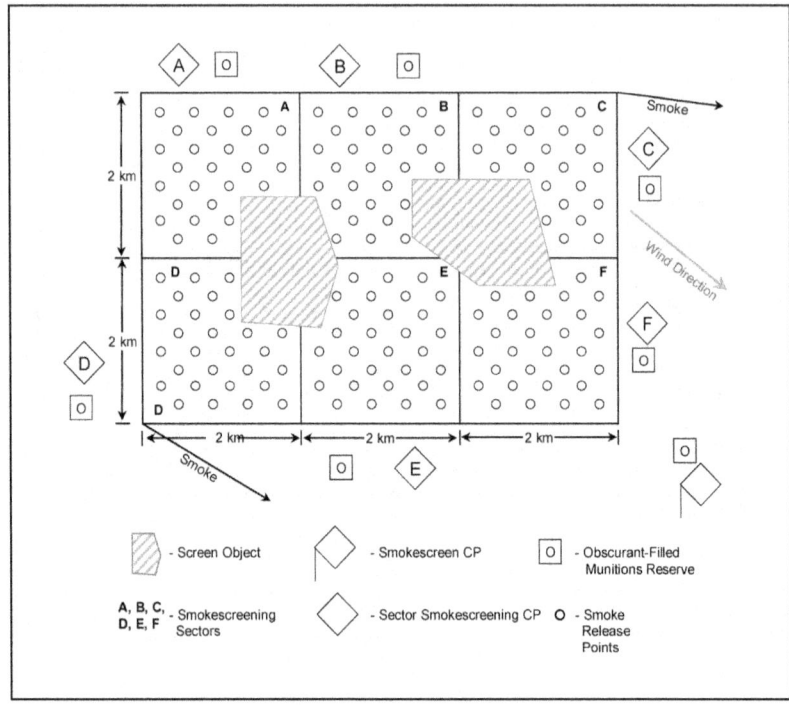

Figure 13-2. Checkerboard area smokescreen (example)

CBRN and Smoke

Ring Area Smokescreen

13-146. A circle or set of concentric rings of smoke release points works well on relatively flat, featureless terrain (see figure 13-3). Generally, the distance between the target and the first obscurant-generation ring is 100 to 250 m. The distance between smoke release points within each ring varies between 20 and 100 m, depending on the obscurant device being used and the meteorological conditions.

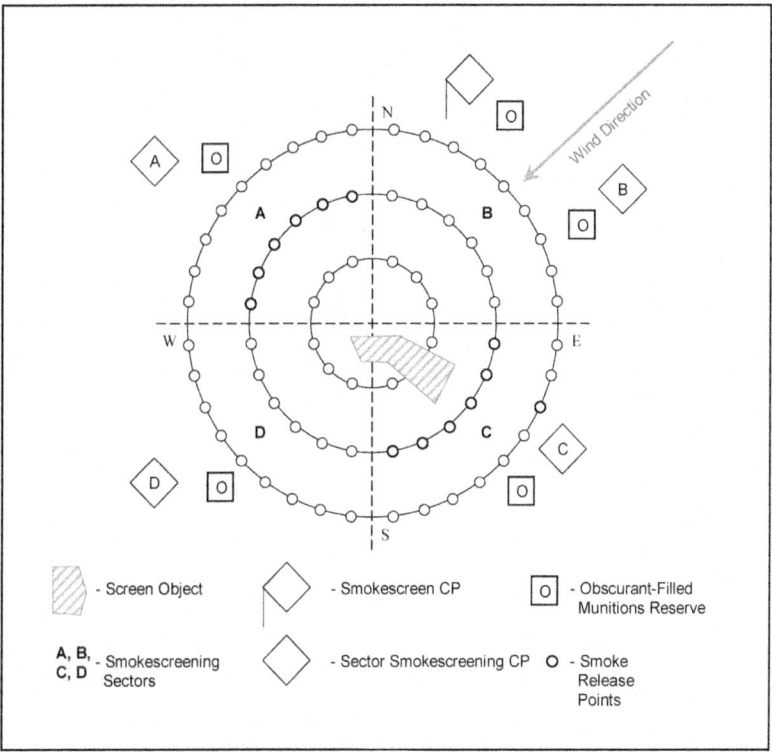

Figure 13-3. Ring area smokescreen (example)

Mixed Area Smokescreen

13-147. The OPFOR uses checkerboard area smokescreens and ring area smokescreens together when objects 3 to 4 km apart must be screened simultaneously. The rings of smoke-generation lines are placed around each object to be screened, and these rings are placed within the squares of a checkerboard. See figure 13-4 for an example of a mixed area smokescreen.

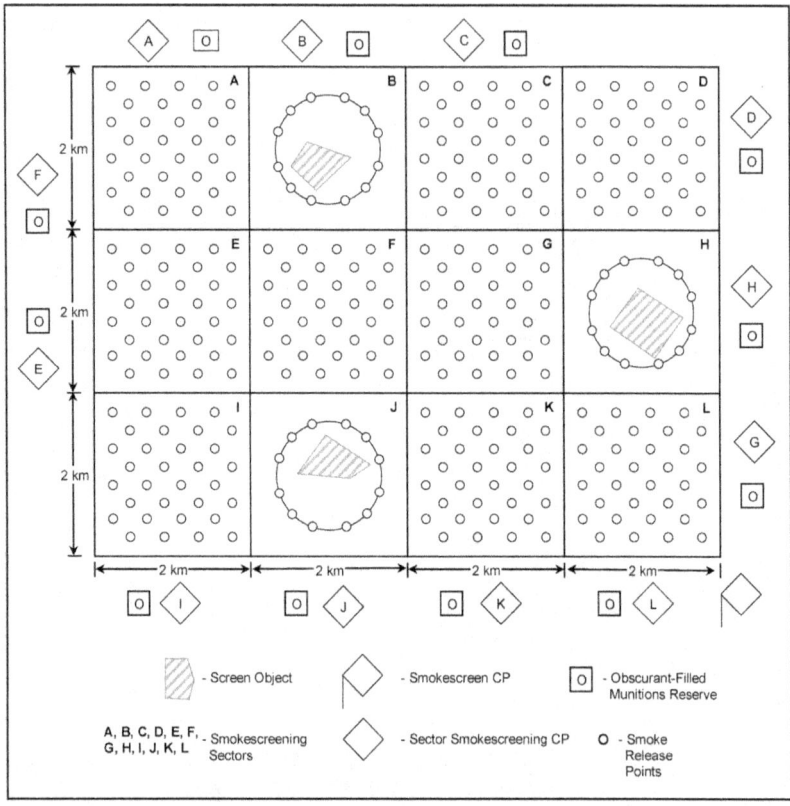

Figure 13-4. Mixed area smokescreen (example)

TACTICAL SMOKESCREEN EMPLOYMENT

13-148. The use of smoke is an important part of tactical C3D efforts. The OPFOR can use smokescreens to blind or deceive enemy forces and to conceal friendly forces from observation and targeting. Smoke can screen units near the battle lines, as well as those in support zones, from direct fire, reconnaissance, and air attack. It has applications in offense and defense, including tactical movement. Smoke is very effective in screening water obstacle crossings. It also has specific applications at night. Table 13-2 shows tactical options for employing smoke and other obscurants.

CBRN and Smoke

Table 13-2. Tactical employment of smoke and other obscurants from various sources

Source	Placement			Uses			
	On Friendly	Between	On Enemy	Blinding	Camouflage	Decoy	Signaling
Smoke Grenade	X	X		X	X	X	X
Smoke Generator	X	X			X	X	
Smoke Pot	X	X			X	X	X
VEESS	X				X	X	
Vehicle Dust	X				X	X	
Helicopter	X	X	X		X	X	
Mortar/Artillery Smoke		X	X	X	X	X	X
Rocket		X	X	X			
Aerial Bomb		X	X	X			
Aircraft Spray	X	X	X	X	X	X	
Mortar/Artillery HE Dust		X	X	X			

Offense

13-149. The OPFOR emphasizes the use of smoke during the offense to help reduce friendly battle losses. However, it understands that smoke may hinder its own C2, battlefield observation, and target engagement capabilities. In addition, the enemy may take advantage of OPFOR smokescreens to shield his own maneuvers or to carry out a surprise attack or counterattack. Thus, a smokescreen is successful when the OPFOR attackers are able to maintain their assigned axis and retain sight of the objective. To prevent the smoke from interfering with friendly maneuver, OPFOR commanders coordinate the planned location and duration of the smoke-generation lines or points with the scheme of maneuver.

13-150. Smoke pots, artillery, mortars, and aircraft are the primary means of smoke dissemination in the offense. Artillery and aircraft are used to spread screening smoke throughout the tactical depth of the enemy's defense. They are also useful in screening the flanks of attacking units.

13-151. The OPFOR uses camouflage, blinding, and decoy smokescreens to conceal the direction and time of attack. The OPFOR can place smoke on enemy firing positions and OPs before and during an attack. Smoke has uses during various types of offensive action, which may be conducted against an enemy occupying defensive positions or an enemy on the move.

Enemy in Defensive Positions

13-152. During the offense, a camouflage smokescreen is typically used to conceal combat formations that are advancing and maneuvering toward the enemy's defensive positions. With a tail wind, the action and enabling forces or elements can generate enough smoke to adequately screen their front. Then they can advance behind the screen as it blows toward the enemy.

13-153. The example in figure 13-5 shows an independent mission detachment (IMD) with two mechanized infantry companies and two tank companies. This IMD is advancing from a wooded area in platoon formations with a flanking wind. In such conditions, the IMD may use its VEESS-equipped tanks and IFVs and smoke grenades. After turning on their VEESS, the tanks and IFVs advance toward the enemy's defensive positions while firing on visible targets. Dismounted infantrymen equipped with smoke grenades follow on foot behind the extended line of tanks and IFVs. As gaps develop in the smokescreen, the infantrymen approach and throw smoke grenades. Infantrymen can also fire incendiary smoke charges from a variant of the encapsulated flamethrower out to a range of up to 1 km. In this example, artillery also delivers blinding smoke where the wind will carry it through the enemy defensive positions.

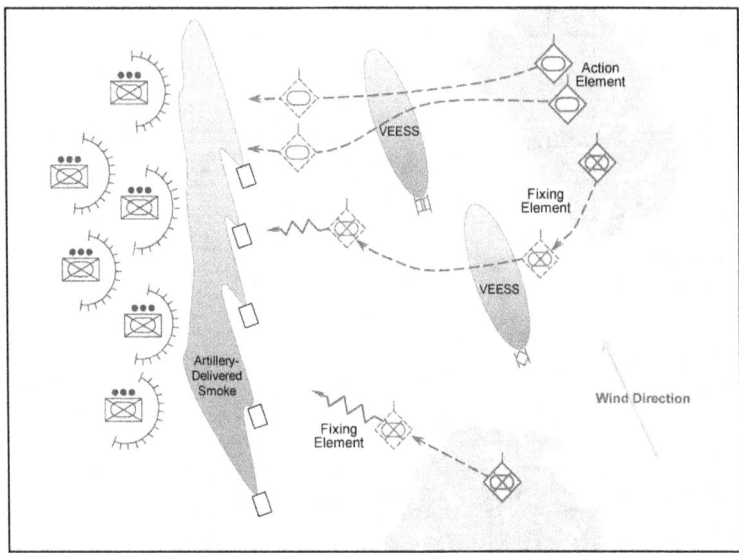

Figure 13-5. Smoke in the offense against an enemy in defensive positions (example 1)

13-154. Figure 13-6 shows another example of smoke employment during an attack against an enemy in defensive positions. In this example, the commander has more opportunity to plan and prepare for a coordinated smokescreen than in the previous example. Again, a camouflage smokescreen is used to prevent observation of advancing fixing and action elements. As the advancing action element nears the rear of units of the fixing element already in contact with the enemy, the unit in contact may set up a camouflage smokescreen using smoke pots and VEESS of forward-deployed armored vehicles. In addition, artillery can deliver blinding smoke on enemy defensive positions while units of the action element negotiate minefields in front of them.

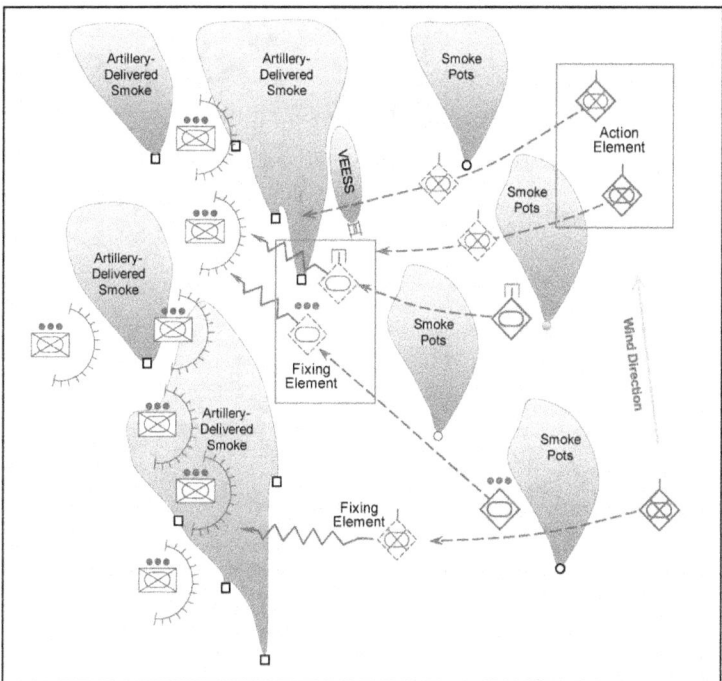

Figure 13-6. Smoke in the offense against an enemy in defensive positions (example 2)

Enemy on the Move

13-155. Figure 13-7 on page 13-28 shows an example of the use of smoke in an offensive action in which the enemy is on the move. This example involves an IMD using two camouflage and one decoy smokescreens. An IMD based on a mechanized infantry battalion with an additional tank company is attacking toward the enemy's left flank (as viewed by the OPFOR). It is camouflaging that attack with a smokescreen laid by VEESS of tanks from the tank company. The tanks move at 100-m intervals to create a continuous smoke cloud. The distance was calculated on the basis of meteorological conditions and the fact that a smokescreen can extend 300 to 400 m from a VEESS and still remain impenetrable to vision. Meanwhile, a mechanized infantry company facing the enemy's right flank lays a decoy smokescreen to divert the enemy's attention from the actual attack. One of the companies on the left flank lays smoke pots along a 1,500-m line at intervals of 20 to 25 m for a total burning time of 6 minutes. The company divides the work among its three mechanized infantry platoons, with each responsible for laying pots along 500 m of the line.

Chapter 13

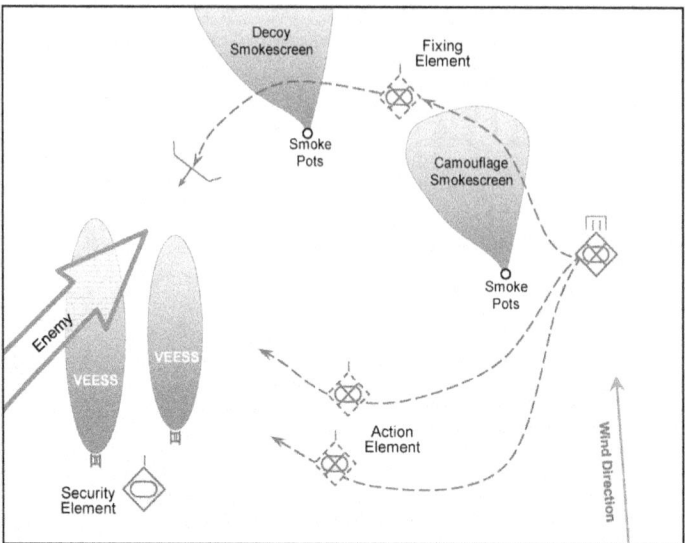

Figure 13-7. Smoke in the offense against an enemy on the move (example)

13-156. Another example of smoke employment in an offensive action against an enemy on the move could be the simultaneous use of frontal camouflage and blinding smokescreens during an enemy counterattack. Following an artillery attack against the enemy advancing for a counterattack, artillery would deliver blinding smoke directly in front of the advancing enemy and camouflage smoke in front of the advancing battalion-size mechanized infantry detachment. As soon as the blinding smokescreen on the enemy dissipates, the detachment's antitank guided missiles (ATGMs) and augmenting tanks would open fire on the enemy.

Defense

13-157. In the defense, the OPFOR may use of smokescreens for—
- Camouflaging the maneuver of friendly units.
- Concealing engineer activities from enemy observation.
- Screening replacements of units under conditions of good visibility.
- Camouflaging the approach of friendly units for a counterattack.
- Screening the movements of defending units between battle positions.
- Providing flank and maneuver security.
- Misleading the enemy on the disposition of reserves and planned counterattack axes.

13-158. Because a completely obscured environment tends to aid the attacker more than the defender, an OPFOR defense uses smoke to minimize the enemy's vision while allowing the defenders a fairly clear view of the enemy's location. Smoke from artillery and mortar shells is the most effective means of blinding an advancing enemy while keeping friendly forces out of the obscured area. The OPFOR would use VEESS, smoke pots, and smoke grenades only to assume the defense while in contact with the enemy, to change positions, or to begin a withdrawal from contact.

13-159. Figure 13-8 shows an example of an IMD using smoke devices in the defense to disrupt and subsequently defeat the attacking enemy. In this example, one of the IMD's mechanized infantry companies disperses its platoons and sends one of those platoons out to lay smoke along three successive smoke-generation lines. As the enemy force approaches, it first encounters smoke from two lines using smoke pots emplaced by the infantrymen and then a final line created by the platoon's IFVs using VEESS. The smoke from those lines disrupts the enemy advance and creates a favorable situation for the defending IMD to launch a counterattack with its tank company from the flank.

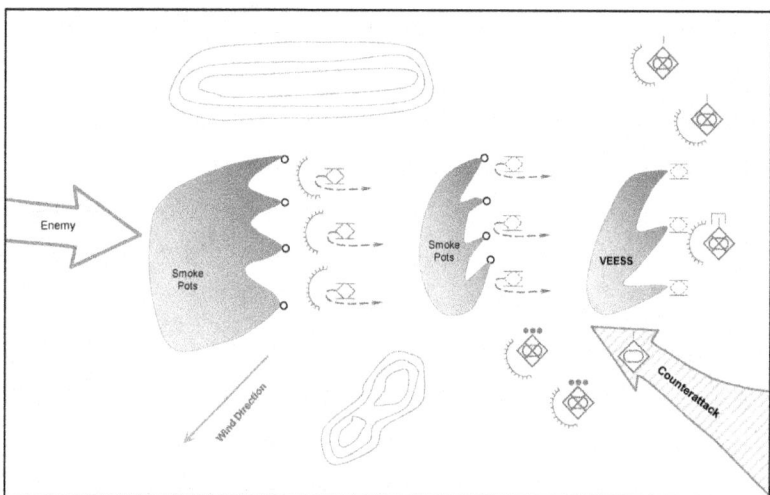

Figure 13-8. Smoke in the defense (example)

Water Obstacle Crossing

13-160. Because of their vulnerability to air attack and direct fires, successful water obstacle crossings require smokescreens for concealment. The OPFOR can place 2 to 3 hours' worth of screening smoke along a wide frontage to cover units conducting water obstacle crossings. It may also place floating smoke pots and barrels in the water. It distinguishes between opposed and unopposed crossings (see chapter 12).

13-161. For *opposed crossings*, OPFOR doctrine emphasizes using all three types of smokescreens (blinding, camouflage, and decoy). An opposed crossing requires greater planning and preparation than an unopposed crossing, because it anticipates contact with the enemy. First, unfavorable meteorological conditions are more difficult to overcome. Friendly forces must have a tail wind or at least a flanking wind in order for smoke generators and smoke pots on the near bank to screen the crossing sites. If the OPFOR faces a head wind, only artillery or aircraft can deliver a blinding smokescreen against enemy positions on the opposite bank. Whenever possible, the OPFOR prefers to lay smoke on both sides of the river. The use of decoy smoke at one or more other likely crossing sites can deceive the enemy as to the actual crossing location. Figure 13-9 on page 13-30 shows an example of an IMD based on a mechanized infantry battalion with augmenting tanks. This example uses smoke delivered by several means to cover friendly forces and deceive and blind enemy forces.

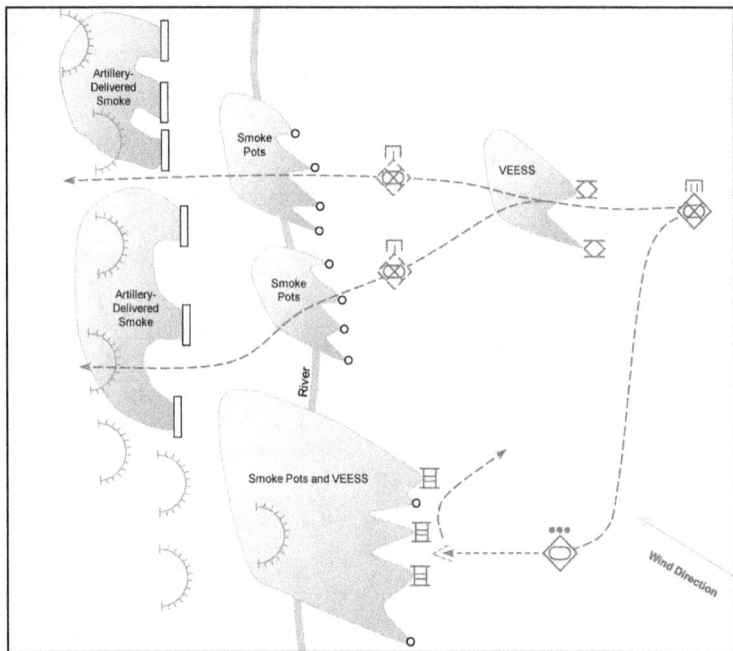

Figure 13-9. Smoke in an opposed water obstacle crossing (example)

13-162. *Unopposed* water obstacle crossings far from the battle line may be crucial for supporting tactical and operational missions. Therefore, they also require the use of smokescreens for concealment whenever feasible. As with smoke use near the battle line, it is important to establish at least one or two decoy smokescreens for every actual crossing site. This is because a smoke cloud in the rear attracts the attention of enemy reconnaissance. Area smokescreens are best for covering crossing sites and surrounding terrain.

Combat at Night

13-163. At night, the OPFOR can conceal its forces from enemy active and passive night-vision and thermal imaging devices by using smoke and other obscurants that are effective in the visible through far-IR portions of the electromagnetic spectrum. The OPFOR uses smoke in three ways to counter various types of enemy EO sensors:

- For active night-vision devices, blind with smoke.
- For passive night-vision devices, blind with illumination or combined use of illuminating and smoke projectiles.
- For thermal imaging devices, camouflage friendly troops with smoke and illuminate enemy targets (to benefit friendly night-vision devices) at the same time.

13-164. In essence, the OPFOR uses smoke when it cannot quickly destroy or neutralize the enemy EO devices, or when the enemy has created high levels of illumination within his defense. However, it can also use smoke in conjunction with its own illumination.

13-165. In night combat, the OPFOR can use smoke to help illuminate enemy vehicles and other targets. The most effective method is to use smoke in conjunction with illuminating rounds to silhouette enemy

CBRN and Smoke

vehicles and other targets. A mechanized infantry battalion can use this method by firing mortar smoke rounds to burst 50 to 100 m beyond the targets, interspersed with illuminating rounds aimed just beyond the screen. This creates a broad, bright background.

13-166. A more elaborate version of the latter method involves the use of artillery-delivered smoke and illuminating rounds. The OPFOR can use smoke and other obscurants to blind the enemy's night-vision equipment. Image intensifiers can be blinded by obscurants and forced to shut down by flares or the flash of artillery shells. In the defense, therefore, OPFOR artillery could use close support fire on advancing enemy forces, alternating blinding smoke with illuminating rounds to blind enemy forces while simultaneously illuminating them for targeting. See figure 13-10 for an example of this technique.

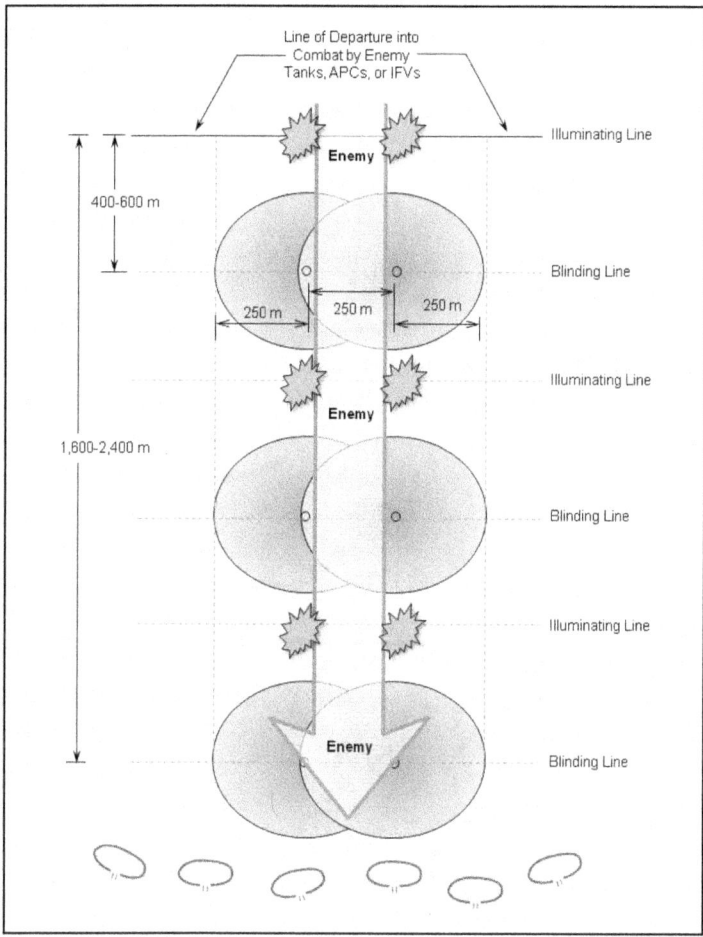

Figure 13-10. Alternating blinding smoke and illuminating lines against the enemy at night (example)

Chapter 13

SIGNALING SMOKE

13-167. Aside from smokescreens, the OPFOR also uses colored smoke for signal purposes. Smoke can mark enemy positions or, occasionally, friendly positions or movement routes for the information of supporting aircraft or artillery. By prearrangement, colored smoke may—
- Identify friendly units.
- Identify targets.
- Control the commencing and lifting of fire.
- Coordinate fire and maneuver of combat units.

Chapter 14
Logistics

Logistics is the process of planning and executing the sustainment of forces in support of military actions. At the tactical level, it focuses on the traditional combat service support functions of materiel support (supply), maintenance, transportation, personnel support, and medical support. These tasks present a challenge in modern combat, where there is not always a clearly defined front line or a relatively secure rear area. Combat can spread over a deep and wide area. Within such an area, combat actions and attrition may not occur evenly or predictably. There may be areas of intense battles and local destruction, while other secondary or defensive sectors have much lighter logistics demands. This requires a flexible logistics system designed to sustain forces throughout conflict, adapting as conditions change.

STRATEGIC AND OPERATIONAL LOGISTICS SUPPORT

14-1. The State's national-level logistics system is designed to provide continuous support to the civilian populace while simultaneously supporting military forces from the strategic level to the individual fighting unit. The OPFOR continues to make major improvements in all aspects of its logistics system. This includes an increased emphasis on support zone security and plans for stockpiling war materiel throughout the country.

14-2. Operational logistics links strategic-level logistics resources with the tactical level of logistics, thus creating the conditions for effective sustainment of a combat force. It covers the support activities required to sustain campaigns and major operations. A dependable logistics system helps commanders seize and maintain the initiative. Operational maneuver and the exploitation of operational or tactical success often hinge on the adequacy of logistics and the ability of the force to safeguard its critical lines of communications (LOCs), materiel, and infrastructure. Operational logisticians interface with tactical-level logisticians in order to determine shortfalls and communicate these shortfalls back to the strategic logistics complex to support operational priorities.

TACTICAL STAFF RESPONSIBILITIES

14-3. At all levels of command, down through division and brigade, the resources section of the primary staff is the principal office for the logistics integration of supply, maintenance, transportation, and services. The resources officer heads this section, with two subsections headed by secondary staff officers who support him: the chief of logistics and the chief of administration. The resources section establishes and controls the sustainment command post (CP) to supervise the execution of sustainment procedures and the movement of support troops. It contains staff officers for fuel supply, medical support, combat equipment repair, ammunition, clothing, and food supply. It is also designed to serve as an alternate CP or to provide multiple sustainment CPs. If an additional sustainment CP is required, the assistant resources officer will control it. The resources section is also structured to accommodate augmentation from the functional staff and liaison teams. See figure 14-1 on page 14-2 for the organization of the resources section.

Chapter 14

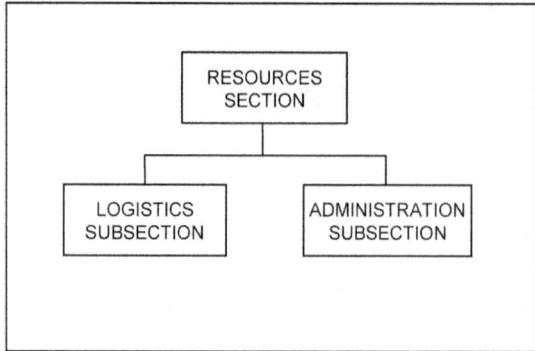

Figure 14-1. Resources section

Note. Throughout this chapter, references to division- and brigade-level logistics support may also apply to a division tactical group (DTG) and brigade tactical group (BTG), unless specifically stated otherwise. Likewise, reverences to a battalion also apply to a battalion-size detachment (BDET).

14-4. On a battalion-level staff, there is a resources officer, who also performs the functions of chief of administration. The battalion's materiel support platoon leader serves as the chief of logistics.

RESOURCES OFFICER

14-5. The resources officer is responsible for the requisition, acquisition, distribution, and care of all of the command's resources, both human and materiel. He executes staff supervision over procedures that ensure the command's logistics and administrative requirements are met. One additional major task of the resources officer is to free the commander from the need to bring his influence to bear on priority logistics and administrative operations. He is also the officer in charge of the sustainment CP.

CHIEF OF LOGISTICS

14-6. The chief of logistics is responsible for—
- Managing the order, receipt, and distribution of supplies to sustain the command.
- Maintaining the condition and combat readiness of armaments and related combat equipment.
- Ensuring the supply, proper utilization, repair, and evacuation or armaments and equipment.
- Overseeing the supply and maintenance of the command's combat and technical equipment.

These responsibilities encompass the essential wartime tasks of organizing and controlling the command's recovery, repair, and replacement system. During combat, he keeps the commander informed on the status of the command's equipment.

CHIEF OF ADMINISTRATION

14-7. The chief of administration supervises all personnel actions and transactions in the command. His subsection—
- Maintains daily strength reports.
- Records changes in table of organization and equipment of units in the administrative force structure (AFS).
- Assigns personnel.

Logistics

- Requests replacements.
- Records losses.
- Administers awards and decorations.
- Collects, records, and disposes of war booty.

TACTICAL LOGISTICS CONCEPTS

14-8. Logistics support must complement the force structure and sustain combat actions. The logistics elements must be ready to provide full support at the start of combat and be capable of rapid movement to keep pace with maneuver forces. A greater quantity of logistics support is concentrated on the combat force assigned the principal mission in a given battle. The OPFOR relies on the following three concepts:

CENTRALIZED PLANNING AND DECENTRALIZED EXECUTION

14-9. To ensure both priority of effort and efficiency in the logistics process, logistics plans are developed at higher levels and executed by units and organizations at lower levels. At division and brigade level, the resources officer has overall responsibility for logistics planning. Centralized planning requires a focal point for logistics planning and resource allocation at all levels. Regardless of whether the focal point is an individual (the resources officer or his secondary staff) or a unit, it must be constantly aware of requirements and capabilities. Decentralized execution enhances the flexibility of lower-level commanders to meet local requirements and to rapidly reprioritize support.

14-10. A careful study of the missions of the total force allows planners to program and measure logistics requirements. This requires concurrent operational, tactical, and logistics planning. Each level of command is responsible for the timely and complete provision of logistics support to subordinate units from available assets. The commander—

- Allocates these assets to support the mission of his units.
- Shifts resources according to the combat situation.
- Retains some emergency reserves to meet unexpected contingencies.

14-11. The bulk of supplies and transport resources are concentrated at the strategic and operational levels. This centralization of logistics resources contributes to operational and tactical flexibility. It enables operational-level commanders to concentrate support where it is needed most, if necessary switching axes rapidly to take advantage of unexpected opportunities. They can quickly strip resources from stalled divisions or brigades and reallocate them to units making better progress. Centralization of resources at the operational level frees divisions or brigades and their subordinates of an unnecessarily large logistics tail, making it easier for them to engage in high-speed maneuver battles.

SUPPORT FORWARD

14-12. Logistics units are organized and deployed to support forward. The guiding principle is that a combat force should retain its organic support resources (such as trucks, recovery equipment, and ambulances) to support its subordinate units. It should not have to use its own resources to go to support areas to pick up supplies or to evacuate resources that can no longer contribute to combat power.

SUSTAINMENT FROM OTHER SOURCES

14-13. Finally, the logistics system may have to rely on sustainment from other than military sources. Supplies may be procured or obtained from social groups, consumer cooperatives, farms, or individual citizens, and by coercion or foraging in the area of responsibility (AOR). Captured enemy supplies and equipment are another source of outside sustainment.

LOGISTICS MISSIONS

14-14. In operational and tactical logistics, three terms describe how the OPFOR provides support to the field. These terms are primary support, area support, and depot support.

14-15. *Primary support* is a mission given to supply, services, transportation, and maintenance units that normally provide support directly to other units. This allows the primary support unit to respond directly to the supported unit's request for assistance or supplies.

14-16. *Area support* is a mission given to supply, services, transportation, and maintenance units that normally provide support to primary support units and other area support units. Lower-priority units may have to rely on area support, rather than receiving supplies and services directly from the next-higher echelon.

14-17. *Depot support* is a mission given to national-level or strategic units that normally provide support to area support units. Depot support missions include the receipt, storage, and issue of war stocks and domestically produced armaments and materiel, and the overhaul and rebuilding of major end items.

TAILORED LOGISTICS UNITS

14-18. The OPFOR concentrates the bulk of logistics units at two levels—theater and operational-strategic command (OSC). This concentration supports the OPFOR philosophy of streamlined, highly mobile combat elements at the tactical level. These higher levels maintain the responsibility and the primary means for logistics support.

14-19. Tailoring allows allocation of logistics resources to the combat elements most essential to mission success. It also allows the OPFOR to assign priorities for logistics support. Subordinate units receive assets according to—

- The importance of their mission.
- The nature of the terrain.
- The level of fighting anticipated.

Commanders can reallocate their own resources in line with changes in the situation. They can also take away a subordinate's organic resources and assign them to another subordinate if the situation warrants.

ADMINISTRATIVE AND FIGHTING FORCE STRUCTURE

14-20. The AFS is the aggregate of military headquarters, facilities, and installations that are designed to man, train, and equip the OPFOR. In wartime, the normal role of an operational-level administrative headquarters is to provide forces for the creation of fighting commands, such as OSCs and DTGs. After transferring control of its major fighting forces to one or more task-organized fighting commands, an administrative headquarters, facility, or installation continues to provide depot and area support-level administrative, supply, and maintenance functions.

14-21. Tailoring of the OPFOR's fighting force structure affects both the number and type of subordinate combat elements and the number and type of assigned logistics units. Divisions and brigades augmented to become DTGs or BTGs in the wartime force structure have increased requirements for logistics support.

INTEGRATED SUPPORT COMMAND

14-22. The integrated support command (ISC) is the aggregate of combat service support units (and perhaps some combat support units) constituent to a division and additional assets allocated from the AFS to a DTG. It contains such units that the division or DTG does not suballocate to lower levels of command in a constituent or dedicated relationship.

14-23. The division (or DTG) further allocates part of its ISC units as an integrated support group (ISG) to support its integrated fires command (IFC). It uses the remainder to support the rest of the division, as a second ISG. For organizational efficiency, combat service support units may be grouped in this ISC and its ISGs, although they may support only one of the major units of the division or IFC. Sometimes, an ISC or ISG might also include units performing combat support tasks that support the division and its IFC. Such tasks may include engineer, chemical warfare; information warfare (INFOWAR); reconnaissance, intelligence, surveillance, and target acquisition (RISTA); or law enforcement. (See chapters 2 and 9 for more detail on the IFC.)

Logistics

14-24. The ISC's mission is to provide command and control (C2), administrative, operations, and support personnel and equipment required for forming the nucleus of the two ISGs. The division resources officer (in consultation with his chiefs of logistics and administration and the ISC commander) task-organizes the ISGs based upon support mission requirements.

14-25. The ISC commander and his staff are the division logisticians. The ISC commander advises the division commander, resources officer, and the rest of the division staff on those logistics matters pertaining to ISG functions. The ISC commander normally receives guidance and direction from the division commander. The overall responsibility for logistics planning belongs to the division resources officer. The division commander tasks the ISC commander to evaluate the logistics supportability of future battle plans or courses of action. The ISC commander tasks and provides guidance to the ISC staff. The ISC staff gives the alternatives and preferred solutions to the commander for a decision. See figure 14-2 for an organizational breakout of the ISC headquarters.

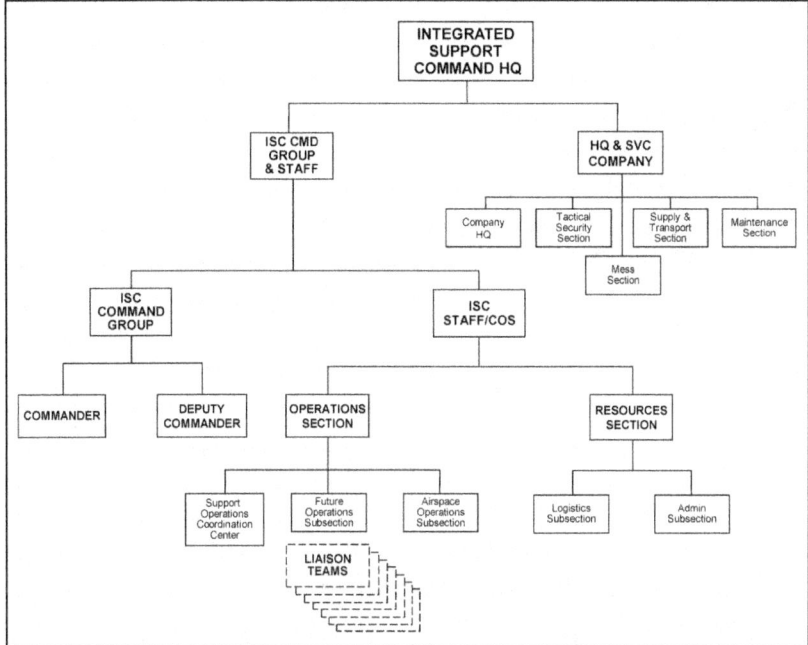

Figure 14-2. Integrated support command headquarters

14-26. The ISC headquarters is composed of the ISC command group and staff and the headquarters and service company. The ISC command group contains the ISC commander and deputy commander. The ISC staff, headed by the chief of staff, is composed of the operations section and resources section. Located within the operations section is the support operations coordination center (SOCC). The SOCC is the staff element responsible for the planning and coordination of support for the division and the IFC. The RISTA and INFOWAR officer resides in the SOCC to ensure all ISC intelligence and INFOWAR requirements are met. The SOCC relies principally on direct liaison among all the ISC subordinate units to ensure the necessary coordination of logistics support for combat actions. Other major components of the operations section are the future operations subsection and airspace operations subsection. For additional details on the organization of the ISC, see FM 7-100.4.

14-27. Liaison teams are not a permanent part of the ISC staff structure. They support the ISC staff with detailed expertise in the mission areas of their own particular branch or service. They also provide direct communications to subordinate and supporting units executing missions in those areas. All liaison teams are under the direct control of the operations section. The operations officer is responsible for ensuring proper placement and utilization of the teams. The ISC staff will also receive liaison teams from multinational and interagency subordinates and from affiliated forces. The number and types of liaison teams is fluid and is determined by many factors. Liaison teams augmenting the ISC staff provide their own equipment.

14-28. The headquarters and service company provides administrative, logistics, and security support to the ISC staff, including general security for the sustainment CP(s). The operations section provides the control, coordination, and communications for the headquarters.

INTEGRATED SUPPORT GROUP

14-29. The ISG is a compilation of units performing various support tasks (primarily logistics,) that support the division and its IFC. Normally, separate ISGs are organized to support the division and the IFC. The ISG has six major functions:

- Supply.
- Maintenance.
- Transportation.
- Medical support.
- Personnel services.
- Field services.

However, the ISG may also perform engineer, chemical warfare, law enforcement, RISTA, or INFOWAR tasks.

14-30. There is no standard ISG organizational structure. The number, type, and mix of subordinate elements vary based on the tactical support situation. For example, an ISG supporting a division composed mainly of tank and mechanized infantry units will differ from an ISG supporting a division composed mainly of infantry or motorized infantry units. Even within a division that receives no augmentation, there can be variations as to which division subordinates may belong to either of the ISGs and which ones are in which ISG.

14-31. In essence, the ISG is tailored to the mission. In the case of a DTG, it is also tailored to the task organization of the DTG. Figure 14-3 is one example of a rather robust ISG that might be appropriate for a DTG not relying on extensive support from a parent OSC. As the number and type of supported units change, the ISGs change the way in which subordinate units are organized to provide support. When the logistics units allocated from the operational level are no longer required for ISG functions, the primary or area support units will do one of the following:

- Revert to control of their original parent units in the AFS.
- Be assigned to other DTGs, as appropriate.

Logistics

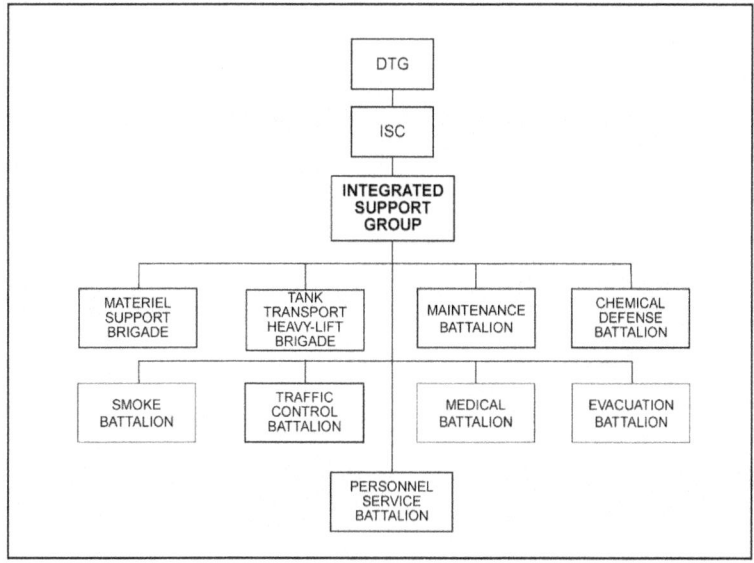

Figure 14-3. DTG ISG (example)

MATERIEL SUPPORT

14-32. The OPFOR materiel support system comprises a mix of very modern and less modern capabilities that vary depending on the priority of the supported units. Generally, high-priority or elite units enjoy the benefits of a robust materiel support system that affords a higher degree of flexibility and responsiveness to rapid changes in plans. For such units, the system may be fully automated to track requirements and control the issue of supplies. Less capable units (including reserve and militia forces) typically have little or no automation support. Both types of materiel support system are based on allocating supplies and services to units in order to accomplish mission objectives. However, the aim is to continue the upgrade of less capable units to a robust supply system capable of sustaining the force in all environments.

14-33. *Supply* includes actions to acquire, manage, receive, store, and issue the materiel required to equip and sustain the force. These actions occur from deployment through combat operations and recovery into OPFOR-controlled territory. The allocation of supplies is based on—

- Unit mission.
- Supply reports.
- Availability of supplies.

14-34. The OPFOR concept of *services* includes all troops, installations, and duty positions that perform logistics support for combat arms units. Such services are not specific to the ground forces, but support other Armed Forces components as well.

14-35. During peacetime, the OPFOR operates under the "pull system" of supply. For example, units in the field may request materiel from a depot where they must pick it up and deliver it to the field.

14-36. During wartime, the OPFOR operates under the forward distribution or "push system" principle, in which the higher echelon directly supplies and services the next-lower echelon. Supplies and services

Chapter 14

are delivered directly to subordinate elements using the transportation assets of the higher headquarters. Lower-priority units may have to rely on area support or even supply point distribution.

METHODS OF DISTRIBUTION

14-37. The three methods by which supplying units distribute supplies to using units are—
- *Supply point distribution*, in which the supplying unit issues supplies from a supply point to a receiving unit. The receiving unit must go to the supply point and use its own transportation to move supplies to where they are needed.
- *Unit distribution*, in which the supplying unit issue supplies and delivers supplies to the receiving unit's area in transportation assets the supplying unit has arranged.
- *Throughput*, in which shipments bypass intermediate supply points or logistics sites. Throughput eliminates the need for double handling, uses transportation assets more efficiently, and is more responsive to the user's needs.

SUPPLY PRIORITIES

14-38. The OPFOR places primary emphasis on maintaining the supply of ammunition, fuel, and weapons. Its logistics system typically operates on the following sequence of priorities:
- Ammunition of all types.
- Petroleum, oils, and lubricants (POL).
- Spare parts and technical supplies (for equipment maintenance and repair).
- Rations, clothing, and medical supplies.

14-39. These priorities can change with the combat situation. For example, during an attack, the principal demand may be for ammunition. On the other hand, a unit advancing rapidly with no opposition might have a greater need for POL than for ammunition. Nonessential supplies may not be delivered if it reduces the ability to provide essential combat supplies. Ammunition and fuel resupply can comprise 80 percent or more of total transportation requirements. Rations may be considered nonessential, for instance, when units can obtain them by foraging.

PLANNING FACTORS

14-40. Essentially, all materiel support assets, from battalion level to the Ministry of Defense, are part of one system. When planning and coordinating division or brigade logistics allocations, the division or brigade resources officer requisitions and allocates supplies according to guidance from the division or brigade commander and pre-established planning factors. Standard tables of logistics planning factors, based on experience and estimated expenditure rates, indicate the amount and type of supplies required by a division or brigade to perform a particular type of combat action. Like his counterparts at higher and lower levels, the division or brigade resources officer refers to these tables when planning for a combat action. Then he must ensure that centralized planning provides adequate amounts of supplies, properly distributed to support the action. Thus, resources officers at all levels coordinate requirements from a common point of reference.

STANDARD UNITS OF ISSUE

14-41. To simplify logistics planning and to standardize ordering and issuing procedures, the OPFOR divides the major classes of materiel supplies into specific quantities or distribution lots. These quantities are called *basic load* for ammunition, *refills* for fuel, *daily rations* for food, and *sets* for spare parts and accessories. Once a standardized planning factor has prescribed a specific quantity as the unit of issue, planners no longer need to refer to the quantity itself, and all future references are given in multiples of the unit of issue.

14-42. Logistics calculations with these standard units of issue normally involve the weight of the unit of issue in metric tons, since this is a key parameter for transport planning. For certain computations, volume is also computed. These figures can then be used for planning transport and storage in connection with similar lots of weights and volumes of standardized units of issue of ammunition, POL, rations, and other lots of combat equipment.

MAINTENANCE

14-43. Maintenance includes actions taken to—
- Keep materiel and equipment in a serviceable condition.
- Return it to service.
- Update and upgrade its capability.

14-44. Since supplies are limited, the OPFOR stresses preventive maintenance, technical inspections, and proper operating methods to extend the life cycle of equipment. The maintenance system is designed to repair vehicles and equipment in the battle zone or as close to it as possible. Repair facilities and units move near the scene of combat rather than waiting for damaged equipment to be evacuated to them. Fixed and mobile repair units extend repair capabilities into the battle zone and provide service to the customer unit. During wartime, the types of repair performed at each level depend on the situation. Generally, they are of a lesser degree than in peacetime. The OPFOR classifies three categories of repair:
- *Routine repairs*—such as replacements, adjustments, or repair of individual components—require a short time to fix. Generally, maintenance personnel do not disassemble major components as part of routine repair.
- *Medium repairs* include the minor overhaul of equipment and the repair of individual components requiring a short time to fix.
- *Capital repairs* are conducted at depot level and involve the major overhaul and/or assembly of equipment.

TRANSPORTATION

14-45. Transportation is a critical function that cannot be viewed in isolation. It is the one element that ties sustainment and all other battlefield activities together. The OPFOR envisions an environment characterized by—
- A rapid tempo of nonlinear operations.
- Wide dispersion of forces.
- The need to concentrate rapidly for battle and disperse quickly.
- The need to conduct a wide range of actions simultaneously.

The mobility of logistics units must match that of the supported force. If the logistics support units fail to achieve this, they may jeopardize the overall success of the combat action.

MOVEMENT PRINCIPLES

14-46. The principles of movement apply to all military transportation services and remain constant throughout peace and war. They apply regardless of the planning level.

Use of All Available Movement Resources

14-47. Military logistics planners base their estimates on the use of all movement resources available. These estimates include tactical combat vehicles as well as civilian transportation assets mobilized to move supplies, equipment, and personnel. For example, during mobilization, civilian trucking and bus companies may be organized as militia truck units to provide transportation of cargo and personnel within friendly or occupied territory. During wartime, civilian personnel, transportation assets (including farm animals, vehicles, aircraft, and water vessels), and materiel-handling equipment are mobilized or commandeered to support the war effort.

Chapter 14

Centralized Planning and Decentralized Execution

14-48. Movement control is centralized at the highest level at which commanders charged with providing total logistics support and monitoring the transportation system and infrastructure can exercise it. However, decentralized execution enhances the flexibility of lower-level commanders to meet local requirements and to rapidly reprioritize support.

Regulated Movement

14-49. All movement is regulated according to command priorities. Movements are not validated, approved, or initiated if any part of the transportation system cannot meet the requirement. Regulating transportation assets and LOCs is required to prevent congestion, confusion, and conflict of movements. Unregulated use of the transportation system can severely hamper the movement of critical cargo and personnel supporting the battle or the overall operation or strategic campaign. Therefore, traffic in the AOR is programmed to provide fluid movement throughout the transportation network.

14-50. OPFOR traffic control units employ a system of measures organized and executed to—
- Ensure convoy and traffic regulation.
- Maintain general order in areas where troops are deployed.

Traffic control units or personnel are responsible for traffic control and law enforcement at the operational and tactical levels. They are responsible for directing military traffic along convoy routes and ensuring that the proper convoy speed and spacing are maintained. Internal Security Forces support movement control through protection of supply routes of movement in the homeland and of key transportation nodes and centers.

14-51. A movement program is a directive that allocates the available transport mode capability to satisfy the movement requirements according to the commander's priorities. The program normally contains detailed information concerning—
- Origins.
- Destinations.
- Weights and volumes of cargo.
- Types and number of personnel to be moved.

Fluid and Flexible Movement

14-52. The transportation system is designed to provide an uninterrupted flow of traffic that adjusts rapidly to changing situations. It is flexible enough to meet the changing priorities of a fluid battlefield and reallocate resources as necessary. Adjustments must be made to meet the variations in combat intensity. For example, when units are in the offense, the transportation system expands to maintain the tempo of the battle. Conversely, when units are in the defense, the system is contracted, the modes change, and differing cargo priorities may be necessary. Changes in the operational environment necessitate adjustments to operate in varying conditions and tactical situations that may dictate the types of convoys and controls established for movement.

14-53. The availability and use of road and rail networks, airfields, inland waterways, ports, and beaches increases flexibility. They not only allow the transportation system to respond to tactical changes, but also provide redundancy within the overall transportation network. For example, if a portion of a road network is destroyed or rendered unusable, the mode could change to rail or inland waterway.

Maximum Use of Carrying Capacity

14-54. The principle of making maximum use of carrying capacity involves more than just loading each transportation asset to its optimum carrying capacity. Transport capability that is not used in one day cannot be stored to provide an increase in capability for subsequent days. Similarly, a situation allowing fully loaded transport to sit idle is just as much a loss of carrying capacity as is a partially loaded vehicle moving through the system. While allowing for sufficient equipment maintenance and personnel rest, planners should keep transportation assets loaded and moving as much as the situation permits.

Logistics

TRANSPORTATION MODES

14-55. Transportation operations may include motor vehicles, rail, aircraft, and waterway transport vessels. The OPFOR generally uses motor vehicles to move large quantities of general cargo, POL, and personnel throughout the AOR. However, waterway transport vessels may be used to move large quantities of supplies and personnel along coastal or inland waterways to remote areas that are not accessible to motor vehicles.

14-56. As requirements for transportation fluctuate, each mode must be properly used to accomplish the commander's objective. For example, air transport is employed if reaction speed is the priority. Motor transport is considered the most flexible surface mode. It provides door-to-door delivery service and an interface with all other transportation modes.

14-57. Motor transport becomes essential as supplies are moved forward from railheads, field depots, or supply points to combat units. After the relocation of supplies from national-level supply bases, the OPFOR distributes them within OSCs and divisions primarily by truck. Within an OSC, the heaviest truck transport requirements are primarily above the division level.

14-58. Under the control of the resources officer at each level, motor transport resources are centralized for operational and tactical employment. This centralized control is especially important in the pre-offensive buildup period and for resupply of advancing columns. It also facilitates the diversion of motor transport assets of reserve forces to support those units engaged in the main effort when necessary.

SUPPLY AND EVACUATION ROUTES

14-59. Within their AORs, divisions and brigades establish and improve supply and evacuation routes, using the network of military roads. Routes are usually as follows:
- **Division routes**: from the ISG deployment area to the deployment areas of the brigades' materiel support units, IFC firing positions, and the brigades' medical points.
- **Brigade routes**: from the deployment area of the brigade's materiel support unit(s) to the deployment areas of battalion-level materiel support units, indirect fire support unit areas (or firing positions), and battalion medical points.

14-60. The division or brigade resources officer, together with the chief of infrastructure management at that level, is responsible for improving supply and evacuation routes and maintaining them in passable condition. At division and brigade levels, subordinate engineer elements perform road maintenance. Engineer units at OSC or division level may form road and bridge construction and repair groups to prepare and maintain these and other movement routes.

14-61. At national level, the Strategic Integration Department (SID) also organizes civil engineering and construction efforts required to sustain military actions. During wartime, civil engineering units from the Ministry of the Interior, as directed by the SID, may be employed at the national, OSC, and division levels. Employed on an area basis, these units are responsible for the upkeep of supply and evacuation routes and for repair of battle-damaged roads and bridges. The chief of infrastructure management at the OSC or division level must coordinate and prioritize the route construction and maintenance functions of both civil and combat engineers within his AOR.

PERSONNEL SUPPORT

14-62. The OPFOR considers people as one of the assets most critical to the success of any military action. Thorough planning and efficient personnel management directly influence mission readiness. During the course of battles, timely personnel replacements are essential.

PERSONNEL MANAGEMENT

14-63. The division or brigade chief of administration is responsible for all personnel actions and transactions in the command. At DTG level, a personnel service company or battalion provides the personnel to operate the personnel operations center. That center's major functions include providing personnel and administrative support, finance support, and legal support.

REPLACEMENT

14-64. Units may maintain strength by piecemeal replacement of casualties during combat, particularly when lightly wounded personnel and damaged equipment can return to parent units quickly. Once casualties are sufficient to threaten total loss of combat effectiveness, the unit withdraws from contact and reconstitutes. Timely replacement of ineffective units is vital to maintaining momentum. The commander may choose to withdraw heavily attrited units and consolidate them to form a smaller number of combat-effective units.

14-65. Personnel replacement is based on unit strength reports and includes the coordinated support and delivery of replacements and soldiers returning from medical facilities. The unit strength report is used to assess a unit's combat power, plan for future battles, and assign replacements on the battlefield.

Individual Replacements

14-66. The OPFOR can use the system of individual replacements in both peacetime and wartime. The sources of replacement personnel are school graduates, reserve assignments, medical returnees, and normal assignments.

Incremental Replacements

14-67. The OPFOR may incrementally replace entire small units such as weapons crews, squads, and platoons. Replacements can be obtained from training units or reserve forces.

Composite Unit Formation

14-68. Composite units may be formed from other units reduced by combat operations. Composite units may be constituted up to division and even OSC level.

Whole-Unit Replacement

14-69. The OPFOR uses whole-unit replacement when massive losses occur as a result of a combat action. Company-level and above units are brought forward from reserve forces to replace combat forces rendered ineffective.

Replacement Training

14-70. OPFOR planners realize that personnel replacement requirements may necessitate any of the aforementioned procedures. Individual and unit replacement exercises are held semiannually to maintain established proficiency standards for personnel units. During these and other training exercises, troops are moved by various modes of transportation such as motor vehicles, waterway, aircraft, or rail.

MEDICAL SUPPORT

14-71. The basic principle of combat medical support is multistage evacuation with minimum treatment by medical personnel at each unit level. They treat the lightly wounded who can return to combat and those casualties who would not survive further evacuation without immediate medical attention. See table 14-1 for the levels of medical care.

Table 14-1. Levels of medical care

Level	Available Care
Platoon	Platoon medic (corpsman) provides basic first aid.
Company	Company medic (paramedic) provides advanced first aid, pain relief, intravenous fluids, and treatment of most common illnesses.
Battalion	Medical assistant (physician's assistant) provides limited medical intervention, minor surgery, and treatment of most common illnesses; limited inpatient capability.
Brigade, BTG, and Division	Medical officers (physicians) provide trauma stabilization and minor surgical intervention.
DTG and Higher	A field hospital provides major surgery and extended care.
OSC or Theater Support Zone	The Central Military Hospital and major civilian hospitals provide definitive care in fixed facilities.

14-72. The OPFOR divides the range of medical treatment into three categories. The first category of procedures includes only mandatory lifesaving measures. The second category includes procedures to prevent severe complications of wounds or injuries. The final category of treatment includes procedures accomplished only when there is a low casualty load and reduced enemy activity.

14-73. In anticipation of an overtaxed combat medical support system, OPFOR doctrine emphasizes the importance of self-help and mutual aid among individual soldiers. This concept extends beyond the battlefield to casualty collection points and unit aid stations. Self-help and mutual aid reduces the demands made on medical personnel, particularly when there is a sudden and massive influx of casualties. Each soldier receives first-aid training.

MEDICAL LOGISTICS

14-74. Medical logistics operates on a "pull system." Personnel in the field request medical materiel (including repair parts for medical equipment) from a medical depot where it must be picked up and delivered to the field. Normally, medical supplies are transported from the support zone to the battle zone on cargo-carrying transport vehicles, water vessels, or aircraft. However, ground ambulances returning to the battle zone may assist in transporting medical supplies. A medical equipment maintenance unit at the medical depot provides all medical equipment maintenance.

CASUALTY HANDLING

14-75. The OPFOR has shown success in handling combat casualties. This success stems from emphasis placed on trauma training and close coordination with the civilian medical sector. Evacuation is based on a higher-to-lower method. The next-higher echelon provides transportation for casualties. Each level has specific responsibilities for the care of the sick and wounded. Besides treating the wounded, medical personnel handle virtually all of their own administration, especially at lower levels. As casualties move through the combat evacuation system, medical personnel at each level make effective use of medical facilities by repeated sorting of the wounded (triage). Helicopters are used for military and civilian search and rescue missions, medical evacuations, and domestic disaster relief flights. During wartime, most casualties arrive at a hospital within 6 to 12 hours after being wounded. The evacuation time is reduced to 2 hours during peacetime.

MEDICAL FACILITIES

14-76. A field hospital is the first level in the evacuation system capable of conducting major surgery and giving extended care. It is mobile and capable of deployment near the battle zone. It constitutes the largest and most extensive military facility with this capability.

14-77. The best medical facilities are permanent (long-term) Army hospitals. During peacetime, military personnel receive treatment at these hospitals, which may also serve as an emergency medical care facilities for foreign diplomats, their families, and tourists.

14-78. During wartime, military personnel are treated at all of the major civilian hospitals and local clinics in addition to field hospitals and Army hospitals. Major university hospitals will also be directed to serve as emergency medical care facilities for the OPFOR. This ensures consistent high-quality medical staffing, care and treatment. A majority of medical facilities or clinics in the outlying areas have sufficient numbers of trained personnel, supplies, and reliable electric power and water. The facilities also contain high-quality, sophisticated medical equipment. The pharmacies are stocked with high-quality domestic and foreign-produced pharmaceuticals.

CBRN TREATMENT

14-79. Treating chemical, biological, radiological, and nuclear (CBRN) casualties is a standard OPFOR trauma protocol. The CBRN medical plan is based on three assumptions:
- Mass casualties will occur.
- Casualties will be similar to those that medical personnel have been trained to treat.
- Medical personnel are able to treat the casualties in a decontaminated environment.

An Army hospital can be converted into a chemical decontamination center within 2 to 6 hours. Most of the remaining major hospitals require up to 30 days to convert to a decontamination center.

SUPPORT OF COMBAT ACTIONS

14-80. During both offense and defense, OPFOR logistics units operate from locations that are protected, concealed, and serviced by good road networks. Commanders emphasize that logistics units make maximum use of urban areas to conduct logistics activities. The dispersion of logistics sites is consistent with support requirements, control, and local security.

14-81. Logisticians must be continuously informed of battle plans and probable changes to those plans. They coordinate logistics preparations with deception plans to avoid giving away the element of surprise. Commanders emphasize passive security measures during the sustainment of combat actions. Logistics unit commanders anticipate that at least 50 percent or more of their work will be done in darkness or under other limited visibility conditions. Therefore, noise and light discipline is a necessity when operating under these conditions.

OFFENSE

14-82. The logistics objective in supporting offensive actions is to maintain the momentum of the attack by supporting in the battle zone or as close to it as possible. Both the battle zone and the support zone can move as the offensive battle progresses.

14-83. Planners must consider the nature of the offensive action as it affects logistics activities. For example, high fuel consumption may dictate making provisions to position substantial quantities in or near the battle zone without signaling the OPFOR's intention to attack to the enemy. Responsive support is made more difficult by lengthening of supply lines and by critical requirements for user resupply vehicles

Logistics

to stay close to their respective units. Planning, coordination, communication, and above all flexibility are key elements to consider. Therefore, planners develop logistics plans flexible enough to meet the changing priorities of a fluid battlefield.

14-84. In considering the attack, materiel support units ensure that all support equipment is ready and that supplies are best located for support. They also ensure that sufficient transportation is available to support maneuver and logistics plans. Normally, ammunition and fuel are the most important supplies in the offense. However, consideration must be given to all supplies, as well as other support procedures, specifically medical and maintenance.

14-85. The following are examples of some specific considerations for planners to use during the development of logistics plans supporting offensive actions:
- Maintenance units should pre-plan maintenance collection points along movement routes, in order to reduce recovery requirements.
- Fuel and ammunition supply points are positioned in the battle zone or as close to it as possible.
- Arrangements are made in advance for aerial resupply of critical items in order to maintain the tempo of combat.
- Planners arrange to throughput obstacle-breaching and bridging material if required.
- Planners must consider potential bypassed enemy units; they must have the latest intelligence on the enemy situation.

DEFENSE

14-86. The logistics objective in supporting defensive actions is to sustain the attrition of enemy attacking forces through support from dispersed sites located in the support zone. A division support zone may be dispersed within the support zones of subordinate brigades, or the division may have a separate support zone site of its own.

14-87. During the defense, supply activity is greatest in the preparation stage. Supplies generally are stockpiled or pre-positioned in initial and subsequent defensive positions. Critical supplies such as ammunition and barrier material should be as mobile as possible to ensure continuous support as combat power is shifted in response to enemy attacks.

14-88. To support stay-behind forces, supply stockage levels may be two to three times higher than normal amounts. This ensures a redundancy of caches and needed equipment that cannot be readily resupplied. Stay-behind forces may require unique maintenance support arrangements to ensure that equipment remains operational.

14-89. Logistics units position themselves in relatively secure positions far enough from maneuver and fire support units to be out of the flow of the battle. However, they should not be so far removed as to render the logistics effort less effective.

14-90. The following are examples of some specific considerations for planners to use during the development of logistics plans supporting defensive actions:
- Maintenance units should position maintenance teams in the battle zone to return the maximum number of weapons systems to the battle as soon as possible.
- Emphasis is on keeping supply and evacuation routes open.
- Nonessential logistics units and operations move into the depth of the support zone as early as possible.
- In a maneuver defense, fuel and ammunition supply points are positioned as far forward as possible and in successive battle positions.

SUPPORT ZONE SECURITY

14-91. The OPFOR expects any enemy to make an effort to conduct reconnaissance, espionage, and diversionary action in its tactical and operational support zones. These enemy actions can be particularly effective in areas where the local population is not sympathetic to the OPFOR's cause. In addition to these

Chapter 14

threats, the OPFOR anticipates attacks on its support zone by airborne and heliborne forces as well as larger-scale attacks by enemy maneuver forces.

14-92. The OPFOR uses a security force or element to counter any threats in its support zones. Each division or DTG deploys a considerable counterintelligence effort. It can assign up to an entire BTG for security tasks. The security force or element is equipped and trained for conventional as well as unconventional warfare. As airborne and amphibious threats grow, there is increasing stress on deploying antilanding reserves, including, or even based on, heliborne units to provide a rapid reaction.

14-93. All logistics and communications units are capable of self-defense. The convalescent sick and wounded provide a reserve of manpower for elements near medical locations or reserve forces.

MISSION SUPPORT SITES

14-94. A mission support site (MSS) is a temporary base used by units that are operating at a considerable distance from their support zone, during an extended mission. The MSS may provide food, shelter, medical support, ammunition, or demolitions. The use of an MSS eliminates unnecessary movement of supplies and allows a force to move more rapidly to and from attack sites or objectives. When selecting an MSS, consideration is given to—

- Cover and concealment.
- Proximity to the objective.
- Proximity to supply routes.
- The presence of enemy security forces in the area.

Security dictates that drop zones or landing zones be a considerable distance from an MSS, cache, or support zone—although this may increase transportation problems.

POST-COMBAT SUPPORT

14-95. OPFOR logisticians are not only focused on supporting units in combat. They are also focused on other post-combat support requirements such as—

- Personnel replacement (see the Personnel Support portion of this chapter).
- Weapon systems replacement.
- Reconstitution.
- Receiving and preparing reinforcements.

WEAPON SYSTEMS REPLACEMENT

14-96. Weapon systems replacement is simply a procedure for providing a weapon system to a combat unit. It involves processing the vehicle or equipment from a storage or transportation configuration to a ready-to-fight condition. It also involves the integration of a completely trained crew with the weapon system.

RECONSTITUTION

14-97. *Reconstitution* is performed in support of all combat actions, in order to restore combat effectiveness. Although it is mainly a command and operations function, the actual refitting, supply, personnel fill, and medical actions are conducted by logistics units. There are two methods for conducting reconstitution: *reorganization* and *regeneration*.

Reorganization

14-98. *Reorganization* is action taken to shift resources internally within a degraded unit to increase its level of combat effectiveness. Reorganization is normally done at unit level and requires only limited external support such as supply replenishment, maintenance assistance, and limited personnel replacement.

When continuity of the mission is of paramount importance, composite units may be formed from other units reduced by combat actions.

Regeneration

14-99. *Regeneration* is action taken to rebuild a unit through large-scale replacement of personnel, equipment, and supplies. Additionally, it is action taken to restore C2 and conduct mission-essential training. Overall, the effort is directed at restoring the unit's cohesion, discipline, and fighting effectiveness.

RECEIVING AND PREPARING REINFORCEMENTS

14-100. OPFOR strategic and operational logisticians prepare contingency plans for the mobilization and reception of reserve forces. Once the unit personnel and equipment are mobilized, they are sustained, configured, and transported to their respective OSC. Normally, strategic-level logistics units provide this type of support. Once the units arrive at the OSC, the OSC assumes responsibility for their further sustainment and transport, and they are available for assignment to appropriate tactical-level missions.

This page intentionally left blank.

Chapter 15
Special-Purpose Forces and Commandos

The OPFOR includes both special-purpose forces (SPF) and commandos, each of which are discussed in their respective sections in this chapter. The chapter also clarifies the relationships between the two types of units in terms of command and control (C2), organization for combat, and tactical capabilities.

SECTION I – SPECIAL-PURPOSE FORCES

15-1. The OPFOR maintains a broad array of SPF as means to carry the battle to the enemy's depth. SPF missions may support national-, theater-, operational-, or tactical-level objectives. They are conducted across the spectrum of military operations either independently or in coordination with regular and/or irregular forces. This section describes the nature of SPF, the roles they can play at various levels of command, and how they reach the tactical level via task-organizing.

COMMAND AND CONTROL

15-2. SPF structure and C2 relationships are significantly different than those normally associated with regular maneuver units. The SPF structure depends primarily on the posture and missions of the OPFOR at any given time.

ADMINISTRATIVE FORCE STRUCTURE

15-3. In the OPFOR's peacetime administrative force structure (AFS), some SPF are national-level forces controlled by the General Staff. Under the General Staff, SPF are subordinate either to the SPF Command or to the service headquarters of the Army, Navy, Air Force, or Internal Security Forces, which have their own SPF. The General Staff can use them against strategic objectives or for power projection in the region. However, some SPF are intended for use at the operational level and thus can be subordinate to operational-level administrative commands even in the AFS. For additional information on SPF organizations at operational and strategic level in the AFS, see FM 7-100.4.

15-4. In peacetime and in garrisons, all SPF are organized administratively into SPF companies, battalions, and brigades. These organizations facilitate peacetime administrative control and training. However, even these administrative groupings do not have a fixed structure in the AFS, and the AFS structure normally differs from the OPFOR's go-to-war (fighting) force structure.

FIGHTING FORCE STRUCTURE

15-5. In wartime or in transition to war, some SPF units from the SPF Command or from the Army, Navy, Air Force, or Internal Security Forces SPF may remain under the C2 of their respective service headquarters. However, some SPF units also might be suballocated to operational- or tactical-level commands during the task-organizing process, in order to perform designated missions. Depending on the situation, command and support relationships may be constituent, dedicated, supporting, or affiliated. (See chapter 2 for a full discussion of these relationships.)

15-6. SPF units generally reach the tactical battlefield via an operational-strategic command (OSC). The OSC is a task organization and, as such, receives SPF and other units from higher headquarters. The SPF units of the Army are more likely to appear in the task organization of an OSC than any of the SPF organizations of the SPF Command or other services.

15-7. In some cases, SPF units may be allocated to an OSC in a supporting relationship, while remaining under the command of their parent SPF organization. Even in a supporting relationship, the commander of the OSC receiving the SPF unit(s) establishes those units' objectives, priorities, and time of deployment.

15-8. In other cases, the OSC commander may receive SPF in a constituent or dedicated status. In that case, he may employ the SPF assets allocated to him as part of his integrated fires command (IFC). However, he may choose to suballocate some or all of them to the task organization of his tactical-level subordinates. The tactical commander may receive these SPF assets in a constituent, dedicated, or supporting relationship. The supporting relationship allows the OSC commander to use the SPF as part of the tactical battle scheme but retain control over these units to ensure that SPF objectives support the overall mission of the OSC.

15-9. No SPF assets are permanently subordinate to tactical-level units at division and below either in the AFS or during war. Therefore, it is necessary to understand their command and support relationships when they arrive at, and support the tactical fight. Depending on the situation, various relationships may exist when SPF detachments or teams operate within a tactical commander's area of responsibility (AOR):

- An SPF unit may operate in that AOR without the approval or even knowledge of the tactical commander. When this occurs, each SPF team or detachment remains in direct communication with a higher headquarters. The controlling headquarters may be an SPF detachment, SPF company, SPF battalion, SPF brigade, or a higher operational or tactical headquarters.
- The SPF may be allocated to the maneuver commander in a constituent, dedicated, or supporting relationship.
- The local maneuver commanders may find themselves in a supporting role to the SPF unit.

However, there are no firm rules governing the C2 of SPF assets. These relationships are determined solely by the situation and mission and what best serves OPFOR interests. See chapter 2 for more detail on command and support relationships. For additional information on the SPF structure and command relationships see FMs 7-100.1 and 7-100.4.

AFFILIATIONS WITH PARAMILITARY OR NONMILITARY ACTORS

15-10. Due primarily to the nature of SPF missions, the relationships between SPF and other paramilitary or nonmilitary personnel or groups are not as quite as firm as SPF relationships with the more regular maneuver units. Any relationship of the SPF unit may be one of affiliation rather than a command relationship. This affiliation may be dependent upon only a single shared or similar goal. These relationships are generally fluctuating and may be fleeting, mission-dependent, or event- or agenda-oriented.

15-11. The nature of the shared goal or interest determines the tenure and type of relationship and the degree of affiliation. For example, the affiliation of an SPF detachment with criminal (or guerrilla) organizations is dependent only on the needs of the criminal (or guerrilla) organization or on the needs of the SPF at a particular time. The relational dynamics of SPF units are very fluid and apt to change from one day to the next. Shifts in affiliations may in turn cause adjustments in the SPF task organization to accommodate these changes. (See chapter 2 for additional information on command and support relationships, including affiliation.)

MISSIONS

15-12. SPF play an important role in support of both the offense and defense. They may perform their missions separately, in support of strategic objectives, or in support of a theater-level campaign, an OSC-level operation, or any tactical-level action. *Regardless of the level of command for which they perform the missions, the SPF always conduct these missions as small SPF teams or detachments.* They perform the same basic forms of tactical action at any level, although the purpose of the action may be dissimilar at different echelons. This chapter focuses on SPF missions at the tactical level. (See FM 7-100.1 for information on strategic- and operational-level missions.)

Special-Purpose Forces and Commandos

15-13. Not all specific missions of the SPF can be forecast. Their missions, like their tactics, techniques, and procedures (TTP) and activities, are adaptable and unpredictable, and lend themselves to improvisation. The SPF do whatever they feel will work.

15-14. SPF allocated to an OSC or division tactical group (DTG) often become part of the disruption force, frequently operating in enemy-held territory before the beginning of an operation or battle. At OSC level and sometimes at DTG level, they may become part of an IFC, to assist in locating and destroying key enemy formations or systems (see chapter 2 and chapter 9).

15-15. SPF are not always employed against military targets for purely military objectives. They will also be used against political, economic, or population centers or tangible targets whose destruction affects intangible centers of gravity. These efforts often place noncombatants at risk and aim to apply diplomatic-political, economic, and psychological pressure. Tactical targets include not only enemy military forces and equipment, but also government agency heads, contractors, private firms, and nongovernmental organizations, and/or personnel involved in transporting troops and materiel into the region or supporting enemy forces in any manner. The goal is to present the enemy with a nonlinear, simultaneous battlefield. Striking such targets will not only deny the enemy sanctuary, but also weaken his national will, particularly if the SPF or affiliated forces can strike targets in the enemy's homeland.

15-16. Depending on the circumstances the OPFOR will completely deny any relationship or affiliation with SPF assets deployed outside the State. This plausible deniability greatly extends the employment options available to the OPFOR and its use of SPF.

BASIC MISSIONS

15-17. SPF actions increase the depth of the battle area. The SPF's simultaneous attack of both forward and rear areas to disrupt or destroy enemy forces includes the following basic missions:
- Conduct strategic and operational reconnaissance.
- Conduct actions in the enemy's tactical, operational, and strategic depth to undermine his morale and to spread panic among the civilian population and the political leadership.
- Conduct surgical strikes and raids.
- Neutralize weapons of mass destruction and precision weapons.
- Attack air defense facilities and airfields. Destroy critical air defense systems and associated radars (especially early warning radars).
- Disrupt lines of communications (LOCs).
- Attack C2 and reconnaissance, intelligence, surveillance, and target acquisition (RISTA) facilities.
- Exploit surprise to disrupt defensive actions.
- Influence the population.
- Support follow-on conventional military actions.
- Support the information warfare (INFOWAR) plan.
- Conduct actions that will cause a flow of refugees, which can hamper enemy deployment, defensive maneuver, and logistics.
- Disrupt enemy power supplies and transportation networks (power utilities, POL transfer and storage sites, and internal transportation).
- Sabotage enemy mobilization and deployment.
- Train agents (affiliated forces or civilians) to operate as political agitators, human intelligence (HUMINT) collectors, and saboteurs.
- Organize local irregular forces (insurgents or guerrillas) or sympathizers.
- Provide communications, liaison, training, and support to stay-behind activities in the defense.
- Serve as part of the disruption force. SPF allocated to an OSC or a tactical-level command often become part of the disruption force, frequently operating in enemy-held territory before the beginning of an operation or battle.

Chapter 15

- Serve as part of an IFC at OSC level and sometimes at DTG level.
- Provide terminal guidance for attacks by aircraft, missiles, and precision weapons.

Note. SPF can recruit, organize, train, advise, and support local irregular forces (insurgents or guerrillas) and possibly even criminal organizations and conduct (or lead) operations in conjunction with them. SPF personnel may fight alongside such affiliates or assist them to prepare for offensive actions, diversionary measures, or other missions. In some cases, the SPF will not only advise and assist but actually control (command) the irregular forces as a surrogate force. SPF missions (and those of their affiliates or surrogates) can include the use of terror tactics.

SPECIAL MISSIONS

15-18. In addition to the basic missions expected of the SPF, there are several requiring specific skills unique to these missions. These missions not fitting under the basic missions are the *special missions*. While there may be a number of special missions conducted by the SPF, the most common special missions are special reconnaissance and direct action.

Special Reconnaissance

15-19. SPF conducting *special reconnaissance* operate deep behind enemy lines. While many of these activities are operational level and above, a large portion is conducted on behalf of the OSC and below. SPF can conduct reconnaissance for future ground force operations or for airborne and/or amphibious landings. SPF are a major source of HUMINT, placing "eyes on target" in hostile, denied, or politically sensitive territory. They can operate in small teams beyond the battle lines of the AOR, conducting long-range reconnaissance. Their reconnaissance priorities include—

- Chemical, biological, radiological, and nuclear (CBRN) delivery systems.
- Precision weapons.
- Headquarters and other C2 installations and centers.
- RISTA systems and centers.
- Rail, road, and air movement routes.
- Airfields and ports.
- Logistics facilities.
- Air defense systems.

Once SPF teams locate such targets, they may simply monitor and report on activity there, or they may conduct direct action or diversionary measures.

15-20. The SPF can train and employ affiliated forces and civilians to perform HUMINT activities. They may also operate in conjunction with HUMINT agents controlled by elements of the General Staff and may or may not operate with the local maneuver commander's knowledge.

Direct Action

15-21. *Direct action* involves an overt, covert, or clandestine attack by armed individuals or groups to damage or destroy high-value targets or to kill or seize a person or persons. Examples of direct-action missions for SPF units are—

- Assassinations of key military and other leaders.
- Abduction and hostage taking.
- Sabotage.
- Capture.
- Ambushes.
- Raids.

- Rescue of hostages (civilian and military).
- Rescue and recovery of downed pilots or aircrews.

Implementation of direct-action missions depends on—
- The size of the enemy's defenses.
- The element of surprise.
- The specific impact and cascading effects of the events.
- The assets available to the SPF unit commander.

15-22. The term *diversionary measures* refers to direct actions of groups or individuals operating in the enemy's rear area. Such measures include the destruction or degradation of key military objectives and the disruption of C2, communications, junctions, transport, and LOCs. Specific measures could include—
- Misdirecting military road movement by moving road markers.
- Jamming or disrupting communications and GPS, and generating false communications.
- Killing personnel.
- Destroying military hardware.
- Spreading disinformation.
- Undermining the morale and will of the enemy by creating confusion and panic.

Other diversionary measures may be part of a larger INFOWAR campaign.

15-23. The OPFOR has trained SPF as an alternate means of delivering CBRN munitions packages it may develop for them. This provides a worldwide strategic means of CBRN delivery that is not limited to the range of the missiles of the Strategic Forces.

ORGANIZATION FOR COMBAT

15-24. No SPF assets are subordinate to tactical level units at division and below in the AFS. All SPF assets employed at the tactical level will have been allocated from higher levels. Therefore, it is important to understand where those SPF teams, companies, battalions, and brigades come from, how they get to the tactical level, and how they are employed to support the tactical fight.

15-25. SPF units reach the tactical level through allocation to an OSC. SPF units from the SPF Command or other services can be allocated in a constituent or dedicated status to an OSC. The OSC commander may then choose to further allocate them to the task organization of a tactical group, again in either a constituent or dedicated relationship. In other cases, SPF units may be allocated in a supporting relationship, while remaining under the command of their parent SPF organization. SPF units, especially at the tactical level, are task-organized to provide the best possible chance of mission success. Regardless of the level from which they come or the level at which they are employed, units at SPF brigade level and lower have similar capabilities, equipment, and manning.

15-26. SPF organizational structures, like the TTP they use and teach, are adaptive, dynamic, and malleable. The malleable structure of SPF units from team to brigade provides the ability to continually adapt and to allow continuous improvisation. Units will be added, deleted, or modified, as the mission and conditions dictate. Every SPF mission is unique and unlike any other, and thus requires forces organized not in a standard fashion but rather adapted into a task organization based on the mission. All of these SPF organizations provide a flexible and capable means of supporting all OPFOR operations.

15-27. **Regardless of the parent organization in the AFS or their deployment location, the SPF normally infiltrate and operate as small SPF teams. When deployed, these teams may operate individually, or they may be task-organized into SPF detachments.** The terms *team* and *detachment* indicate the temporary nature of the groupings. In the course of an operation, teams can leave a detachment and join it again. Each team may in turn break up into smaller teams (of as few as two men) or, conversely, come together with other teams to form a larger team, depending on the mission. At a designated time, teams can join up and form a detachment (for example, to conduct a raid), which can at any moment split up again. This process can be planned before the operation begins, or it can evolve during the course of an operation, often at differing levels.

Chapter 15

> *Note.* In most OPFOR organizations, a *detachment* is a battalion or company designated to perform a specific mission and allocated the forces necessary to do so. However, SPF are organized into companies and battalions for administrative purposes only, not for performing missions. Therefore, a *detachment* in SPF terms is a temporary grouping of SPF teams that is task-organized for a particular mission but is not necessarily linked to a company or battalion organization.

15-28. For a very small SPF detachment, the senior team leader may act as the detachment commander. When a number of SPF teams from an SPF company join efforts for a particular mission, the original SPF company headquarters may act as the temporary detachment headquarters. Alternatively, the SPF company commander may designate his deputy commander or another individual as the detachment commander and provide him assets to form the temporary detachment headquarters. When a large number of teams join to form a larger detachment (especially if the detachment includes teams from more than one SPF company), the SPF battalion headquarters may temporarily serve as the detachment headquarters or provide personnel and equipment to form that headquarters.

SPF ASSETS

15-29. SPF assets allocated to a DTG or brigade tactical group (BTG) are drawn from an SPF brigade, normally one belonging to the Army SPF. However, SPF units generally reach the tactical level via the OSC. SPF units in an OSC may come from Army SPF, but can also come from the national-level SPF Command or from Air Force or Navy SPF, since an OSC is the lowest level of joint command. If an OSC has received SPF units, it may further allocate some of these units to supplement the long-range reconnaissance assets of a subordinate DTG or division (which would thus become a DTG).

OTHER ORGANIZATIONS AND ASSETS

15-30. Based on the mission, SPF units can be task-organized into any maneuver unit. Likewise, any type of other unit (maneuver or otherwise) can be task-organized as part of an SPF detachment.

15-31. The mission and available resources drive the task-organizing of the SPF unit. For example, if the mission requires an affiliation with irregular forces, those units can be included in the task organization with the appropriate command and support relationships. This may also apply to mechanized infantry, air defense, or chemical defense units; elements providing air support and/or transport; or any other type of military or nonmilitary organization. All of these malleable SPF organizations (task-organized or otherwise) provide the OPFOR a flexible and capable means of attaining its military goals. If specific units or organizations are not available at the next higher level, the OPFOR commander requests up the chain that they be provided.

SPF BRIGADES, BATTALIONS, COMPANIES, AND TEAMS

15-32. SPF brigades, battalions, companies, and teams exist in the AFS, but are not organized as uniformly as regular ground forces. Although FM 7-100.4 provides basic structures for such organizations, there is actually no fixed structure. Examples in figures 15-1 through 15-4 (on pages 15-7, 15-8, 15-10, and 15-11, drawn from FM 7-100.4) indicate this by depicting some subordinates in dashed boxes, indicating optional structures. The size and composition of SPF organizations at each level in the AFS can vary greatly, depending on the situation and the assets available. When actually employed, the units will be further task-organized to fit specific missions.

SPF BRIGADE

15-33. The majority of SPF assets reside originally in SPF brigades. This is true whether they are in the national-level SPF Command or in Army, Air Force, and Navy SPF. An SPF brigade can also be part of an operational-level command (such as a corps, army, or army group) in the AFS. SPF units from the SPF brigade can be allocated in a constituent or dedicated status to be task-organized as part of an OSC or a tactical group or to another SPF unit. In other cases, SPF units may be allocated in a supporting relationship, while remaining under the command of the parent SPF brigade.

15-34. Figure 15-1 shows a typical example of an SPF brigade. The structure of individual SPF brigades will vary depending on mission requirements and conditions. Had the mission or the environment dictated, the organizational example in figure 15-1 could easily have included an engineer battalion (for civil affairs), an INFOWAR company, affiliated local insurgents or guerrilla unit(s), or even a helicopter company. The structure of each SPF brigade is situational, adapted to the specific conditions.

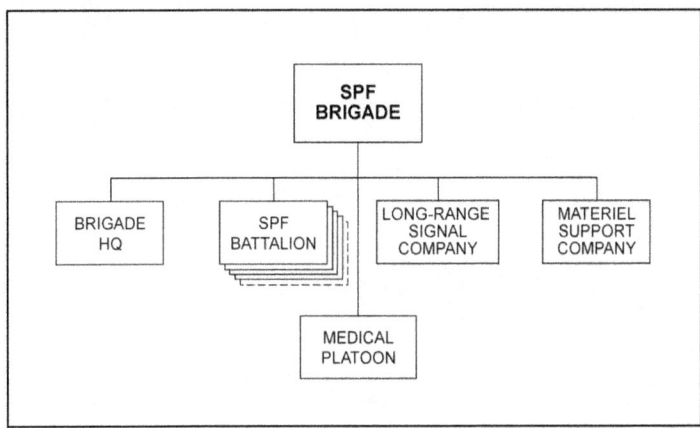

Figure 15-1. SPF brigade (example)

15-35. The SPF battalions and their companies are addressed in some detail below. The long-range signal company, materiel support company, and medical platoon provide typical support functions that are not unique to the SPF. For further information on these organizations, see FM 7-100.4.

SPF BATTALION

15-36. In the AFS, an SPF battalion may be part of an SPF brigade or directly subordinate to an operational-level command (a corps, army, or perhaps an army group). SPF battalions are generally assigned a geographic AOR. This AOR may be quite large depending on the terrain, population, and other factors. The battalion is likely to be allocated to the task organization of an OSC operating in a fairly large geographic area. However, an SPF battalion might also be allocated to a tactical group. Individual SPF companies can be further suballocated depending on the circumstances.

15-37. The SPF battalion organization shown in figure 15-2 on page 15-8 is an example only. It provides a typical example of the capabilities associated with an SPF battalion. The structure of SPF battalions can vary to meet mission requirements.

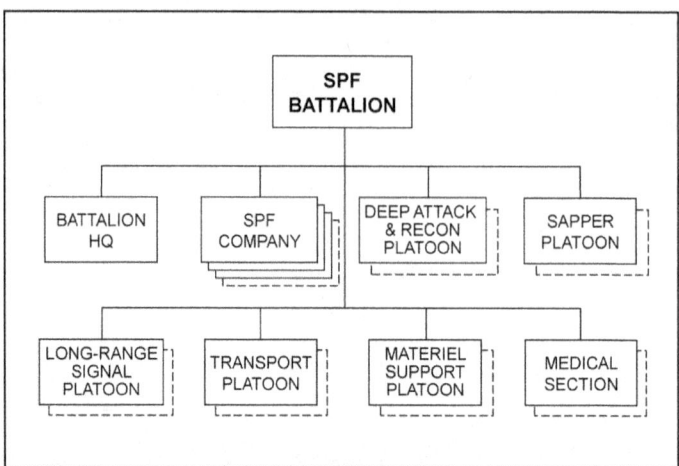

Figure 15-2. SPF battalion (example)

15-38. All SPF battalions have a battalion headquarters that may vary in size depending on any number of factors. The primary mission of the battalion headquarters is to direct, support, and sustain its deployed SPF companies. The required degree of each function is influenced by numerous variables. The capabilities and functions of some of the more unique SPF battalion assets are explained in detail below. For simplicity, SPF company functions and capabilities are addressed separately below, under SPF Company.

Deep Attack and Reconnaissance Platoon

15-39. The SPF deep attack and reconnaissance (DAR) platoon of an SPF battalion provides a unique deep, fast, attack (or reconnaissance) capability to the SPF battalion. The platoon is extremely versatile and is used in many different roles, either behind enemy lines or in the first line of combat. Although it is small (less than 40 personnel and 14 vehicles), it is extremely mobile and lethal. In order to provide the appropriate lethality, mobility, and also the stealth required for mission success, the equipment and weapons mix typically includes—

- Fast attack vehicles or light strike vehicles.
- Tactical unmanned aerial vehicles (UAVs).
- Air defense and antiarmor systems.

15-40. The name of the DAR platoon is indicative of the type of missions it is designed to accomplish. Missions include (but are not limited to)—

- Conduct deep (strategic and operational) reconnaissance.
- Conduct operations along tactical, operational, or strategic axes, including deep attack raids and sophisticated ambushes.
- Destroy critical air defense systems and associated radars, and communications and intelligence systems.
- Disrupt and/or destroy enemy rear area operations.
- Assist local guerrillas or insurgents in offensive operations.
- Provide communications, liaison, and support to stay-behind and guerrilla activities in the defense.

- In the strategic and operational depth, undermine the enemy's morale and spread panic among the civilian population and the political leadership (This may cause a flow of refugees which will hamper enemy employment, defensive maneuver, and logistics.)
- Serve as part of a disruption force operating in enemy-held territory prior to the beginning of an operation.
- Serve as an alternate means of delivering CBRN munitions.
- Serve as target designation and forward observer teams.
- Establish and resupply or restock caches and mission support sites (MSSs).

15-41. All members of the SPF DAR platoon are cross-trained in the use of all equipment, weapons, and vehicles assigned to the platoon. The equipment and weapons mix is determined by the mission. Some may be left in the vehicles until required or not carried on the mission at all. For example, if the mission is target designation, the laser target designator will be carried and employed. If not it may be either left in the vehicle or left behind at an MSS, cache, or at battalion.

15-42. The ease of operation, size, and simple design of tactical UAV used by the SPF DAR platoon lends itself to field expedient modification. Converting this UAV into a munitions delivery system (improvised attack UAV) is not difficult and offers several tactical advantages and extends the attack range of the platoon.

Sapper Platoon

15-43. The mission of the sappers (raiders) is to serve as the lead or primary (assault) element in an assault on fixed installations or military field positions. Sappers are SPF (or infantry or guerrillas) trained to perform some typically raider, engineer, and ranger functions. Sappers are not engineers. The SPF also train local civilians and affiliated forces to be sappers.

15-44. A sapper platoon in an SPF battalion may consist of—
- A platoon headquarters.
- An infiltration and scout squad.
- A mine warfare and demolition squad.
- An improvised explosive/booby traps squad.
- A general support squad.

However, all squads are cross-trained to perform all functions. (See Sapper Team, under SPF Company for more detail on the nature of sappers and the types of functions they can perform.) Sapper squads can be task-organized into teams.

15-45. SPF battalions using sappers in an assault and/or demolition role may form several sapper platoons. Most if not all of the additional sapper platoons will be manned by affiliated guerrillas or insurgents. A sapper platoon may also be a mixture of SPF and locals.

15-46. The sappers may or may not be in a uniform. The affiliated forces or civilians (probably trained and lead by the SPF) may be men, women, and children. Women and children may be used as runners, messengers, scouts, guides, drivers, porters, fighters, suicide bombers, lookouts, or in several other roles. They may also emplace and/or detonate improvised explosive devices (IEDs) and mines.

15-47. The SPF sapper platoon not only emplaces and detonates IEDs and booby traps. It also manufactures them and teaches others how to manufacture, emplace, and detonate them. (Booby traps are actually a subcategory of IEDs.)

15-48. Sappers accompany and/or augment standard SPF teams, direct action teams, and guerrillas. They may also work completely independent of other operations. The platoon may also serve as a stay-behind or independent unit to conduct disruption operations. The sapper platoon also provides general engineer support to the SPF battalion.

Long-Range Signal Platoon

15-49. The long-range signal platoon of an SPF battalion is responsible for all communications (satellite and non-satellite) in the SPF battalion. It also performs unit-level maintenance on communications and electronic equipment. It also coordinates retransmission requirements when necessary. All SPF communications are encrypted.

15-50. This platoon can be augmented with personnel and equipment from the long-range signal company of the parent SPF brigade. The platoon can be employed together or broken up to support subordinate SPF companies and activities. This platoon also assists in the training of affiliated irregular forces and civilians on how to set up, operate, maintain, and transport communications equipment.

SPF COMPANY

15-51. The AOR for SPF companies may be quite large. The SPF company headquarters serves as the lifeline for the SPF teams. The teams are designed to operate independently (if necessary) throughout the company's AOR, or to support regular military, irregular forces, or other SPF units and/or activities. The company headquarters and staff provide C2 and coordinate activities of individual teams and their mission support requirements. Examples of types of support span the spectrum from logistics, to intelligence, to coordinating external fire support, to acquiring additional medical attention for locals. The company's primary task is to facilitate the success of individual teams.

15-52. The SPF company does not have a fixed structure. The structure, and the equipment and weapons mix, all depend on specific mission requirements and conditions. The SPF company, therefore, can have any number of teams. However, it typically has 10 to 12 teams (as shown in the example in figure 15-3). The number of teams deployed generally depends on the team size required for specific missions. The mission, environment, geographic factors, and other conditions determine the configuration and composition of each company and team.

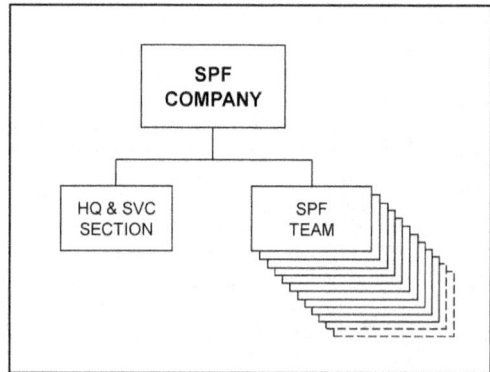

Figure 15-3. SPF company (example)

15-53. The company may be composed of several different teams, each with a different focus, or it may be composed of teams with a single focus. The single focus may be air infiltration (all teams of air infiltration), or divers (all teams of divers), or medical, or sniper, or any of the other types of teams listed below. More often, however, the company will be composed of a mix of teams with specialized capabilities and several multifunction (standard SPF) teams, or all teams will be the standard SPF team structure. Figure 15-4 shows an example of a company that has at least one of each of the types of specialized teams. Another example might be four standard SPF teams, one direct action team, one air infiltration team, one sniper team, one medical team, one signal team, and one sapper team. Each SPF company normally has a full range of these specialty capabilities, whether in standard teams or specialized teams. See SFP Teams below.

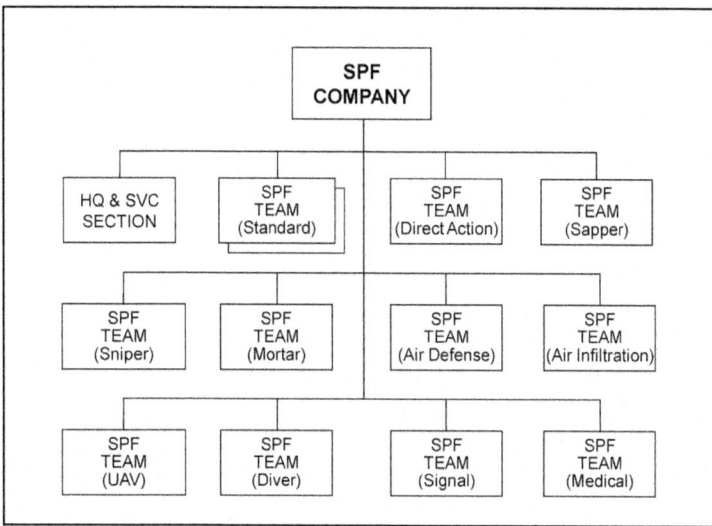

Figure 15-4. SPF company with specialized teams (example)

15-54. The SPF company has sufficient assets to transport munitions and materiel for the immediate fight. It is dependent upon support from higher (transport platoon at SPF battalion, or materiel support company at SPF brigade) or external sources (caches or civilian augmentation) to transport materiel for a sustained fight. The organic vehicles provide the company a degree of operational autonomy and may be employed separately from a battalion.

SPF TEAMS

15-55. Regardless of the parent administrative organization, SPF normally infiltrate and operate as small SPF teams. Each team may break up into smaller teams (of as few as two men) or temporarily join with other teams to form a larger team or an SPF detachment, depending on the mission. These teams contain the majority of actual fighters and shooters in all SPF organizations. They perform the sabotage, assassinations, IED (and suicide bomber) emplacement and detonation, extortion, kidnapping, taking of hostages, ambushes, sniping, firing of mortars, and other SPF functions. The SPF teams also provide instruction to guerrillas, insurgents, or other indigenous personnel. In some cases, the SPF team also serves as the planning and command element for these forces.

15-56. SPF team composition is not fixed and varies from team to team, mission to mission, and environment to environment. This includes the personnel, equipment, weapons mixes, and language abilities. Within a given SPF company, SPF teams can be—
- All multifunctional.
- Each with a different focus.
- All teams composed of a single focus or specialty.
- Any mix of these.

Chapter 15

The single focus may be a multifunction direct action mission, or any of the other functions represented by specialized teams in figure 15-4. Each of these may also be the focus of one or two teams (possibly more). Specific composition of each type of SPF team can be found in FM 7-100.4.

15-57. The mix of weapons and equipment varies with the mission. Each SPF team can be equipped with—

- State-of-the-art communications gear including satellite telephones and satellite communications transceivers.
- Man-portable GPS jammers.
- Handheld thermal viewers.
- Communications intercept and direction finding sets.
- Sensor sets.
- Automatic grenade launchers (AGLs).
- Sniper and antimateriel rifles.
- General purpose machineguns.
- Antitank guided missiles (ATGMs), antitank grenade launchers (ATGLs), and/or antitank disposable launchers (ATDLs).
- Assorted remotely detonated mines, side-attack mines, and/or IEDs.

For additional information on personnel, weapons, and equipment, see FM 7-100.4.

Standard SPF Team

15-58. The standard SPF teams are the default for all SPF organizations not requiring unique specialties. For additional detail see FM 7-100.4. All SPF teams are based on this standard structure. SPF commanders begin with the standard team and tailor (task-organize) it to provide a high degree of success for specialized functions. Standard teams possess all of the specialized capabilities to a lesser degree and are more multifunctional. Any specialized team can serve as a standard SPF team when not engaged in specialized missions. SPF team members are cross-trained in the use of all equipment, weapons, and vehicles assigned to the SPF company.

Direct Action SPF Team

15-59. The structure and weapons mix of the direct action teams provides extreme lethality and is ideal for dispersed combat such as fighting in built-up areas, especially urban combat. The teams specialize in infantry antiarmor style attacks; antiarmor ambushes; and hit-and-run attacks against armored and/or hardened or rear area targets.

15-60. SPF direct action teams can work completely independently. However, several direct action teams (typically 12-men each) usually work together, with each team breaking down into smaller teams. For example, 3 or 4 direct action teams, each broken into 3 smaller teams, might act together—for a total of 9 to 12 smaller teams, with perhaps 4 persons each. The teams will probably be supported or augmented by different types of SPF specialty teams such as sniper teams or sapper teams. They may also be supported by guerrillas or other local affiliated forces. One or more SPF UAV teams may be used to acquire reconnaissance information on targets or facilities.

15-61. When fighting in close terrain, SPF direct action teams are also the masters of the "defenseless defense" in which no defense or opposition is observed or suspected. Therefore, the enemy enters into the kill zone completely unaware. If teams are not in fixed defensive positions, they can remain totally mobile and hard to find. The teams prefer to trap vehicle columns in city streets where destruction of the first and last vehicles will trap the column and allow its total destruction. Normally, several direct action teams (often broken into several smaller teams) simultaneously attack a single armored vehicle. Kill shots are generally made against the top, rear, and sides of vehicles. Targets are engaged simultaneously to maximize effectiveness and confusion. Once isolated, single vehicles are easily defeated.

15-62. While the direct action teams may fight from ground level, they prefer a three-dimensional attack. That means that they also attack from basements or sewers (spider holes) and from trees or the upper

Special-Purpose Forces and Commandos

stories and tops of buildings. The elevation and depression angles of gun barrels on many armored vehicles are incapable of engaging the teams fighting from basements and second- or third-story positions. Multiple antiarmor rounds fired from different heights and directions limit a vehicle commander's ability to respond. The SPF predetermine (and target) enemy escape routes. SPF can use thermobaric flame weapons as "pocket artillery" to provide the firepower equal to that of a 122-mm artillery shell or 120-mm mortar shell, both of which are especially lethal in the close environment.

15-63. Direct action teams are generally equipped with a mix of infantry-type weapons (including man-portable antiarmor weapons); sniper and antimateriel rifles; and an assortment of sensors, designators, and other gear. The teams use command-detonated, controllable, and side-attack mines (antitank [AT], anti-vehicle, and antipersonnel). Side-attack (off-route) mines may be placed out of sight, such as inside windows and alleys. Tripwires for claymore-type antipersonnel mines can be strung at antenna height to destroy troops riding on the top of armored vehicles. The OPFOR also uses AT grenades and will attempt to drop bottles filled with gasoline or jellied fuel and other improvised flammables or demolitions on top of vehicles. See the *Worldwide Equipment Guide* for specific information and detail on equipment and weapons.

Sapper Team

15-64. Sappers are SPF (or infantry or guerrillas) trained to perform some typically raider and engineer functions. The SPF sappers also train local irregular forces and civilians to be sappers. Sappers are not engineers. An SPF sapper team can—

- Infiltrate enemy installations and areas.
- Scout (making accurate diagrams for future attacks).
- Guide other sappers, guerrillas, or SPF through enemy lines and obstacles to perform their missions.
- Conduct route reconnaissance.
- Conduct and/or assist in raids, assaults, and ambushes.
- Breach obstacles.
- Facilitate water obstacle crossing.
- Conduct assault breaching.
- Emplace field expedient fortifications and obstacles (such as cratering).
- Lay mines (especially nuisance minefield, IEDs, and booby traps).
- Lay controlled minefields.
- Emplace obstacles (including anti-vehicle wire obstacles).
- Set side-attack mines.
- Deliver mines using man-portable mine-scattering systems.
- Support AT and countermobility operations.
- Conduct general demolition.
- Provide general engineer support (including minor construction).
- Provide water purification.
- Employ limited smoke or expedient obscurants.

The sapper team may use an SPF UAV team to acquire reconnaissance information on targets, geography, or facilities.

15-65. In the raider role, sapper teams may serve as the lead or primary (assault) element in an assault on fixed installations or military field positions. Armed primarily with explosives charges, the sapper attempts to breach the defensive perimeter and neutralize tactical and strategic positions in advance of the main body or behind enemy lines. The sapper (raiders) can serve as an independent combat force making deep thrusts from different directions simultaneously, or they can support guerrilla or infantry operations.

15-66. The sapper team may be broken into three, four, or six smaller teams, which may work in concert or may each be employed individually. Sappers may accompany and/or augment standard SPF teams,

Chapter 15

direct action teams, and guerrillas. They may also work completely independent of other operations. The team may serve as a stay-behind element to conduct disruption operations. This team may serve with and train local irregular forces and civilians how to infiltrate, set demolitions, and assault enemy installations.

15-67. The sapper team manufactures IEDs and booby traps (booby traps are actually a subcategory of IEDs). The team may also emplace and/or detonate the IEDs (booby traps are triggered by the unsuspecting) or they may give them to other trained sappers, SPF, or guerrillas to emplace and/or detonate. The OPFOR commonly uses IEDs as "secondary devices" to detonate on the arrival of responding personnel. IEDs can be detonated by a variety of means, including remote, command, electrical, tripwire, pressure, time, and others. The sapper team trains local irregular forces and civilians in the manufacture and employment of improvised munitions and booby traps.

15-68. Members of the sapper team do not serve as suicide bombers. However, they do recruit suicide bomber prospects, and plan and coordinate such bombings. They may also control the bombers.

Sniper Team

15-69. The primary mission of the SPF sniper team is to serve in the conventional offensive sniper-countersniper role and to extend the lethal range of the team or the supported organization. Other missions include assassination, attrition, interdiction and disruption, providing covering fires, and supporting the INFOWAR plan. Depending on mission requirements, however, the SPF sniper team may also serve as a reconnaissance element, a stay-behind element, or part of a hunter-killer (HK) team. Sniper teams may also serve as laser designators for indirect fire or air-delivered ordnance.

15-70. These teams may also work completely independent of other operations. Several SPF sniper teams may saturate an area with snipers to serve in an INFOWAR role, or stay behind to conduct disruption or direct action operations.

15-71. The composition of the sniper team fluctuates with the mission, conditions, and available resources. However, the SPF sniper team typically consists of 12 members and can break down into four smaller sniper teams, each with three people. In this case, the original SPF team leader serves as the leader of one small sniper team. Each small sniper team generally consists of a team leader/observer (spotter), sniper (shooter)/target designator, and a driver/assistant sniper/gunner.

15-72. Each small team may have its own vehicle (possibly amphibious). This may be a tactical utility vehicle, all-terrain vehicle, or light strike vehicle. However, the mission may dictate that no vehicles be used. In such cases, even teams with vehicles would leave their vehicle behind and carry only mission-critical equipment. Teams with no vehicles would depend completely on caches, porters, or other transportation and supply means.

15-73. The sniper carries either a sniper rifle or an antimateriel rifle, depending on the mission. The other rifle may remain behind with the vehicle, at an MSS or cache, or at company. On some occasions (when the vehicle is left behind), the driver/assistant sniper may carry the other rifle.

15-74. The sniper in each small team provides covering fires. The other team members may use AGLs or machineguns to engage enemy personnel when they exit the armored vehicles, and pin down the supporting infantry, allowing an ATGM or ATGL gunner to engage the armored vehicle. If the sniper uses the antimateriel rifle, it can provide accurate fires 1,800 meters or more. The infantry ATGM or ATGL can engage and penetrate armor from 600 meters or more, depending on the weapons used.

15-75. SPF snipers may accompany and/or augment other SPF teams, direct action teams and guerrillas. They may also work completely independent of other operations. The SPF sniper team can also train and equip local irregular forces and civilians to serve as snipers or marksmen. See chapter 16 for more detail on sniper teams and their employment.

Mortar Team

15-76. The SPF mortar teams normally use the light, mobile 60-mm "commando"-style mortar. More lethal 81- or 82-mm mortars are used when the mission dictates.

15-77. SPF mortar teams may accompany and/or augment other (standard) SPF teams, direct action teams, or affiliated irregular forces. They may be employed as a whole team (typically 12-men) or smaller individual mortar teams. The small teams may be employed in concert or may be broken up and assigned to support other SPF teams. SPF mortar teams may also work completely independent of other operations. Several teams may saturate an area with mortars to serve in an INFOWAR role, or stay behind to conduct disruption operations.

15-78. The SPF mortar team can train and equip local irregular forces and civilians in the proper use, logistics, and employment of mortars. It will also provide mortar support to these forces whenever necessary.

15-79. Weapons and ammunition may be cached. Local sympathizers may volunteer to serve as ammunition bearers. Others may be forced to serve as ammunition bearers or porters. Local burden animals such as donkeys, mules, or camels may also be used to transport ammunition and materiel. See Logistics later in this section.

Air Defense Team

15-80. The primary mission of the SPF air defense team is to provide security from air attack. Other missions include—

- Air ambushes.
- Disruption of enemy air operations.
- Destruction of key aircraft.
- Support the INFOWAR plan (denial of airspace and enemy claims of air supremacy and/or superiority).

15-81. SPF air defense teams may accompany and/or augment other SPF teams, direct action teams, and affiliated irregular forces. The team may be employed as a whole team (typically 12 men) or as smaller air defense teams with individual air defense weapons. These smaller teams may be employed in concert or broken up either to support other SPF teams or to work completely independent of other operations. Several SPF air defense teams may saturate an area with air defense weapons to serve in an INFOWAR role, or stay behind to conduct disruption operations. The team is extremely effective in conducting air defense ambushes. The mobility and low signature of the team enhances its lethality and overall effect in disruption of enemy air operations and planning.

15-82. The SPF air defense teams are usually equipped with state-of-the-art air defense equipment, such as the air defense/antiarmor systems (ADAAS) or man-portable air defense systems (MANPADS). Aside from the ADAAS or MANPADS, equipment usually includes a mix of machineguns and air defense acquisition and warning receivers.

Note. An ADAAS can also be used against ground targets such as light armored vehicles, and snipers in bunkers or buildings.

15-83. When dismounted, each ADAAS or MANPADS gunner carries a shoulder-fired launcher with one missile. The ADAAS or MANPADS assistant gunner carries one additional missile and can bring more from the vehicle, as required. The SPF air defense team vehicles carry extra missiles for each ADAAS or MANPADS launcher. The assistant gunner also carries the electronic plotting board and enters location and direction data of approaching targets.

15-84. The SPF air defense team machinegunner remains with the heavy machinegun (HMG), mounted or dismounted. Typically the ADAAS or MANPADS gunner and assistant gunner dismount and prepare launch positions. The HMG gunner dismounts and prepares an air defense firing position (assisted by the driver). The driver then hides the vehicle and either stays with the vehicle (ready to pick up the others) or assists the HMG gunner with the electronic plotting board. The HMG gunner and/or the ADAAS or MANPADS gunner may stay with the vehicle if the terrain and situation allow.

Chapter 15

15-85. The SPF air defense team can train and equip local irregular forces and civilians in the proper use, logistics, and employment of air defense weapons. It will also provide air defense support to these forces whenever necessary. See chapter 11 for more information on ADAAS and MANPADS employment.

Air Infiltration Team

15-86. The SPF air infiltration team either infiltrates by air and/or assists other SPF teams in infiltrating by air and in conducting other air operations. It is responsible for all insertion, supply, and planning and coordination of other air operations for the SPF company. The team also performs tests, diagnostic checks, repairs, and maintenance on all parachutes, ultralight aircraft, support systems, and related equipment. All team members are qualified riggers and provide individual and cargo parachute packing and rigging, unit maintenance of air delivery items, and individual rigger support for the SPF company.

15-87. The ultralight aircraft may, or may not, be equipped with light and general-purpose machineguns; AGLs, ATDLs, or other weapons systems. They may also be fitted with radios, cameras, sensor sets, or laser target designators. The ultralights are dismantled for trailer transport, two aircraft per trailer. All team members are qualified to operate the ultralight aircraft (single- or two-seaters) or to serve as gunners/equipment operators.

Note: Armed ultralight aircraft can also be used in attack, reconnaissance, or multimission roles, not just in infiltration.

15-88. The air infiltration team also has powered parachutes available. These are extremely effective for surreptitious infiltration and exfiltration. Once successfully infiltrated, the team hides the parachutes in a cache. Upon successful completion of the mission, the parachutes are retrieved and the team exfiltrates to a more secure location, possibly for further transport out of the area. For additional information on the SFP air-infiltration team, see FM 7-100.4.

UAV Team

15-89. The primary mission of tactical UAVs is reconnaissance. However, the ease of operation, size, and simple design of some tactical UAVs lend themselves to field expedient modification. Converting a UAV into a munitions delivery system (improvised attack UAV) is not difficult and offers several tactical advantages. Off-the-shelf remote-controlled aircraft can also provide this capability (remote-controlled IEDs).

15-90. SPF UAV teams may accompany and/or augment other SPF teams (especially sapper and direct action teams) or affiliated irregular forces. The UAV team may be employed as a whole or may break down into smaller teams. For example, a 12-man team could break into 4 smaller (3-man) teams. Either the large team or individual smaller teams can be assigned to support other SPF teams. They may also work completely independent of other operations.

15-91. When the SPF UAV team breaks into four small teams, each small team would generally carry two UAVs on each mission: one is used as a backup. In dismounted operations, the driver carries the appropriate radio and the UAV maintenance and repair kit. The UAV operator carries one UAV while the UAV assistant operator carries the extra UAV, batteries, and the computer-laptop UAV ground station. Almost any laptop computer can serve as a ground station coupled with the proper software and flight controls. Specialized training is not required for this system. Range, reconnaissance capability, and flight time vary depending on the type of tactical UAV used.

Diver Team

15-92. The primary missions of the SPF diver team are water infiltration, reconnaissance, and demolition. All members of the SPF diver team are qualified special-purpose combat divers. The teams specialize in operating in a riverine, swamp, wetland, or coastal environment. The team is responsible for all waterborne insertion, infiltration, supply, attack, beach, river crossing, and surf/tide studies. It is also responsible for other waterborne operations, planning, and coordination for the SPF company. It maintains a sufficient amount of equipment to support the SPF company's diving and waterborne infiltration requirements.

15-93. The size and number of diver teams employed depends on the mission and geographic AOR of the SPF company. For example, companies and/or teams operating in a coastal or river environment perform more diving missions. SPF diver teams may accompany and/or augment other SPF teams (especially sappers and direct action teams) or affiliated irregular forces. They may be employed as a whole team (typically 12-men) or individual smaller diving teams and suballocated to supported units. They may also work completely independent of other operations and serve in various roles.

15-94. The SPF diver team either infiltrates by water and/or assists other SPF teams in infiltrating and exfiltrating by water and in conducting other waterborne operations. The team also performs tests, diagnostic checks, repairs, and maintenance on all diving gear, diving support equipment, and water craft. Surface craft and other specialized craft and equipment such as rigid inflatable boats, light patrol boats, submersibles, semi-submersible infiltration landing craft, and specialized open water craft will be provided by the supporting naval organization depending on mission requirements.

Signal Team

15-95. The SPF signal team provides state-of-the-art secure long- and short-range communications for the SPF company and its deployed teams. This team can also train local guerrilla or insurgent forces and civilians on how to set up, operate, maintain, and transport communications equipment.

15-96. SPF signal teams may accompany and/or augment other SPF teams, insurgents, or guerrillas. They may be employed as a whole team (typically 12-men) or individual small signal teams and suballocated to supported units. They may also work completely independent of other operations and serve in roles such as retransmission sites.

15-97. A single small SPF signal team can provide long-range communications support for guerrilla units up to battalion size. A full SPF signal team can do the same for a brigade-size unit. Teams can also support insurgent operations. This team may also serve in a signals reconnaissance collection role. In the collection role, the signal equipment is exchanged one-for-one with communications intercept and direction finding equipment. Each team then becomes a communications intercept and direction finding unit.

Medical Team

15-98. All personnel in the SPF medical team are qualified medics. (All other types of SPF team also have a dedicated medic assigned.) SPF medical teams may accompany and/or augment other SPF teams or affiliated irregular forces. They may be employed as a whole team (typically 12-men) or broken down into small medical teams or individual personnel and suballocated to supported units, villages, or sectors.

15-99. Each SPF medical team is designed to provide medical support to the SPF company and for guerrilla units up to brigade size. A small medical team can support a battalion-size force. SPF medical teams can also support insurgent operations.

15-100. This team also trains local irregular forces and civilians on how to perform emergency medicine, battlefield medical procedures, and evacuation. The SPF medical team members provide limited medical intervention, minor surgery, and treatment of most common illnesses. Supported irregular force units receive litters and medical supplies from the medical section to transport and treat wounded. The supported unit provides its own litter bearers. The SPF teams in the field have a very limited and very temporary inpatient capability. Severe and longer-term care relies on evacuation to civilian, military, or other medical facilities. More routine and excess ill and wounded are backhauled in general-purpose cargo and civilian vehicles.

15-101. Some local medical support may be available. SPF medical support is coupled with local sympathizers and/or irregular forces' medical assets in the area. Irregular forces may mobilize local medical personnel to assisting in treat and evacuating ill and wounded. Maximum use is made of local medical assistance and facilities regardless of capability. Local civilian sympathizers may volunteer their homes, equipment, vehicles, and services. They may also assist in the evacuation of wounded to civilian, irregular force, or military facilities. Whenever possible, medical functions are performed in tents, tunnels, caves, or local accommodations. In some instances, the SPF medical teams will attempt to colocate with a village clinic.

15-102. Cargo trailers transport medical equipment and supplies. In emergencies they may transport wounded. These trailers may be dropped at the aid station when the team's light trucks serve as ambulances. Vehicles may be a mix of military and civilian, or all civilian. Carts may also be used to transport wounded. Depending on the situation, the trailer transporting equipment and supplies may be cached or dropped to be recovered later. It may also be abandoned (concealed) when no longer needed.

15-103. SPF medical personnel are combatants. When necessary—they fight. A medical aid station is usually set up at base camp while other medics accompany other SPF teams or affiliated irregular forces in the fight. Medical team personnel may be a mixture of men and women. Local civilian augmentation, including women, may make up a large percentage of the total strength of people serving as assistant medics, litter bearers, and in other roles.

TACTICS, TECHNIQUES, AND PROCEDURES

15-104. The point of the SPF spear—the SPF team—is malleable and extremely lethal. The teams lend themselves to adapting to the situation and will adopt any TTP, weapons, or equipment that may prove successful or place them at advantage.

15-105. There are no standard prescribed TTP used or taught by the SPF. Once on the ground, the SPF are amorphous. Their ability to continually adapt to all aspects of their environment is directly relational not only to mission success but also to their survivability and lethality. These lethal organizations, and their TTP and actions, manifest themselves primarily by their—

- Unpredictability.
- Sharp learning curve.
- Continuous improvisation.
- Adaptive and unpredictable TTP.
- Mobility.
- Ability to influence and to blend in with the population.
- Ability to use local culture and agendas to their advantage.
- Ability to shape local perceptions and alliances.
- Shifting architectures, affiliations, alliances, behaviors, and players.
- Spectrum of lethality—ranging from a single precise rifle shot to IEDs to weapons of mass destruction (WMD).

15-106. SPF missions may support national-, theater-, operational-, or tactical-level objectives. They are conducted across the spectrum of military operations independently or in coordination with regular and/or irregular forces. However, the TTP used by all SPF units are similar and are actually performed at a tactical level, as small teams or detachments. Therefore, SPF would use the same types of tactical action (such as raids and ambushes) regardless of the level of command from which they come or the level at which they for which they perform the missions. The basic TTP used by SPF to conduct these missions are not unique to the SPF, nor are the SPF limited to those basic TTP. Special missions may require specialized skills and TTP that are unique to those missions. For either basic or special missions, the TTP the SPF select for a particular mission or purpose may be unexpected.

15-107. One mission of SPF is to advise, train, and assist irregular forces. Therefore, the SPF must be able to use the same types of TTP as the irregular forces they train, support, or fight alongside. As SPF train and work with affiliated irregular forces, they may also learn some effective TTP (such as terror tactics) from those forces. If, on the other hand, the SPF teach terror tactics to the irregular forces, they must have the expertise and capability to use terror tactics themselves.

INFILTRATION

15-108. The success of SPF land, air, and amphibious operations in support of regular military and/or irregular forces or independent SPF missions is primarily dependent upon detailed planning and preparation. Infiltration techniques include—
- Air infiltration.
- Water infiltration.
- Land infiltration.
- Stay-behind forces.

Land

15-109. Land infiltration involves the use of various modes of transportation or techniques such as commercial vehicles, railway trains, or infiltration on foot, possibly along with refugees. It is conducted in a manner similar to that of a long-range reconnaissance patrol infiltrating into enemy territory. Generally, guides are required. If guides are not available, the SPF team or detachment must have detailed intelligence of the route, particularly if it is to cross borders. Routes are selected to take maximum advantage of cover and concealment and to avoid enemy outposts, patrols, and installations.

15-110. Before the mission, the team or detachment is briefed on the known locations of selected individuals who will furnish assistance and on the established means of contacting them. These individuals may be used as local guides and sources of information, food, and shelter. Since there are local sources for survival items, the SPF team or detachment can restrict the equipment and supplies to be carried to mission-essential items (individual arms, equipment, and communications gear).

15-111. A very successful infiltration method used by the SPF is to infiltrate under the guise of reconnaissance probes. This is especially successful when the SPF is either guiding or using affiliated irregular forces, from a team of 3 to 4 men to a squad, or a platoon, or even a company of approximately 200 men. The SPF and/or irregular forces conduct small probes along the enemy defensive positions. If the enemy does not respond to these probes, the SPF and/or irregular forces infiltrate in small numbers and spread out. This permits larger numbers to penetrate. Once behind enemy lines, one team may cut off the escape route of the enemy, while the other teams conduct a coordinated assault on both the front and flanks. The attacks will continue on all sides until the defenders are destroyed or forced to withdraw. The SPF and/or irregular forces will then move stealthily forward to the open flank of the next enemy position and repeat the tactics.

15-112. The OPFOR also conducts another very successful variation of this infiltration and subsequent action. In this variant, the SPF and/or irregular forces do not immediately attack as soon as they are successfully behind enemy lines. Once behind the enemy, they may wait a few hours or up to 3 days or more and may number as much as a full irregular company or even a battalion, depending on the circumstances. Once emplaced either behind the enemy or more likely behind and on both flanks of the enemy, the infiltrated force then waits for the main OPFOR attack. If the main OPFOR attack is successful, the enemy will either retreat or fall back. At that time, the infiltrated SPF and/or irregular forces will ambush and destroy the remaining enemy forces. If the OPFOR main attack if faltering or appears as if it may fail, the infiltrated SPF and/or irregular forces simultaneously attack from both the rear and flanks, ensuring victory.

Air

15-113. Air delivery by parachute is one of the principal means available for the infiltration of SPF personnel. However, they can also use ultralight aircraft as well as powered parachutes. In preparing a team for air infiltration, the team leader or detachment commander considers the following:
- Aircraft capabilities.
- Timing (day/night).
- Weather and geographic effects.
- Detection by enemy personnel, which may compromise the mission.
- Equipment and supplies.

Chapter 15

- C2.
- Reception personnel in the drop zone (DZ).
- Ground assembly.
- Emergency plans.
- Exfiltration.

15-114. The Air Force fields light transport aircraft for insertion of its own SPF or those belonging to other service components. The SPF Command also has some tactical transport for use in inserting SPF units. Commercial aircraft may also be used to support high-altitude air drops.

15-115. All OPFOR SPF companies normally contain at least one air infiltration team equipped with six ultralight aircraft as well as an assorted number of powered parachutes.

High-Altitude Air Drop

15-116. When enemy air defense discourages normal infiltration by air, parachute entry from very high altitudes may be necessary. This may involve either high-altitude low-opening (HALO) or high-altitude high-opening (HAHO) techniques. Whenever this type of drop is planned in denied areas protected by enemy radar and other detection devices, a system of jamming or disruption of these systems should be established.

15-117. An important consideration is the availability of aircrews trained in operating under arduous conditions in depressurized aircraft at high jump altitudes. The team leader or detachment commander must devise a system for freefall assembly of personnel after they have exited the aircraft, but before opening the parachutes. This is particularly important at night or when conditions preclude visual contact with DZ markings. Assembly aids include special marking devices and materials, visible at night, applied to pack trays, backpacks, and other designated equipment.

Blind Drop

15-118. Selected SPF personnel may be air-dropped during the initial infiltration phase on DZs devoid of reception personnel. This technique is referred to as a "blind drop" and may be employed when an area is known to contain a local irregular force of sufficient size and nature to warrant cultivating as an affiliated force. In all probability, the irregular force will be receptive to outside support. Other SPF, regular OPFOR, or interested interagency partners were either unable or did not have time and means to train the irregular force as DZ reception personnel. Additionally, the enemy situation might preclude normal DZ markings and recognition signals.

15-119. Once on the ground, SPF personnel move to the selected assembly area and establish security. The SPF team or detachment attempts to make contact with the local irregular forces.

Water

15-120. Water offers another practical means for infiltration into areas having exposed coastlines or riverbanks. Water infiltration normally terminates in a land movement phase. Considerations for water infiltration include the following:
- Watercraft capabilities.
- Detection by enemy personnel, which may compromise the mission.
- Reception personnel.
- Equipment and supplies.
- C2.
- Ship-to-shore movement.
- Land assembly and movement.
- Emergency plans.
- Exfiltration.

Special-Purpose Forces and Commandos

15-121. SPF companies normally contain at least one SPF diver team equipped with Zodiac-type inflatable boats with motors, and assorted scuba and rebreathing gear. The SPF diver team either infiltrates by water and/or assists other SPF teams in infiltrating and/or exfiltrating by water. This team is responsible for all waterborne insertion, infiltration, and supply. For additional information on this team, see Diver Team, above, or FM 7-100.4.

15-122. Depending on mission requirements, SPF units may require surface craft and other specialized craft and equipment to be provided by a supporting naval organization. This can include rigid inflatable boats, light patrol boats, mini-submarines, submersibles, semi-submersible infiltration landing craft, and specialized open water craft.

15-123. Infiltration by means of amphibious aircraft landing on large lakes, rivers, or coastal waters may be possible. In such a case, infiltration planning by the team or detachment considers the ship-to-shore and subsequent land movement characteristics of water infiltration.

SWARMING

15-124. Swarming is a tactic that results in the convergent attack(s), from multiple directions, and possibly multiple dimensions, by numerous elements on a single target(s). The SPF generally use this type of attack when accompanied by affiliated irregular forces. SPF teams plan or otherwise facilitate the attack and may or may not accompany their surrogate forces conducting the swarming attack.

15-125. There are two basic types of swarming: the massed swarm and the dispersed swarm. In the massed swarm, the elements begin as a massed (assembled) unit. On command, the elements then disassemble and conduct a convergent attack(s) to swarm the enemy from numerous directions. In the dispersed swarm, the elements are geographically dispersed from the beginning. On command, the elements infiltrate. Once prepared, they attack (from their respective directions), converging on the enemy without forming a single massed unit. Swarms are equally effective in both the offense and the defense.

15-126. Of the two types of swarming, the dispersed swarm is the most difficult to defend against because the attacking elements never present a massed target. The OPFOR prefers to use the dispersed swarm attack where the attackers are initially dispersed, then converge on the target(s). It is more appropriate to the dispersed fight the OPFOR, and especially the SPF, prefers. Once the attack is complete, the attacking elements can either dissipate into the local population, exfiltrate back to where they came from, or move to hide positions or sanctuary (possibly cross-border).

STAY-BEHIND

15-127. SPF teams may be pre-positioned in proposed AORs or may remain in areas formerly under OPFOR control, before the enemy occupies these areas. This provides them the opportunity to organize the nucleus of an affiliated irregular force, conduct surveillance, or conduct direct action. Stringent precautions are taken to preserve security, particularly that of the refuge areas or other safe sites to be used during the initial period of enemy occupation. Information concerning locations and identities within the indigenous organization is kept on a need-to-know basis. Contacts among various elements use clandestine communications.

15-128. Dispersed caches, to include radio equipment, are pre-positioned when possible. SPF personnel have a better chance of survival in small towns, villages, and rural areas. However, when stay-behind operations are attempted in heavily populated urban areas, the SPF team or detachment is completely dependent upon the indigenous organization for security, the contacts required for expansion, and the buildup effort.

EQUIPMENT

15-129. SPF personnel generally use the best equipment and weapons available to accomplish their missions, typically high-end tier 1 systems. (See FM 7-100.4 for additional information on the tier system of OPFOR weapons and equipment and specific information on SPF equipment and its allocation.) The OPFOR routinely uses the SPF as an avenue for introducing new or improved weapons and equipment technology onto the battlefield. This often includes advanced weaponry such as MANPADS and ATGMs, extremely lethal warhead updates for older AT weapons, and communications, targeting, and reconnaissance systems. At the tactical level, this often provides asymmetric advantage out of proportion to the capabilities of the individual piece of equipment. Depending on the type of equipment or weapon introduced, it may carry a significant psychological impact at the strategic level. In some instances, the OPFOR uses the SPF to train both regular and irregular troops in the use of new or high-technology niche weapons and equipment.

15-130. SPF team members are cross-trained in the use of all equipment, weapons, and vehicles assigned to the team or company. Some weapons and/or equipment may be left in the vehicles until required or not carried on the mission at all. These may be left behind at an MSS, in a cache, or left at the company. In this case, the SPF teams carry only mission-critical equipment and weapons.

15-131. The environment and the mission determine the categories and types of equipment and weapons mix used by the SPF battalion and below. For example, in a swamp, wetland area, or other difficult and marginal terrain, the SPF may use all-terrain vehicles (ATVs), amphibious ATVs, and/or similar amphibious vehicles in lieu of tactical utility vehicles or light strike vehicles. In short, all vehicles and weapons are interchangeable and/or tailored to fit the specific mission and conditions.

15-132. Vehicles used by the SPF are a mix of military and civilian. An SPF brigade, battalion, company, or team may be augmented by military and/or civilian vehicles. Depending on the mission, civilian vehicles may include—

- Motorcycles.
- Agricultural trucks.
- Commercial trucks.
- Liquid cargo carriers (for POL or water).
- Flat beds.
- Busses.
- Farm trailers.
- Tractors.
- Cars.
- Half-ton civilian trucks.
- Bicycles.
- Carts.
- High-mobility vehicles.
- ATVs.

15-133. Local sympathizers may volunteer their equipment and services. The SPF commander may requisition or confiscate local civilian transportation assets and materiel. This includes the use of civilian personnel for porters or labor. Draft animals may also serve as bearers or porters.

PERSONNEL

15-134. All SPF soldiers are airborne qualified. Most speak several languages (which also enhances their ability to blend in), and many are qualified combat divers.

15-135. SPF personnel are trained in blending in and therefore may, or may not, be in uniform, depending on the circumstances. They may attempt to be indistinguishable from the local population (other than weaponry, which they may conceal or discard). Some, or all, may completely melt into the civilian population when not engaged in military operations.

15-136. The ability of the SPF teams to blend in with the population usually is in direct relationship to their ability to survive and also facilitates bonding with locals. When weapons are required, they may attempt to appear as if they are local militia, police, security guards, or other acceptable armed groups.

15-137. One of the primary missions of the SPF is to train, direct, use, and fight alongside of, or assist affiliated irregular forces to prepare for offensive actions, diversionary measures, or other missions. The irregular forces may be a mixture of men, women, and children. Local women and children may be used as runners, messengers, scouts, guides, suicide bombers, drivers, porters, snipers, lookouts, or in other roles. They may also emplace and/or detonate IEDs, booby traps, and mines. Women (and possibly children) may be fighters and participate in "drive-bys," assassinations, ambushes, and/or assaults.

LOGISTICS

15-138. Secrecy during movement or delivery complicates resupply of the deployed SPF team or detachment. This is complicated by the fact that the AOR for SPF units may be quite large geographically and cover a significant amount of terrain.

15-139. Detection by enemy personnel is always a priority concern. Once detected, the supplies can then be intercepted and destroyed, or the SPF personnel can be attacked, killed, or captured. The ability to escape detection while successfully resupplying the deployed SPF teams is susceptible to many factors, including time, distance, terrain, weather, and support of the local populace. All these factors pose challenges for SPF logistics planners at all levels. The surreptitious nature of resupplying what are usually covert activities is amplified by emergency requests and some on-call requests.

15-140. An SPF company or battalion usually has sufficient assets to transport munitions and materiel to meet immediate needs. It is dependent upon external support (brigade materiel support company or battalion transport platoon) or other sources (caches or civilian augmentation) to transport and/or stage materiel for a sustained fight. Depending on the mission and other variables, however, SPF teams and even their parent SPF companies or battalions may have no vehicles at all and depend on caches, porters, or other transportation or supply means.

15-141. A team leader or detachment commander requesting extensive logistics support from the outside should limit his request to essential items not readily obtainable in the AOR. This could include major items such as weapons, ammunition, demolitions, communications equipment, medical supplies, or other items that are normally denied to the local population by the enemy. The team leader or detachment commander has several techniques available that will give him the supplies required when he needs them.

ACCOMPANYING SUPPLIES

15-142. Accompanying supplies are items taken into the AOR by the SPF team during insertion or infiltration. These supplies are issued at the staging base in the final briefing stages and rigged by the team for delivery. When he plans his accompanying supplies, the team leader or detachment commander considers automatic resupply of survival and mission-essential items that he will receive. The accompanying supplies, plus the automatic resupply, will constitute the supply level of items required for the mission.

15-143. Some SPF units do not possess sufficient assets to transport munitions and materiel to meet immediate needs. This is especially true of the smaller teams or those with a mission of longer distance and/or duration.

AUTOMATIC RESUPPLY

15-144. Automatic resupply is prearranged for time, location, and content during the team's or detachment's final preparation stage at the staging base. The automatic resupply gives the team or detachment flexibility by allowing it to include backup communications equipment, weapons, ammunition, demolitions, medical supplies, and other items to support small-scale tactical actions and/or training of local irregular forces that may be affiliated with the OPFOR.

15-145. The supplies designated for the automatic resupply are selected on the basis of available intelligence, indicating items essential to complete or continue the mission, conduct exfiltration, or support affiliated forces. Once they are issued and received, they are rigged and prepared and stored for delivery in accordance with the predetermined schedule.

Chapter 15

ON-CALL SUPPLIES

15-146. After commitment into an AOR, and once it has established communications with its higher headquarters, the team or detachment is ready to begin requesting supplies based on mission needs and the capability to receive and store them.

15-147. In order to expedite supply requests, ensure accurate identification of needs, and minimize communication transmission time, the team leader or detachment commander uses a logistics brevity code system. The code includes the general category, unit designation, unit weight, total bundle weight, and number of individual man-loads per package. The logistics brevity code system is used to request three categories of supplies:
- Survival items (medical supplies, blankets, clothing, and food).
- Mission-essential items (weapons, ammunition, and communications equipment).
- Bulk items (the aforementioned items in bulk quantities to support extended missions, or the rapid expansion of affiliated forces).

Each load is prepackaged and self-contained. For example, a weapon will be packaged with ammunition, tools, POL, batteries, medical supplies, cleaning equipment, and spare parts. Unused weight or space will be used for additional survival items.

EMERGENCY RESUPPLY

15-148. The emergency resupply procedure is used to restore the combat capability of a team or detachment. This procedure is initiated when requested or after sustaining losses from enemy actions, missing a scheduled radio contact, or discovering faulty equipment. It may also be triggered by other incidents such as medical emergencies (usually of the supported population). Items delivered normally consist of communications equipment and other mission-essential equipment.

15-149. Local sympathizers may volunteer their equipment and services. The commander may requisition or confiscate local civilian transportation assets and materiel, to include POL and food. Civilian personnel may volunteer or be forced to serve as porters, farmers, or as general labor.

RECONSTITUTION AND REORGANIZATION

15-150. Restoring combat effectiveness of subordinates is one of the most important duties of SPF commanders. It includes—
- Determining the degree of combat effectiveness of subordinates.
- Assigning missions to subordinates that are still combat-effective.
- Withdrawing units from areas of destruction or contamination.
- Providing units with replacement personnel, weapons, ammunition, fuel, and other supplies.
- Restoring disrupted C2.

15-151. The OPFOR makes an effort to keep some units at full strength rather than all units at an equally reduced level. Usually, the unit with the fewest losses is the first to receive replacement personnel and equipment. However, once the casualties or equipment losses are sufficient to threaten the total loss of combat effectiveness, the commander may apply the concept of composite unit replacement. The composite unit concept involves a unit formed from other units reduced by combat action.

MISSION SUPPORT SITES

15-152. A mission support site (MSS) is a temporary base used by units and personnel who are away from their base camp, during an extended mission. The MSS may provide food, shelter, medical support, ammunition, or demolitions. Often some weapons and/or equipment may not be required and therefore not carried on the mission. These may be left behind at a MSS, in a cache, or left at the company until needed for the next mission.

15-153. The use of an MSS eliminates unnecessary movement of supplies and allows a force to move more rapidly to and from objectives. When selecting an MSS, consideration is given to cover and concealment, proximity to the objective, proximity to supply routes, and the presence of enemy security forces in the area. Security dictates that DZs or landing zones (LZs) be a considerable distance from an MSS, cache, or base camp—although this may increase transportation problems.

SECTION II – COMMANDOS

15-154. The SPF Command includes elite commando units. There are no commando units constituent to tactical level units at division and below in the AFS. Therefore, all commando assets employed at the tactical level will have been allocated from higher levels during task-organizing. This section addresses where those commando teams, squads, platoons, companies, battalions, and brigades come from, how they get to the tactical level, and how they are employed.

15-155. Most of what is addressed in this chapter applies to SPF and commando units alike. Like SPF units, commandos normally operate in enemy-controlled territory. In addition to proficiency in various infantry-type tactics, these elite units receive training for more specialized commando missions.

15-156. Every commando mission is unique and unlike any other, and thus requires forces organized not in a standard fashion but rather adapted into a task organization based on the mission. All of these commando organizations provide the OPFOR a flexible, capable, and lethal means of achieving its military goals.

COMMAND AND CONTROL

15-157. In the AFS, commando battalions are subordinate to the SPF Command. For administrative purposes, these battalions may be grouped under a commando brigade headquarters. However, commandos are employed as battalions, companies, platoons, and squads or as small teams, depending on the type of mission. The primary fighting element of the commandos is the company and platoon, more similar to regular infantry than SPF. Therefore, commandos are usually employed as battalions or companies depending on the mission, geographic area, and other conditions. In many cases, the commando companies and/or platoons are task-organized into teams. This is especially the case when operating in urban environments or other complex terrain.

15-158. Commando units can be allocated in a constituent or dedicated status to be task-organized as part of an OSC or of a tactical group based on a regular ground forces organization. Even in such cases, however, the reason for incorporating a commando unit into such an organization normally would be to perform specialized commando missions that contribute to the overall mission for which that task organization was created. In other cases, commando units may be allocated in a supporting relationship, while remaining under the command of their parent commando unit or the SPF Command. (See chapter 2 for a thorough discussion of the various command and support relationships.)

MISSIONS

15-159. Commandos are elite units, specially trained for missions in enemy territory. When assigned such missions, the commando units may disperse into small teams. These small teams are harder to detect during infiltration and provide the ability to strike many targets simultaneously to achieve maximum effect. If necessary, once they infiltrate, they can re-form into platoon- to company-size units or into task organizations to perform attacks against subsequent targets.

INFANTRY

15-160. Sometimes, commandos may be called on to perform regular infantry missions. This may occur particularly in defensive situations, if the defensive mission is more important than reconnaissance or security. Commando units may fill gaps between the battle positions of dispersed regular forces. When performing such infantry-type missions, commandos typically fight as companies or battalions, using tactics similar to those of regular infantry or motorized infantry units.

Chapter 15

COMMANDO

15-161. Commando units generally conduct various types of reconnaissance and combat missions in the disruption zone or deep in enemy territory, during larger operations or tactical actions that are either offensive or defensive. The reconnaissance missions include actions such as surveillance, monitoring, and searches. Commando units are expected to conduct reconnaissance within the context of any combat mission. Conversely, when employed as reconnaissance elements, the commando units' activities are not limited to reconnaissance. They are also tasked with assaulting and destroying military or civilian targets.

15-162. Commandos provide the OPFOR with flexible, lethal forces capable of employment in a variety of roles. Typical missions that are assigned to the commandos include but are not limited to—

- Collecting information on deployment of enemy forces and reserve unit movement.
- Collecting information on logistics facilities and seaports.
- Collecting information on enemy aircraft operating from forward airfields.
- Conducting reconnaissance of terrain and enemy forces, in support of the offense.
- Locating and destroying enemy WMD.
- Conducting platoon-size or smaller raids and ambushes and destroy critical military or civilian targets in enemy territory.
- Conducing larger-scale (company- or battalion-size) raids and ambushes in the disruption zone or in enemy territory.
- Clearing LOCs for use by supported units during the offense or defense.
- Clearing or emplacing obstacles.
- Acting as an antilanding reserve.
- Conducting surprise attacks on enemy forces.
- Creating disturbances after infiltrating into enemy territory.
- Acting as a functional force or element—or part of one—in a combined arms tactical action (see Offense and Defense below).

Offense

15-163. Commandos are employed as infiltration units during the offense. Following land, air, or water infiltration, commandos—operating independently—may perform various reconnaissance and combat missions described above. However, they may also act in conjunction with regular ground forces. In the latter role, commandos can conduct the following missions to ensure the success of the overall offensive action.

15-164. Commandos can act as a *disruption force or element*, or as part of such a force or element. In addition to reconnaissance missions, they can be tasked with creating confusion in the disruption zone or in enemy territory by—

- Removing or emplacing obstacles.
- Raiding and destroying headquarters, LOCs, and tactical missile firing locations.
- Occupying key terrain features (in advance of regular ground forces).
- Occupying ambush positions on enemy withdrawal routes.

15-165. Commandos can act as a *fixing force or element*. In this role, they can set up ambushes or emplace obstacles to prevent further enemy forces from coming to the aid of the target of the regular forces' attack. They can occupy key terrain features that control choke points that hinder enemy reserve unit movements. Such choke points may be valleys, bridges, and crossroads that are critical for the enemy movement.

15-166. Commandos can act as part of an *action force or element*. In this role, they can conduct raids and surprise attacks against C2 sites, logistics elements, fire support units (to include attack helicopter units), and other high-priority civilian and military targets. They also conduct attacks against other objectives or seize terrain that hinders enemy reserve unit movements or hampers his withdrawal.

15-167. Commandos may attack a withdrawing enemy force from his flank and rear. Commando units can be air-inserted ahead of the withdrawing enemy force to establish ambush positions along the enemy's withdrawal route.

Defense

15-168. During a defensive action conducted by tactical group or detachment based on a regular ground force unit, commando units allocated to that task organization can support the action primarily in reconnaissance and security roles. Commando units can conduct reconnaissance in the disruption zone or deep in enemy territory. They may also act as a security force in the support zone. When acting as a security force, commandos are normally employed as companies or battalions. The commando unit can be augmented with vehicles and/or additional forces (such as tank or mechanized units, fire support, or aviation) to act as an action force in limited-objective attacks against enemy airborne, air assault, or special operations forces units.

15-169. When regular maneuver forces are forced to withdraw from an area, commando units can remain deployed in the original disruption zone and battle zone to perform reconnaissance, raids, and ambushes. The stay-behind commandos attempt to maneuver in small teams to conduct reconnaissance and limited-objective attacks against enemy targets such as C2 sites, isolated combat units, LOCs, and logistics units.

ORGANIZATION FOR COMBAT

15-170. Commando brigades and battalions, along with their subordinate companies, platoons, and squads exist in the AFS. However, they are not organized as uniformly as regular ground forces. Examples in figures 15-5 through 15-12 (on pages 15-28 through 15-32) indicate the fact that there is actually no fixed structure by depicting some subordinates in dashed boxes, indicating optional structures. The size and composition of commando organizations at each level in the AFS can vary greatly, depending on the situation and the assets available. When actually employed, the units may be further task-organized to fit specific missions.

15-171. No commando assets are constituent to tactical level units at division and below in the AFS. All commando assets employed at the tactical level will have been allocated from higher levels. Therefore, it is necessary to understand where those commando platoons, companies, battalions, and brigades come from, how they get to the tactical level, and how they are employed to support the tactical fight.

15-172. Every mission performed by commandos is unique and unlike any other, and thus requires task organization based on the mission. All of these commando organizations provide the OPFOR a flexible and capable means of achieving its military goals.

15-173. Although some commando units are intended for use at the operation level, the TTP used by all commando units are similar and are actually performed at a tactical level. Regardless of the level from which they come or the level at which they are employed, all units at commando brigade level and lower have similar capabilities, equipment, manning, and TTP.

15-174. Commando units from the SPF Command can be allocated in a constituent or dedicated status to be task-organized as part of a tactical group. In other cases, commando units may be allocated in a supporting relationship, while remaining under the command of their parent organization.

COMMANDO BRIGADES AND BATTALIONS

15-175. As with SPF units, commandos generally reach the tactical battlefield via the OSC. The OSC commander may employ the commando assets allocated to him as constituent or dedicated as part of the overall tactical battle scheme. Commando units, especially at the tactical level, are task-organized to provide the best possible chance of mission success.

15-176. The malleable structure of commando units from squad to brigade provides the ability to continually adapt and to allow continuous improvisation. Units will be added, deleted, or modified, as the mission and conditions dictate. For additional information on commando organizations at operational and strategic level, see FM 7-100.1 and FM 7-100.4. See chapter 2 for specifics on task organizations.

COMMANDO BRIGADE

15-177. Figure 15-5 shows a typical example of a commando brigade in the AFS. A commando brigade will generally need to be task-organized to meet specific mission requirements and conditions, in which case it becomes a commando BTG. Its AFS organization is limited to those units required to perform most missions. The brigade is structured in a manner that easily accepts appropriate task-organizing to meet specific or unique mission requirements. Had the mission or the environment dictated, the organizational example in figure 15-5 could easily have included an engineer battalion (for civil affairs), an artillery and/or MRL battalion or battery, an SPF battalion or company, affiliated irregular forces, or even a helicopter company. The structure of each commando brigade is situational, adapted to the specific conditions.

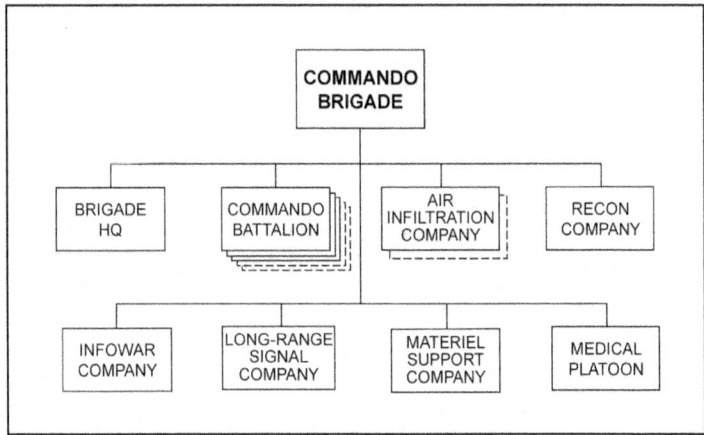

Figure 15-5. Commando brigade (example)

Air Infiltration Company

15-178. Assets of the commando air infiltration company either infiltrate by air and/or assist other units in air infiltration and in conducting other air operations. It is responsible for all insertion, supply, and other air operations planning and coordination for the commando brigade. The company also performs tests, diagnostic checks, repairs, and maintenance on all parachutes, ultralight aircraft, support systems, and related equipment. All members of this company are qualified riggers. See figure 15-6 for organization.

Special-Purpose Forces and Commandos

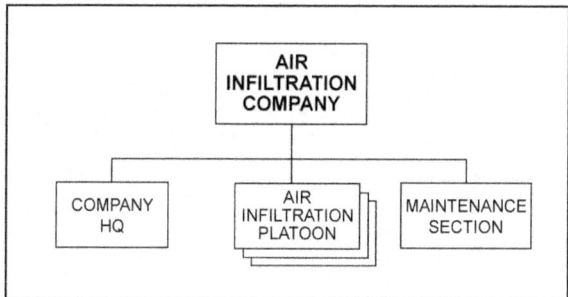

Figure 15-6. Air infiltration company (example)

15-179. The capabilities and techniques of the commando air infiltration company are very similar to those of the SPF counterpart air infiltration team, except that this is a much larger organization. The ultralight aircraft may, or may not, be equipped with light and general-purpose machineguns, AGLs, ATDLs, or other weapons systems. They may also be fitted with radios, cameras, sensor sets, or laser target designators. The ultralights are dismantled for trailer transport, two aircraft per trailer. All members of the air infiltration company are qualified to operate the ultralight aircraft (single- or two-seaters) or to serve as gunners/equipment operators.

Note: Armed ultralight aircraft can also be used in attack, reconnaissance, or multimission roles, not just in infiltration.

15-180. The company also has powered parachutes available. These are extremely effective for surreptitious infiltration and exfiltration. Once successfully infiltrated, the team hides the parachutes in a cache. Upon successful completion of the mission, the parachutes are retrieved and the company exfiltrates to a more secure location, possibly for further transport out of the area. For additional information on the commando air infiltration company, see FM 7-100.4.

Reconnaissance Company and INFOWAR Company

15-181. Two other units peculiar to a commando brigade are the reconnaissance company and the INFOWAR company (see figures 15-7 and 15-8 on page 15-30). Other brigade subordinates are organized and function similar to their counterparts in regular ground forces organizations.

Chapter 15

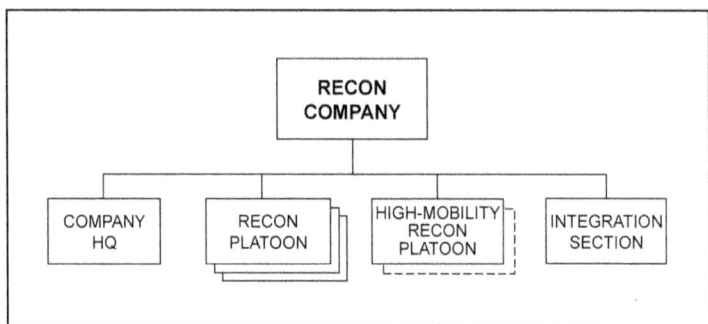

Figure 15-7. Reconnaissance company (example)

15-182. The high-mobility reconnaissance platoon(s) in the reconnaissance company of a commando brigade would likely be very similar to the one found in a commando battalion. (See figure 15-13 on page 15-32.)

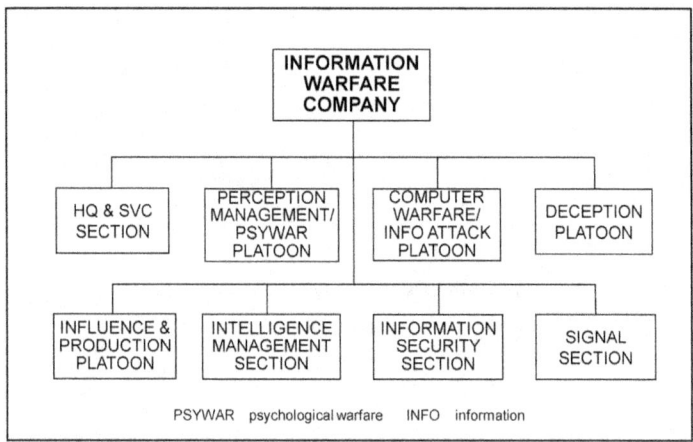

Figure 15-8. INFOWAR company (example)

COMMANDO BATTALION

15-183. The commando battalion and its subordinates contain organic transportation assets sufficient for it to move several hundred kilometers and sustain itself for as long as 7 days without resupply. Quite often, however, the battalion or its companies are required to quickly deploy as specialized light infantry units. In these events, all of the organic transport may be left in garrison, and support is provided by either higher commando organizations or outside assistance. Without organic transportation, the battalion can only be expected to sustain itself for 3 days without resupply, caches, or external support.

15-184. See figure 15-9 for the structure of an example commando battalion. See figures 15-10 through 15-13 (on pages 15-31 and 15-32) for the composition of the commando company, weapons company,

INFOWAR platoon, and high-mobility reconnaissance platoon subordinate to the commando battalion. Other subordinates are organized and function similar to their counterparts in regular ground forces organizations.

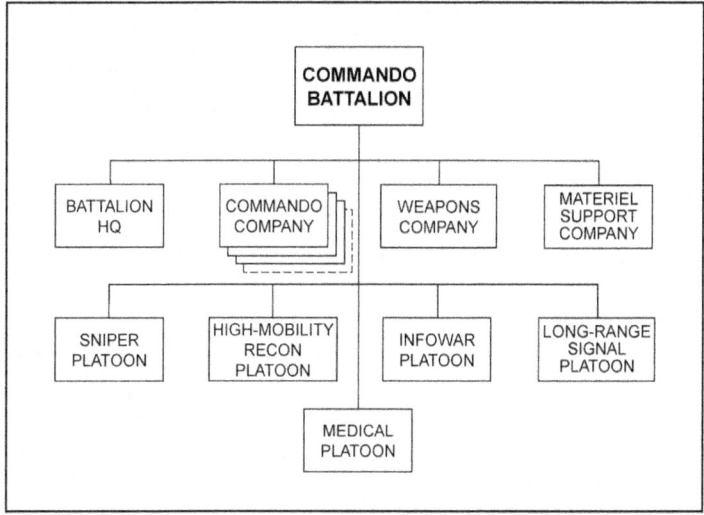

Figure 15-9. Commando battalion (example)

15-185. A commando company typically consists of a company headquarters, three commando platoons, and a weapons platoon. (See figure 15-10.) In the AFS, the commando platoons typically have a headquarters and weapons squad, and three commando squads. In organization for combat, however, these squads may form into task-organized teams of various sizes. These teams may resemble the HK teams formed by regular infantry or perhaps some of the types of teams found in SPF organizations.

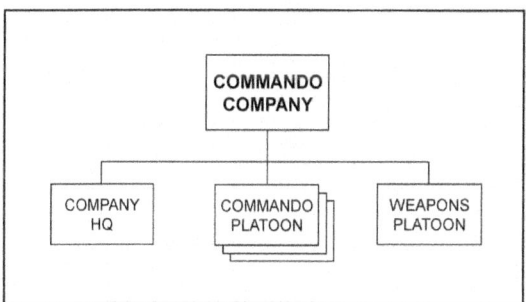

Figure 15-10. Commando company (example)

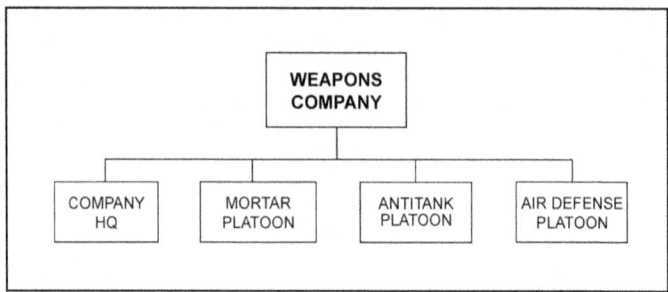

Figure 15-11. Weapons company (example)

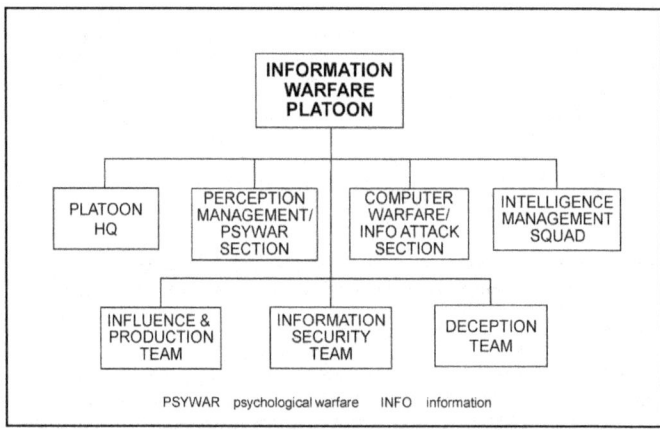

Figure 15-12. INFOWAR platoon (example)

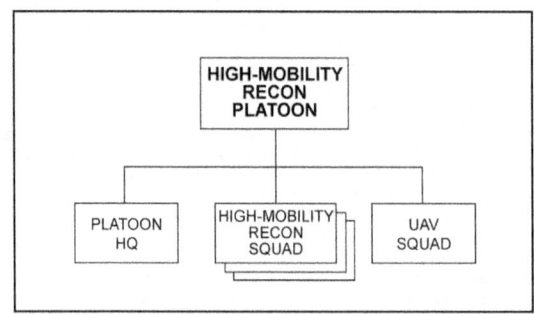

Figure 15-13. High-mobility reconnaissance platoon, commando battalion (example)

TACTICS, TECHNIQUES, AND PROCEDURES

15-186. Regardless of the level from which commando units come or the level at which they are employed, the TTP used by all commando units are similar and are actually performed at a tactical level. Their TTP, like their organization structures, are adaptive, dynamic, and malleable.

INFILTRATION

15-187. Commandos infiltrate during the offense and defense. Although commandos prefer to conduct infiltration at night or under the cover of reduced visibility, they will infiltrate whenever and wherever they are least expected. The method they use will depend on the mission, situation, condition, terrain, time available, distance and ease of infiltration, available transportation means, geography, or many other factors.

Land

15-188. Although commando units may have vehicles and are capable of infiltrating using the vehicles, the most common method of infiltration by far, is by foot. The infiltrating units are dispersed to one or more areas, depending on the size of the infiltrating force. Infiltration is accomplished as a unit or as small teams, using different routes. After careful consideration, routes are selected in—

- Difficult or complex terrain the enemy may consider impassable.
- Areas having sufficient camouflage to cover movement.
- Areas where streams or mountains form natural corridors.
- Areas where the enemy is not using night observation or surveillance equipment.
- Gaps between enemy units.

15-189. A very successful infiltration method used by the commandos is to infiltrate under the guise of reconnaissance probes. Small units, from a team of 3 to 4 men to a squad, or a platoon, or even a company of over 100 men, conduct small probes along the enemy defensive positions. If the enemy does not respond to these probes, the commandos infiltrate in small numbers and spread out. This permits larger numbers to penetrate. Once behind enemy lines, one team may cut off the escape route of the enemy, while the other teams conduct a coordinated assault on both the front and flanks. The attacks will continue on all sides until the defenders are destroyed or forced to withdraw. The commandos will then move stealthily forward to the open flank of the next enemy position, and repeat the tactics.

15-190. The OPFOR also conducts another very successful variation of this infiltration. In this variant, the commandos do not immediately attack as soon as they are successfully behind enemy lines. Once behind the enemy, the commandos may wait a few hours or up to 3 days or more and may number as much as a full company or even a battalion, depending on the circumstances. Once emplaced either behind the enemy or more likely behind and on both flanks of the enemy, the infiltrated force then awaits for the main OPFOR attack. If the main OPFOR attack is successful, the enemy will either retreat or fall back. At that time, the infiltrated commandos will ambush and destroy the remaining enemy forces. If the OPFOR main attack is faltering or appears as if it may fail, the infiltrated commandos simultaneously attack from both the rear and flanks, ensuring victory.

Air

15-191. When time is limited and air transport is available, commandos may be airdropped or air-landed from helicopters or fixed-wing transport aircraft for deeper penetration. The enemy situation and terrain features in the area selected as the DZ or LZ are carefully studied before airborne or heliborne infiltration. DZs are generally mountainous valleys, hills, and clear areas in or adjacent to a forest.

Chapter 15

Water

15-192. Commando units may use naval vessels (to include mini-submarine and semi-submersible insertion craft) for seaborne infiltration. The landing sites are selected in coastal areas far from any hostile naval bases, fishing villages, and coastal defense units. The landing time is determined by considering—
- The time required to arrive at the landing site.
- Maritime and coastal security measures employed by the enemy forces.
- Tidal conditions.
- Hours of moonrise and moonset.
- Weather conditions.

Commandos can also infiltrate using watercraft on inland waterways.

SWARMING AND STAY-BEHIND

15-193. Commando units can use the same types of swarming and stay-behind TTP as SPF. (See section I of this chapter.)

PERSONNEL

15-194. Commando organizations are elite units, specially trained for missions in the disruption zone or in enemy territory. Normally, personnel selected for commando units come from soldiers who have already served 3 to 7 years in other combat arms. In addition to proficiency in various infantry-type tactics, they receive training for more specialized commando missions, with emphasis on infiltrating and fighting in complex terrain and at night.

Chapter 16

Marksmen and Snipers

The OPFOR recognizes both marksmen and snipers as significant combat multipliers at the tactical level. It also realizes that their successes (especially those of snipers) may have an impact at the operational and strategic levels as well. Marksmen and snipers each fill a unique niche and are an integral part of virtually all OPFOR tactical actions. This chapter primarily addresses the use of marksmen and snipers subordinate to an OPFOR regular military force, since that is the focus of this TC.

SIMILARITIES AND DIFFERENCES

16-1. Some of the missions of marksmen and snipers may overlap, but their roles and impact (psychological and physical) on the battlefield are different. (See Missions under Marksmen and Snipers below.) While the marksman operates as part of the unit, snipers often operate autonomously, either individually or as part of a sniper team. The proper use on the battlefield of both marksmen and snipers often causes the enemy to modify his tactics, techniques, and procedures (TTP) and/or to operate in a more tentative fashion.

Note. Marksmen and snipers assigned to regular OPFOR units share few similarities with what are commonly referred to as "irregular force snipers," but there are many differences. (See Differences between OPFOR Regular Military Snipers and Irregular Force Snipers, below.)

MARKSMEN

16-2. A *marksman* is normally a skilled rifleman assigned to and providing direct support to a military organization—usually an infantry squad. Using his assigned, specially prepared (usually scoped) weapon, he engages and dispatches targets at a greater distance than the infantry rifleman.

16-3. The terms *marksman, designated marksman,* and *sharpshooter* are generally interchangeable. While some of their missions may overlap those of a *sniper*, these shooters are completely different from a military sniper. (See the Snipers portion of this chapter.)

MISSION

16-4. The mission of the OPFOR marksman is similar to that of marksmen in armies around the world. He provides accurate, precision fire support to his assigned unit, thereby extending the lethality of the unit. This includes firing, maneuvering, and providing support for the other members of his unit. He may also provide precision fires as a member of a hunter-killer team. The marksmen also provide—

- Overwatch.
- Covering fires.
- Unit security and sniper watch.

They can also assist in the control of key terrain. However, whenever a marksman is not performing marksman or security duties, he serves as a rifleman assigned to the squad.

16-5. Marksmen work at closer ranges than snipers and are required to place several aimed shots in a short period of time, usually at multiple fleeting targets. Marksmen usually engage targets, using a scoped 7.62-mm rifle, in the gap between those engaged by OPFOR soldiers (up to 300 m) and trained snipers (500 m and beyond).

16-6. Many of the marksman's targets may be 100 m or less in an urban or close environment. However, marksmen also easily engage enemy soldiers with precision fires out to 500 m and area fires often well beyond that. Accuracy has increased with the common use, lower costs, and availability of laser rangefinders and scoped rifles. Quite often, talented OPFOR marksmen are selected for advanced training to become snipers.

ORGANIZATION

16-7. Each OPFOR infantry squad (in an infantry, motorized infantry, or mechanized infamtry company) has one marksman and one 7.62-mm sniper/marksman rifle assigned. An infantry or motorized infantry company also has a weapons platoon with two sniper/marksman rifles and two 12.7-mm/.50 cal antimateriel rifles assigned. Thus, an infantry or motorized infantry company has a total of 11 sniper/marksman rifles and 2 antimateriel rifles. This gives an infantry or motorized infantry battalion a total of 33 sniper/marksman rifles and 6 antimateriel rifles. Each mechanized infantry company has 9 sniper/marksman rifles (1 per squad), and the weapons platoon of a mechanized infantry battalion has 3 antimateriel rifles, for a total of 27 sniper/marksman rifles and 3 antimateriel rifles in the battalion. For additional information on OPFOR organizations and equipment, see FM 7-100.4.

16-8. OPFOR marksmen are not limited only to maneuver units. Many supporting units have their own marksmen. The marksmen in the supporting units are usually not infantrymen. They are merely assigned soldiers trained and equipped to be skilled marksmen. The OPFOR does not limit marksmen to males. It also trains women soldiers to serve as markswomen.

EQUIPMENT

16-9. Unlike other armies, which may use the smaller-caliber 5.45-mm (.215 cal.) or 5.56-mm (.223 cal.) rifle for the marksman, the OPFOR prefers the marksman's rifle be at least 7.62x54R-mm (.308 cal.). The 7.62x39-mm cartridge used in many assault rifles is a capable marksman cartridge at ranges 300 m and less, especially when the marksman is equipped with a scoped rifle. Often, however, in some of the support units, a 7.62x54R-mm rifle may not be available. In these cases, the marksman uses his or/her assigned weapon. However, in the marksman role, the 7.62x54R-mm is a much more effective and capable round. It has significantly more penetration and has a greater effective range than the standard issue OPFOR tier-2 rifle (5.45-mm AK-74M) of the infantryman (less than 300 m). The 7.62x54R-mm round is available in a wide range of ammunition, some of which can penetrate lightly armored vehicles.

Note: The Russian 7.62x54R-mm rifle cartridge was developed in 1891. It has been in continuous active military service for over 120 years—longer than any other standard-issue cartridge. Ballistically similar to the 7.62x51-mm NATO round, it is still used in the PK series of machineguns and SVD sniper rifles.

16-10. OPFOR marksmen generally use a mix of sniper, ball, and armor-penetrating ammunition. The marksman using a 7.62x54R-mm rifle provides the squad with an organic weapon capable of penetrating approximately 40 inches of pine at 200 m and 10 inches of cinder block at 100 m. Not many standard assault rifles possess either the range or penetration abilities of the 7.62-mm rifles of the marksman and/or sniper. For additional information on the capabilities of OPFOR marksman rifles and equipment, see the *Worldwide Equipment Guide*.

16-11. The OPFOR marksman carries all the equipment typically carried by a rifleman. The scoped marksman rifle is usually the marksman's only assigned weapon, although he can also be assigned an assault rifle. In addition to the typical infantry equipment, specific mission equipment may include, but is not limited to—

Marksmen and Snipers

- 5.45-mm/7.62x39-mm scoped assault rifle.
- 7.62x54R-mm, 7.62x51-mm NATO (.308 cal.) scoped sniper rifle.
- Optical scope (of various types).
- Binoculars.
- Sound suppressor.
- Laser rangefinder.
- Night vision goggles.
- Night rifle scope.
- Laser pointer.

16-12. The OPFOR also completely understands the advantages of converting standard infantry weapons into precision weapons for a mimimal cost outlay. Almost every rifle, automatic grenade launcher (AGL), or machinegun can mount a day or night scope, which has resulted in a proliferation of scopes on such weapons (especially vehicle-mounted machineguns). A machinegun has a heavier barrel and sturdier mount (bipod, tripod, or pintle) than a rifle. When combined with a mounted scope (and a laser rangefinder), it makes the perfect platform for a skilled marksman. Sound suppresson devices have also proliferated and assist in concealing the shooter's location by masking the direction of the sound.

TARGETS

16-13. Targets for marksmen may be predetermined by leaders prior to execution, depending on the mission. However, marksmen generally concentrate on engaging targets of opportunity such as—

- Battlefield leaders (such as officers and NCOs).
- Key persons (such as communications personnel, couriers, reconnaissance, or small unit leaders).
- Very important persons.
- Medical personnel and religious leaders.
- Enemy crew-served weapons and crews.
- Equipment of crew-served weapons.
- Vehicle crewmen.
- Enemy bunkers, caves, and concealed positions.
- Enemy unit security.
- Possible landmines, roadside bombs, and improvised explosive devices.
- Enemy armored vehicles (causing them to button up and become more vulnerable to antiarmor weapons).
- Vulnerable areas of armored vehicles (such as periscopes, infrared and thermal sensors, or external fuel).
- Other materiel targets.

TACTICS, TECHNIQUES, AND PROCEDURES

16-14. A marksman can serve either as a member of his assigned unit or as a member of a task-organized hunter-killer (HK) team. The vast majority of the time the marksman supports his assigned unit. However, when fighting in complex terrain, the OPFOR prefers to use marksmen as part of HK teams.

Assigned Unit

16-15. As a member of his assigned unit, the OPFOR marksman employs those TTP that enable him to best support his unit with precision fires. This usually places him in the best position to provide those covering or precision fires. While the marksman may be temporately separated from his squad as he maneuvers for the location with the best field of fire and/or cover, he is never very far from his unit. He accompanies his unit whenever and wherever it goes.

Chapter 16

Hunter-Killer Team

16-16. While snipers can also be used to provide precision fires, the marksman is the preferred choice for manning HK teams. HK teams are task-organized using infantry platoons or companies (regular or irregular) as the base organization. The HK team structure is extremely lethal, especially against armored vehicles, and is ideal for dispersed combat such as fighting in urban areas and other complex terrain.

16-17. When fighting an armored unit in complex terrain such as urban areas, forests, and jungles, the OPFOR commander forms infantry antiarmor HK teams controlled by an infantry company or platoon or group. An infantry company can form as many as three of these infantry antiarmor groups or platoons. Each group or platoon consists of six or seven infantry antiarmor HK teams. Each team is generally composed of four or more personnel, which can include—

- A marksman (with scoped rifle) and machinegunner. These pin down the supporting infantry allowing the antiarmor gunner to engage the armored vehicle.
- An antiarmor gunner armed with a shoulder-fired antitank or antiarmor weapon. The task of this gunner is to immobilize or destroy the armored vehicle. (Weapons with reduced back blast will be used if firing from confined spaces such as a room or sewer.)
- An RPO-A gunner. The RPO-A is a shoulder-fired thermobaric weapon that is especially effective in neutralizing enemy troops that have escaped into, or occupied, buildings, sewers, bunkers, or other enclosed areas.
- Additional personnel may serve as ammunition bearers, security, and assistant gunners. Each team member is equipped with a handheld, very-low-power radio for communicating within the team and one level up.

Note: Often referred to as "pocket artillery," the RPO-A provides the HK team and infantry squad organic firepower equal to that of an 122-mm high-explosive artillery shell. (See the *Worldwide Equipment Guide*.)

16-18. Often four or five of these teams work together in a "pack" to attack a single armored vehicle. In some cases, the HK team may be composed of only three members: the marksman, the machinegunner, and the antiarmor gunner.

16-19. Depending on the target, antimateriel weapons and/or AGLs may be used to disable or destroy the vehicles. Some AGL fire can penetrate lightly armored vehicles. All AGLs can be used to "button up" the vehicles, destroy any external optics and electronics, and kill enemy soldiers exiting the vehicles. For additional information on the organization of the HK team, see FM 7-100.4.

SNIPERS

16-20. Technology and information have elevated the potential of OPFOR snipers' tactical capabilities to have strategic effects. This influence may be magnified by an effective information warfare (INFOWAR) campaign. The mere presence of a sniper can also have a tremendous psychological effect on enemy forces, instilling fear and demoralizing the enemy as well as influencing his decisions and actions, especially at the tactical level.

16-21. The skills and abilities of the snipers set them apart from the marksmen. OPFOR snipers have all received centralized advanced marksmanship training (precision long-range fires). They are also intensively trained to master field craft, stalking, stealth, concealment, infiltration, and exfiltration. These skills, while handy, are not necessarily required for marksmen. All OPFOR snipers may or may not be in uniform depending on the presence (and physical similarity) of supporting local population, the mission, and other factors.

16-22. While the marksman is employed as part of his unit, the sniper may operate autonomously either individually or as part of a sniper team. Although the action of the sniper is a tactical task, the result may easily have operational and/or strategic impact. In fact, the OPFOR may design some sniper missions specifically for the operational and/or strategic effect.

Marksmen and Snipers

16-23. The OPFOR is aware of the lethality that snipers bring to the battlefield. It is also aware of the emphasis placed on snipers by foreign armies, some of which have several sniper squads assigned to each maneuver battalion. The OPFOR does not limit snipers to males only; women can also serve as snipers.

Note: In World War II, the Soviet Army had over 100,000 snipers, the top 20 of whom accounted for over 7,400 confirmed kills. In 1943, it had over 2,000 women snipers 1,000 of whom accounted for over 12,000 confirmed kills.

MISSION

16-24. The OPFOR sniper's primary combat mission is to deliver precision long-range fires. Effective long-range sniper fire can—
- Create casualties.
- Impede movement.
- Instill fear.
- Influence enemy decisions, actions, and TTP.
- Lower morale.
- Damage or destroy materiel.
- Disrupt enemy tempo.

In contrast to automatic weapons and indirect fire support, the sniper is especially effective in avoiding collateral damage due to the proximity of civilians.

16-25. The missions of the sniper are as varied as the terrain and conditions in which he operates and are not unique to the OPFOR. Generally, the missions are to conduct long-range, precision fires to produce casualties and damage materiel (attrition). Below are several typical examples of specific sniper missions:
- Conduct countersniper actions.
- Provide security.
- Support checkpoints and roadblocks.
- Cover avenues of approach.
- Support reconnaissance and/or counterreconnaissance (eliminate enemy observers and reconnaissance).
- Provide overwatch.
- Target medical personnel and religious or political leaders.
- Target materiel (see Antimateriel Role below).
- Serve as forward observer, controller, or laser designator for precision-guided munitions.
- Provide blocking or screening.
- Support or conduct raids, ambushes, or patrols.
- Support other offensive and defensive operations.
- Support INFOWAR (see Role in Information Warfare below).

See Targets below.

ORGANIZATION

16-26. Each OPFOR infantry squad in an infantry company (of a motorized infantry battalion) or mechanized infantry company has one sniper (or marksman) with a 7.62-mm sniper rifle. Each infantry company (in a motorized infantry battalion) has a sniper section in its weapons platoon. That sniper section has two snipers, each with both a 7.62-mm sniper rifle or a .50 cal (or 12.7-mm) antimateriel rifle available. However, the sniper carries only one of these weapons, depending on the mission, and the other rifle remains behind with the vehicle or headquarters element. A mechanized infantry company has a weapons platoon, in which each of three weapons squads has one antimateriel sniper with an antimateriel rifle (and a 5.45-mm carbine as an alternate weapon).

Chapter 16

16-27. A motorized infantry battalion has no weapons platoon and, therefore, no snipers or sniper weapons above company level. Each OPFOR mechanized infantry battalion has three antimateriel rifles in the weapons platoon for use by snipers. The snipers at battalion level can also choose to use the smaller 7.62-mm sniper rifle. The OPFOR routinely creates task organizations that include sniper teams or sniper units.

16-28. OPFOR motorized and mechanized infantry brigades have a sniper platoon with 12 snipers assigned, divided into 3 squads. Each sniper has both a 7.62-mm sniper rifle and an antimateriel rifle available. However, the sniper carries only one of these weapons, depending on the mission, and the other rifle remains behind with the vehicle or headquarters element. The sniper squad typically consists of four 2-person sniper teams. The squad leader serves as the team leader of one sniper team. Each sniper team consists of a team leader/observer (spotter) and a sniper (shooter)/target designator. The duties of spotting and shooting are interchangeable between the two team members. Two teams share a vehicle when necessary.

16-29. All OPFOR mechanized and motorized infantry divisions have an organic sniper company composed of 36 snipers. Each sniper at division level has both a 7-62-mm sniper rifle and an antimateriel rifle available. For additional information see FM 7-100.4.

Single Sniper

16-30. A skilled, experienced sniper is basically a lone hunter who fires one shot from distance (with suppressor) and disappears either into the local population or into the environment. For example, the lone sniper may cache his weapon and hide in a spider hole for a day or more, prior to exfiltrating, or he may hide his weapon and "rejoin" a band of "other" shepherds or goat herders. It is often crucial for the single sniper to wear civilian clothing and blend in with the local population.

16-31. The single sniper also serves as the spotter and is responsible for his/her own security. Even the lone sniper may have a large psychological impact.

Sniper Teams

16-32. There is no set number or organization of OPFOR sniper teams. The composition of the OPFOR sniper team can vary greatly, as can sniper operations. The OPFOR organizes and equips each team depending on the situation, conditions, and mission. The size of team (number of snipers) also determines its TTP.

16-33. Individual snipers or multiple-person teams can support offensive or defensive actions and/or conduct independent operations. Separate sniper teams (observer and shooter) often work together. They may support each other or may even share the same target(s). Depending on the mission, numerous sniper teams may converge on an area.

16-34. Multiple-member teams may act as a single unit or break out into two or more mutually supporting subelements. As a single unit, the multiple-member team can provide security or early warning, or serve as spotter, recorder, and sniper. As multi-action subelements, the team may split into two or more shooters. This provides multiple fields of fire on the intended target(s) and increases the snipers' chances of success.

16-35. In some cases, if the OPFOR feels it has proper situational awareness and combat superiority, the sniper unit may have enough combat power to become decisively engaged. This could be the case when supporting a larger OPFOR offensive or defensive action. In other cases, the element will conduct a phased withdrawal using bounds within the support subelement to provide cover for the actual sniper.

Two-Member Sniper Team

16-36. The typical OPFOR sniper team is a two-person team. Each team consists of a team leader/observer (spotter) and a sniper (shooter)/target designator. The duties are interchangeable, as both members are qualified snipers and spotters. One member shoots while the second member performs the functions of spotting, security, and recording the attack. For additional information, see FM 7-100.4.

Marksmen and Snipers

Three-Member Sniper Team

16-37. One member serves as the observer/spotter, one is the shooter/target designator, and the third provides security, transports equipment, and may be a driver. This team is similar to the two-member team above except that it provides 360-degree security and an extra man to carry ammunition, radios, water, and other items. The three-member team is more appropriate for longer-range (deeper) missions requiring infiltration and exfiltration, and/or longer duration.

Four-Member and Larger Sniper Team

16-38. One member serves as the sniper, one is the shooter/target designator, and two or more provide security and/or support and may serve as driver. Teams of four persons and above may be broken into two or more subelements. One of these subelements is the basic sniping element with observer and shooter, while the other personnel (or subelements) provide support and/or security.

16-39. Often the mission requires snipers to operate in environments where the expectation of close contact with the enemy is high and fleeting targets are the norm. On these missions, it is quite normal for some or all snipers to carry an assault rifle in addition to their sniper weapons. Often a member of larger teams will also be equipped with a light (5.45-mm) or medium (7.62-mm) machinegun to provide security to the team.

EQUIPMENT

16-40. Military capabilities and equipment have a large impact on sniper TTP. The equipment is selected to best fit the mission, conditions, and environment. For example, in some environments, targets may appear within 200 m. Therefore, the sniper may choose small arms versus a larger, difficult-to-conceal, sniper rifle as his weapon of choice or a semiautomatic rather than a bolt action rifle if the sniper expects multiple fleeting targets within a short distance. If precision fire at longer distance is required, then a sniper rifle would likely be preferred. If the target is materiel, generally a more capable, larger-caliber weapon will be selected. Optics, night vision sights, and laser rangefinders are sniper enablers and, while not required, are available to the OPFOR sniper, greatly enhancing his effectiveness. Generally, the OPFOR sniper carries either a 7.62-mm or .50 cal. sniper rifle depending on the mission. The other rifle remains behind with the tactical utility vehicle (TUV) or headquarters element. Two teams share a TUV, when the unit is so equipped.

16-41. The OPFOR sniper-specific mission equipment may include, but is not limited to—

- 7.62x54R-mm or 7.62x51-mm NATO (.308 cal.) sniper rifle.
- 8.58-mm (.338 cal.) (or larger) sniper rifle.
- .50 cal./12.7-mm antimateriel rifle.
- Assault rifle (for missions in complex terrain).
- Light or medium machineguns (for missions in complex terrain).
- Detachable sound suppressor.
- Under-barrel grenade launcher.
- Optical (or electro-optical) scope (of various types).
- Bullet drop compensator integrated into scope.
- Binoculars (possibly digitial with transmission capability).
- Digital camera (possibly digitial with transmission capability).
- Ballistic computer.
- Day/night observation scope (60X).
- Handheld GPS receiver.
- Laser rangefinder.
- Laser target designator.
- Night vision goggles.
- Handheld thermal viewer.

Chapter 16

- Night rifle scope (image intensifier or thermal imager).
- Laser pointer.
- Tactical periscope.
- Ghillie suits and/or other special camouflage items.
- Manpack, low-power radio or toher tactical radios.
- Satellite radio/telephone.
- Organic or provided transportation.
- Hide material and equipment.

Some of this equipment may not be carried on all missions.

TARGETS

16-42. The mission of the sniper generally defines the target set available to the sniper. Since the types of missions the sniper supports are virtually limitless, so therefore are the targets the sniper can engage. The sniper can engage all of the targets listed in the Marksmen portion of this chapter, above, and all of the targets listed in the Antimateriel Role portion, below.

16-43. When engaging very high-value targets, the OPFOR may assign several snipers or sniper teams the mission of eliminating a single target. In order to ensure the highest probability of hit and subsequent kill, all snipers fire at the target simultaneously. The teams all fire on command, at a prearranged signal (such as from a laser pointer). Simultaneous firing is extremely difficult to achieve under tactical conditions. Due to the level of planning, coordination, and integration required, only the very best snipers and sniper teams can attain this level of proficiency.

TACTICS, TECHNIQUES, AND PROCEDURES

16-44. The TTP used by the OPFOR sniper are not unique to the OPFOR. However, the OPFOR may not be hindered by traditional values and laws of warfare. For example, the OPFOR sniper has no problem using civilian noncombatants as shields or lookouts. While the OPFOR sniper exploits successful, time-proven TTP, he/she will use whatever TTP provides the best chance of mission success and survival.

Snipers in Urban Areas and Other Complex Terrain

16-45. Snipers are extremely effective in complex terrain, the nature of which provides cover and concealment to the sniper's ingress; surveillance and/or firing position; and egress. Urban structures reverberate sound. When combined with the everyday noise of urban activity, these structures can mask rifle sounds. This benefits the sniper by adding a degree of difficulty to the enemy in determining the true direction of fire and/or identifying the sniper's firing position.

16-46. Complex terrain also enables the sniper to move undetected and establish surveillance and/or firing positions relatively close to his intended kill zone. Therefore, urban settings and other complex terrain may allow for shots of 200 m or less.

16-47. Most sniper rifles are designed to engage at a distance. Therefore, missions in complex terrain often require weapons better designed to compensate for conditions unique to close combat. This is usually in the form of the sniper carrying an assault rifle and possibly a machinegun assigned to the sniper team. (For additional information see Four-Member and Larger Sniper Team, above.)

Snipers in Open Terrain and Rural Areas

16-48. In contrast to urban settings, rural settings may lack man-made structures but provide natural settings that afford cover and concealment. Distance by itself may provide sufficient concealment, since the sniper can fire much farther than the average enemy soldier can discriminate a concealed sniper. Rural areas may dictate greater distance between the shooter and his target, thus requiring shots at ranges greater than 500 m. However, the snipers' long-range precision fire can engage targets at a distance, and their advanced optics can easily discriminate individuals or point targets.

Marksmen and Snipers

16-49. Open and rural areas may also dictate a greater distance the sniper must travel to obtain an optimum surveillance and/or firing position. These conditions require that snipers have a higher level of proficiency in using stealth in moving, communicating, conducting surveillance, and placing precision fire on the intended target.

Note: During combat against the Russian Army, Chechen snipers in rural areas usually operated in conjunction with a four- to six-person support (overwatch) element armed with assault rifles. The sniper would fire one or two shots at the Russians from about 1,000 m and then change firing positions. The support element usually positioned itself approximately 500 m behind the sniper. Should the Russian soldiers either fire at or advance toward the sniper, the support element would open fire to draw their fire. This allowed the higher-value sniper to escape. The well-equipped Chechens snipers generally did not deploy without a support element of four to six guerrillas.

Exfiltration

16-50. The OPFOR expends significant capital on training and fielding effective and efficient snipers and sniper teams. Therefore the OPFOR expects to use its snipers for numerous missions, not just one time. While the OPFOR may decide to sacrifice a sniper and/or team for a very high-value target, this is not normally the case. OPFOR planners put as much time and resources into exfiltrating and extracting the sniper and team as they do in the initial planning, infiltration, and execution of the mission. Snipers, having a vested interest, go into extreme detail and countermeasures to ensure their survival and subsequent extraction.

ANTIMATERIEL ROLE

16-51. The OPFOR views the antimateriel role from several perspectives. The term *antimateriel* can refer to a target, a mission, or a specific category of weapons or rifles. Generally, there is no difference between how an OPFOR sniper or marksman targets materiel or personnel. The same TTP and the same caliber and type of rifle may be used for both missions, depending on the nature of the target, range, and other factors.

16-52. Generally, however, the weapon used in the antimateriel role is more capable and larger caliber with greater penetration, and often with a greater range. Some of the antimateriel rifles can damage or destroy targets at ranges in excess of 2,000 m.

MISSION

16-53. The mission of the sniper engaged in the antimateriel role is to destroy, damage, and/or attrit enemy materiel. Examples of types of materiel targets may be found below. Antimateriel rifles (and associated missions) are integral to any modern battlefield. Due to their low price, easy availability, and capability as a combat multiplier, they are proliferating very rapidly in most armies worldwide, modern and otherwise. Although not snipers, marksmen will also engage materiel targets of opportunity as they are presented.

16-54. All OPFOR snipers have antimateriel rifles available for use as the mission dictates. Typical missions involving use of antimateriel rifles would be—
- An ambush on a lightly armored column.
- A raid on a command and control (C2) facility.
- An attack on critical equipment.

ORGANIZATION

16-55. The OPFOR does not have dedicated antimateriel sniper organizations. However, if needed, commanders could task-organize (and equip) their existing snipers for the mission. Team configuration is the same as for the standard sniper. See Organization in the Sniper portion of this chapter, above, for details. For additional information see FM 7-100.4.

Chapter 16

EQUIPMENT

16-56. The sniper with an antimateriel mission has the same equipment as for a standard mission. The primary difference may be in the weapon selection, determined by mission, availability, and conditions. Antimateriel rifles may be single-shot, bolt action, or semiautomatic. Calibers range from 5.45-mm (with limited antimateriel capability) to 20-mm (in limited numbers). The most prevalent caliber for antimateriel rifles is .50 cal./12.7-mm. Some example typical calibers used in the antimateriel role are—

- 7.62-mm (.308 cal.).
- 8.58-mm (.338 cal.).
- 12.7-mm (.50 cal.).
- 14.5-mm to 20-mm.

16-57. Although the larger-caliber rifles (12.7-mm and above) may also be categorized as sniper rifles, they are generally employed as antimateriel rifles. Armor-piercing and incendiary ammunition or specialty ammunition is generally used in the antimateriel role. For additional information on the capabilities (ammunition, penetration, and ranges) of antimateriel weapons, see the *Worldwide Equipment Guide*.

TARGETS

16-58. The sniper is the soldier that services the antimateriel mission and engages the enemy materiel. Although the OPFOR may use larger-caliber weapons (12.7-mm and above) against personnel, their primary use is against materiel. The OPFOR has no set target list for these weapons. The range of targets the OPFOR will engage with these weapons is unlimited. The following are some typical antimateriel targets:

- Lightly armored vehicles.
- Fixed-wing aircraft (ground or taxiing).
- Rotary-wing aircraft (ground, taxiing, hovering, or in flight).
- Radars.
- C2 assets.
- Unmanned aerial vehicle support systems (such as launch and control vehicles).
- Missiles (all types).
- POL storage facilities, flexible storage tanks, and vehicles.
- Transformers, generators, and other electrical systems.
- Water purification units.
- Other critical infrastructure.

TACTICS, TECHNIQUES, AND PROCEDURES

16-59. The TTP used by OPFOR snipers to engage enemy materiel targets are no different than the TTP used for engaging human targets. While the range, munitions, and equipment may be different, the basic TTP remain. Some TTP may have to be modified, however, due to the heavier weight of antimateriel weapons and ammunition and other factors.

ROLE IN INFORMATION WARFARE

16-60. The OPFOR realizes that the nature of sniping carries a significant (and very personal) terror (psychological) element. Single acts of sniping (other than the elimination of key military, national, or religious leaders) rarely have a widespread psychological impact. However, cumulative attacks, especially over time, may have an extensive psychological impact. This impact can cascade to several levels and ranges from—

- Influencing individual behaviors (such as instilling fear and creating a tentative soldier).
- To a tactical influence (modifying enemy TTP).
- To completely demoralizing the enemy.
- To affecting civilian support of a cause.

MODIFYING INDIVIDUAL AND GROUP BEHAVIORS

16-61. The goal of using precision sniper fires as a tool of OPFOR INFOWAR is to influence both military and civilian populations. While the methods used to affect both groups may differ in application, they still depend on the manipulation and exploitation of human behaviors.

Military

16-62. Precision sniper fire instills fear and creates a tentative soldier. Some individual soldiers may be terrorized to the point they are combat ineffective, may be a casualty, and must be withdrawn from the fight. Sniper fire may also cause entire units to modify their TTP.

Civilian Population

16-63. The terror generated by sniper fires demonstrates and reinforces OPFOR dominance. Targeting the local population leads the population to believe the OPFOR's enemy cannot protect them and is not the dominant force in the region. The same applies when a sniper fires a single round and the locals see all the enemy soldiers scramble right and left.

DISSEMINATION OF MISSION RESULTS

16-64. The OPFOR plans precision sniper fires to support the INFOWAR plan, and disseminates the results of such fires in accordance with that plan. Usually an OPFOR combat photographer or videographer records the successful sniper event and then passes the data (usually digital) either through the battalion intelligence officer or through OPFOR INFOWAR channels. The information will be screened, analyzed, processed, and then disseminated by the INFOWAR personnel in accordance with the plan. The results of the incident may be disseminated via a sympathetic media provider or other means. The single act of a sniper may easily reach an international audience within hours of his making the shot.

DIFFERENCES BETWEEN REGULAR MILITARY SNIPERS AND IRREGULAR FORCE SNIPERS

16-65. Snipers at all levels rely on a system to support and sustain operations. To one degree or another, logistics (including weapons, ammunition, and sights), intelligence, and an INFOWAR capability (such as camera, video, and media) are all facets of the sniper system. These sniper support and sustainment functions also emphasize a number of differences between the regular military sniper and irregular force sniper. Table 16-1 on page 16-12 contains several points of comparison illustrating the differences between the two. For additional information on OPFOR insurgent and other irregular force snipers, see other parts of the TC 7-100 series.

Table 16-1. Differences between regular military snipers and irregular force snipers

Regular Military Sniper	Points of Comparison	Irregular Force Sniper
Ability and skill high. Field craft, long-range marksmanship, stalking, stealth, concealment, and infiltration/exfiltration high.	Ability and Skill Level (Includes Discipline)	Low to high. Limited (but some) field craft. May be limited by equipment. Initially, skill levels may vary widely but will improve with experience.
Tier 1 with niche technology. The trend among modern armies is to more capable weapons (range and accuracy) and more 12.7-mm antimateriel rifles in addition to 7.62-mm sniper rifles.	Equipment	Varies tier 1 to 4. Generally limited to what is available. Trend is to more sophisticated and capable sniper systems and support systems. Increasing numbers of 12.7-mm antimateriel rifles and systems.
600 to 1,000 m with 7.62-mm rifles and well beyond (out to 2,000 + m) with 12.7-mm rifles.	Effective Range	Most capable of neck and/or head shots out to 300+ m. A few may be effective out to 800 to 1000 m (beyond with 12.7-mm) depending on training and equipment.
Professional occupation. Highly trained in marksmanship, countersniper, field craft, stalking, infiltration/exfiltration, stealth, concealment, communications, and targeting.	Training	Varies from none to formal sniper training provided in another country.
Selected from existing force. Elite shooters recruited from trained troops, usually from the pool of designated marksmen.	Selection and Recruiting	Persons with previous military or sniper experience and/or hunters preferred.
Same as other regular military.	Sustainment/Logistics	Usually reliant on external funding and the local population. This link is vulnerable.
Same as other regular military.	Media Manipulation	Major tool. Often primary reason for attack.
Standard military sniper targets (see above) and targeting.	Targeting and Reconnaissance	Population assists in target identification. Not restricted by laws on whom they target.
Desired but not critical.	Local Support	Critical. May receive food, concealment, targeting information, transportation, and other support.
Same as other regular military.	Incentive	Varies (such as money, religious, cultural, ethnic, or revenge).
Same as other regular military.	Finance	Critical, vulnerable.
Same as other regular military.	Life Cycle Vulnerability	Extremely vulnerable.
From individual sniper to a sniper team of 4 or more (see Organization under Snipers, above.)	Team Composition	Individual or team of 2 to 4 (shooter, driver/ spotter/ security, support, video, possibly cell leader/trainer) or more. (Leader/trainer is usually an experienced sniper.)
May or may not be restricted by laws of ground warfare. Proliferating into organizational structure of most modern armies.	Other	Ability to blend in with the population (usually because part of the population). Not restricted by laws of warfare. Cost-effective combat multiplier.

Notes:
1. Snipers in guerrilla units are usually much more skilled than insurgent snipers but may not be as proficient as those in regular military units.
2. For equipment tiers, see the *Worldwide Equipment Guide.*

Glossary

SECTION I – ACRONYMS AND ABBREVIATIONS

µm	micrometer(s)
AA	antiaircraft
ABNCS	airborne control station
AC	hydrogen cyanide
ACC	assistant crossing commander
ACO	airspace coordination order
ACP	air control point
ADAAS	air defense/antiarmor system
admin	administration
AFCS	automated fire control system
afl	affiliated
AFS	administrative force structure
AGL	above ground level; automatic grenade launcher
AIRCP	airborne command post
AKO	Army Knowledge Online
ALR	antilanding reserve
ALTCP	alternate command post
AOP	aerial observation post; air observer platoon
AOR	area of responsibility
AOS	airspace operations subsection
AP	antipersonnel; attack position
APC	armored personnel carrier
APOD	aerial port of debarkation
AR	Army regulation
ARNG	Army National Guard
ARNGUS	Army National Guard of the United States
ASP	aviation support plan
AT	antitank
ATDL	antitank disposable launcher
ATGL	antitank grenade launcher
ATGM	antitank guided missile
ATO	air tasking order
ATR	antitank reserve
ATV	all-terrain vehicle
AUXCP	auxiliary command post
AWACS	airborne warning and control system

Glossary

bde	brigade
BDET	battalion-size detachment
bn	battalion
BP	battle position
BTG	brigade tactical group
btry	battery
C2	command and control
C3D	camouflage, concealment, cover and deception
cal	caliber
CAO	chief of airspace operations
CBP	complex battle position
CBRN	chemical, biological, radiological, and nuclear
CC	crossing commander
CCD	charged-coupled device
CDET	company-size detachment
cdr	commander
CELT	chief of engineer liaison teams
CFSC	chief of fire support coordination
CG	phosgene (chemical agent)
CGS	Chief of the General Staff
cm	centimeter(s)
cmd	command
co	company
COA	course of action
COE	Contemporary Operational Environment
combo	combination
COP	command and observation post
COR	chief of reconnaissance
COS	chief of staff
COTS	commercial off-the-shelf
CP	command post
CR	counterreconnaissance
CRD	counterreconnaissance detachment
CRZ	counterreconnaissance zone
CS	combat support; tear gas
CSOP	combat security outpost
CSS	combat service support
CTC	combat training center
CTID	COE and Threat Integration Directorate
DAR	deep attack and reconnaissance
DAS	direct air support

Glossary

DC	deputy commander
decon	decontamination
ded	dedicated
dep	deputy
DF	direction-finding
div	division; divisional
DOD	Department of Defense
DP	diphosgene (chemical agent)
DTG	division tactical group
DZ	drop zone
EA	electronic attack
ECP	engineer checkpoint
ELINT	electronic intelligence
EO	electro-optical
ERP	engineer reconnaissance patrol
EW	electronic warfare
EXJAM	expendable jammer
FAC	forward air controller
FARC	Fuerzas Armadas Revolutionarias de Colombia
FARP	forward arming and refueling point
FCP	forward command post
FDC	fire direction center
FG	field group
FM	field manual
FOP	forward observation post
FP	fighting patrol
FSCC	fire support coordination center
ft	feet
G	a type of nerve agent
GB	sarin (nerve agent)
GHz	gigahertz
GPS	global positioning system
H	mustard gas
h	hour (in km/h)
HAHO	high-altitude high-opening
HALO	high-altitude low-opening
HC	hexachloroethane
HCB	hexachlorobenzene
HE	high-explosive
HF	high-frequency
HK	hunter-killer

Glossary

HMG	heavy machinegun
HPT	high-payoff target
HQ	Headquarters
HT	Hybrid Threat for training
HUMINT	human intelligence
HVT	high-value target
IA	information attack
IADS	integrated air defense system
IED	imitative electronic deception; improvised explosive device
IFC	integrated fires command
IFF	identification, friend or foe
IFV	infantry fighting vehicle
IMD	independent mission detachment
inf	Infantry
info	Information
INFOWAR	information warfare
intel	Intelligence
IP	initial point
IR	Infrared
IRP	independent reconnaissance patrol
ISC	integrated support command
ISG	integrated support group
JP	joint publication
km	kilometer(s)
km/h	kilometers per hour
L	lewisite (chemical agent)
LLL	low-light-level (television)
LOC	line of communications
LOP	lateral observation post
LOR	limit of responsibility
LRR	long-range reconnaissance
LTD	laser target designator
LZ	landing zone
m	meter(s)
m/s	meters per second
MANPADS	manportable air defense system
MCP	main command post
MED	manipulative electronic deception
mm	millimeter(s)
MMW	millimeter wave
MOD	Ministry of Defense

Glossary

MRL	multiple rocket launcher
MRP	mobile reconnaissance post
MSD	movement support detachment
MSR	main supply route
MSS	mission support site
NATO	North Atlantic Treaty Organization
NCA	National Command Authority
NCO	noncommissioned officer
ND	Neodymium
NGO	nongovernmental organization
nm	nanometer(s)
NOE	nap-of-the earth
NRT	near real time
OD	obstacle detachment
OE	operational environment
OP	observation post
OPFOR	opposing force
OR	operational readiness
OSC	operational-strategic command
PEL	predicted enemy location
plt	Platoon
PMESII-PT	political, military, economic, social, information, infrastructure, physical environment, and time (memory aid for operational variables)
POL	petroleum, oils, and lubricants
PSYWAR	psychological warfare
PWP	plasticized white phosphorus
PZ	pick-up zone
QRF	quick response force
RCS	radar cross section
RD	reconnaissance detachment
RDD	radiological dispersal device ("dirty bomb")
RDL	General Dennis J. Reimer Training and Doctrine Digital Library
recon	reconnaissance
RED	radiological exposure device
RISTA	reconnaissance, intelligence, surveillance, and target acquisition
ROZ	restricted operations zone
RP	reconnaissance patrol; red phosphorus
RPV	remotely piloted vehicle
RTO	radio telephone operator
RZ	reference zone
SACLOS	semiautomatic command-to-line-of-sight (guidance)

SAM	surface-to-air missile
SATCOM	satellite communications
SBP	simple battle position
SCP	strategic campaign plan; sustainment command post
SEAD	suppression of air enemy defenses
SED	simulative electronic deception
sep	Separate
SHC	Supreme High Command
SHF	super-high-frequency
SID	Strategic Integration Department
SIGINT	signals intelligence
SOCC	support operations coordination center
SP	self-propelled
SPF	special-purpose forces
SPOD	sea port of debarkation
spt	Supporting
SSM	surface-to-surface missile
SUSCP	sustainment command post
svc	service; services
TC	training circular
TCP	traffic control point
tech	Technical
TIC	toxic industrial chemical
TOE	table of organization and equipment
TRADOC	U.S. Army Training and Doctrine Command
TRISA	TRADOC G-2 Intelligence Support Activity
TTP	tactics, techniques, and procedures
TUV	tactical utility vehicle
TV	Television
U.S.	United States
UAV	unmanned aerial vehicle
UD	urban detachment
UHF	ultra-high-frequency
USAR	United States Army Reserve
V	a type of nerve agent
VEESS	vehicle engine exhaust smoke system
VHF	very-high-frequency
VO	visual observer
VTDP	vectoring and target designation post
WMD	weapons of mass destruction
WP	white phosphorus

Glossary

ZORR   zone of reconnaissance responsibility

SECTION II – TERMS

operational environment

A composite of the conditions, circumstances, and influences that affect the employment of capabilities and bear on the decisions of the commander (JP 3-0).

opposing force

A plausible, flexible military and/or paramilitary force representing a composite of varying capabilities of actual worldwide forces, used in lieu of a specific threat force for training and developing U.S. forces (AR 350-2).

hybrid threat

The diverse and dynamic combination of regular forces, irregular forces, terrorist forces, and/or criminal elements unified to achieve mutually benefitting effects.

This page intentionally left blank.

References

DOCUMENTS NEEDED
These documents must be available to the intended users of this publication.
JP 1-02. *Department of Defense Dictionary of Military and Associated Terms.* Available online at http://www.dtic.mil/doctrine/jel/doddict/.
FM 1-02. *Operational Terms and Graphics.* 21 September 2004.

READINGS RECOMMENDED
These sources contain relevant supplemental information.
FM 7-100.1. *Opposing Force Operations.* 27 December 2004.
FM 7-100.4. *Opposing Force Organization Guide.* 3 May 2007. Associated online organizational directories, volumes I-IV, available on TRADOC G2-TRISA Web site at https://www.us.army.mil/suite/files/19296289 (AKO access required). Associated *Worldwide Equipment Guide*, volumes 1-3, available on TRADOC G2-TRISA Web site at https://www.us.army.mil/suite/files/21872221 (AKO access required).
TC 7-100. *Hybrid Threat.* 26 November 2010.
TC 7-101. *Exercise Design.* 26 November 2010.

DEPARTMENT OF ARMY FORMS
DA forms are available on the APD Web site <www.apd.army.mil>.
DA Form 2028. *Recommended Changes to Publications and Blank Forms.*

This page intentionally left blank.

Index

Entries are by paragraph number unless page (p.) or pages (pp.) is specified. After a page reference, the subsequent use of paragraph reference is indicated by the paragraph symbol (¶).

A

action element, 2-52–2-53, 3-44–3-45
   in antilanding reserve, 6-34
   in reconnaissance attack, 3-199
   in urban detachment, 6-46
action force, 2-52–2-53
   in counterattack, 3-103–3-104
   in dispersed attack, 3-84
   in integrated attack, 3-73
   in offense, 3-36–3-40
actions on contact, 5-4–5-15
adaptive operations, 1-10, 1-23–1-26, 1-54–1-55
administrative force structure, 1-6–1-7
   logistics functions, 14-20–14-21
   tactical-level organizations, 2-10–2-30
aerial minelaying, 12-162–12-168
aerial reconnaissance, 8-31–8-37
   Air Force, 10-73–10-74
   army aviation, 10-104–10-106
affiliated forces, 2-9
   in disruption force, 6-4
   in reconnaissance roles, 8-28
air defense, pp. 11-1–11-30
   air defense ambush, 11-137–11-145
   air defense and CBRN warning communications net, 2-161
   air defense command post, 11-31
   Air Defense Forces, 1-5
   air defense jammers, 11-179–11-180
   air defense system, 11-1–11-20. See also integrated air defense system.
   air defense SPF team, 15-80–15-85
   air surveillance, 11-43–11-59
   all arms air defense, 11-72–11-178
   command and control of, 11-21–11-42
   communications, 11-33–11-37
   deployment, 11-99–11-100
   engagement procedures, 11-132–11-135
   in defense, 11-111–11-117
   in offense, 11-101–11-110
   integrated air defense system, 11-14–11-19, 11-21, 11-33–11-34
   in urban areas, 11-162–11-163
   innovative techniques, 11-172–11-176
   missions, 11-81
   movement of, 11-119–11-122
   nonlethal measures, 11-177–11-189
   of assembly areas, 11-167–11-171
   of tactical movement, 11-125–11-131
   phases, 11-20
   planning, 11-82–11-84
   radars, 11-47–11-50
   roving air defense units, 11-146
   support of ambush, 3-169
   support of assault, 3-128–3-131
   support of counterreconnaissance detachment, 6-20
   support of defense, 4-136, 4-166
   support of raid, 3-189
   support of reconnaissance attack, 3-209
   support of urban detachment, 6-52–6-53
   support of water obstacle crossing, 12-103–12-105
   tactical assets, 11-60–11-78
Air Defense Forces, 1-5
Air Force, 1-5
   Air Defense Forces, 1-5
   missions, 10-62–10-88
air infiltration
   of commandos, 15-178–15-180, 15-191
   of SPF, 15-86–15-88, 15-113–15-119
air interdiction, 10-75–10-77
air parity, 10-71
airspace control net, 2-159
airspace management, 10-21–10-59, 11-38–11-42
airspace operations subsection, 10-27–10-46
air superiority, 10-68–10-69. See also local air superiority.
air supremacy, 10-67
air surveillance, 11-43–11-59
air transport, 10-88
all arms air defense, 11-72–11-178
ambush, 3-133–3-173
   air defense ambush, 11-137–11-145
   ambush team, 4-113, 6-4
   annihilation ambush, 3-115–3-157
   containment ambush, 3-163–3-164
   harassment ambush, 3-158–3-162
   reconnaissance (by) ambush, 8-105–8-107
ambush element, 3-106, 3-138
ambush team, 4-113
   in disruption zone, 6-4
annihilation ambush, 3-115–3-157
antiarmor hunter-killer team. See hunter-killer team.
antihelicopter mines, 11-136, 11-147–11-148, 11-173, 12-1
antilanding minefields, 12-144

Index

Entries are by paragraph number unless page (p.) or pages (pp.) is specified. After a page reference, the subsequent use of paragraph reference is indicated by the paragraph symbol (¶).

antilanding operations, 6-31–6-37
antilanding reserve, 4-20, 4-28, 4-34, 4-150, 6-33–6-34
　in disruption force, 6-4
antimateriel role of snipers, 16-51–16-59
antipersonnel minefields, 12-137–12-140
antitank reserve, 4-33
antitank minefields, 12-133–12-136
AOR. See area of responsibility.
area defense, 4-85–4-104
　air defense in 11-116–11-117
area of responsibility, 2-33–2-35
　linear AOR, 2-23
　nonlinear AOR, 2-36
Armed Forces, p. xii, ¶ 1-4–1-7. See also opposing force.
armor,
　support of counterreconnaissance detachment, 6-27
　support of defense, 4-134, 4-164
　support of raid, 3-187
　support of reconnaissance attack, 3-205–3-206
　support of urban ddetachment, 6-49
Army, 1-5
army, 1-6
army aviation, 9-30. See also aviation.
　missions, 10-89–10-116
　tactics, 10-174–10-206
army group, 1-6
artillery. See also fire support and indirect fire support.
　artillery component of IFC, 9-28–9-29
artillery reconnaissance, 8-27, 9-89, 9-92
biological weapon delivery, 13-49
chemical delivery, 13-5, 13-11
　in antilanding reserve, 6-33

in counterreconnaissance detachment, 6-16, 6-22
in disruption force, 4-20, 6-4
in reconnaissance elements and detachments, 8-69, 8-92
in urban detachment, 6-51, 6-62
mine delivery, 12-158
nuclear delivery, 13-60
smoke delivery, 13-127, 13-130, 13-132–13-133, 13-150, 13-153–13-154, 13-156, 13-158, 13-161, 13-166
assembly area,
　air defense of, 11-167–11-171
　communications, 2-165
　deception assembly areas, 7-48
　force protection, 5-20
　in ambush, 3-146–3-147
　in antilanding actions, 6-36
　in assault, 3-121
　in urban combat, 6-39
　of artillery units, 9-141
　of reconnaissance elements, 8-15
　of reserves, 4-84, 4-104
　of SPF, 15-119
　preparation of, 12-177, 12-179–12-180
　smoke cover of, 13-135, 13-142
assault, 3-106–3-132
assault element,
　in assault, 3-109, 3-115–3-117
assault force,
　in counterattack, 3-101
　in dispersed attack, 3-82
　in integrated attack, 3-71
　in offense, 3-32
attack, 3-60–3-84. See also dispersed attack; integrated attack.
attack helicopters, 10-91–10-107
　employment, 10-179–10-196
　engagement techniques, 10-187–10-196
　formations, 10-184–10-185
　movement techniques, 10-186

attack missions, 10-78–10-81. See also air strike.
attack zone, 2-47, 3-70, 3-93, 6-32, 6-37
autonomous weapon attack, 9-136–9-139
aviation, pp. 10-1–10-31
　aerial observation post, 9-93
　aerial minelaying, 12-162–12-168
　aerial reconnaissance, 8-31–8-37, 10-73–10-74, 10-104–10-106
　airspace management, 10-21–10-59
　army aviation. See main entry.
　aviation support plan, 10-148–10-164
　command and control of, 10-1–10-20
　flight tactics. 10-171–10-206
　in integrated fires command, 10-10–10-11
　missions, 10-60–10-123
　planning and preparation, 10-124–10-170
　request process, 10-130–10-137
　readiness conditions, 10-165–10-166
　risk management, 10-167–10-170
　sortie generation, 10-138–10-147
　support of counterreconnaissance detachment, 6-21
　support of reconnaissance attack, 3-208
　target selection, 10-126–10-129
　unmanned aerial vehicles. See main entry.
　weather and night capabilities, 10-202–10-206

B

battalion, 2-23. See also detachment.
　command section, 2-108–2-109
　staff, 2-110–2-114
　types of tactical defensive action, 4-105–4-171

Index

Entries are by paragraph number unless page (p.) or pages (pp.) is specified. After a page reference, the subsequent use of paragraph reference is indicated by the paragraph symbol (¶).

battalion *(continued)*
  types of tactical offensive action, 3-105–3-211
battalion-size detachment, 2-25. *See also* battalion.
battery (fire support), 2-24
battle drills, 2-3, 3-55, 4-57–4-58, pp. 5-1–5-10
battle line, 2-36, 2-40
battle position, 4-106–4-108. *See also* complex battle position; simple battle position.
battle zone, 2-36–2-37, 2-42–2-45, 3-142, 4-119–4-125, 4-128, 4-156
  air defense in, 11-94–11-97
BDET. *See* battalion-size detachment.
biological warfare, 13-39–13-52
  biological agents, 13-39–13-46
  delivery means, 13-47–13-49
  release authority, 13-51–13-52
  targets, 13-50
breaching. *See* obstacle breaching; situational breach.
breaching element,
  in obstacle breach, 12-51
  in situational breach, 5-30
breaking contact, 5-16–5-24
brigade, 2-17–2-18
  separate brigades, 1-6, 2-17–2-18
  command group, 2-100–2-103
  headquarters, 2-99–2-106
  staff, 2-1-4–2-106
  types of tactical defensive action, 4-61–4-104
  types of tactical offensive action, 3-59–3-104
brigade tactical group, 2-19–2-22. *See also* brigade.
BTG. *See* brigade tactical group.

C

C3D. *See* camouflage, concealment, cover, and deception.
camouflage, concealment, cover, and deception, 7-69–7-70, 12-181–12-189
  air defense, 11-183–11-185
  deception. *See* main entry.
  smoke. *See* main entry.
CBRN (chemical, biological, radiological, and nuclear), pp. 13-1–13-17.
  air defense and CBRN warning communications net, 2-161
  biological warfare. *See* main entry.
  CBRN detection and warning reports, 13-97–13-103
  CBRN protection, 13-79–13-119. *See also* chemical defense.
  CBRN reconnaissance, 8-26, 13-86–13-96
  chemical warfare. *See* main entry.
  decontamination, 13-104–13-114
  nuclear warfare. *See* main entry.
  radiological weapons. *See* main entry.
CDET. *See* company-size detachment.
chemical, biological, radiological, and nuclear. *See* CBRN. See also biological warfare; chemical warfare; nuclear warfare; radiological weapons.
chemical defense. *See also* CBRN protection.
  chemical defense reserve, 4-35
  chemical defense units, 13-81–13-82
  chemical reconnaissance. *See* CBRN reconnaissance.
  equipment, 13-84–13-85
chemical warfare, 13-10–13-38
  chemical agents, 13-12–13-21

chemical defense. *See* main entry; *see also* CBRN protection.
chemical munitions, 9-17, 13-11
  defensive employment, 13-38
  delivery means, 13-11
  offensive employment, 13-33–13-37
  release authority, 13-30–13-32
  toxic industrial chemicals, 13-23–13-28
chief of administration, 14-3, 14-7
  division or DTG staff, 2-88
chief of airspace operations, 10-24–10-26, 11-26–11-27
  division or DTG staff, 2-80
  airspace control net, 2-159
chief of communications,
  battalion of BDET, 2-116
  division or DTG staff, 2-85
chief of current operations,
  division or DTG staff, 2-78
chief of engineer liaison teams, 12-16
chief of fire support coordination,
  battalion of BDET, 2-116, 9-44
  brigade or BTG staff, p. 2-25n, ¶ 9-36, 9-40–9-41
chief of force protection,
  brigade or BTG staff, 2-105
  CBRN protection, 13-9
  division or DTG staff, 2-92
chief of future operations,
  division or DTG staff, 2-79
chief of information warfare,
  deception nets, 2-164
  division or DTG staff, 2-84, 2-92, 2-95
chief of infrastructure management,
  brigade or BTG staff, 2-105
  division or DTG staff, 2-96
chief of integrated fires,
  division or DTG staff, 2-91, 9-36

Index

Entries are by paragraph number unless page (p.) or pages (pp.) is specified. After a page reference, the subsequent use of paragraph reference is indicated by the paragraph symbol (¶).

chief of logistics, 14-3, 14-6
　battalion of BDET, 2-116
　division or DTG staff, 2-87
chief of population management,
　brigade or BTG staff, 2-105
　division or DTG staff, 2-95
chief of reconnaissance,
　battalion of BDET, 2-116
　brigade or BTG staff, 8-52–8-54
　division or DTG staff, 2-83, 8-52–8-54
　intelligence net, 2-158
chief of special-purpose operations,
　brigade or BTG staff, 2-105
　division or DTG staff, 2-93
chief of staff,
　battalion or BDET, 2-108, 2-113
　brigade or BTG, 2-103, 2-124–2-125
　division or DTG, 2-68–2-69, 2-124–2-125
　reconnaissance planning, 8-46–8-49
chief of weapons of mass destruction, 13-8
　brigade or BTG staff, 2-105
　division or DTG staff, 2-94, 2-92
clearing element, 2-54, 12-52. See also obstacle-clearing element.
　in urban detachment, 6-46–6-47, 6-54
close support fire, 9-98
combat security outpost, 4-112
combat service support helicopters, 10-108–10-110, 10-197–10-201
combat support helicopters, 10-111–10-114, 10-197–10-201
combat system, 1-63–1-68, 1-71. See also systems warfare.
command and control, pp. 2-1–2-34
　C2 systems, 2-142–2-172
　C2 systems survivability, 2-167–2-172

command and support relationships, 2-5–2-9, 9-19
　of air defense, 11-21–11-42
　of aviation, 10-1–10-20
　of commandos, 15-157–15-158
　of disruption force, 3-23–3-24
　of engineers, 12-14–12-24
　of functional forces, 3-23–3-24
　of indirect fire support, 9-18–9-56
　of SPF, 15-2–15-11
command and support relationships, 2-5–2-9
　affiliated, 2-9
　dedicated, 2-7
　constituent, 2-6
　fire support units, 9-19
　supporting, 2-8
commander,
　battalion or BDET, 2-109
　brigade or BTG, 2-101
　company or CDET, 2-118
　division or DTG, 2-63–2-66
　reconnaissance planning, 8-41–8-45
commander's reconnaissance group, 8-70–8-71
command group,
　brigade or BTG, 2-100–2-103
　division or DTG, 2-62–2-69
command net, 2-154
commandos, 15-154–15-194
　air infiltration company of commando brigade, 15-178–15-180
　command and control of, 15-157–15-158
　commando battalion, 15-183–15-185
　commando brigade, 15-177–15-182
　in administrative force structure, 15-157, 15-170
　infiltration, 15-159, 15-163, 15-178–15-180, 15-187–15-192, 15-194
　information warfare company of commando brigade, 15-181–15-182
　in OSC, 15-175
　missions, 15-159–15-169

organization for combat, 15-170–15-174
　part of SPF Command, 1-5, 15-154, 15-157
　personnel, 15-194
　reconnaissance company of commando brigade, 15-181
　tactics, techniques, and procedures, 15-186–15-193
command posts, 2-119–2-141
　airborne command post, 2-131
　alternate command post, 2-132
　auxiliary command post, 2-133
　command post location, 2-138–2-139
　command post movement, 2-135–2-137
　command post security, 2-140–2-141
　deception command post, 2-134, 7-46
　forward command post, 2-127–2-129
　IFC command post, 2-126
　main command post, 2-124–2-125
　sustainment command post, 2-130
command section,
　battalion or BDET, 2-108–2-109
command team,
　company or CDET, 2-118
company, 2-24. See also detachment.
　command team, 2-118
　supply and transport team, 2-118
　support team, 2-118
　types of tactical defensive action, 4-105–4-171
　types of tactical offensive action, 3-105–3-211
company-size detachment, 2-25. See also company.
computer warfare, 7-87–7-91
communications, 2-142–2-166
communications subsection,
　division or DTG staff, 2-81–2-85

Index-4　　　　　TC 7-100.2　　　　　9 December 2011

Index

Entries are by paragraph number unless page (p.) or pages (pp.) is specified. After a page reference, the subsequent use of paragraph reference is indicated by the paragraph symbol (¶).

complex battle position, 4-105, 4-108
  defense of, 4-145–4-171
complex terrain, p. 3-25n, ¶ 3-121–3-124, 3-173, 4-49
  snipers in, 16-45–16-47
constituent command relationship, 2-6
contact force, 2-53, 4-25, 4-70–4-72, 4-81
containment ambush, 3-163–3-164
contour flight, 10-177
corps, 1-6
counterair missions,
  Air Force 10-72
  army aviation, 10-107
counterattack, 3-95–3-104
  air defense in, 11-110
counterattack force, 4-29, 4-42
counterbattery fire, 9-101
counterfire, 9-101
countermobility, 12-107–12-168
countermortar fire, 9-101
counterreconnaissance, 6-8–6-30, 7-67–7-68
  counterreconnaissance detachment, 4-20, 4-42, 6-4, 6-15–6-27
  counterreconnaissance zone, 6-10
  kill zone, 6-13
  predicted enemy locations, 6-12
  reference zone, 6-11
criminal elements
  affiliation with regular military forces, 1-76, 2-9, 4-20, 6-4
  affiliation with SPF, 15-11, p. 15-4n
  as part of hybrid threat, p. xi
  as part of Hybrid Threat, for training, pp. xi, xii, 1-1n, ¶ 1-31
  as part of OPFOR, p. xii

D

deception, 7-28–7-58
deception battery, 9-132–9-133
deception command post, 7-46
deception element, 2-55, 4-150, 7-34–7-35
deception firing position, 9-120
deception force, 2-55, 3-34, 4-32, 4-36, 4-42, 4-75–4-79, 7-34–7-35
electronic deception, 7-51–7-58
in defense of simple battle position, 4-130
decontamination, 13-104–13-114
decoys, 7-42–7-51, 12-184–12-189
decoy minefields, 12-143
decoy smokescreens, 13-140
dedicated command relationship, 2-7
deep attack and reconnaissance platoon of SPF battalion, 15-39–15-42
defense, pp. 4-1–4-35
  air defense in, 11-111–11-117
  area defense, 4-85–4-104
  defense of complex battle position, 4-145–4-171
  defense of simple battle position, 4-109–4-144
  defensive arrays, 4-67–4-69
  defensive maneuver, 4-70–4-74
  executing, 4-53–4-60
  maneuver defense, 4-64–4-84
  planned defense, 4-10
  planning for, 4-8–4-42
  preparing for, 4-43–4-52
  purpose, 4-1–4-7
  situational defense, 4-11–4-13
  types of operational defensive action, 1-57
  types of tactical defensive action, 4-61–4-171
demonstration, 7-40
deputy commander,
  battalion or BDET, 2-108–2-109
  brigade or BTG, 2-102
  company or CDET, 2-108–2-109
division or DTG, 2-67, 2-126
detachment, 2-25–2-28
  battalion-size detachment, 2-25
  company-size detachment, 2-25
  counterreconnaissance detachment, 4-20, 4-42, 6-4, 6-15–6-27
  independent mission detachment, 2-26
  movement support detachment, 3-47, 12-40–12-47
  obstacle detachment, 4-33, 4-35, 4-48, 4-78, 4-96, 12-110–12-119
  reconnaissance detachment, 8-92–8-95
  urban detachment, 6-43–6-55
  types of tactical defensive action, 4-105–4-171
  types of tactical offensive action, 3-105–3-211
direct air support,
  Air Force, 10-82–10-87
  army aviation, 10-99–10-103
dirty bomb. See radiological dispersal devices.
dispersed attack, 3-74–3-84
  air defense in, 11-104–11-106
dispersed (firing)platoon, 9-126–9-127
disruption, 6-1–6-7
disruption element, 2-55, 4-38–4-39
  in antilanding reserve, 6-34
  in defense of CBP, 4-150
  in defense of SBP, 4-111–4-113
disruption force, 2-55, 6-1–6-7
  C2 of, 3-23–3-24
  commandos in 15-164
  in area defense, 4-92–4-101
  in defense, 4-18–4-24
  in dispersed attack, 3-80
  in integrated attack, 3-69
  in maneuver defense, 4-70, 4-75–4-79
  in offense, 3-28
  SPF in, 15-17, 15-40

Index

Entries are by paragraph number unless page (p.) or pages (pp.) is specified. After a page reference, the subsequent use of paragraph reference is indicated by the paragraph symbol (¶).

disruption zone, 2-36, 2-38–2-41
air defense in, 11-90–11-93
and disruption element in defense, 4-39–4-112, 4-115
and disruption force, 6-1–6-7
and disruption force in defense, 4-17–4-24, 4-76, 4-78, 4-92–4-101, 4-103
offense, 3-28
in area defense, 4-87, 4-92–4-101, 4-103
in defense of CBP, 4-154–4-155
in defense of SBP, 4-118, 4-124, 4-127, 4-143
in maneuver defense, 4-63–4-64, 4-73, 4-76, 4-78
in urban combat, 6-41
diver team of SPF company, 15-92–15-94
division, 2-13–2-16
separate divisions, 1-6, 2-13
command group, 2-62–2-69
headquarters, 2-60–2-61
staff, 2-70–2-98
types of tactical defensive action, 4-61–4-104
types of tactical offensive action, 3-59–3-104
division tactical group, 2-19–2-22. *See also* division.
DTG. *See* division tactical group.

E

electronic attack, 7-22–7-24
electronic deception, 7-51–7-58
electronic warfare, 7-14–7-27. *See also* information warfare.
  electronic attack, 7-22
  signals reconnaissance, 7-18–7-21
  support of counterreconnaissance detachment, 6-26
enabling element, 2-54–2-55, 3-46
enabling force, 2-54–2-55

in counterattack, 3-99–3-102
in dispersed attack, 3-80–3-83
in integrated attack, 3-69–3-72
in offense, 3-26–3-35
engineer support, pp. 12-1–12-32
  chief of engineer liaison teams, 12-16
  C2 of, 12-14–12-24
  countermobility, 12-107–12-168
  engineer reconnaissancel, 8-25, 12-27–12-36, 12-75–12-77
  engineer reserve, 4-35
  fortification, 12-171–12-180
  minelaying, 12-147–12-168
  missions and tasks, 12-2–12-13
  mobility, 12-37–12-106
  movement support detachment, 3-47, 12-40–12-47
  obstacle breaching, 12-48–12-60
  obstacle detachment, 4-33, 4-35, 4-48, 4-78, 4-96, 12-110–12-119
  support of ambush, 3-170
  support of counterreconnaissance detachment, 6-24
  support of defense, 12-9–12-13
  support of defense of CBP, 4-167
  support of defense of SBP, 4-137–4-142
  support of information warfare, 12-25–12-26
  support of offense, 12-6–12-8
  support of raid, 3-190
  support of reconnaissance attack, 3-210
  support of urban detachment, 6-54–6-55
  survivability, 12-169–12-189
  water obstacle crossing, 12-61–12-106
exfiltration,
  of complex battle position, 4-168

of MANPADS team, 11-140
of snipers, 16-21, 16-30, 16-37, 16-50
of SPF, 15-88, 15-94, 15-113, 15-120–15-121, 15-126, 15-145, 15-180
of support zone, 4-160
exploitation element, 3-44
exploitation force, 2-53, 3-38, 3-73, 3-84, 3-94, 3-103–3-104
explosive obstacle breaching, 12-49–12-58
explosive obstacles, 12-25–12-127. *See also* mines; minefields.

F

false battery, 9-134
false deployment, 7-47–7-50
feint, 7-39
field group, 1-39–1-40
fighting patrol, 8-89–8-91
final protective fire, 9-102
fire and decoy, 9-129–9-134
fire and maneuver, 5-32–5-39
fire support. *See also* indirect fire support.
  chief of fire support coordination, p. 2-25n, ¶ 2-116, 9-36, 9-40–9-41, 9-44
  coordination measures, 9-64
  fire missions, 9-65
  fire requests, 9-45–9-46
  fire support coordination center, 9-27, 9-38, 9-41
  fire support observers, 9-52–9-56
  naval fire support, 9-47–9-51
  support of ambush, 3-166
  support of assault, 3-127
  support of counterreconnaissance detachment, 6-22
  support of defense, 9-108–9-111
  support of defense of CBP, 4-165
  support of defense of SBP, 4-135

Index-6　　　　　　　　　　　　　　　TC 7-100.2　　　　　　　　　　　　　　　9 December 2011

Index

Entries are by paragraph number unless page (p.) or pages (pp.) is specified. After a page reference, the subsequent use of paragraph reference is indicated by the paragraph symbol (¶).

fire support *(continued)*
  support of offense, 9-106–9-107
  support of raid, 3-188
  support of reconnaissance attack, 3-207
  support of strike, 9-105
  support of urban detachment, 6-51
  targeting, 9-66–9-78
fixing, 5-40–5-46. *See also* fixing element; fixing force.
  countermobility actions, 5-46
  fires, 5-44
fixing element, 2-54–2-55, 3-176, 3-198, 5-11, 5-13–5-14, 5-22–5-23, 5-36–5-37
fixing force, 2-54–2-55
  in counterattack, 3-100
  in dispersed attack, 3-81
  in integrated attack, 3-70
  in offense, 3-29–3-31
fortification, 12-171–12-180
forward air controller, 10-42–10-45
  role in preplanned missions, 10-157
  role in immediate missions, 10-160–10-162
forward arming and refueling points, 10-115–10-116
functional organization, 2-49–2-58
action element. *See* main entry.
action force. *See* main entry.
ambush element. *See* main entry.
assault element. *See* main entry.
assault force. *See* main entry.
clearing element. *See* main entry; *see also* obstacle-clearing element.
contact force *See* main entry.
counterattack force *See* main entry.
deception element. *See* main entry.
deception force. *See* main entry.

disruption element. *See* main entry.
disruption force. *See* main entry.
enabling element. *See* main entry.
enabling force. *See* main entry.
exploitation element. *See* main entry.
exploitation force. *See* main entry.
fixing element. *See* main entry.
fixing force. *See* main entry.
  for ambush, 3-137–3-141
  for assault, 3-108–3-111
  for counterattack, 3-98–3-104
  for defense, 4-14–4-42
  for defense of complex battle position, 4-149–4-153
  for defense of simple battle position, 4-110–4-116
  for dispersed attack, 3-79–3-84
  for integrated attack, 3-68–3-73
  for offense, 3-22–3-47
  for raid, 3-176–3-183
  for reconnaissance attack, 3-196–3-200
  for spoiling attack, 3-94
main defense element. *See* main entry.
main defense force. *See* main entry.
mission force. *See* main entry.
obstacle-clearing element. *See* main entry.
  of battalions, 3-42–3-47
  of brigades, 3-22–3-41
  of companies, 3-42–3-47
  of detachments, 3-42–3-47
  of divisions, 3-22–3-41
  of tactical groups, 3-22–3-41
protected force. *See* main entry.
raiding element. *See* main entry.
raiding force. *See* main entry.
reconnaissance element. *See* main entry.

reconnaissance force. *See* main entry.
reserve element. *See* main entry.
reserve force. *See* main entry.
security element. *See* main entry.
security force. *See* main entry.
shielding force. *See* main entry.
specialist elements. *See* main entry.
strike force. *See* main entry.
support element. *See* main entry.
support force. *See* main entry.
functional staff,
  brigade or BTG staff, 2-105
  division or DTG staff, 2-89–2-96

G

General Staff, 1-4
  Chief of the General Staff, 1-34–1-35
  military strategic campaign plan, 1-32
global positioning system. *See* GPS.
GPS jammers, 7-15, 7-25, p. 10-18n, ¶ 10-122, 11-159, 11-181, 15-22, 15-57
ground reconnaissance, 8-21–8-28
guerrillas,
  affiliated with an OSC, 1-18
  affiliated with SPF, 15-11, 15-17, 15-34, 15-40, 15-45, 15-48, 15-60, 15-65–15-67, 15-75, 15-96–15-97
  relation to nation-state forces, 1-18, p. 3-16n, ¶ 4-20, 6-4, 9-80, 10-120, 13-11
  trained by SPF, 15-17, 15-55, 15-95

H

harassment ambush, 3-158–3-162

9 December 2011 TC 7-100.2 Index-7

Index

Entries are by paragraph number unless page (p.) or pages (pp.) is specified. After a page reference, the subsequent use of paragraph reference is indicated by the paragraph symbol (¶).

headquarters. *See* command group and staff (by level of command).
heliborne landing, 10-109, 10-112
helicopter. *See* army aviation, attack helicopters, combat service support helicopters, combat support helicopters, and rotary-wing tactics.
helicopter attack, 10-94–10-98
hunter-killer team, 3-154–3-157
and SPF snipers, 15-69
and marksmen, 16-14, 16-16–16-19
hybrid threat, p. xi
Hybrid Threat for training, p. xi

I

IMD. *See* independent mission detachment.
imitative electronic deception, 7-56–7-57
improvised explosive devices, and SPF, 15-67, 15-105
independent mission detachment, 2-26
independent reconnaissance patrol, 8-22, 8-83–8-86
indierct fire support, pp. 9-1–9-152. *See also* fire support and aviation.
aerial observation post, 9-93
autonomous weapon attack, 9-136–9-139
battalion firing position areas, 9-113–9-116
battery deployment, 9-121–9-139
battery firing positions, 9-117–9-120
close support fire, 9-98
command and observation post, 9-83–9-86
counterbattery fire, 9-101
counterfire, 9-101
countermortar fire, 9-101
deception battery, 9-132–9-133
dispersed platoon, 9-126–9-127
false battery, 9-134
final protective fire, 9-102

fire and decoy, 9-129–9-134
forward opbervation post, 9-87–9-89
integrated with information warfare, 9-1, 9-3, 9-66
lateral observation post, 9-90
logistics, 9-145–9-152
methods of fire, 9-94–9-103
mobile reconnaissance post, 9-91–9-92
reconnaissance by fire, 8-29–8-30, 9-103
reconnaissance fire, 9-95–9-97
roving gun, 9-130
roving unit, 9-131
shoot and move, 9-135
split battery, 9-125
systems warfare, 9-6–9-7
tactical deployment, 9-112–9-144
tactical movement, 9-140–9-144
target acquisition and reconnaissance, 9-79–9-93
target damage criteria, 9-11–9-15
weapons, 9-16–9-17
infantry
support of
counterreconnaissance detachment, 6-23
infantry antiarmor hunter-killer team. *See* hunter-killer team.
infiltration. *See also* air infiltration.
by observation post, 8-74
by assault force, 3-71, 3-82
by breach teams, 12-54
by commandos, 15-159, 15-163, 15-178–15-180, 15-187–15-192, 15-194
by elements for reconnaissance attack, 3-202
by raiding element, 3-177
by SPF teams, 3-177, 11-174–11-175, 15-27, 15-55, 15-64, 15-86–15-88, 15-92, 15-94, 15-108–15-123, 15-125, 15-142

by water obstacle crossing, 12-63, 12-91
engineer support of, 12-37
in urban combat, 6-57
through minefield, 12-50
to conduct spoiling attack or ambush, 4-9
information attack, 7-82–7-86
information security, 7-71–7-72
information warfare, pp. 7-1–7-16
C2 survivability (protection and security measures), 2-168
computer warfare *See* main entry.
deception. *See* main entry.
electronic warfare. *See* main entry.
elements of, 7-9–7-91
in fixing, 5-45
information attack *See* main entry.
information warfare company of commando brigade, 15-181
information warfare platoon of commando battalion, 15-184
in strategic operations, 1-14
perception management *See* main entry.
physical destruction. *See* main entry.
protection and security measures. *See* main entry.
sniper role, 16-60–16-63
support of ambush, 3-172
support of assault, 3-132
support of defense of CBP, 4-171
support of defense of SBP, 4-144
support of raid, 3-192
support of reconnaissance attack, 3-212
support of urban combat, 6-58
information warfare subsection, division or DTG staff, 2-81, 2-84
INFOWAR. *See* information warfare.

Index

Entries are by paragraph number unless page (p.) or pages (pp.) is specified. After a page reference, the subsequent use of paragraph reference is indicated by the paragraph symbol (¶).

instruments of national power, p. xii, ¶ 1-1
 in strategic campaign, 1-27
 in strategic operations, 1-13–1-14
insurgent forces,
 affiliated with an OSC, 1-18
 affiliated with SPF, 15-11, 15-17, 15-34, 15-40, 15-45, 15-48, 15-60, 15-65–15-67, 15-75, 15-96–15-97
 relation to nation-state forces, 1-14, 1-18, 2-9, p. 3-16n, ¶ 4-20, 6-4, 9-80, 10-120, 13-11
 trained by SPF, 15-17, 15-55, 15-95
integrated attack, 3-64–3-73
 air defense in, 11-103
integrated fires command, 2-14, 9-20–9-35
 artillery component, 9-28–9-29
 army aviation component, 9-30
 IFC commander, 2-67, 2-126, 9-22–9-23
 IFC headquarters, 9-27
 IFC in division or DTG, 9-58
 IFC in OSC, 9-47–9-48, 9-58
 IFC planning, 9-62–9-63
integrated support group, 9-35
 long-range reconnaissance component,
 missile component, 9-31
integrated support command, 2-15–2-16, 14-22–14-28
 ISC headquarters, 14-25
integrated support group, 14-29–14-31
 in IFC, 2-15–2-16, 9-35
intelligence and information section,
 division or DTG staff, 2-81
intelligence net, 2-158
intelligence officer,
 battalion or BDET, 2-111
 division or DTG, 2-81–2-82
 reconnaissance planning, 8-50
interagency forces, 2-13, 2-18
interdiction. See air interdiction.

Internal Security Forces, 1-5, 1-75–1-76, 2-95, 3-33, 4-28, 14-50, 15-3, 15-5
irregular forces. See also guerrillas and insurgents.
 as part of hybrid threat, p. xi
 as part of Hybrid Threat for training, p. xi, ¶ 1-31
 irregular force snipers, 16-65
 role in operations, 1-75–1-77

J

jamming. See electronic attack.

K

kill zone, 2-48, 6-10

L

liaison teams,
 battalion or BDET staff, 2-114
 brigade or BTG staff, 2-106
 division or DTG staff, 2-97–2-98
limited-objective attack, 3-85–3-104. See also counterattack; spoiling attack.
limit of responsibility, 2-33
local air superiority, 10-70
local populace, 3-49, 6-65, 7-73, 7-77–7-78, 7-80, 14-1, 15-139
logistics, pp. 14-1–14-17
 integrated support command. See main entry.
 integrated support group. See main entry.
 logistics units, 14-18–14-31
 maintenance, 14-43–14-44
 materiel support, 14-32–14-42
 medical support, 14-71–14-79
 missions, 14-14–14-17
 mission support site, 3-146, 14-94, 15-152–15-153
 personnel support, 14-62–14-70
 post-combat support, 14-95–14-100
 staff responsibilities, 14-3–14-4. See also chief of

administration; chief of logistics; resources officer.
 support of ambush, 3-171
 support of defense, 14-86–14-90
 support of defense of CBP, 4-170
 support of defense of SBP, 4-143
 support of indirect fire support, 9-145–9-152
 support of offense, 14-82–14-85
 support of raid, 3-191
 support of reconnaissance attack, 3-211
 support of SPF, 15-138–15-153
 tactical concepts, 14-8–14-13
 transportation, 14-45–14-61
long-range reconnaissance patrol, 6-4
long-range reconnaissance component of IFC, 9-32–9-34
long-range reconnaissance unit, 6-19
long-range signal platoon, SPF battalion, 15-49–15-50
LOR. See limit of responsibility.
low-level flight, 19-178

M

main defense element, 2-53, 4-38–4-39
 in defense of CBP, 4-151
 in defense of SBP, 4-114
main defense force, 2-53, 4-25
 in area defense, 4-102–4-103
 in maneuver defense, 4-80–4-83
maintenance, 14-43–14-44
maneuver brigades. See brigade.
maneuver defense, 4-64–4-84
 air defense in, 11-112–11-115
maneuver reserve, 4-31–4-32
manipulative electronic deception, 7-52–7-54

9 December 2011 TC 7-100.2 Index-9

Index

Entries are by paragraph number unless page (p.) or pages (pp.) is specified. After a page reference, the subsequent use of paragraph reference is indicated by the paragraph symbol (¶).

marksmen, 16-2–16-19
  and snipers, p. 16-1
  equipment, 16-9–16-12
  in hunter-killer team, 16-14, 16-16–16-19
  mission, 16-4–16-6
  organization, 16-7–16-8
  tactics, techniques, and procedures, 16-14–16-19
  targets, 16-13
materiel support, 14-32–14-42
media, 6-60, 7-12, 7-33, 7-74, 7-76–7-77, 7-79–7-80, 16-64–16-65
medical support, 14-71–14-79
  medical team of SPF company, 15-98–15-103
military district, 1-6
military region, 1-6
militia forces,
  Army, 1-5
mines,
  antihelicopter mines. *See* main entry.
  antitank minefields, 12-133–12-136
  antipersonnel minefields, 12-137–12-140
  minefields, 12-129–12-146
  minefield breaching. *See* explosive obstacle breaching.
  minelaying, 12-147–12-168
  mixed minefields, 12-141–12-142
Ministry of Defense, 1-2–1-4
  military strategic campaign plan, 1-32
  national strategic campaign plan, 1-30
Ministry of the Interior, 1-2, 1-5, 2-95
  Internal Security Forces, 1-5, 1-75–1-76, 2-95, 3-33, 4-28, 14-50, 15-3, 15-5
Ministry of Public Information, 2-95
mission force, 3-40, 3-94
mission support site, 3-146, 14-94
mission tactics, 2-2

movement,
  air defense of, 11-125–11-131
  logistics, 14-46–14-54
  movement routes, 12-33–12-36, 12-39
  movement support detachment, 3-47, 12-40–12-47
  of air defense units, 11-119–11-122
  of command posts, 2-135–2-137
  of indirect fire support units, 9-140–9-144
  to ambush site, 3-146
movement support detachment, 3-47, 12-40–12-47

N

nap-of-the-earth flight, 10-176
National Command Authority, 1-2–1-3
national security strategy, 1-8–1-26
national strategic campaign. *See* strategic campaign.
national strategic goals, 1-9, 1-27–1-29
Navy, 1-5
  naval fire support, 9-47–9-51
NCA. *See* National Command Authority.
night,
  air defense, 11-164
  aviation capability, 10-202–10-206
  smoke employment, 13-163–13-166
nuclear warfare, 13-59–13-78
  defensive employment, 13-78
  delivery means, 13-60
  nuclear munitions, 9-17, 13-60
  offensive employment, 13-69–13-77
  release authority, 13-67–13-68

O

observation, 8-99–8-100
  aerial observation post, 9-93
  air defense observation posts, 11-54–11-59
  command and observation post, 9-83–9-86
  fire support observers, 9-52–9-56
  fire support observation posts, 9-81–9-82
  forward opbervation post, 9-87–9-89
  lateral observation post, 9-90
  observation post, 8-73–8-74
  observer, 8-72
obstacle breaching, 12-48–12-60
obstacle-clearing element, 3-47, 3-124, 4-40, 12-41–12-44
obstacle detachment, 4-33, 4-35, 4-48, 4-78, 4-96, 12-110–12-119
obstacles. *See* mines; obstacle breaching; obstacle detachment.
offense, pp. 3-1–3-43. *See also* ambush; assault; attack; limited-objective attack; raid; reconnaissance attack; strike.
  air defense in, 11-101–11-110
  executing, 3-53–3-58
  fire support of, 9-106–9-107
  planned offense, 3-17
  planning for, 3-16–3-47
  preparing for, 3-48–3-52
  purpose, 3-1–3-15
  situational offense, 3-18–3-21
  strike, 1-56
  types of operational offensive action, 1-56
  types of tactical offensive action, 3-59–3-211
officer reconnaissance patrol, 8-87–8-88

Index-10   TC 7-100.2   9 December 2011

Index

Entries are by paragraph number unless page (p.) or pages (pp.) is specified. After a page reference, the subsequent use of paragraph reference is indicated by the paragraph symbol (¶).

operational designs, 1-44–1-55
  adaptive operations, 1-54–1-55
  regional operations, 1-45–1-48
  transition operations, 1-49–1-53
operational environments, pp. ix–x
  operational variables, p. x
  training applications, pp. ix–x
operational-strategic command, 1-41–1-43
operational variables, p. x
opetrations net, 2-155
operations officer,
  battalion or BDET, 2-111
  division or DTG, 2-75–2-77
operations security, 7-71–7-72
OPFOR. *See* opposing force.
opposing force, pp. x–xiv
  adaptability, p. xiii
  administrative force structure, 1-6–1-7, 2-10–2-30
  baseline OPFOR, p. xii
  fighting force structure, 1-6–1-7
  Hybrid Threat for training, p. xi
  strategic and operational framework, pp. 1-1–1-16
  tactical-level organizations, 2-10–2-30
  terminology, p. xiii
organizing the tactical battlefield, 2-31–2-48
OSC. *See* operational-strategic command.

P

paramilitary forces. *See also* guerrilla, insurgent, irregular forces.
  affiliated with regular military forces, p. xi, ¶2-9
  patrol, p. 2-17n
  fighting patrol, 8-89–8-91
  independent reconnaissance patrol, 8-22, 8-83–8-86
  long-range reconnaissance patrol, 6-4

officer reconnaissance patrol, 8-87–8-88
  reconnaissance patrol, 8-78–8-82
patrol squad, 8-75–8-76
perception management, 7-73–7-80
personnel support, 14-62–14-70
physical destruction, 7-59–7-64. *See also* information warfare.
planning
  for ambush, 3-142–3-147
  for offense, 3-16–3-47
platoons, 2-29–2-30
predicted enemy locations, 6-12
protected force, 2-57, 4-26
protection and security measures, 7-65–7-72. *See also* information warfare.
psychological warfare, 7-75–7-77. *See also* perception management.
public affairs, 7-78. *See also* perception management.

R

radiological weapons, 13-53–13-58
  radiological dispersal devices, 13-55–13-57
  radiological exposure devices, 13-58
raid, 3-174–3-192
  for reconnaissance, 8-101–8-104
  reconnaissance raid, 8-102–8-104
raiding element, 2-53, 3-177–3-179
raiding force, p. 37n
reconnaissance, pp. 8-1–8-18
  aerial reconnaissance, 8-31–8-37, 10-73–10-74, 10-104–10-106
  CBRN reconnaissance, 8-26, 13-86–13-96
  chief of reconnaissance, 2-83, 2-116, 2-158, 8-52–8-54

commander's reconnaissance group, 8-70–8-71
deep attack and reconnaissance platoon of SPF battalion, 15-39–15-42
engineer reconnaissance, 8-25, 12-27–12-36, 12-75–12-77
fighting patrol, 8-89–8-91
ground reconnaissance, 8-21–8-28
high-mobility reconnaissance platoon of commando battalion, 15-184
independent reconnaissance patrol, 8-22, 8-83–8-86
information flow and communications, 8-60–8-66
intelligence officer, 8-50
long-range reconnaissance patrol, 6-4
long-range reconnaissance component of IFC, 9-32–9-34
observation, 8-99–8-100
observation post, 8-73–8-74
observer, 8-72
officer reconnaissance patrol, 8-87–8-88
patrol squad, 8-75–8-76
raid, 8-101–8-104
reconnaissance attack, 3-8, 3-193–3-211, 8-108–8-111
reconnaissance (by) ambush, 8-105–8-107
reconnaissance by fire, 8-29–8-30
reconnaissance company of commando brigade, 15-181–15-182
reconnaissance detachment, 8-92–8-95
reconnaissance element, 3-197, p. 8-1n, ¶ 8-67–8-95
reconnaissance force, 8-67
reconnaissance patrol, 8-78–8-82, 12-29–12-30
reconnaissance plan, 8-55–8-59
reconnaissance raid, 8-102–8-104

9 December 2011   TC 7-100.2   Index-11

Index

Entries are by paragraph number unless page (p.) or pages (pp.) is specified. After a page reference, the subsequent use of paragraph reference is indicated by the paragraph symbol (¶).

reconnaissance (continued)
reconnaissance report, 8-64
reconnaissance subsection, division or DTG staff, 2-81, 2-83
reconnaissance summary, 8-66
reconnaissance team, 8-77
reporting, 8-62–8-63
signals reconnaissance, 8-24
special reconnaissance, 15-19–15-20
support of ambush, 3-167
support of assault, 3-126
support of counterreconnaissance detachment, 6-19
support of defense of CBP, 4-163
support of defense of SBP, 4-133
support of raid, 3-186
support of reconnaissance attack, 3-204
unmanned aerial vehicles, 8-33–8-37
zone of reconnaissance responsibility, 8-39–8-40
reconnaissance and obstacle-clearing element, 12-42–12-44
reconnaissance attack, 3-8, 3-193–3-211, 8-108–8-111
reconnaissance by fire, 8-29–8-30, 9-103
reconnaissance element, 3-197, p. 8-1n, ¶ 8-67–8-95
reconnaissance force, 8-67
reconnaissance fire, 9-95–9-97
reconnaissance, intelligence, surveillance, and target acquisition (RISTA), p. 8-1n. *See also* reconnaissance.
RISTA and INFOWAR section, IFC, 9-27
RISTA assets in IFC, 9-20, 9-23, 9-25
target acquisition. *See* indirect fire support: target acquisition.
reconstitution, 14-97–14-99, 15-150–15-151
reference zone, 6-11

regional operations, 1-10, 1-15–1-19, 1-45–1-48
reserve (reserve element, reserve force), 2-56
in area defense, 4-104
in defense, 4-30–4-35, 4-38–4-39, 4-42
in defense of CBP, 4-152
in defense of SBP, 4-115
in maneuver defense, 4-84
in offense, 3-41
reserve forces, Army, 1-5
resources officer, 14-3–14-5
battalion or BDET, 2-111
division or DTG staff, 2-83
support net, 2-157
RISTA. *See* reconnaissance, intelligence, surveillance, and target acquisition (RISTA). *See also* reconnaissance.
river crossing. *See* water obstacle crossing.
road and bridge construction and repair element, 12-45–12-46
rotary-wing aircraft. *See* army aviation, aviation,and helicopters.
roving gun, 9-130
roving unit, 9-131
ruse, 7-41

S

sappers, in assault, 3-124
in urban combat, 6-54
sapper platoon of SPF battalion, 15-43–15-48
sapper team, SPF company, 15-64–15-68
security element, 2-55
as fixing element, 5-11
in actions on contact, 5-9
in ambush, 3-139–3-140
in antilanding reserve, 6-34
in assault, 3-110, 3-118
in breaking contact, 5-19, 5-21–5-22
in fire and maneuver, 5-36–5-38
in offense, 3-46
in raid, 3-180–3-181

in reconnaissance attack, 3-198
in situational breach, 5-28
in urban detachment, 6-46–6-47
security force, 2-55
in defense, 4-27–4-28, 4-42
in offense, 3-33
SHC. *See* Supreme High Command.
shielding force, 2-53, 4-25, 4-70–4-72, 4-74, 4-81
shoot and move, 9-135
SID. *See* Strategic Integration Department.
signal. *See also* communications.
long-range signal platoon, SPF battalion, 15-49–15-50
signal support of counterreconnaissance detachment, 6-25
signal team of SPF company, 15-95–15-97
signature reduction, 7-50
signals reconnaissance, 7-18–7-21, 8-24
simple battle position, 4-105, 4-107
defense of, 4-109–4-144
simulative electronic deception, 7-55
situational breach, 5-25–5-31
smoke, 12-120–13-167
delivery means, 13-127–13-130
defensive employment, 13-157–13-159
offensive employment, 13-149–13-156
signalling smoke, 13-167
smoke agents, 13-122–13-126
smokescreens, 13-131–13-147
smoke units, 13-121
support of water obstacle crossing, 12-106, 13-160–13-162
tactical employment, 13-148–13-166, 16-20–16-59

Index

Entries are by paragraph number unless page (p.) or pages (pp.) is specified. After a page reference, the subsequent use of paragraph reference is indicated by the paragraph symbol (¶).

sniper, 16-20–16-59
  and marksmen, p. 16-1, 16-21–16-22
  antimateriel role, 16-51–16-59
  equipment, 16-40–16-41, 16-56–16-57
  exfiltration, 16-21, 16-30, 16-37, 16-50
  information warfare role, 16-60–16-63
  irregular force snipers, 16-65
  mission, 16-24–16-25, 16-53–16-54
  organization, 16-26–16-39, 16-55
  single sniper, 16-30–16-31
  sniper team, 16-32–16-39
  iper team in SPF, 15-69–15-75
  tactics, techniques, and procedures, 16-44–16-50, 16-59
  targets, 16-42–16-43, 16-58
specialist elements, 3-47, 4-38, 4-40
special mission net, 2-160
Special-Purpose Forces, 1-5, 15-1–15-153
  affiliations with irregular forces, 15-10–15-11, 15-17, 15-34, 15-40, 15-45, 15-48, 15-60, 15-65–15-67, 15-75, 15-96–15-97, 15-107
  air defense team, 15-80–15-85
  air infiltration team, 15-86–15-88
  allocation in fighting force structure, 15-5–15-9. See also SPF: organization for combat.
  C2 of, 15-2–15-11
  direct action missions, 15-21–15-23
  direct action team, 15-59–15-63
  diver team, 15-92–15-94
  equipment, 15-129–15-133
  exfiltration, 15-88, 15-94, 15-113, 15-120–15-121, 15-126, 15-145, 15-180
  in administrative force structure, 15-3–15-4

infiltration, 3-177, 11-174–11-175, 15-27, 15-55, 15-64, 15-86–15-88, 15-92, 15-94, 15-108–15-123, 15-125, 15-142
in LRR component of IFC, 9-34, 15-8
in OSCs, 1-18, 9-34, 15-6
in strategic operations, 1-14
logistics, 15-138–15-153
medical team, 15-98–15-103
missions, 15-12–15-23
mortar team, 15-76–15-79
organization for combat, 15-24–15-31
personnel, 15-134–15-137
sapper team, 15-64–15-68
signal team, 15-95–15-97
sniper team, 15-69–15-75
special reconnaissance, 15-19–15-20
SPF battalion, 15-36–15-50
SPF brigade, 15-33–15-35
SPF Command, 1-5
SPF company, 15-51–15-103
SPF teams, 15-55–15-103
standard SPF team, 15-58
stay-behind tactics, 15-127–15-128
swarming tactic, 15-124–15-126
tactics, techniques, and procedures, 15-104–15-128
terror tactics, 15-107
UAV team, 15-89–15-91
Special-Purpose Forces Command, 1-5
special reserves, 4-35
SPF. See Special-Purpose Forces.
split battery, 9-125
spoiling attack, 3-90–3-94
  air defense in, 11-109
squads, 2-29–2-30
staff,
  battalion or BDET, 2-110–2-114
  brigade or BTG, 2-104–2-106
  division or DTG, 2-70–2-98
  staff command, 2-115–2-117

State, the, p. xii
  administrative force structure, 1-6–1-7, 2-10–2-30
  Armed Forces, p. xii, ¶ 1-4–1-7. See also opposing force.
  instruments of national power, p. xii, ¶1-1
  National Command Authority, 1-2–1-3
  national-level organization, 1-1–1-7
  national security strategy, 1-8–1-7, ¶ 1-26
  national strategic campaign plan, 1-30
  national strategic goals, 1-9, 1-27–1-29
  strategic and operational framework, pp. 1-1–1-16
  strategic-level courses of action, 1-10–1-26
  strategic and operational framework, pp. 1-1–1-16
  strategic campaign, 1-27–1-37
  strategic campaign plan, in adaptive operations, 1-24
  military strategic campaign plan, 1-32–1-37
  national strategic campaign plan, 1-30
  strategic courses of action, 1-10–1-26
  Strategic Forces, 1-5.
  strategic goals, 1-9, 1-27–1-29
  Strategic Integration Department, 1-3
  strategic operations, 1-10–1-14
  strike, 1-56, 3-39, p. 9-4n, ¶ 9-105
  strike force, 3-39
  supply. See materiel support.
  supply and transport team, company or CDET, 2-118
  support element, 2-55, 4-39
    in ambush, 3-141
    in antilanding reserve, 6-34
    in assault, 3-111, 3-119
    in defense, 4-38–4-39
    in defense of CBP, 4-153
    in defense of SBP, 4-116
    in offense, 3-46
    in raid, 3-182–3-183

Entries are by paragraph number unless page (p.) or pages (pp.) is specified. After a page reference, the subsequent use of paragraph reference is indicated by the paragraph symbol (¶).

support element (*continued*)
  in reconnaissance attack, 3-200
  in situational breach, 5-29
support force, 2-55
  in counterattack, 3-102
  in dispersed attack, 3-83
  in integrated attack, 3-72
  in offense, 3-35
supporting command relationship, 2-8
support line, 2-36
support net, 2-157
support team, company or CDET, 2-118
support zone, 2-36, 2-46
  air defense in, 11-98
  and protected force, 4-26
  in offense, 3-33
  in defense, 4-27, 4-126–4-127, 4-129, 4-157, 4-160
support zone security, 14-91–14-93
suppression of enemy air defenses, 10-169–10-170
Supreme High Command, 1-4–1-6
  military strategic campaign plan, 1-35
surface-to-surface missiles, 9-2, 9-4, 9-16, 9-31
survivability, 12-169–12-189
swarming tactic, 15-124–15-126, 15-193
systems warfare, 1-58–1-73
  combat system, 1-64–1-65
  indirect fire support, 9-6–9-7
  operational level, 1-63–1-70
  tactical level, 1-71
  strategic level, 1-60–1-62

T
tactical groups, 2-19–2-22
  brigade tactical group, 2-19–2-22
  division tactical group, 2-19–2-22
tactical movement. *See* movement.
tanks. *See* armor.
target damage criteria, 9-11–9-15
terror tactics,
  by SPF, 15-107
  in strategic operations, 1-14
  transition operations, 1-10, 1-20–1-22, 1-49–1-53
toxic industrial chemicals, 3-23–13-28
transportation, 14-45–14-61

U
UAV. *See* unmanned aerial vehicles.
unit symbols,
  detachments, 2-28
  platoons (task-organized), 2-30
  squads (task-organized), 2-30
  tactical groups, 2-22
unmanned aerial vehicles, 10-19–10-20
  airspace management, 11-41–11-42
  capabilities, 10-118–10-121
  defense against, 11-149–11-161
  in SPF, 15-42, 15-64, 15-89–15-91
  missions, 10-122–10-123
  reconnaissance, 8-33-8-37, 9-80

urban combat, 6-38–6-64
  air defense in, 11-162–11-163
  snipers, 16-45–16-47
urban detachment, 6-43–6-55

V
vehicle engine exhaust system, 13-127, 13-129, 13-142, 13-153–13-155, 13-158–13-159

W
water obstacle crossing, 12-61–12-106
  opposed crossing, 12-86–12-88
  smoke support, 13-160–13-162
  unopposed crossing, 12-89–12-102
weapons of mass destruction, 13-1–13-2. *See* CBRN.
  chief of weapons of mass destruction, 2-94, 2-92, 2-105, 13-8
  delivery by SPF, 15-105
  in strategic operations, 1-14

Z
zones, 2-36–2-48
  attack zone. *See* main entry.
  battle zone, 2-36–2-37, 2-42–2-45
  disruption zone *See* main entry.
  kill zone, 2-48, 6-10
  support zone. *See* main entry.
zone of reconnaissance responsibility, 8-39–8-40

TC 7-100.2
9 December 2011

By order of the Secretary of the Army:

RAYMOND T. ODIERNO
General, United States Army
Chief of Staff

Official:

JOYCE E. MORROW
Administrative Assistant to
  Secretary of the Army
1125005

DISTRIBUTION:
Active Army, Army National Guard, and United States Army Reserve: Not to be distributed; electronic media only.

PIN: 102339-000

www.ingramcontent.com/pod-product-compliance
Lightning Source LLC
Chambersburg PA
CBHW050045230526
45470CB00004B/1411